GCSE Biology

The Complete Course for AQA

How to get your free Online Edition

Go to **cgpbooks.co.uk/extras** and enter this code...

0099 3770 0697 7703

This code will only work once. If someone has used this book before you, they may have already claimed the Online Edition.

Contents

Introduction

Working Scientifically

Topic 1 — Cell Biology

Topic 1a — Cell Structure and Cell Division

Topic 1b — Transport in Cells

Topic 2 — Organisation

Topic 2a — Tissues, Organs and Organ Systems

Topic 2b — Health and Disease

Topic 2c — Enzymes and Digestion

Topic 3 — Infection and Response

Topic 4 — Bioenergetics

Topic 5 — Homeostasis and Response

Topic 5a — The Nervous System

Topic 5b — The Endocrine System

Topic 5c — Animal and Plant Hormones

Topic 6 — Inheritance, Variation and Evolution

Topic 6a — DNA and Reproduction

Published by CGP

From original material by Richard Parsons.

Editors:
Charlotte Burrows, Katherine Faudemer, Emily Howe, Chris McGarry, Sarah Pattison, Sean Walsh.

Contributor:
Paddy Gannon.

ISBN: 978 1 78294 595 6

With thanks to Sophie Anderson and Camilla Simson for the proofreading.
With thanks to Jan Greenway for the copyright research.

Clipart from Corel®

How to use this book

Learning Objectives:
- Know that in 1859 Darwin published "On the Origin of Species" to propose his theory.
- Understand why Darwin's ideas caused controversy when they were published, and why his ideas were only gradually accepted.
- Understand that there have been other hypotheses about evolution, including Lamarck's — which was based on the idea that 'acquired characteristics' could be inherited.
- Understand that we now know Lamarck's hypothesis was wrong.

Specification Reference 4.6.3.1

3. Ideas About Evolution

Lamarck had different ideas from Darwin, but they were eventually rejected — leaving Darwin's ideas to form the theory of evolution.

Controversy

Darwin's theory of evolution by natural selection is widely accepted today. But when he proposed his theory in his book "On the Origin of Species" in 1859, his idea was very controversial for various reasons...

- It went against common religious beliefs about how life on Earth developed — it was the first plausible explanation for the existence of life on earth without the need for a "Creator" (God).
- Darwin couldn't explain why these new, useful characteristics appeared or how they were passed on from individual organisms to their offspring. But then he didn't know anything about genes or mutations — they weren't discovered 'til 50 years after his theory was published.
- There wasn't enough evidence to convince many scientists, because not many other studies had been done into how organisms change over time.

For these reasons, Darwin's idea was only accepted gradually, as more and more evidence came to light.

Figure 1: Darwin.

Lamarck

Darwin wasn't the only person who tried to explain evolution. There were different scientific hypotheses about evolution around at the same time, such as Lamarck's:

Jean-Baptiste Lamarck (1744-1829) argued that changes that an organism acquires during its lifetime will be passed on to its offspring — e.g. he thought that if a characteristic was used a lot by an organism, then it would become more developed during its lifetime, and the organism's offspring would inherit the acquired characteristic.

Using Lamarck's theory, if a rabbit used its legs to run a lot (to escape predators), then its legs would get longer. The offspring of that rabbit would then be born with longer legs.

Figure 2: Lamarck.

Accepting or rejecting hypotheses

Often scientists come up with different hypotheses to explain similar observations. Scientists might develop different hypotheses because they have different beliefs (e.g. religious) or they have been influenced by different people (e.g. other scientists and their way of thinking)... or they just think differently.

290 Topic 6c Evolution and Classification

Learning Objectives

- These tell you exactly what you need to learn, or be able to do, for the exam.
- There's a specification reference at the bottom that links to the AQA specification.

Examples

These are here to help you understand the theory.

Maths Skills

- There's a range of maths skills you could be expected to apply in your exams. The section on pages 376-383 is packed with plenty of maths that you'll need to be familiar with.
- Examples that show these maths skills in action are marked up with this symbol.

Working Scientifically

- Working Scientifically is a big part of GCSE Biology. There's a whole section on it at the front of the book.
- Working Scientifically is also covered throughout this book wherever you see this symbol.

WORKING SCIENTIFICALLY

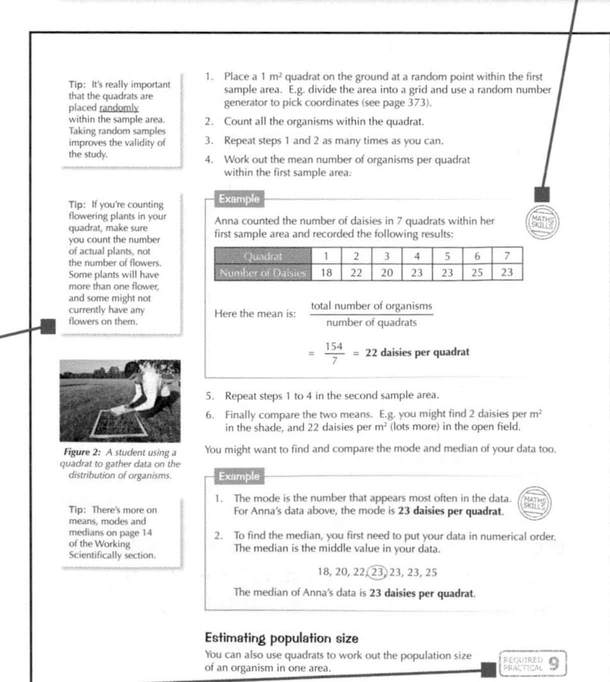

Tip: It's really important that the quadrats are placed randomly within the sample area. Taking random samples improves the validity of the study.

1. Place a 1 m^2 quadrat on the ground at a random point within the first sample area. E.g. divide the area into a grid and use a random number generator to pick coordinates (see page 373).
2. Count all the organisms within the quadrat.
3. Repeat steps 1 and 2 as many times as you can.
4. Work out the mean number of organisms per quadrat within the first sample area:

Tip: If you're counting flowering plants in your quadrat, make sure you count the number of actual plants, not the number of flowers. Some plants will have more than one flower, and some might not currently have any flowers on them.

Example

Anna counted the number of daisies in 7 quadrats within her first sample area and recorded the following results:

Quadrat	1	2	3	4	5	6	7
Number of Daisies	18	22	20	23	23	25	23

Here the mean is: $\frac{\text{total number of organisms}}{\text{number of quadrats}}$

$= \frac{154}{7} =$ **22 daisies per quadrat**

5. Repeat steps 1 to 4 in the second sample area.
6. Finally compare the two means. E.g. you might find 2 daisies per m^2 in the shade, and 22 daisies per m^2 (lots more) in the open field.

Figure 2: A student using a quadrat to gather data on the distribution of organisms.

You might want to find and compare the mode and median of your data too.

Example

1. The mode is the number that appears most often in the data. For Anna's data above, the mode is **23 daisies per quadrat**.
2. To find the median, you first need to put your data in numerical order. The median is the middle value in your data.

18, 20, 22, (23), 23, 23, 25

The median of Anna's data is **23 daisies per quadrat**.

Tip: There's more on means, modes and medians on page 14 of the Working Scientifically section.

Estimating population size

You can also use quadrats to work out the population size of an organism in one area.

REQUIRED PRACTICAL 9

Example

Students used 0.25 m^2 quadrats to randomly sample daisies on an open field. The students found a mean of 10.5 daisies per quadrat. The field had an area of 800 m^2. Estimate the population of daisies on the field.

326 Topic 7a Organisms and Their Environment

Tips and Exam Tips

- There are tips throughout this book to help you understand the theory.
- There are also exam tips to help you with answering exam questions.

Required Practical Activities

There are some Required Practical Activities that you'll be expected to do throughout your course. You need to know all about them for the exams. They're all marked with stamps like this:

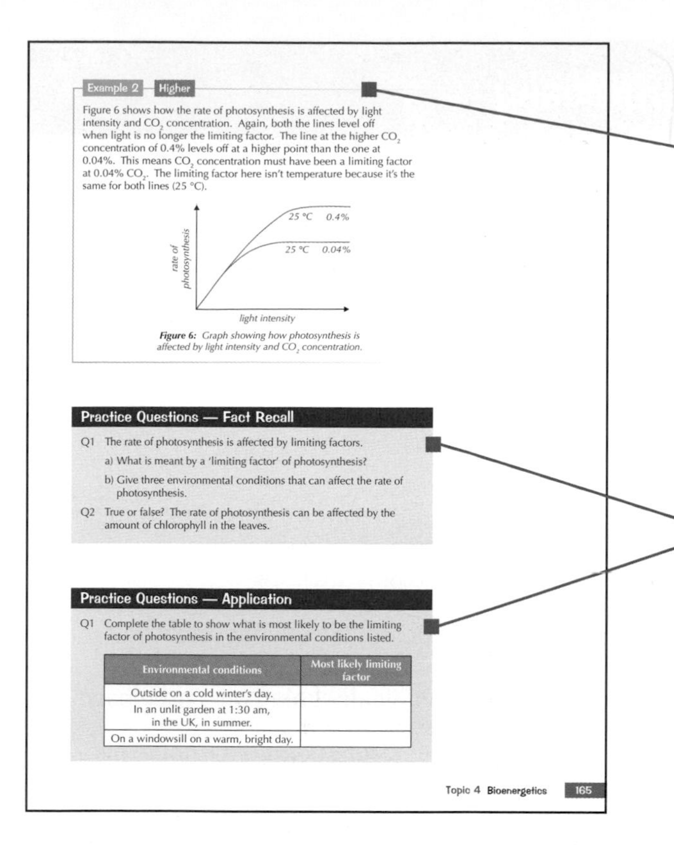

Example 2 Higher

Figure 6 shows how the rate of photosynthesis is affected by light intensity and CO_2 concentration. Again, both the lines level off when light is no longer the limiting factor. The line at the higher CO_2 concentration of 0.4% levels off at a higher point than the one at 0.04%. This means CO_2 concentration must have been a limiting factor at 0.04% CO_2. The limiting factor here isn't temperature because it's the same for both lines (25 °C).

Figure 6: Graph showing how photosynthesis is affected by light intensity and CO_2 concentration.

Practice Questions — Fact Recall

Q1 The rate of photosynthesis is affected by limiting factors.
a) What is meant by a 'limiting factor' of photosynthesis?
b) Give three environmental conditions that can affect the rate of photosynthesis.

Q2 True or false? The rate of photosynthesis can be affected by the amount of chlorophyll in the leaves.

Practice Questions — Application

Q1 Complete the table to show what is most likely to be the limiting factor of photosynthesis in the environmental conditions listed.

Environmental conditions	Most likely limiting factor
Outside on a cold winter's day.	
In an unlit garden at 1:30 am, in the UK, in summer.	
On a windowsill on a warm, bright day.	

Topic 4 Bioenergetics 165

Higher Exam Material

- Some of the material in this book will only come up in the exam if you're sitting the higher exam papers.
- This material is clearly marked with boxes that look like this:

Practice Questions

- Fact recall questions test that you know the facts needed for GCSE Biology.
- Annoyingly, the examiners also expect you to be able to apply your knowledge to new situations — application questions give you plenty of practice at doing this.
- All the answers are in the back of the book.

Practical Skills

There's also a whole section on pages 368-375 with extra details on practical skills you'll be expected to use in the Required Practical Activities, and apply knowledge of in the exams.

Exam-style Questions

- Practising exam-style questions is really important — this book has some at the end of every topic to test you.
- They're the same style as the ones you'll get in the real exams.
- All the answers are in the back of the book, along with a mark scheme to show you how you get the marks.
- Higher-only questions are marked like this: 

Topic Checklist

Each topic has a checklist at the end with boxes that let you tick off what you've learnt.

Glossary

There's a glossary at the back of the book full of definitions you need to know for the exam, plus loads of other useful words.

Exam Help

There's a section at the back of the book stuffed full of things to help you with the exams.

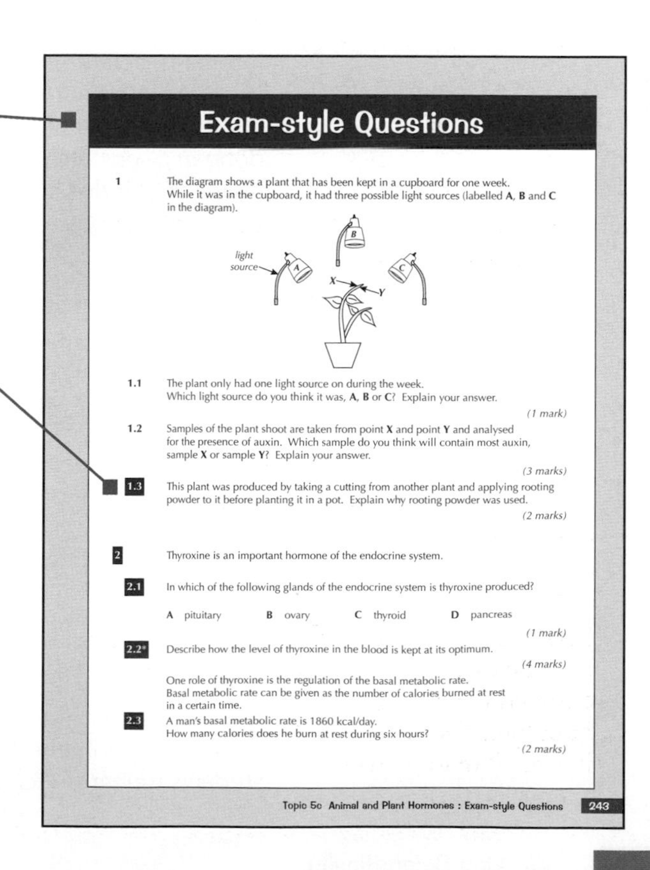

Exam-style Questions

1 The diagram shows a plant that has been kept in a cupboard for one week. While it was in the cupboard, it had three possible light sources (labelled **A**, **B** and **C** in the diagram).

1.1 The plant only had one light source on during the week. Which light source do you think it was, **A**, **B** or **C**? Explain your answer.
(1 mark)

1.2 Samples of the plant shoot are taken from point **X** and point **Y** and analysed for the presence of auxin. Which sample do you think will contain most auxin, sample **X** or sample **Y**? Explain your answer.
(3 marks)

1.3 This plant was produced by taking a cutting from another plant and applying rooting powder to it before planting it in a pot. Explain why rooting powder was used.
(2 marks)

2 Thyroxine is an important hormone of the endocrine system.

2.1 In which of the following glands of the endocrine system is thyroxine produced?

A pituitary B ovary C thyroid D pancreas
(1 mark)

2.2* Describe how the level of thyroxine in the blood is kept at its optimum.
(4 marks)

One role of thyroxine is the regulation of the basal metabolic rate. Basal metabolic rate can be given as the number of calories burned at rest in a certain time.

2.3 A man's basal metabolic rate is 1860 kcal/day. How many calories does he burn at rest during six hours?
(2 marks)

Topic 5c Animal and Plant Hormones : Exam-style Questions 243

1. The Scientific Method

Science is all about finding things out and learning things about the world we live in. This section is all about the scientific process — how a scientist's initial idea turns into a theory that is accepted by the wider scientific community.

Hypotheses

Scientists try to explain things. Everything. They start by observing something they don't understand — it could be anything, e.g. planets in the sky, a person suffering from an illness, what matter is made of... anything.

Then, they come up with a **hypothesis** — a possible explanation for what they've observed. (Scientists can also sometimes form a model — a description or a representation of what's physically going on — see page 3).

The next step is to test whether the hypothesis might be right or not. This involves making a **prediction** based on the hypothesis and testing it by gathering evidence (i.e. data) from investigations. If evidence from experiments backs up a prediction, you're a step closer to figuring out if the hypothesis is true.

Tip: Investigations include lab experiments and studies.

Testing a hypothesis

Normally, scientists share their findings in peer-reviewed journals, or at conferences. **Peer-review** is where other scientists check results and scientific explanations to make sure they're 'scientific' (e.g. that experiments have been done in a sensible way) before they're published. It helps to detect false claims, but it doesn't mean that findings are correct — just that they're not wrong in any obvious way.

Once other scientists have found out about a hypothesis, they'll start basing their own predictions on it and carry out their own experiments. They'll also try to reproduce the original experiments to check the results — and if all the experiments in the world back up the hypothesis, then scientists start to think the hypothesis is true.

However, if a scientist somewhere in the world does an experiment that doesn't fit with the hypothesis (and other scientists can reproduce these results), then the hypothesis is in trouble. When this happens, scientists have to come up with a new hypothesis (maybe a modification of the old hypothesis, or maybe a completely new one).

Tip: Sometimes it can take a really long time for a hypothesis to be accepted.

Accepting a hypothesis

If pretty much every scientist in the world believes a hypothesis to be true because experiments back it up, then it usually goes in the textbooks for students to learn. Accepted hypotheses are often referred to as **theories**.

Our currently accepted theories are the ones that have survived this 'trial by evidence' — they've been tested many, many times over the years and survived (while the less good ones have been ditched). However... they never, never become hard and fast, totally indisputable fact. You can never know... it'd only take one odd, totally inexplicable result, and the hypothesising and testing would start all over again.

Example

Over time scientists have come up with different hypotheses about how illnesses are caused:

- Hundreds of years ago, we thought demons caused illness.
- Then we thought it was caused by 'bad blood' (and treated it with leeches).
- Now we've collected more evidence, we know that illnesses that can be spread between people are due to microorganisms.

Figure 1: *Historical artwork of a woman using leeches to treat disease.*

Models

Models are used to describe or display how an object or system behaves in reality. They're often based on evidence collected from experiments, and should be able to accurately predict what will happen in other, similar experiments. There are different types of models that scientists can use to describe the world around them. Here are just a few:

- A **descriptive model** describes what's happening in a certain situation, without explaining why. It won't necessarily include details that could be used to predict the outcome of a different scenario. For example, a graph showing the measured rate of an enzyme-catalysed reaction at different temperatures would be a descriptive model.
- A **representational model** is a simplified description or picture of what's going on in real life. It can be used to explain observations and make predictions. E.g. the lock and key model of enzyme action is a simplified way of showing how enzymes work (see page 115). It can be used to explain why enzymes only catalyse particular reactions.
- **Spatial models** are used to analyse the way that data is arranged within a physical space. For example, a spatial model could be used to look for relationships between the distribution of a species and physical aspects of its environment.
- **Computational models** use computers to make simulations of complex real-life processes, such as climate change. They're used when there are a lot of different variables (factors that change) to consider, and because you can easily change their design to take into account new data.
- **Mathematical models** can be used to describe the relationship between variables in numerical form (e.g. as an equation), and therefore predict outcomes of a scenario. For example, an equation can be written to predict how quickly molecules will diffuse into cells based on factors such as how large they are. On a larger scale, a mathematical model can be used to predict the way that a disease might spread through a population, e.g. by estimating the number of people that will become infected with the disease based on the number of people that are currently infected.

Tip: Like hypotheses, models have to be tested before they're accepted by other scientists. You can test models by using them to make a prediction, and then carrying out an investigation to see whether the evidence matches the prediction.

Tip: Mathematical models are made using patterns found in data and also using information about known relationships between variables.

Tip: Like hypotheses, models are constantly being revised and modified, based on new data.

All models have limitations on what they can explain or predict. Climate change models have several limitations — for example, it's hard to take into account all the biological and chemical processes that influence climate. It can also be difficult to include regional variations in climate.

Communicating results

Some scientific discoveries show that people should change their habits, or they might provide ideas that could be developed into new technology. So scientists need to tell the world about their discoveries.

Tip: New scientific discoveries are usually communicated to the public in the news or via the internet. They might be communicated to governments and large organisations via reports or meetings.

Example

Gene technologies make use of discoveries that scientists have made about genes and DNA. Some of these technologies are used in genetic engineering, to produce genetically modified crops. Information about these crops needs to be communicated to farmers who might benefit from growing them, and to the general public so they can make informed decisions about the food they buy and eat.

Reports about scientific discoveries in the media (e.g. newspapers or television) aren't peer-reviewed. This means that, even though news stories are often based on data that has been peer-reviewed, the data might be presented in a way that is over-simplified or inaccurate, leaving it open to misinterpretation.

Tip: If you're reading an article about a new scientific discovery, always think about how the study was carried out. It may be that the sample size was very small, and so the results aren't representative of the whole situation (see page 11 for more on sample sizes).

It's important that the evidence isn't presented in a **biased** way. This can sometimes happen when people want to make a point, e.g. they overemphasise a relationship in the data. (Sometimes without knowing they're doing it.) There are all sorts of reasons why people might want to do this.

Tip: An example of bias is a newspaper article describing details of data supporting an idea without giving any of the evidence against it.

Examples

- They want to keep the organisation or company that's funding the research happy. (If the results aren't what they'd like they might not give them any more money to fund further research.)
- Governments might want to persuade voters, other governments or journalists to agree with their policies about a certain issue.
- Companies might want to 'big up' their products, or make impressive safety claims.
- Environmental campaigners might want to persuade people to behave differently.

There's also a risk that if an investigation is done by a team of highly-regarded scientists it'll be taken more seriously than evidence from less well known scientists. But having experience, authority or a fancy qualification doesn't necessarily mean the evidence is good — the only way to tell is to look at the evidence scientifically (e.g. is it repeatable, valid, etc. — see page 9).

2. Scientific Applications and Issues

New scientific discoveries can lead to exciting new ways of using science in our everyday lives. Unfortunately, these developments may also come with social, economic, environmental or personal problems that need to be considered.

Using scientific developments

Lots of scientific developments go on to have useful applications.

Examples

- When it was discovered that penicillin acted as an antibiotic, drugs were developed that used penicillin to treat bacterial infections.
- As scientists have investigated the role of reproductive hormones in the body, treatments that use these hormones have been developed to allow women to become pregnant who wouldn't be able to otherwise.

Tip: There's lots more about antibiotics on page 140.

Issues created by science

Scientific developments (e.g. new technologies or new advice) can create issues. For example, they could create political issues, which could lead to developments being ignored, or governments being slow to act if they think responding to the developments could affect their popularity with voters.

Example

Some governments were pretty slow to accept the fact that human activities are causing global warming, despite all the evidence. This is because accepting it means they've got to do something about it, which costs money and could hurt their economy. This could lose them a lot of votes.

Tip: See page 343 for more on global warming.

Scientific developments can cause a whole host of other issues too.

Examples

- Economic issues: Society can't always afford to do things scientists recommend (e.g. investing heavily in alternative energy sources) without cutting back elsewhere.
- Social issues: Decisions based on scientific evidence affect people — e.g. should junk food be taxed more highly (to encourage people to be healthy)? Should alcohol be banned (to prevent health problems)? Would the effect on people's lifestyles be acceptable?
- Environmental issues: Human activity often affects the natural environment — e.g. genetically modified crops may help us produce more food, but some people think they could cause environmental problems.
- Personal issues: Some decisions will affect individuals. For example, someone might support alternative energy, but object if a wind farm is built next to their house.

Figure 1: *Solar farms can be used to generate 'clean' electricity, but some people think that the money needed to invest in them should be spent elsewhere.*

3. Limitations of Science

Science has taught us an awful lot about the world we live in and how things work — but science doesn't have the answer for everything.

Questions science hasn't answered yet

We don't understand everything. And we never will. We'll find out more, for sure — as more hypotheses are suggested, and more experiments are done. But there'll always be stuff we don't know.

***Figure 1:** Global warming could cause weather patterns to change — which may result in longer, hotter droughts in some areas.*

Examples

- Today we don't know as much as we'd like about the impacts of global warming. How much will sea levels rise? And to what extent will weather patterns change?
- We also don't know anywhere near as much as we'd like about the universe. Are there other life forms out there? And what is the universe made of?

In order to answer scientific questions, scientists need data to provide evidence for their hypotheses. Some questions can't be answered yet because the data can't currently be collected, or because there's not enough data to support a theory. But eventually, as we get more evidence, we probably will be able to answer these questions once and for all. By then there'll be loads of new questions to answer though.

Questions science can't answer

There are some questions that all the experiments in the world won't help us answer — for example, the "should we be doing this at all?" type questions.

Example

Think about embryo screening (which lets you choose an embryo with certain characteristics). It's possible to do it, but does that mean we should?

- Some people say it's good... couples whose existing child needs a bone marrow transplant, but who can't find a donor, will be able to have another child selected for its matching bone marrow. This would save the life of their first child — and if they want another child anyway... where's the harm?
- Other people say it's bad... they say it could have serious effects on the new child. For example, the new child might feel unwanted — thinking they were only brought into the world to help someone else. And would they have the right to refuse to donate their bone marrow (as anyone else would)?

Tip: Some experiments have to be approved by ethics committees before scientists are allowed to carry them out. This stops scientists from getting wrapped up in whether they can do something, before anyone stops to think about whether they should do it.

The question of whether something is morally or ethically right or wrong can't be answered by more experiments — there is no "right" or "wrong" answer. The best we can do is get a consensus from society — a judgement that most people are more or less happy to live by. Science can provide more information to help people make this judgement, and the judgement might change over time. But in the end it's up to people and their conscience.

4. Risks and Hazards

A lot of things we do could cause us harm. But some things are more hazardous than they at first seem, whereas other things are less hazardous. This may sound confusing, but it'll all become clear...

What are risks and hazards?

A **hazard** is something that could potentially cause harm. All hazards have a **risk** attached to them — this is the chance that the hazard will cause harm.

The risks of some things seem pretty obvious, or we've known about them for a while, like the risk of causing acid rain by polluting the atmosphere, or of having a car accident when you're travelling in a car.

New technology arising from scientific advances can bring new risks. These risks need to be thought about alongside the potential benefits of the technology, in order to make a decision about whether it should be made available to the general public.

Example

Radiotherapy is a cancer treatment that uses radiation to kill cancer cells. However, radiation itself carries a risk of causing cancer in the patient.

The risk of cancer caused by radiotherapy has to be weighed against the risk of the existing cancer that needs to be treated.

In some cases, the risk from the original cancer far outweighs the risk from radiotherapy, which is why it is available to be used as a treatment.

Tip: The application of a new technology can change over time if new risks are discovered. For example, radiation was widely used as a medical treatment before it was known that it can cause cancer. These days, its use as a treatment is a lot more limited and techniques have been improved to reduce the risk.

Estimating risk

You can estimate the risk based on how many times something happens in a big sample (e.g. 100 000 people) over a given period (e.g. a year). For example, you could assess the risk of a driver crashing by recording how many people in a group of 100 000 drivers crashed their cars over a year.

To make a decision about an activity that involves a hazardous event, we don't just need to take into account the chance of the event causing harm, but also how serious the consequences would be if it did.

The general rule is that, if an activity involves a hazard that's very likely to cause harm, with serious consequences if it does, that activity is considered high-risk.

Example 1

If you go for a walk, you may sprain an ankle. But most sprains recover within a few weeks if they're rested, so going for a walk would be considered a low-risk activity.

Figure 1: *BASE jumping is considered a high-risk activity.*

Example 2

If you go BASE jumping, you could die or get a serious injury that could cause further complications later on in life, and the chance of this happening is high. So BASE jumping would be considered a high-risk activity.

Perceptions of risk

Not all risks have the same consequences, e.g. if you chop veg with a sharp knife you risk cutting your finger, but if you go scuba-diving you risk death. You're much more likely to cut your finger during half an hour of chopping than to die during half an hour of scuba-diving. But most people are happier to accept a higher probability of an accident if the consequences are short-lived and fairly minor.

People tend to be more willing to accept a risk if they choose to do something (e.g. go scuba diving), compared to having the risk imposed on them (e.g. having a nuclear power station built next door).

Tip: Risks people choose to take are called 'voluntary risks'. Risks that people are forced to take are called 'imposed risks'.

People's perception of risk (how risky they think something is) isn't always accurate. They tend to view familiar activities as low-risk and unfamiliar activities as high-risk — even if that's not the case. For example, cycling on roads is often high-risk, but many people are happy to do it because it's a familiar activity. Air travel is actually pretty safe, but a lot of people perceive it as high-risk. People may over-estimate the risk of things with long-term or invisible effects.

Reducing risk in investigations

Part of planning an investigation is making sure that it's safe. To make sure your experiment is safe you must identify all the **hazards**. Hazards include:

Tip: You can find out about potential hazards by looking in textbooks, doing some internet research, or asking your teacher.

- Microorganisms: e.g. some bacteria can make you ill.
- Chemicals: e.g. sulfuric acid can burn your skin and alcohols catch fire easily.
- Fire: e.g. an unattended Bunsen burner is a fire hazard.
- Electricity: e.g. faulty electrical equipment could give you a shock.

Once you've identified the hazards you might encounter, you should think of ways of reducing the risks from the hazards.

Figure 2: *Scientists wearing safety goggles to protect their eyes during an experiment.*

Examples

- If you're working with sulfuric acid, always wear gloves and safety goggles. This will reduce the risk of the acid coming into contact with your skin and eyes.
- If you're using a Bunsen burner, stand it on a heat proof mat. This will reduce the risk of starting a fire.

5. Designing Investigations

To be a good scientist you need to know how to design a good experiment.

Making predictions from a hypothesis

Scientists observe things and come up with hypotheses to explain them. To decide whether a **hypothesis** might be correct, you need to do an investigation to gather evidence, which will help support or disprove the hypothesis. The first step is to use the hypothesis to come up with a **prediction** — a statement about what you think will happen that you can test.

Example

If your hypothesis is "eating a diet containing a large amount of saturated fat causes a high blood cholesterol level", then your prediction might be "people who eat large amounts of saturated fats will have a high level of cholesterol in their blood".

Once a scientist has come up with a prediction, they'll design an investigation to see if there are patterns or relationships between two variables. For example, to see if there's a pattern or relationship between the variables 'amount of saturated fats eaten' and 'blood cholesterol level'.

Tip: A variable is just something in the experiment that can change.

Repeatable and reproducible results

Results need to be **repeatable** and **reproducible**. Repeatable means that if the same person does an experiment again using the same methods and equipment, they'll get similar results. Reproducible means that if someone else does the experiment, or a different method or piece of equipment is used, the results will be similar. Data that's repeatable and reproducible is **reliable** and scientists are more likely to have confidence in it.

Example

In 1998, a scientist claimed to have found a link between the MMR vaccine (for measles, mumps and rubella) and autism. This meant many parents stopped their children from being vaccinated, leading to a rise in the number of children catching measles. However, the results have never been reproduced. Health authorities have now concluded that the vaccine is safe.

Figure 1: *The MMR vaccine.*

Ensuring the test is valid

Valid results are repeatable, reproducible and answer the original question.

Example

Do power lines cause cancer? — Some studies have found that children who live near overhead power lines are more likely to develop cancer. What they'd actually found was a **correlation** (relationship) between the variables "presence of power lines" and "incidence of cancer". They found that as one changed, so did the other.

But this data isn't enough to say that the power lines cause cancer, as there might be other explanations. For example, power lines are often near busy roads, so the areas tested could contain different levels of pollution. As the studies don't show a definite link they don't answer the original question.

Tip: Peer review (see page 2) is used to make sure that results are valid before they're published.

Tip: See page 18 for more on correlation.

Ensuring it's a fair test

> **Tip:** For the results of an investigation to be valid the investigation must be a fair test.

In a lab experiment you usually change one variable and measure how it affects another variable. To make it a fair test, everything else that could affect the results should stay the same (otherwise you can't tell if the thing you're changing is causing the results or not — the data won't be valid).

Example

You might change only the temperature of an enzyme-controlled reaction and measure how it affects the rate of reaction. You need to keep the pH the same, otherwise you won't know if any change in the rate of reaction is caused by the change in temperature, or the change in pH.

The variable you change is called the **independent variable**. The variable you measure when you change the independent variable is called the **dependent variable**. The variables that you keep the same are called **control variables**.

Example

In the enzyme-controlled reaction example above, temperature is the independent variable, the rate of the reaction is the dependent variable and the control variables are pH, volume and concentration of reactants, etc.

Control experiments and control groups

> **Tip:** Control experiments let you see what happens when you don't change anything at all.

To make sure no other factors are affecting the results, you also have to include a **control experiment** — an experiment that's kept under the same conditions as the rest of the investigation, but doesn't have anything done to it.

Example

You can investigate antibiotic resistance in bacteria by growing cultures of bacteria on agar plates, then adding paper discs soaked in antibiotic.

If the bacteria are resistant to the antibiotic they will continue to grow. If they aren't resistant a clear patch will appear around the disc where they have died or haven't grown.

A disc that isn't soaked in antibiotic is included to act as a control. This makes sure any result is down to the antibiotic, not the presence of a paper disc.

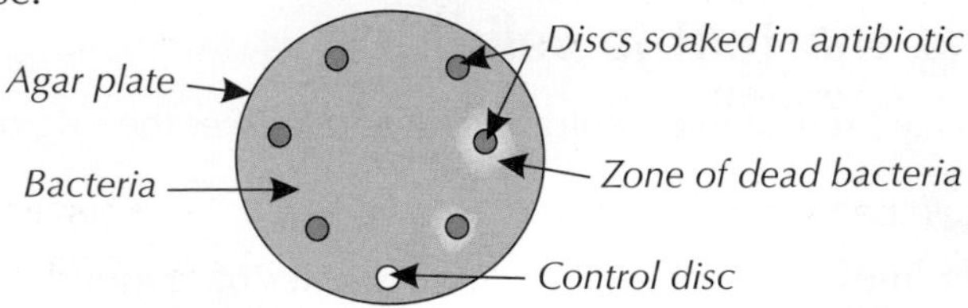

Figure 2: *An investigation into antibiotic resistance.*

> **Tip:** A study is an investigation that doesn't take place in a lab.

It's important that a study is a fair test, just like a lab experiment. It's a lot trickier to control the variables in a study than it is in a lab experiment though. Sometimes you can't control them all, but you can use a **control group** to help. This is a group of whatever you're studying (people, plants, lemmings, etc.) that's kept under the same conditions as the group in the experiment, but doesn't have anything done to it.

Example

If you're studying the effect of pesticides on crop growth, pesticide is applied to one field but not to another field (the control field). Both fields are planted with the same crop, and are in the same area (so they get the same weather conditions).

The control field is there to try and account for variables like the weather, which don't stay the same all the time, but could affect the results.

Tip: A pesticide is a chemical that can be used to kill insects.

Sample size

Data based on small samples isn't as good as data based on large samples. A sample should be representative of the whole population (i.e. it should share as many of the various characteristics in the population as possible) — a small sample can't do that as well.

The bigger the sample size the better, but scientists have to be realistic when choosing how big.

Tip: It's hard to spot anomalies if your sample size is too small.

Example

If you were studying how lifestyle affects people's weight it'd be great to study everyone in the UK (a huge sample), but it'd take ages and cost a bomb. Studying a thousand people with a mixture of ages, gender and race would be more realistic.

Tip: Samples should be selected randomly. There's an example of how to do this on page 373.

Trial runs

It's a good idea to do a **trial run** (a quick version of your experiment) before you do the proper experiment. Trial runs are used to figure out the range (the upper and lower limits) of independent variable values used in the proper experiment. If you don't get a change in the dependent variable at the upper values in the trial run, you might narrow the range in the proper experiment. But if you still get a big change at the upper values you might increase the range.

Tip: If you don't have time to do a trial run, you could always look at the data other people have got doing a similar experiment and use a range and interval values similar to theirs.

Example

For an experiment into the rate of an enzyme-controlled reaction, you might do a trial run with a temperature range of 10-50 °C. If there was no reaction at the upper end (e.g. 40-50 °C), you might narrow the range to 10-40 °C for the proper experiment.

Trial runs can be used to figure out the appropriate intervals (gaps) between the values too. The intervals can't be too small (otherwise the experiment would take ages), or too big (otherwise you might miss something).

Example

If using 1 °C intervals doesn't give much change in the rate of reaction each time, you might use 5 °C intervals instead, e.g. 10, 15, 20, 25, 30, 35, 40 °C.

Trial runs can also help you figure out whether or not your experiment is repeatable — e.g. if you repeat your experiment three times and the results are all similar, the experiment is repeatable.

Tip: Consistently repeating the results is crucial for checking that your results are repeatable.

6. Collecting Data

Once you've designed your experiment, you need to get on and do it. Here's a guide to making sure the results you collect are good.

Getting good quality results

When you do an experiment you want your results to be **repeatable**, **reproducible** and as **accurate** and **precise** as possible.

Tip: Sometimes, you can work out what result you should get at the end of an experiment (the theoretical result) by doing a bit of maths. If your experiment is accurate there shouldn't be much difference between the theoretical result and the result you actually get.

To check repeatability you need to repeat the readings and check that the results are similar — you should repeat each reading at least three times. To make sure your results are reproducible you can cross check them by taking a second set of readings with another instrument (or a different observer).

Your data also needs to be accurate. Really accurate results are those that are really close to the true answer. The accuracy of your results usually depends on your method — you need to make sure you're measuring the right thing and that you don't miss anything that should be included in the measurements. For example, estimating the amount of gas released from a reaction by counting the bubbles isn't very accurate because you might miss some of the bubbles and they might have different volumes. It's more accurate to measure the volume of gas released using a gas syringe (see p. 369).

Tip: For more on means see page 14.

Your data also needs to be precise. Precise results are ones where the data is all really close to the mean (average) of your repeated results (i.e. not spread out).

Example

Look at the data in this table. Data set 1 is more precise than data set 2 because all the data in set 1 is really close to the mean, whereas the data in set 2 is more spread out.

Repeat	Data set 1	Data set 2
1	12	11
2	14	17
3	13	14
Mean	13	14

Figure 1: Different types of measuring cylinder and glassware — make sure you choose the right one before you start an experiment.

Choosing the right equipment

When doing an experiment, you need to make sure you're using the right equipment for the job. The measuring equipment you use has to be sensitive enough to measure the changes you're looking for.

Example

If you need to measure changes of 1 cm^3 you need to use a measuring cylinder that can measure in 1 cm^3 steps — it'd be no good trying with one that only measures 10 cm^3 steps, it wouldn't be sensitive enough.

The smallest change a measuring instrument can detect is called its **resolution**. For example, some mass balances have a resolution of 1 g, some have a resolution of 0.1 g, and some are even more sensitive.

Also, equipment needs to be **calibrated** by measuring a known value. If there's a difference between the measured and known value, you can use this to correct the inaccuracy of the equipment.

Example

If a known mass is put on a mass balance, but the reading is a different value, you know that the mass balance has not been calibrated properly.

Tip: Calibration is a way of making sure that a measuring device is measuring things accurately — you get it to measure something you know has a certain value and set the device to say that amount.

Errors

Random errors

The results of an experiment will always vary a bit due to **random errors** — unpredictable differences caused by things like human errors in measuring.

Example

Errors made when reading from a measuring cylinder are random. You have to estimate or round the level when it's between two marks — so sometimes your figure will be a bit above the real one, and sometimes a bit below.

You can reduce the effect of random errors by taking repeat readings and finding the mean. This will make your results more precise.

Systematic errors

If a measurement is wrong by the same amount every time, it's called a **systematic error**.

Tip: If there's no systematic error, then doing repeats and calculating a mean can make your results more accurate.

Example

If you measured from the very end of your ruler instead of from the 0 cm mark every time, all your measurements would be a bit small.

Just to make things more complicated, if a systematic error is caused by using equipment that isn't zeroed properly it's called a **zero error**. You can compensate for some of these errors if you know about them though.

Tip: A zero error is a specific type of systematic error.

Example

If a mass balance always reads 1 gram before you put anything on it, all your measurements will be 1 gram too heavy. This is a zero error. You can compensate for this by subtracting 1 gram from all your results.

Tip: Repeating the experiment in the exact same way and calculating a mean won't correct a systematic error.

Anomalous results

Sometimes you get a result that doesn't seem to fit in with the rest at all. These results are called **anomalous results** (or outliers).

Example

Look at the data in this table. The entry that has been circled is an anomalous result because it's much larger than any of the other data values.

Experiment	A	B	C	D	E	F
Rate of reaction (cm^3/s)	10.5	11.2	10.8	85.4	10.6	11.1

Tip: There are lots of reasons why you might get an anomalous result, but usually they're due to human error rather than anything crazy happening in the experiment.

You should investigate anomalous results and try to work out what happened. If you can work out what happened (e.g. you measured something totally wrong) you can ignore them when processing your results.

7. Processing Data

Once you've collected some data, you might need to process it.

Organising data

It's really important that your data is organised. Tables are dead useful for organising data. When you draw a table use a ruler, make sure each column has a heading (including the units) and keep it neat and tidy.

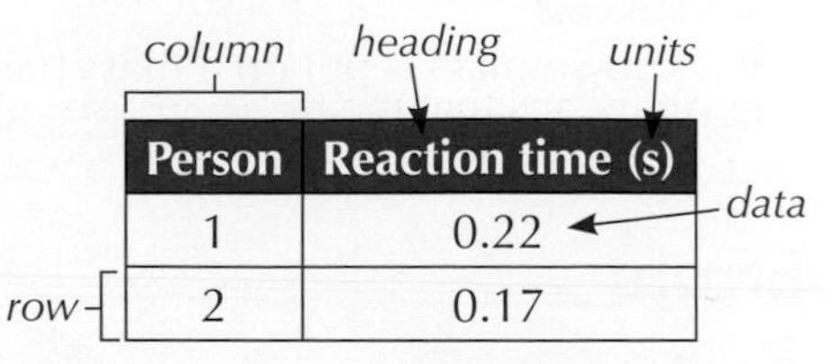

Person	Reaction time (s)
1	0.22
2	0.17

Figure 1: *Table showing the time taken to react to a stimulus for two people.*

Tip: If you're recording your data as decimals, make sure you give each value to the same number of decimal places.

Tip: Annoyingly, it's difficult to see any patterns or relationships in detail just from a table. You need to use some kind of graph or chart for that (see pages 16-18).

Processing your data

When you've collected data from an experiment, it's useful to summarise it using a few handy-to-use figures.

Mean and range

When you've done repeats of an experiment you should always calculate the **mean** (a type of average). To do this add together all the data values and divide by the total number of values in the sample.

You might also need to calculate the **range** (how spread out the data is). To do this find the largest number and subtract the smallest number from it.

Example

Look at the data in the table below. The mean and range of the data for each test tube has been calculated.

Test tube	Repeat (g)			Mean (g)	Range (g)
	1	2	3		
A	28	36	32	(28 + 36 + 32) ÷ 3 = 32	36 – 28 = 8
B	47	52	60	(47 + 52 + 60) ÷ 3 = 53	60 – 47 = 13

Median and mode

There are two more types of average, other than the mean, that you might need to calculate. These are the **median** and the **mode**.

- To calculate the median, put all your data in numerical order — the median is the middle value.
- The number that appears most often in a data set is the mode.

Tip: You should ignore anomalous results when calculating the mean, range, median or mode — see page 13 for more on anomalous results.

Example

The results of a study investigating how many minutes it took for people's heart rates to return to resting after sprinting 100 m are shown below:

3, 6, 5, 4, 6, 4, 6, 5, 7

MATHS SKILLS

First put the data in numerical order: 3, 4, 4, 5, 5, 6, 6, 6, 7

There are 9 values, so the median is the 5th number, which is **5**.

6 comes up 3 times. None of the other numbers come up more than twice. So the mode is **6**.

Tip: If you have an even number of values, the median is halfway between the middle two values.

Uncertainty

When you repeat a measurement, you often get a slightly different figure each time you do it due to random error. This means that each result has some **uncertainty** to it. The measurements you make will also have some uncertainty in them due to limits in the resolution of the equipment you use. This all means that the mean of a set of results will also have some uncertainty to it. Here's how to calculate the uncertainty of a mean result:

uncertainty = range ÷ 2

The larger the range, the less precise your results are and the more uncertainty there will be in your results. Uncertainties are shown using the '±' symbol.

Example

The table below shows the results of a respiration experiment to determine the volume of carbon dioxide produced.

Repeat	1	2	3	mean
Volume of CO_2 produced (cm^3)	20.2	19.8	20.0	20.0

1. The range is: 20.2 – 19.8 = 0.4 cm^3
2. So the uncertainty of the mean is: range ÷ 2 = 0.4 ÷ 2 = 0.2 cm^3.
   You'd write this as **20.0 ± 0.2 cm^3**

Measuring a greater amount of something helps to reduce uncertainty. For example, in a rate of reaction experiment, measuring the amount of product formed over a longer period compared to a shorter period will reduce the percentage uncertainty in your results.

Tip: There's more about errors on page 13.

Tip: Since uncertainty affects precision, you'll need to think about it when you come to evaluating your results (see page 22).

Rounding to significant figures

The first **significant figure** (s.f) of a number is the first digit that isn't a zero. The second, third and fourth significant figures follow on immediately after the first (even if they're zeros). When you're processing your data you may well want to round any really long numbers to a certain number of significant figures.

Example

0.6874976 rounds to **0.69** to **2 s.f.** and to **0.687** to **3 s.f.**

When you're doing calculations using measurements given to a certain number of significant figures, you should try to give your answer to the lowest number of significant figures that was used in the calculation. If your calculation has multiple steps, only round the final answer, or it won't be as accurate.

Example

For the calculation: 1.2 ÷ 1.85 = 0.648648648...

1.2 is given to 2 significant figures. 1.85 is given to 3 significant figures. So the answer should be given to 2 significant figures.

Round the final significant figure (0.6<u>4</u>8) up to 5: 1.2 ÷ 1.85 = **0.65 (2 s.f.)**

The lowest number of significant figures in the calculation is used because the fewer digits a measurement has, the less accurate it is. Your answer can only be as accurate as the least accurate measurement in the calculation.

Exam Tip
If a question asks you to give your answer to a certain number of significant figures, make sure you do this, or you might not get all the marks.

Tip: Remember to write down how many significant figures you've rounded to after your answer.

Tip: When rounding a number, if the next digit after the last significant figure you're using is less than 5 you should round it <u>down</u>, and if it's 5 or more you should round it <u>up</u>.

8. Graphs and Charts

It can often be easier to see trends in data by plotting a graph or chart of your results, rather than by looking at numbers in a table.

Plotting your data on a graph or chart

One of the best ways to present your data after you've processed it is to plot your results on a graph or chart. There are lots of different types you can use. The type you use depends on the type of data you've collected.

Bar charts

If either the independent or dependent variable is **categoric** you should use a bar chart to display the data.

Tip: Categoric data is data that comes in distinct categories, for example, blood type, eye colour, sex (i.e. whether you're male or female).

You also use a bar chart if one of the variables is **discrete** (the data can only take certain values, where there's no in-between value, e.g. number of people is discrete because you can't have half a person).

There are some golden rules you need to follow for drawing bar charts:

- Draw it nice and big (covering at least half of the graph paper).
- Leave a gap between different categories.
- Label both axes and remember to include the units.
- If you've got more than one set of data include a key.
- Give your graph a title explaining what it is showing.

Have a look at Figure 1 for an example of a pretty decent bar chart.

Tip: These golden rules will make sure that your bar chart is clear, easy to read and easy to understand if someone else looks at it.

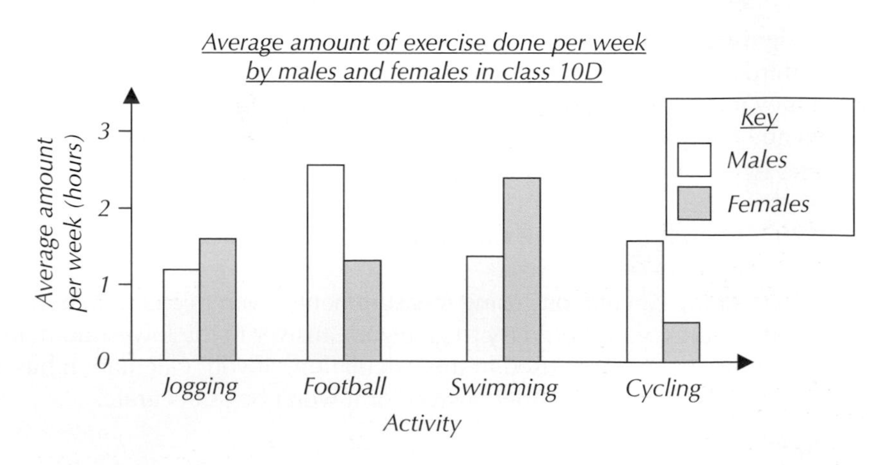

Figure 1: *An example of a bar chart.*

Histograms

Histograms are a useful way of displaying frequency data when the independent variable is **continuous** (numerical data that can have any value within a range, e.g. length, volume, temperature). They may look like bar charts, but it's the area of the bars that represents the frequency (rather than the height). The height of each bar is called the **frequency density**.

Tip: Frequency is just the number of times that something occurs.

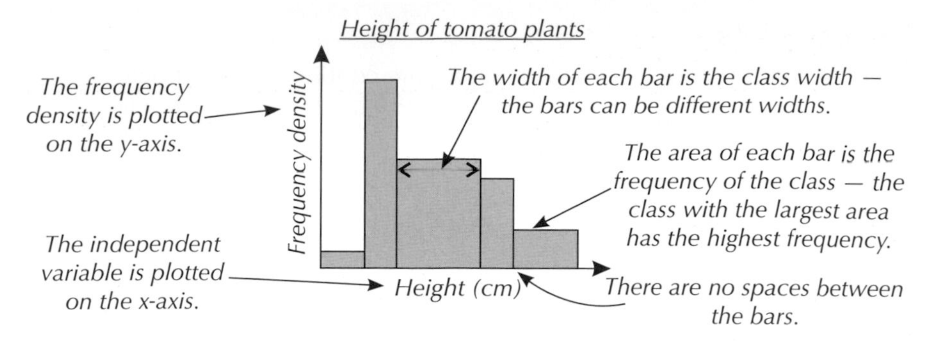

Figure 2: *An example of a histogram.*

You calculate the frequency density using this formula:

frequency density = frequency ÷ class width

Example

The table on the right shows the results of a study into variation in pea plant height. The heights of the plants were grouped into four classes.

Height of pea plant (cm)	Frequency
$0 \le x < 5$	5
$5 \le x < 10$	14
$10 \le x < 15$	11
$15 \le x < 30$	3

1. To draw a histogram of the data, you first need to work out the width of each class. Write the class width in a new column.

Class width
5 – 0 = **5**
10 – 5 = **5**
15 – 10 = **5**
30 – 15 = **15**

2. Use the formula above to calculate the frequency density for each class and write it in another new column.
3. Work out a suitable scale for each axis, then plot the histogram. It should look something like this:

Frequency density
5 ÷ 5 = **1**
14 ÷ 5 = **2.8**
11 ÷ 5 = **2.2**
3 ÷ 15 = **0.2**

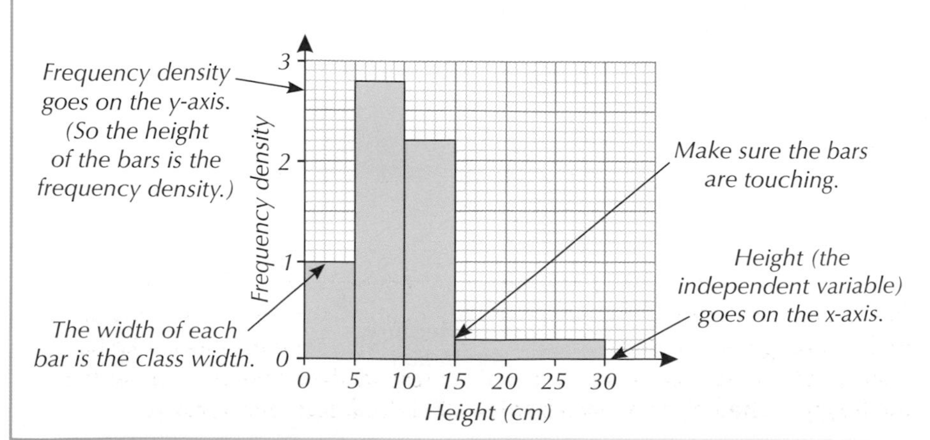

Tip: Don't be fooled by the height of the bars in a histogram — the tallest bar doesn't always belong to the class with the greatest frequency.

Tip: If all the classes are the same width, you can just plot frequency on the *y*-axis, e.g.

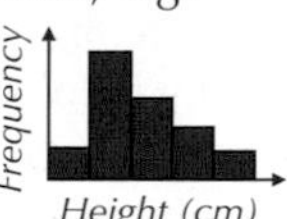

This type of graph is called a frequency diagram.

Tip: The continuous data here has been split into classes. $0 \le x < 5$ means the data in the class is more than or equal to 0 and less than 5.

Tip: The table in the example is a frequency table — it shows how many plants are in each height category.

Tip: You might have to round the frequency density — if so, choose a sensible number of decimal places that will be possible to plot using your graph's scale.

Tip: The width of the whole histogram shows the range of results (how spread out they are).

Tip: You can calculate the frequency of a class from a histogram by rearranging the formula: frequency = frequency density × class width.

Plotting points

If the independent and dependent variables are continuous you should plot points on a graph to display the data. Here are the golden rules for plotting points on graphs:

- Draw it nice and big (covering at least half of the graph paper).
- Put the independent variable (the thing you change) on the *x*-axis (the horizontal one).
- Put the dependent variable (the thing you measure) on the *y*-axis (the vertical one).
- Label both axes and remember to include the units.
- To plot the points, use a sharp pencil and make a neat little cross.
- In general, don't join the dots up. If you're asked to draw a line of best fit (or a curve of best fit if your points make a curve), try to draw the line through or as near to as many points as possible, ignoring anomalous results.
- If you've got more than one set of data include a key.
- Give your graph a title explaining what it is showing.

Exam Tip
You could be asked to plot points for a graph in your exam. If so, make sure you follow the golden rules or you could end up losing marks.

Tip: Use the biggest data values you've got to draw a sensible scale on your axes. Here, the highest rate of reaction is 22 cm³/s, so it makes sense to label the *y*-axis up to 25 cm³/s.

Tip: If you're not in an exam, you can use a computer to plot a graph and draw the line of best fit for you.

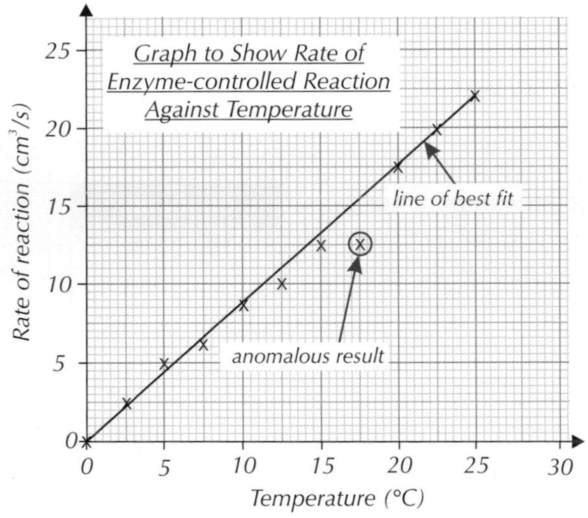

Figure 3: *An example of a graph with a line of best fit.*

Correlations

Graphs are used to show the relationship between two variables. Data can show three different types of **correlation** (relationship):

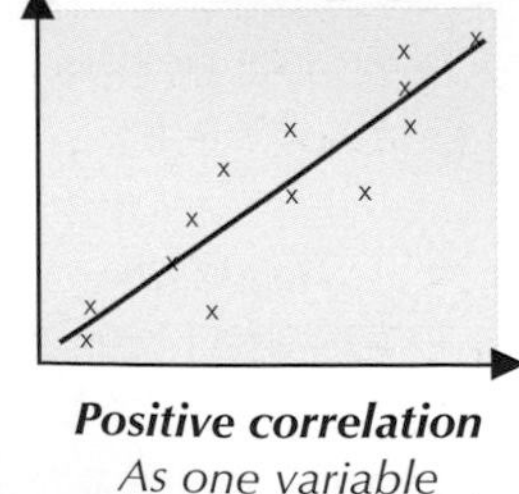

Positive correlation
As one variable increases the other increases.

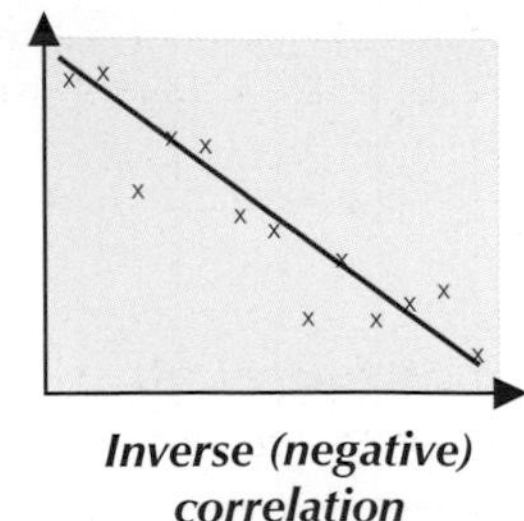

Inverse (negative) correlation
As one variable increases the other decreases.

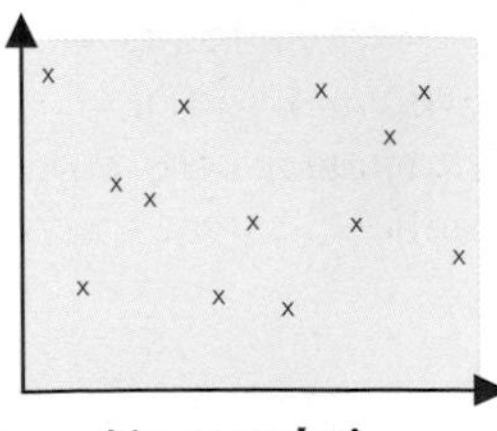

No correlation
There's no relationship between the two variables.

Tip: Just because two variables are correlated doesn't mean that the change in one is causing the change in the other. There might be other factors involved, or it may just be down to chance — see pages 21-22 for more.

9. Units

Using the correct units is important when you're drawing graphs or calculating values with an equation. Otherwise your numbers don't really mean anything.

SI units

Lots of different units can be used to describe the same quantity. For example, volume can be given in terms of cubic feet, cubic metres, litres or pints. It would be quite confusing if different scientists used different units to define quantities, as it would be hard to compare people's data. To stop this happening, scientists have come up with a set of standard units, called **SI units**, that all scientists use to measure their data. Figure 1 shows some SI units you'll see in biology:

Quantity	SI Base Unit
mass	kilogram, kg
length	metre, m
time	second, s

Figure 1: *Some common SI base units used in biology.*

Tip: SI stands for 'Système International', which is French for 'international system'.

Scaling prefixes

Quantities come in a huge range of sizes. For example, the volume of a swimming pool might be around 2 000 000 000 cm^3, while the volume of a cup is around 250 cm^3. To make the size of numbers more manageable, larger or smaller units are used. Figure 2 shows the prefixes which can be used in front of units (e.g. metres) to make them bigger or smaller:

prefix	tera (T)	giga (G)	mega (M)	kilo (k)	deci (d)	centi (c)	milli (m)	micro (μ)	nano (n)
multiple of unit	10^{12}	10^{9}	1 000 000 (10^{6})	1000	0.1	0.01	0.001	0.000001 (10^{-6})	10^{-9}

Figure 2: *Scaling prefixes used with units.*

These prefixes are called **scaling prefixes** and they tell you how much bigger or smaller a unit is than the base unit. So one kilometre is one thousand metres.

Converting between units

To swap from one unit to another, all you need to know is what number you have to divide or multiply by to get from the original unit to the new unit — this is called the **conversion factor** and is equal to the number of times the smaller unit goes into the larger unit.

- To go from a bigger unit to a smaller unit, you multiply by the conversion factor.
- To go from a smaller unit to a bigger unit, you divide by the conversion factor.

Tip: If you're going from a smaller unit to a larger unit, your number should get smaller.
If you're going from a larger unit to a smaller unit, your number should get larger.
This is a handy way to check you've done the conversion correctly.

There are some conversions that'll be particularly useful for GCSE biology. Here they are...

- Mass can have units of kg and g.

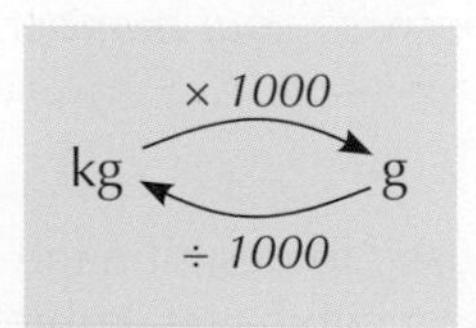

- Length can have lots of units, including mm, μm and nm.

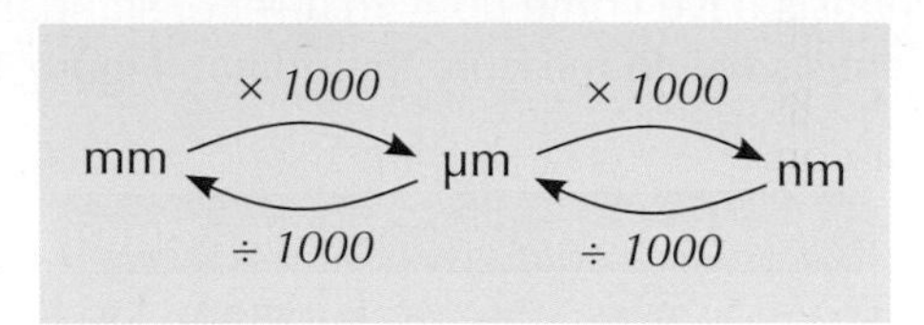

Exam Tip
Being familiar with these common conversions could save you time when it comes to doing calculations in the exam.

- Time can have units of min and s.

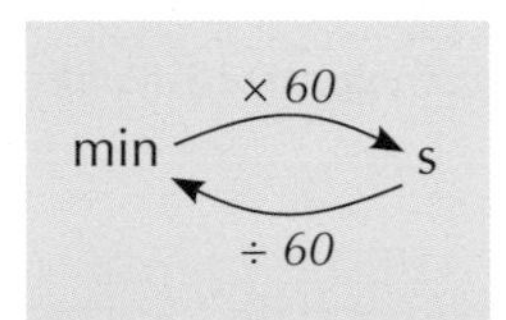

- Volume can have units of m^3, dm^3 and cm^3.

m^3 (× 1000 →, ← ÷ 1000) dm^3 (× 1000 →, ← ÷ 1000) cm^3

Tip: Volume is also often given in L (litres) or ml (millilitres). To convert, you'll need to remember that 1 ml = 1 cm^3 and so 1 L =1000 cm^3.

Examples

- To go from metres to centimetres, you'd multiply by the conversion factor (which is 100).

  2 m is equal to 2 × 100 = **200 cm**

- To go from grams to kilograms, you'd divide by the conversion factor (which is 1000).

  3400 g is equal to 3400 ÷ 1000 = **3.4 kg**

10. Conclusions and Evaluations

So... you've planned and carried out an amazing experiment, got your data and have processed and presented it in a sensible way. Now it's time to figure out what your data actually tells you, and how much you can trust what it says.

How to draw conclusions

Drawing conclusions might seem pretty straightforward — you just look at your data and say what pattern or relationship you see between the dependent and independent variables.

But you've got to be really careful that your conclusion matches the data you've got and doesn't go any further. You also need to be able to use your results to justify your conclusion (i.e. back up your conclusion with some specific data).

When writing a conclusion you need to refer back to the original hypothesis and say whether the data supports it or not.

Example

The table shows the heights of pea plant seedlings grown for three weeks with different fertilisers.

Fertiliser	Mean growth (mm)
A	13.5
B	19.5
No fertiliser	5.5

You could conclude that fertiliser B makes pea plant seedlings grow taller over a three week period than fertiliser A.

The justification for this conclusion is that over the three week period, fertiliser B made pea plants grow 6 mm more on average than fertiliser A.

You can't conclude that fertiliser B makes any other type of plant grow taller than fertiliser A — the results could be totally different.

The hypothesis for this experiment might have been that adding fertiliser would increase the growth of plants and that different types of fertiliser would affect growth by different amounts. The data supports this hypothesis.

Figure 1: *A pea plant seedling.*

Correlation and causation

If two things are correlated (i.e. there's a relationship between them) it doesn't necessarily mean that a change in one variable is causing the change in the other — this is really important, don't forget it. There are three possible reasons for a correlation:

Tip: Graphs can be useful for seeing whether two variables are correlated (see page 18).

Tip: Causation just means one thing is causing another.

1. Chance

Even though it might seem a bit weird, it's possible that two things show a correlation in a study purely because of chance.

Example

One study might find a correlation between the number of people with breathing problems and the distance they live from a cement factory. But other scientists don't get a correlation when they investigate it — the results of the first study are just a fluke.

Tip: Lots of things are correlated without being directly related. E.g. the level of carbon dioxide (CO_2) in the atmosphere and the amount of obesity have both increased over the last 100 years, but that doesn't mean increased atmospheric CO_2 is causing people to become obese.

2. They're linked by a third variable

A lot of the time it may look as if a change in one variable is causing a change in the other, but it isn't — a third variable links the two things.

Example

There's a correlation between water temperature and shark attacks. This isn't because warm water makes sharks crazy. Instead, they're linked by a third variable — the number of people swimming (more people swim when the water's hotter, and with more people in the water shark attacks increase).

3. Causation

Sometimes a change in one variable does cause a change in the other.

Example

There's a correlation between smoking and lung cancer.
This is because chemicals in tobacco smoke cause lung cancer.

You can only conclude that a correlation is due to cause if you've controlled all the variables that could be affecting the result. (For the smoking example, this would include age and exposure to other things that cause cancer.)

Evaluation

An evaluation is a critical analysis of the whole investigation. Here you need to comment on the following points about your experiment and the data you gathered:

- **The method**: Was it valid? Did you control all the other variables to make it a fair test?
- **The quality of your results:** Was there enough evidence to reach a valid conclusion? Were the results repeatable, reproducible, accurate and precise?
- **Anomalous results**: Were any of the results anomalous? If there were none then say so. If there were any, try to explain them — were they caused by errors in measurement? Were there any other variables that could have affected the results? You should comment on the level of uncertainty in your results too.

Tip: When you're commenting on the quality of your results, think about things like whether you took enough repeats and whether other people's results were similar to your own.

Once you've thought about these points you can decide how much confidence you have in your conclusion. For example, if your results are repeatable, reproducible, accurate and precise and they back up your conclusion then you can have a high degree of confidence in your conclusion.

You can also suggest any changes to the method that would improve the quality of the results, so that you could have more confidence in your conclusion. For example, you might suggest changing the way you controlled a variable, or increasing the number of measurements you took. Taking more measurements at narrower intervals could give you a more accurate result.

You could also make more predictions based on your conclusion, then further experiments could be carried out to test them.

Tip: When suggesting improvements to the investigation, always make sure that you say why you think this would make the results better.

1. Cell Structure

All living things are made of cells — they're the building blocks of every organism on the planet. But different organisms have different cell structures...

Learning Objectives:
- Know that plant and animal cells are eukaryotic cells, and that bacterial cells are prokaryotic cells.
- Understand that prokaryotic cells are much smaller than eukaryotic cells.
- Know that most animal cells have a nucleus, cytoplasm, cell membrane, mitochondria and ribosomes, and know the function of each of these parts.
- Know that plant and algal cells have the same parts as animal cells, plus a cell wall.
- Know that plant cells also usually have a permanent vacuole and chloroplasts, and know the function of each of these parts.
- Know that bacterial cells have a cell wall, cell membrane, cytoplasm, a single loop of DNA and may have plasmids, and know the function of each of these parts.

Specification References
4.1.1.1, 4.1.1.2

Prokaryotes and eukaryotes

Cells can be either **prokaryotic** or **eukaryotic**. Eukaryotic cells are complex and include all animal and plant cells. Prokaryotic cells are smaller and simpler, e.g. bacteria (see next page).

Eukaryotes are organisms that are made up of eukaryotic cells.
A **prokaryote** is a prokaryotic cell (it's a single-celled organism).

Both eukaryotic and prokaryotic cells contain various cell parts called **subcellular structures**.

Animal cells

Most animal cells have the following subcellular structures — make sure you know them all. The parts are labelled in Figure 1.

- **Nucleus** — contains genetic material that controls the activities of the cell.
- **Cytoplasm** — a gel-like substance where most of the chemical reactions happen. It contains enzymes (see p. 115) that control these chemical reactions.
- **Cell membrane** — holds the cell together and controls what goes in and out.
- **Mitochondria** — these are where most of the reactions for aerobic respiration take place (see p. 173). Respiration transfers energy that the cell needs to work.
- **Ribosomes** — these are where proteins are made in the cell.

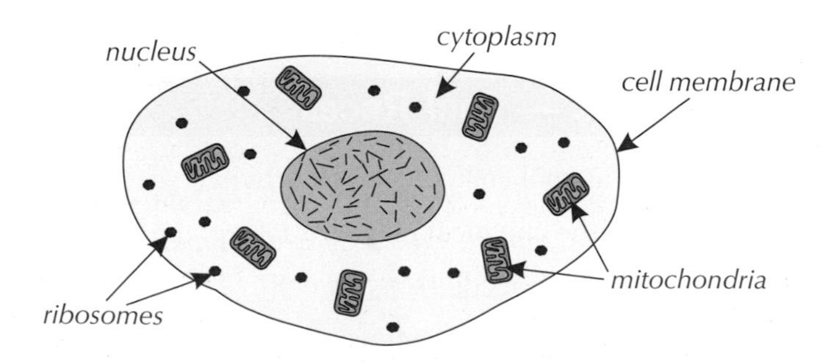

Figure 1: *The structure of a typical animal cell.*

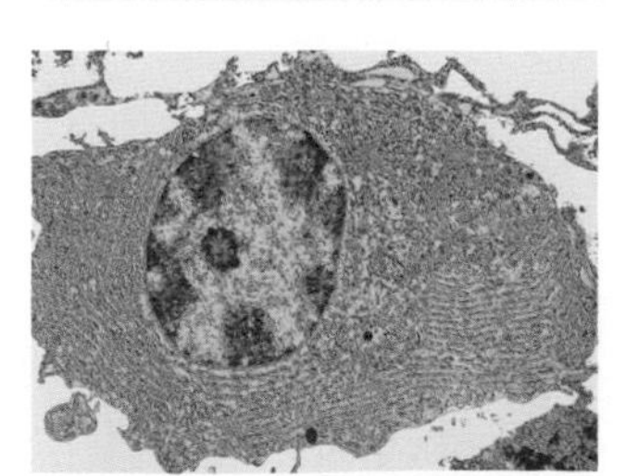

Figure 2: *A human cell seen under a microscope — the blue and yellow oval is the nucleus.*

> **Tip:** The diagrams on this page and the previous one all show 'typical' cells.
> In reality the structure of a cell varies according to what job it does, so most cells won't look exactly like these.

Plant cells

Plant cells usually have all the bits that animal cells have, plus a few extra:

- **Cell wall** — a rigid structure made of cellulose. It supports and strengthens the cell. The cells of algae (e.g. seaweed) also have a rigid cell wall.
- **Permanent vacuole** — contains cell sap, a weak solution of sugar and salts.
- **Chloroplasts** — these are where photosynthesis occurs, which makes food for the plant (see page 158). They contain a green substance called **chlorophyll**, which absorbs the light needed for photosynthesis.

The subcellular structures of a typical plant cell are shown in Figure 4.

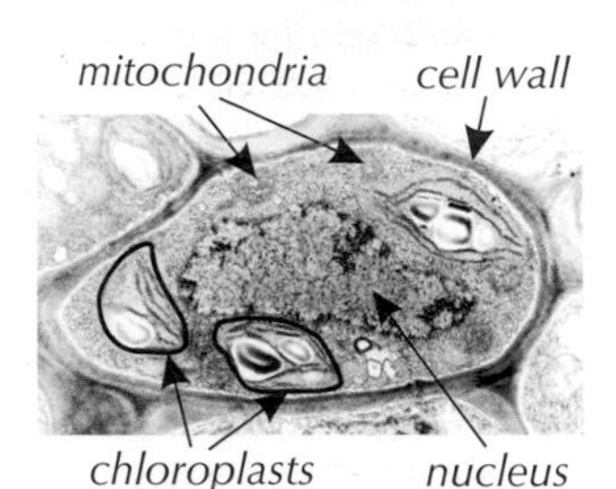

Figure 3: *A cross-section of a plant cell seen under a microscope.*

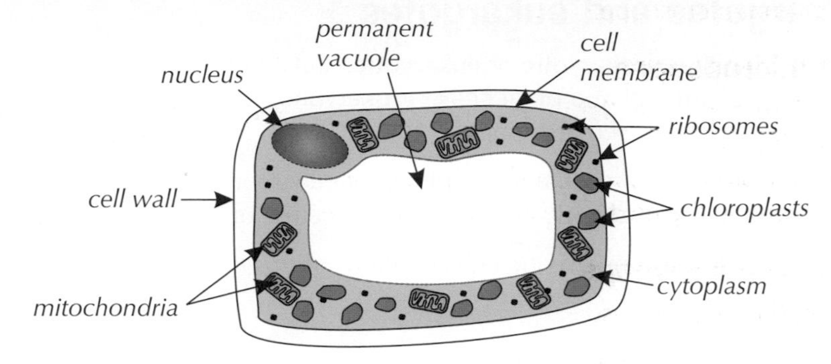

Figure 4: *The structure of a typical plant cell.*

Bacterial cells

Bacteria are prokaryotes. A bacterial cell has cytoplasm and a cell membrane surrounded by a cell wall. The cell doesn't have a 'true' nucleus — instead it has a single circular strand of DNA that floats freely in the cytoplasm (see Figure 5). Bacterial cells may also contain one or more small rings of DNA called **plasmids**.

> **Tip:** Bacterial cells contain ribosomes, but they don't contain any chloroplasts or mitochondria.

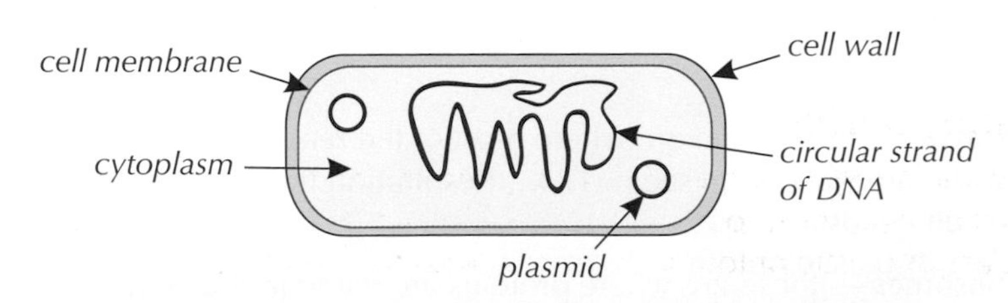

Figure 5: *The structure of a typical bacterial cell.*

Practice Questions — Fact Recall

Q1 Which part of an animal cell controls its activity?

Q2 Where do most of the chemical reactions take place in a cell?

Q3 What are mitochondria needed for in a cell?

Q4 Name three things that a plant cell usually has, that an animal cell doesn't.

Q5 What is a plasmid?

> **Exam Tip**
> You need to learn the functions of the subcellular structures, not just their names.

2. Microscopy

Microscopy is the study of very small objects (such as cells) using an instrument called a microscope. It's an essential part of Biology, really.

Microscopes

Microscopes let us see things that we can't see with the naked eye, such as individual cells and their subcellular structures. The microscopy techniques we can use have developed over the years as technology and knowledge have improved. Two common types of microscope are the light microscope and the electron microscope.

Light microscopes

Light microscopes use light and lenses to form an image of a specimen and magnify it (make it look bigger). They let us see individual cells and large subcellular structures, like nuclei. You can read all about how to use a light microscope on the next page.

Electron microscopes

Electron microscopes use electrons instead of light to form an image.

They have a much higher **magnification** than light microscopes.
They also have a higher **resolution**. (Resolution is the ability to distinguish between two points, so a higher resolution gives a sharper image.)

Electron microscopes let us see much smaller things in more detail, like the internal structure of mitochondria and chloroplasts. They even let us see tinier things like ribosomes and plasmids.

Using a light microscope

REQUIRED PRACTICAL 1

You need to know the ins and outs of using a light microscope, including how to prepare your specimen, how to focus using the microscope and how to make accurate drawings of what you observe.

Preparing a slide

If you want to look at a specimen (e.g. plant or animal cells) under a light microscope, you need to put it on a microscope slide first. A slide is a strip of clear glass or plastic onto which the specimen is mounted.

For example, here's how to prepare a slide to view onion cells:

1. Add a drop of water to the middle of a clean slide.
2. Cut up an onion and separate it out into layers. Use tweezers to peel off some epidermal tissue from the bottom of one of the layers.
3. Using the tweezers, place the epidermal tissue into the water on the slide.
4. Add a drop of iodine solution. Iodine solution is a **stain**. Stains are used to highlight objects in a cell by adding colour to them.

Examples

- Iodine solution is used to stain starch in plant cells.
- Eosin is used to make the cytoplasm show up.

Learning Objectives:

- Understand that, over the years, microscopy techniques have improved.
- Know the difference in magnification and resolving power between light microscopes and electron microscopes.
- Understand how the ability to see cells in more detail under an electron microscope has allowed biologists to have a better understanding of subcellular structures.
- Be able to use a light microscope to look at plant and animal cells, and to correctly draw and label observations (Required Practical 1).
- Be able to use the magnification formula and express answers in standard form.
- Be able to understand the size and scale of cells.
- Understand how to use estimations to work out the relative size or area of subcellular structures.

Specification References
4.1.1.1, 4.1.1.2, 4.1.1.5

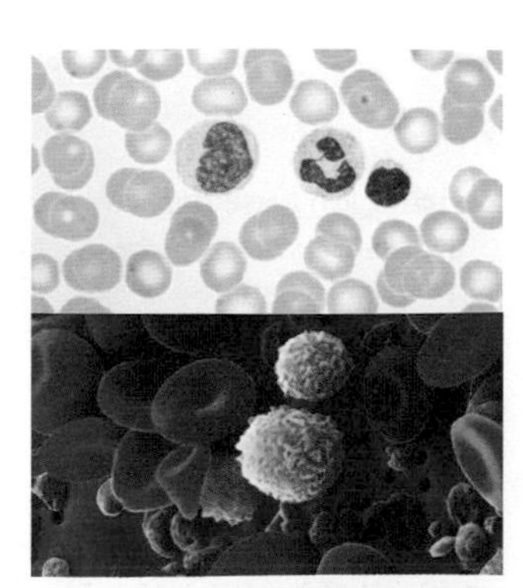

Figure 1: *Blood cells seen under a light microscope (top) and an electron microscope (bottom).*

Tip: Make sure you carry out a risk assessment so that you're aware of any hazards, particularly for any stains you're using, before you start preparing your slide.

Tip: Sometimes you can place the specimen on the slide without it being suspended in liquid — it depends on what you're looking at.

Tip: In light microscopes, the beam of light passes through the object being viewed. An image is produced because some parts of the object absorb more light than others. Sometimes the object being viewed is completely transparent, so the whole thing looks white because the light rays just pass straight through. To get round this, stains are used to highlight parts of the object.

5. Place a cover slip (a square of thin, transparent plastic or glass) on top. To do this, stand the cover slip upright on the slide, next to the water droplet. Then carefully tilt and lower it so it covers the specimen. Try not to get any air bubbles under there — they'll obstruct your view of the specimen. Steps 1-5 are shown in Figure 2.

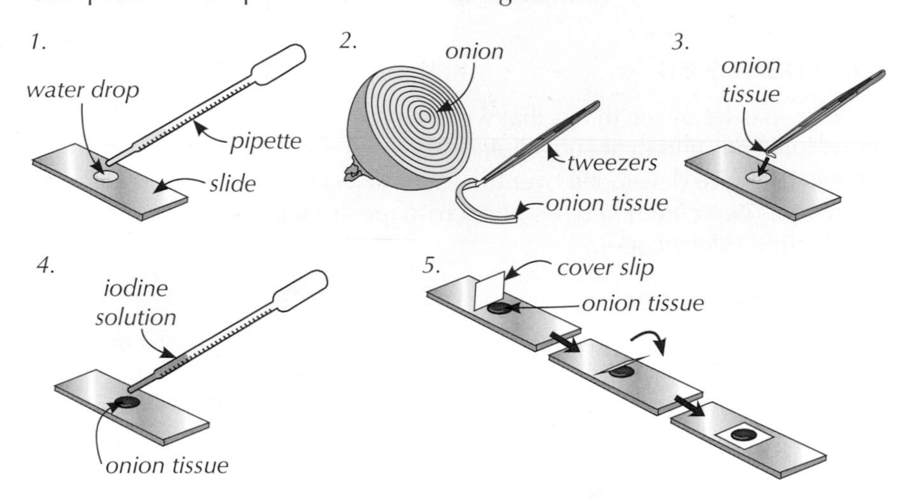

Figure 2: *Preparation of a microscope slide.*

Observing the specimen

You need to know how to set up and use a light microscope (see Figure 3) to observe your prepared slide:

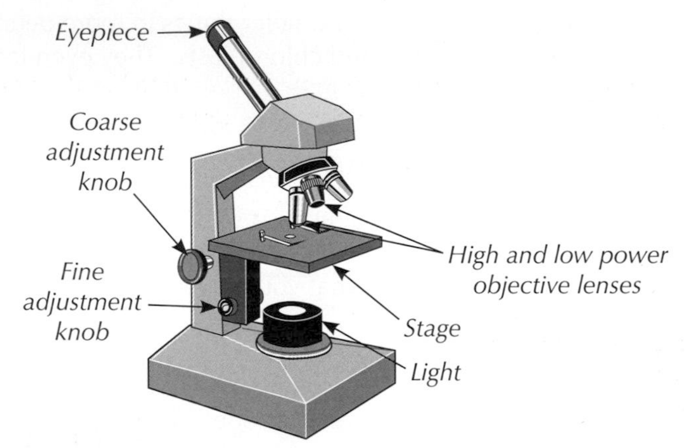

Figure 3: *A light microscope.*

Tip: Don't use the eyepiece while you're moving the stage up with the coarse adjustment knob, as you could cause the lens and stage to collide.

1. Start by clipping the slide you've prepared onto the stage.
2. Select the lowest-powered objective lens (i.e. the one that produces the lowest magnification).
3. Use the coarse adjustment knob to move the stage up to just below the objective lens.
4. Look down the eyepiece. Use the coarse adjustment knob to move the stage downwards until the image is roughly in focus.
5. Adjust the focus with the fine adjustment knob, until you get a clear image of what's on the slide.
6. If you need to see the slide with greater magnification, swap to a higher-powered objective lens and refocus.

Drawing observations

When making a drawing of your specimen from under the microscope, make sure you use a pencil with a sharp point and that you draw with clear, unbroken lines. Your drawing should not include any colouring or shading. If you are drawing cells, the subcellular structures should be drawn in proportion (the correct sizes relative to each other). If you're asked to make your drawing in a certain amount of space (e.g. on a worksheet), it needs to take up at least half of the space available.

You also need to include a title of what you were observing and label the important features of your drawing (e.g. nucleus, chloroplasts) using straight, uncrossed lines. Finally, you need to work out and write down the magnification of your drawing.

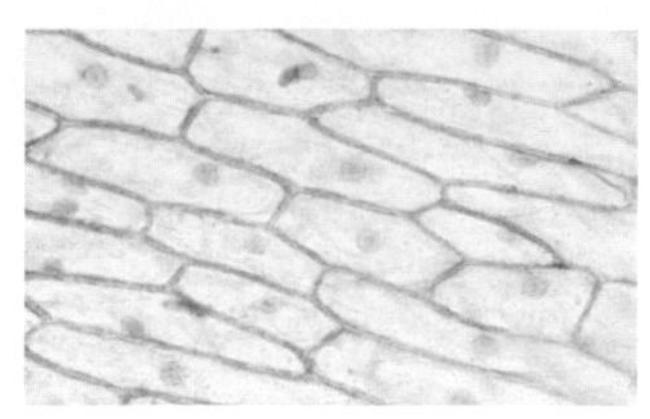

Figure 4: *Onion cells stained with iodine, viewed under × 400 magnification.*

Tip: You can work out the magnification of your drawing using this formula: magnification = length of drawing of cell ÷ real length of cell. For example, in Figure 5, 40 mm ÷ 0.3 mm = 133

Tip: Make sure you also know the magnification that your specimen was observed under.

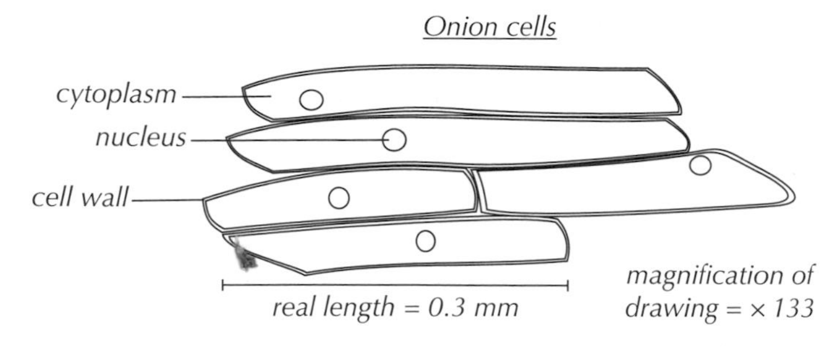

Figure 5: *A drawing of onion cells as viewed under a light microscope.*

Tip: You can work out the real size of a cell by counting the number of cells in 1 mm of the sample — see page 372.

Magnification

Magnification is how much bigger the image is than the real object that you're looking at. It's calculated using this formula:

$$\text{magnification} = \frac{\text{image size}}{\text{real size}}$$

Example

If you have a magnified image that's 2 mm wide and your specimen is 0.02 mm wide the magnification is:

$$\text{magnification} = \frac{\text{image size}}{\text{real size}}$$

$$= \frac{2}{0.02} = \mathbf{\times\ 100}$$

Tip: Both the image size and the real size should have the same units. If they don't, you'll need to convert them first (see page 19).

If you want to work out the image size or the real size of the object, you have to rearrange the equation.

Example

Calculating image size

If your specimen is 0.3 mm wide and the magnification of the microscope is × 50, then the size of the image is:

$$\text{image size} = \text{magnification} \times \text{real size}$$

$$= 50 \times 0.3 = \mathbf{15\ mm}$$

Exam Tip
If you find rearranging formulas hard you can use a formula triangle to help:

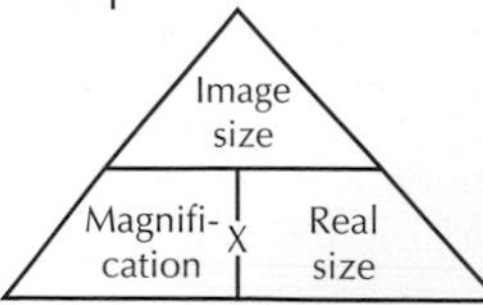

All you do is put your finger over the one you want and read off the formula. E.g. if you want the real size of the object, you put your finger over that and it leaves behind image size ÷ magnification.

Tip: All cells are really small, so you might see their sizes written in standard form. You can find out more about standard form on p. 376.

Calculating the real size of an object

If you have a magnified image that's 10 mm wide and the magnification is × 400, then the real size of the specimen you're looking at is:

$$\text{real size} = \frac{\text{image size}}{\text{magnification}} = \frac{10}{400} = \mathbf{0.025\ mm}$$

Also, you might need to write the real size of a specimen in **standard form**.

Example

To write 0.025 mm in standard form:

1. Move the decimal point to after the '2' so that the first number is between 1 and 10.
2. Write the number of places the decimal point has moved as the power of 10. Don't forget the minus sign because the decimal point has moved right.

$$0.025 \longrightarrow \mathbf{2.5 \times 10^{-2}\ mm}$$

Converting between units

When you're calculating magnification you need to make sure that all measurements have the same units, e.g. all in millimetres. When dealing with microscopes these units can get pretty tiny. The table below shows common units and how to convert between them:

To convert	Unit	How many millimetres it is	To convert
	Millimetre (mm)	1 mm	
× 1000	Micrometre (μm)	0.001 mm	÷ 1000
× 1000	Nanometre (nm)	0.000001 mm	÷ 1000

Examples

MATHS SKILLS

- To convert from a smaller unit to a bigger unit you have to divide. So to convert 5 micrometres to millimetres you divide 5 by 1000. 5 ÷ 1000 = **0.005 mm**
- To go from a bigger unit to a smaller unit you have to multiply. So to convert 0.025 micrometres to nanometres you multiply 0.025 by 1000. 0.025 × 1000 = **25 nm**
- If you have a specimen that's 5 μm wide and an image that's 5 mm wide, you need to put both measurements in the same units before you can work out the magnification using the formula:

$$5\ \text{mm} = 5 \times 1000 = 5000\ \mu\text{m}$$

$$\text{magnification} = \frac{\text{image size}}{\text{real size}} = \frac{5000}{5} = \mathbf{\times\ 1000}$$

Tip: It doesn't matter which units you choose to convert to. In the calculation on the right, if you converted 5 μm to mm (5 ÷ 1000 = 0.005 μm) instead, you'd get the same answer of × 1000 magnification.

Estimating size and area of cell structures

You can estimate the size of a subcellular structure by comparing its size to that of the cell.

> **Tip:** Estimates can come in handy if you don't have a small enough scale to measure something accurately.

Example

Figure 6 shows a cheek cell.

Estimate the width of the nucleus of the cell.

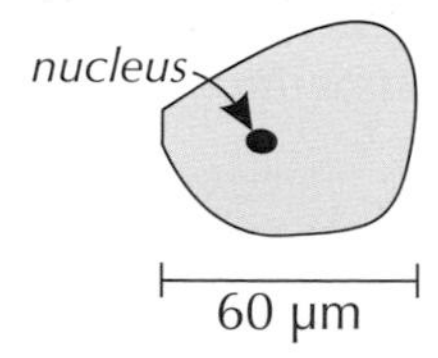

Figure 6: *Cheek cell with scale.*

1. Estimate the number of times that the nucleus could fit across the width of the cell.

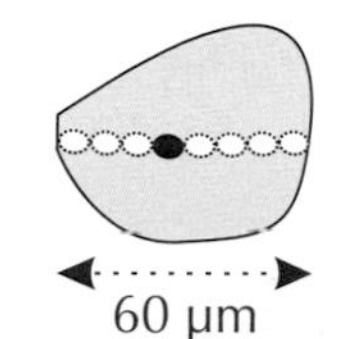

The nucleus would fit across the cell about 8 times.

2. Divide the width of the cell by the number of times the nucleus would fit across it.

width of nucleus = 60 μm ÷ 8 = **7.5 μm**

> **Exam Tip**
> In the exam, you could be given a photomicrograph (a picture taken down a light microscope) of a cell and its subcellular structures rather than a diagram like this. You can estimate the size of the subcellular structures in a photomicrograph in the same way.

You can estimate the area of a subcellular structure by comparing it to a regular shape.

Example

Figure 7 shows a diagram of a mitochondrion measured under an electron microscope.

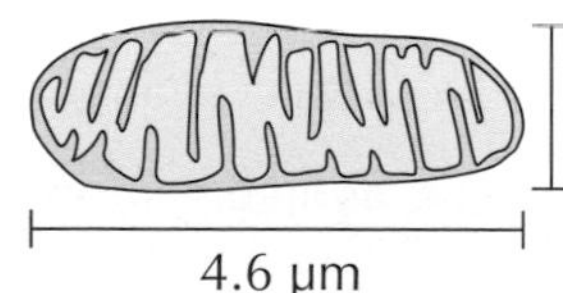

1.5 μm

Figure 7: *A mitochondrion.*

The shape of the mitochondrion is close to that of a rectangle.

So to estimate its area, use the formula for the area of a rectangle:
area = length × width

Area = 4.6 μm × 1.5 μm = **6.9 μm²**

> **Tip:** Don't forget to include the units of area in your answer.

Practice Questions — Application

Q1 A bacterial cell is 0.002 mm long. A magnified image of the cell is 18 mm long. What is the magnification?

Q2 A plant cell is 0.08 mm wide. It is examined under a × 400 microscope lens. Calculate the size of the magnified image.

Q3 A specimen is observed under × 1500 magnification. The magnified image is 10.5 mm wide. What is the real width of the specimen? Give your answer in μm.

> **Tip:** If it helps you to work out what size you're looking for, you could draw a diagram of what the question is describing. So for Q2 it might be like this:
>
> 0.08 mm → × 400 → ? mm
> real size — image

Learning Objectives:
- Understand that many cells are specialised — they perform a specific function.
- Understand that cells differentiate to become specialised as an organism develops, and that differentiation involves the cells developing different subcellular structures, which enable the cells to carry out their specific functions.
- Know that most animal cells lose the ability to differentiate at an early stage but lots of plant cells never lose the ability to differentiate.
- Know that mature animals mainly use cell division to replace and repair cells.
- Be able to relate the structure of different types of cell to their function, including sperm cells, nerve cells and muscle cells in animals and root hair cells, xylem and phloem cells in plants.

Specification References
4.1.1.3, 4.1.1.4

3. Cell Differentiation and Specialisation

Not all cells in an organism do the same job. A cell's structure is related to the job it does, so cell structure can vary. Different cell structures are the result of differentiation.

What is a specialised cell?

A specialised cell is one that performs a specific function. Most cells in an organism are specialised. A cell's structure (e.g. its shape and the parts it contains) helps it to carry out its function — so depending on what job it does, a specialised cell can look very different to the cells you saw on pages 23-24.

Cell differentiation

Differentiation is the process by which a cell changes to become specialised for its job. As cells change, they develop different subcellular structures and turn into different types of cells (see Figure 1).

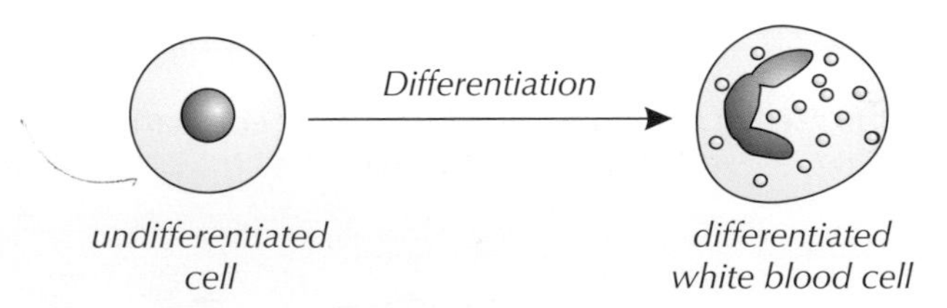

Figure 1: *Diagram showing cell differentiation.*

Most differentiation occurs as an organism develops. In most animal cells, the ability to differentiate is then lost after they become specialised. However, lots of plant cells don't ever lose this ability. The cells that differentiate in mature animals are mainly used for repairing and replacing cells, such as skin or blood cells.

Some cells are undifferentiated cells — they're called stem cells. There's more about them on pages 44-46.

Examples of specialised cells

In the exam, you could be asked to relate the structure of a cell to its function. The idea is that you apply your knowledge of cell structure (see pages 23-24) to the information you're given about the role of the cell in an organism. Don't panic — it's easier than it sounds. Here are a few examples to help you understand how cell structure and function are related:

1. Sperm cells (in animals)

Sperm cells are specialised for reproduction. The function of a sperm is to get the male DNA to the female DNA. It has a long tail and a streamlined head to help it swim to the egg. There are also lots of **mitochondria** (see p. 23) in the cell to provide the energy it needs to do this. It also carries enzymes in its head to digest through the egg cell membrane. The structure of a sperm cell is shown by Figures 2 and 3 on the next page.

Exam Tip
Energy is transferred by respiration, which mostly takes place in the mitochondria. So any cell that needs lots of energy to do its job will have lots of mitochondria.

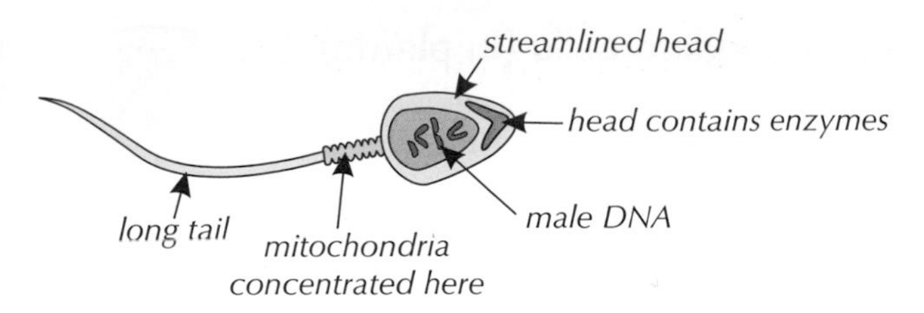

Figure 2: *A sperm cell.*

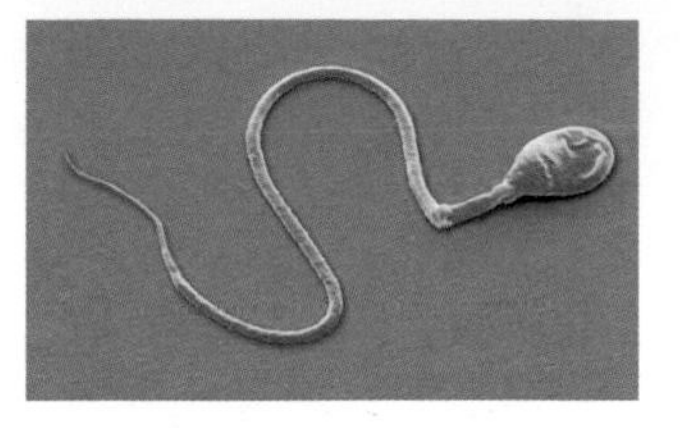

Figure 3: *A microscope image of a sperm cell. It's easy to see the streamlined head and tail.*

2. Nerve cells (in animals)

Nerve cells are specialised for rapid signalling. The function of nerve cells is to carry electrical signals from one part of the body to another. These cells are long (to cover more distance) and have branched connections at their ends to connect to other nerve cells and form a network throughout the body.

Tip: Nerve cells join together to form the nervous system (see page 186).

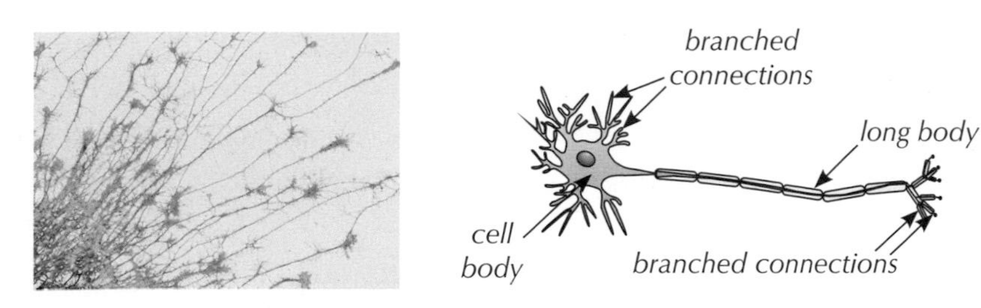

Figure 4: *An electron microscope image of the branched connections from the cell body of a nerve cell (left) and the whole structure of a nerve cell (right).*

Tip: The electrical signals are slowed down when they pass between two nerve cells, so a few long cells are better for fast signal transmission than lots of short cells.

3. Muscle cells (in animals)

Muscle cells are specialised for contraction. The function of a muscle cell is to contract quickly. These cells are long (so that they have space to contract) and contain lots of mitochondria to transfer the energy needed for contraction.

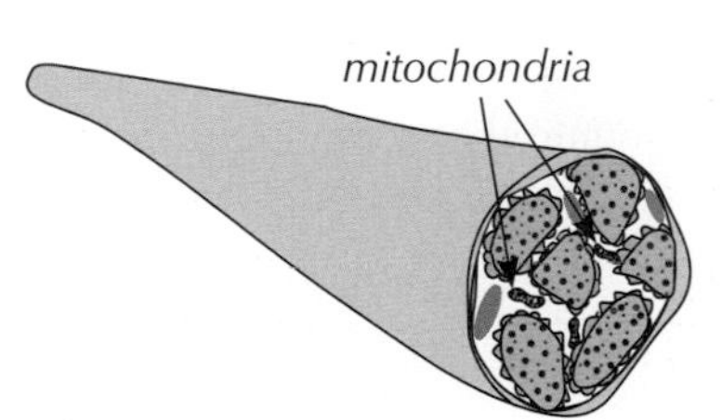

Figure 5: *A muscle cell.*

4. Root hair cells (in plants)

Root hair cells are cells on the surface of plant roots that are specialised for absorbing water and minerals. They grow into long "hairs" that stick out into the soil. This gives the plant a big surface area for absorbing water and mineral ions from the soil.

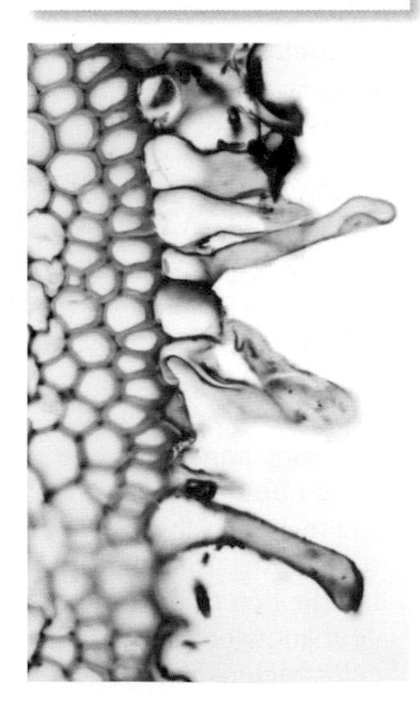

Figure 6: *Root hair cells as seen under a microscope.*

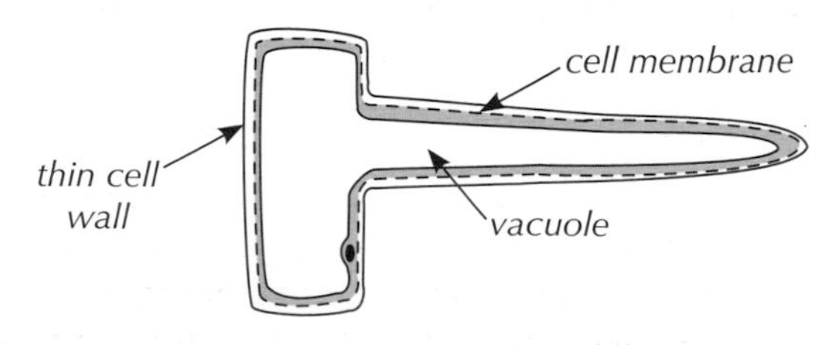

Figure 7: *A root hair cell.*

Tip: Root hair cells don't contain any chloroplasts — this is because they are underground and so they don't carry out photosynthesis.

5. Phloem and xylem cells (in plants)

Phloem and xylem cells are specialised for transporting substances. They form phloem and xylem tubes, which transport substances such as food and water around plants.

To form the tubes, the cells are long and joined end to end. Xylem cells are hollow in the centre and phloem cells have very few subcellular structures, so that substances can flow through them.

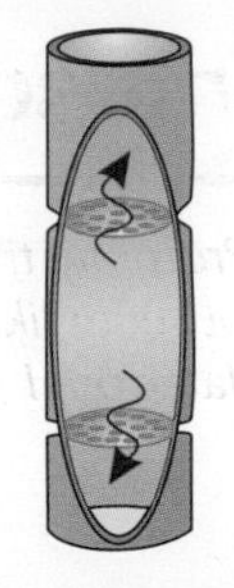

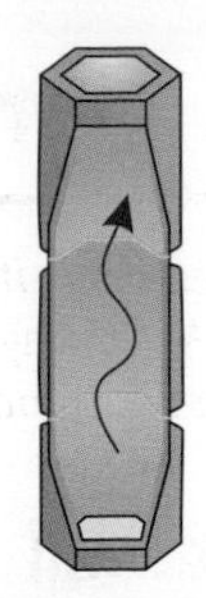

Figure 8: *A phloem tube (left) and xylem tube (right).*

Tip: There's more about phloem and xylem on page 91.

Practice Questions — Fact Recall

Q1 What is cell differentiation?

Q2 Describe the differences in the ability of plant and animal cells to differentiate.

Q3 a) What is the function of a nerve cell?

b) Describe two features of a nerve cell that make it specialised for its function.

Q4 Muscle cells need lots of energy for contraction. What feature of muscle cells make them specialised for this?

Practice Questions — Application

Tip: Think about the function of ribosomes — see page 23.

Q1 The function of gastric chief cells is to secrete enzymes (proteins) into the stomach during digestion. Comment on the amount of ribosomes you'd expect to find in a gastric chief cell.

Q2 A diagram of an epithelial cell from the small intestine is shown here:

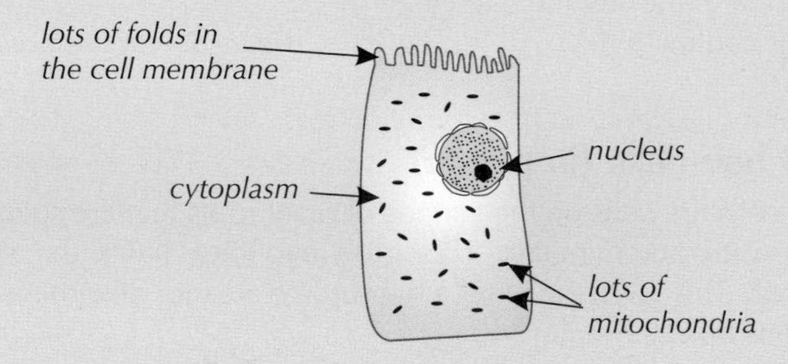

These cells line the inner surface of the small intestine and their function is to absorb food molecules as they move through the intestine. Sometimes this process requires energy.

Use this information and the diagram to suggest two ways in which the structure of these cells helps them to carry out their function.

Tip: You're not expected to actually <u>know</u> the answer to these questions — just make sensible suggestions using the information given and your knowledge about cell structure from pages 23-24.

Tip: Look back at the examples on the last couple of pages if you're struggling with this one — the function of some of those cells involves absorption too.

4. Chromosomes

Well, this is it. Probably the most important topic in the whole of Biology — DNA. It may not seem like a big deal right now, but this stuff is going to come in really handy later on, I promise. So crack on and learn it.

Learning Objectives:
- Know that chromosomes are found in the cell nucleus.
- Know that chromosomes are usually found in pairs in body cells.
- Know that chromosomes are made of DNA and carry genes.

Specification Reference 4.1.2.1

The cell nucleus

Most cells in your body have a structure called a **nucleus**. The nucleus contains your genetic material (the instructions you need to grow and develop). This material is stored in the form of **chromosomes** — see Figure 1.

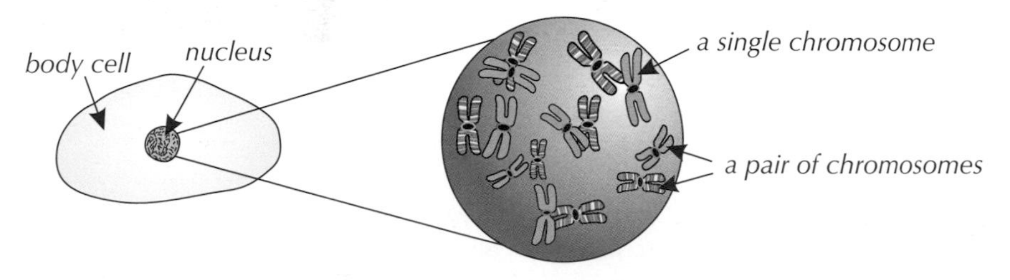

Figure 1: *Diagram to show that a cell nucleus contains chromosomes.*

Body cells normally have two copies of each chromosome — one from the organism's 'mother', and one from its 'father'. So, humans have two copies of chromosome 1, two copies of chromosome 2, etc.

Figure 2 shows the 23 pairs of chromosomes from a human cell. (The 23rd pair are a bit different — see page 261).

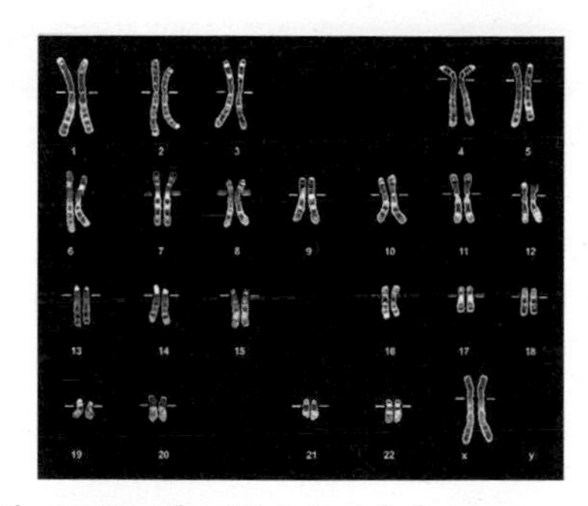

Figure 2: *The 23 pairs of chromosomes from a female human body cell.*

Tip: You might sometimes see chromosomes drawn like this:

That's because chromosomes only look like this:

just before a cell divides for reproduction and growth (see p. 35).

Chromosomes and DNA

Chromosomes are long lengths of a molecule called **DNA**.
The DNA is coiled up to form the arms of the chromosome (see Figure 4).

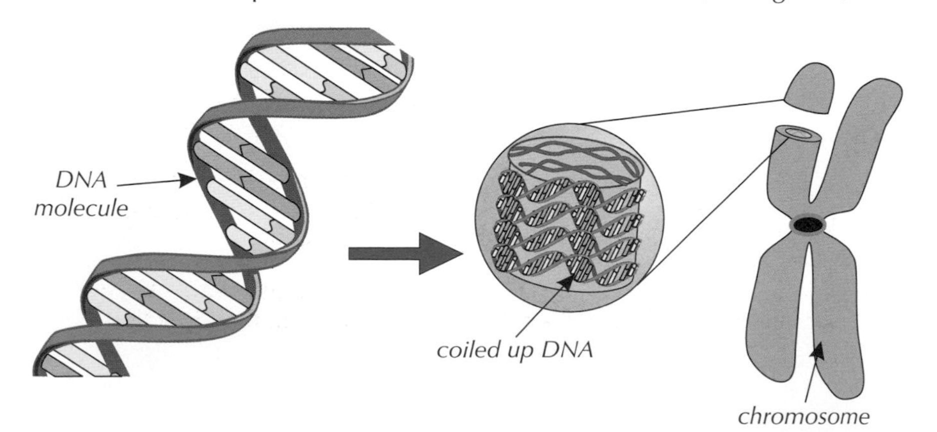

Figure 3: *Diagram to show how DNA coils up to form a chromosome.*

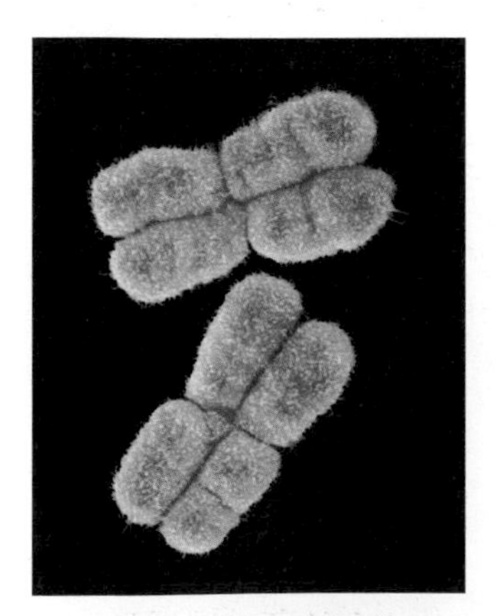

Figure 4: *A pair of human chromosomes seen under a microscope. (In fact, it's pair No. 3.)*

Chromosomes and genes

Each chromosome carries a large number of genes (see Figure 5). These are short sections of DNA. Genes control our characteristics. Different genes control the development of different characteristics, e.g. hair colour, eye colour. There's more about genes on page 244.

Tip: A chromosome contains thousands of genes, not just one. This diagram is just to help show you what's going on.

gene for brown hair colour

gene for green eye colour

brown hair colour

green eye colour

Figure 5: *Diagram showing relationship between chromosomes, genes and characteristics.*

Practice Questions — Fact Recall

Q1 Where in the cell are chromosomes found?

Q2 What molecule are chromosomes made up of?

Q3 a) What are genes?

b) What do we need different genes for?

Practice Questions — Application

The photograph on the right was taken under a microscope. It shows a cell, its nucleus and chromosomes.

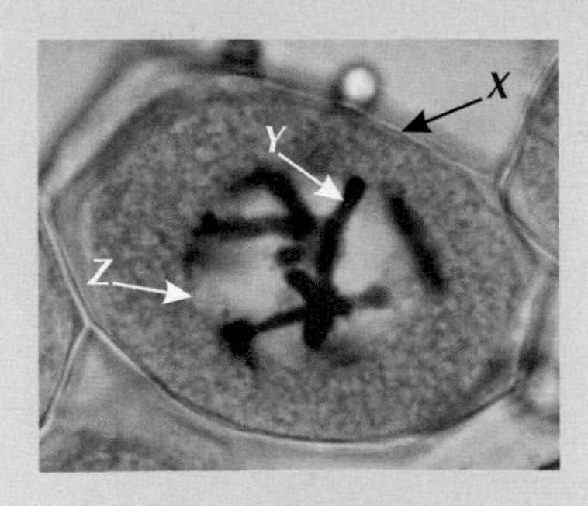

Q1 Which arrow is pointing to:

a) the cell?

b) the nucleus?

c) a chromosome?

Q2 Suggest why it is not possible to see individual genes in this photograph.

Exam Tip
Make sure you don't get genes and chromosomes mixed up in the exam if you have to label them — chromosomes are long lengths of DNA and they carry genes.

5. Mitosis

Our body cells are able to make copies of themselves so that we can grow or repair any damaged tissue. It's pretty clever stuff...

Learning Objectives:

- Know that body cells divide by mitosis, which is a stage of the cell cycle.
- Know that mitosis allows organisms to grow and develop.
- Be able to describe how, during the cell cycle, body cells grow and increase their amount of subcellular structures, then copy their genetic material before undergoing a single division.
- Understand how a cell divides by mitosis to form two genetically identical cells.
- Be able to describe mitosis in a given context and recognise when mitosis is occurring.

Specification Reference 4.1.2.2

Mitosis and the cell cycle

Body cells in multicellular organisms divide to produce new cells as part of a series of stages called the cell cycle. The cell cycle starts when a cell has been produced by cell division and ends with the cell dividing to produce two identical cells. It can be summarised as a stage of cell growth and DNA replication, followed by a stage of cell division (see Figure 1).

The stage of the cell cycle when the cell divides is called **mitosis**:

> Mitosis is when a cell reproduces itself by splitting to form two identical offspring.

Multicellular organisms use mitosis to grow and develop or replace cells that have been damaged. The end of the cell cycle results in two new cells identical to the original cell, with the same number of chromosomes.

cell growth and DNA replication

mitosis

***Figure 1**: The cell cycle*

As a cell goes through the cell cycle, it has to undergo some key changes.

What happens during the cell cycle?

Here are the steps involved in the cell cycle:

Growth and replication

1. In a cell that's not dividing, the DNA is all spread out in long strings.
2. Before it divides, the cell has to grow and increase the amount of subcellular structures such as mitochondria and ribosomes.
3. It then duplicates its DNA — so there's one copy of each chromosome for each new cell. The DNA is copied and forms X-shaped chromosomes. Each 'arm' of the chromosome is an exact duplicate of the other.

   Now the cell is ready for mitosis...

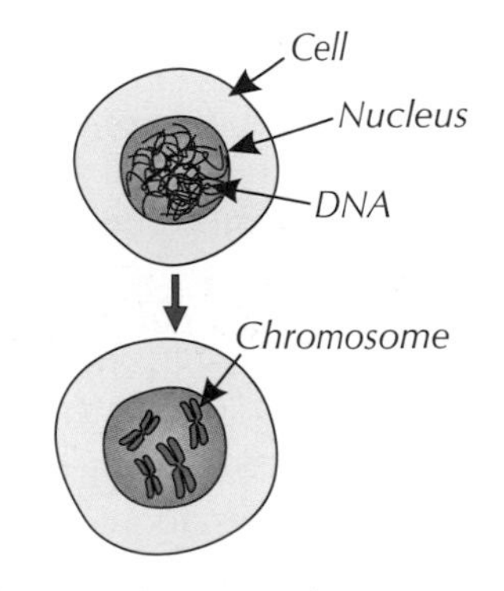

***Figure 2:** Diagram showing the growth and replication stage of the cell cycle.*

Tip: When a cell has copied its DNA, the left arm of an X-shaped chromosome has the same DNA as the right arm.

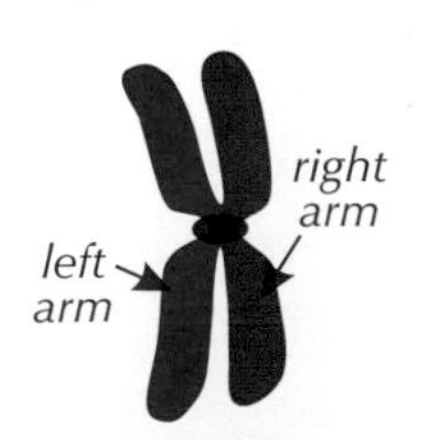

Mitosis

4. The chromosomes line up at the centre of the cell and cell fibres pull them apart. The two arms of each chromosome go to opposite ends of the cell.
5. Membranes form around each of the sets of chromosomes. These become the nuclei of the two new cells — the nucleus has divided.
6. Lastly, the cytoplasm and cell membrane divide.

The cell has now produced two new daughter cells. The daughter cells contain exactly the same DNA — they're identical. Their DNA is also identical to the parent cell.

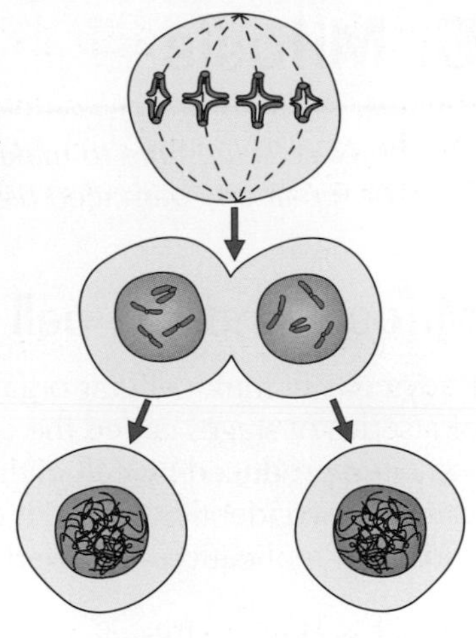

Figure 3: *Diagram showing the steps of mitosis in the cell cycle.*

Tip: Mitosis is really one continuous process, but it's often described as a series of stages.

Tip: The key things to remember about mitosis are that the dividing cell already contains two copies of its DNA and it splits into two, to form two genetically identical daughter cells.

Recognising mitosis

In your exam, you might have to identify or describe what's going on in photos of real cells undergoing mitosis.

Example

The photos below show onion cells in different stages of the cell cycle. Describe what is happening to each cell and identify which cells are going though mitosis.

A

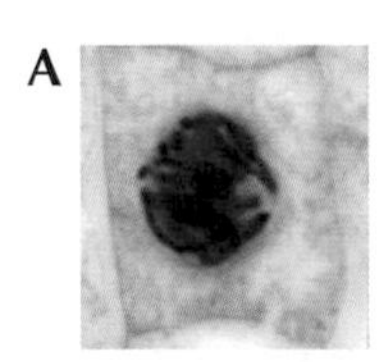

B

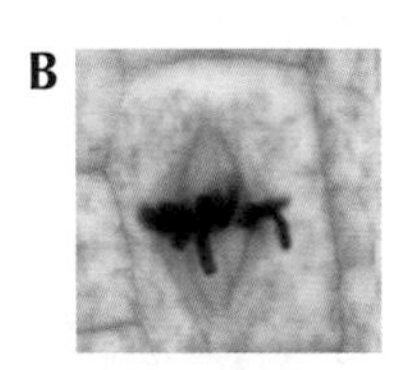

C

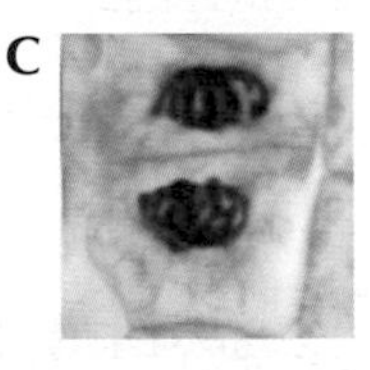

In cell A, the chromosomes are visible in the nucleus. The DNA has formed X-shaped chromosomes because it's being duplicated. So this cell must be in the replication stage of the cell cycle — it is not yet going through mitosis.

In cell B, the chromosomes are lined up across the centre of the cell. You can just about see the cell fibres, which will pull the two arms of each chromosome apart. So this cell is undergoing mitosis.

In cell C, the chromosomes have been pulled apart to opposite ends of the cell. They have formed two separate nucleii. A cell membrane is visible between the two halves of the cell, dividing the cytoplasm, so the cell must be at the end stage of mitosis. It is about to form two identical daughter cells.

Tip: If the cell isn't going through the cell cycle (so it's not dividing or preparing to divide), you won't be able to see the chromosomes in the nucleus.

How long is each stage of the cell cycle?

The time taken for a stage of the cell cycle varies depending on the cell type and the environmental conditions. You can calculate how long a stage lasts if you're given the right information.

Example

MATHS SKILLS

A scientist observes a section of growing tissue under a microscope. He counts 120 cells in one field of view. Of those, 42 cells are in the replication stage of the cell cycle. One complete cell cycle of the cells in this tissue lasts 24 hours. How long do the cells spend in the replication stage of the cell cycle? Give your answer in minutes.

1. The scientist has observed that 42 out of 120 cells are in the replication stage of the cell cycle. This suggests that the proportion of time the cells spend in this stage must be 42/120th of the cell cycle.

2. You're told that the cell cycle in these cells lasts 24 hours. That's (24 × 60 =) 1440 minutes.

3. So the cells spend 42/120th of 1440 minutes in the replication stage, which you can work out like this:

   $\frac{42}{120} \times 1440 =$ **504 minutes** in the replication stage of the cell cycle.

Tip: The fraction of cells in any one stage of the cell cycle is proportional to the time taken for that stage. So if you know the number of cells in a particular stage, the total number of cells and the total time for the cell cycle, you can work out an estimate for how long that particular stage takes.

Practice Questions — Fact Recall

Q1 Give two uses of mitosis in multicellular animals.

Q2 How many times does a body cell divide during mitosis?

Practice Questions — Application

This photograph shows the last stage in mitosis — two new daughter cells are formed.

Q1 a) Will these new cells be genetically identical or genetically different to the parent cell?

b) How many sets of chromosomes will these cells have?

Q2 In a tissue sample of 198 cells, 6 cells are observed to be in the last stage in mitosis. One complete cell cycle in the tissue cells takes 1500 minutes. How long does the last stage in mitosis take for an average cell in this tissue? Give your answer in hours.

Exam Tip
You don't have to know anything to do with the specific cells to answer a mitosis question like this in the exam.

6. Binary Fission

It's not just cell structure that sets prokaryotic and eukaryotic cells apart — they have different methods of dividing too.

Learning Objectives:

- Know that bacteria multiply by binary fission.
- Know that binary fission can occur as quickly as every 20 minutes.
- Understand that the speed at which bacteria multiply is affected by temperature and the availability of nutrients.
- Be able to use the mean division time to calculate the number of bacteria in a population after a certain time.
- H Be able to write the number of bacteria in standard form.

Specification Reference 4.1.1.6

Prokaryotic cell replication

Prokaryotic cells, such as bacteria, replicate by a type of simple cell division called **binary fission**. In binary fission, the cell makes copies of its genetic material, before splitting into two daughter cells.

The process of binary fission

Step 1

The circular DNA and plasmid(s) replicate.

Step 2

The cell gets bigger and the circular DNA strands move to opposite 'poles' (ends) of the cell.

Step 3

The cytoplasm begins to divide and new cell walls begin to form.

Step 4

The cytoplasm divides and two daughter cells are produced. Each daughter cell has one copy of the circular DNA, but can have a variable number of copies of the plasmid(s).

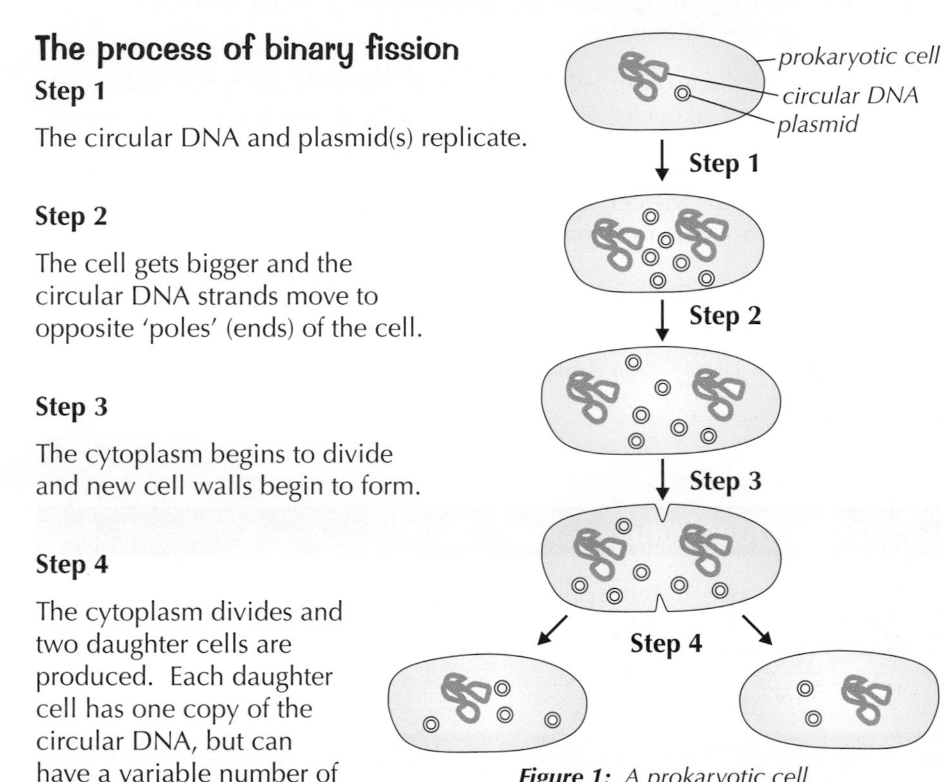

Figure 1: *A prokaryotic cell undergoing binary fission.*

Tip: During Step 1, the circular DNA is copied once, but the plasmid(s) can be copied multiple times.

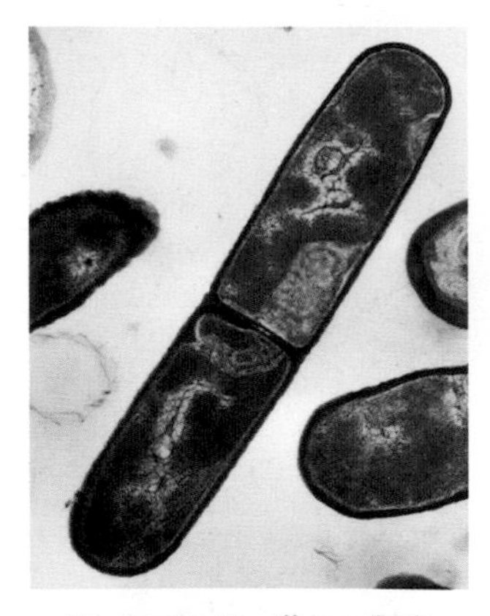

Figure 2: *Bacillus subtilis bacteria replicating by binary fission.*

Bacteria can divide very quickly if given the right conditions (e.g. a warm environment and lots of nutrients). Some bacteria, such as *E. coli*, can take as little as 20 minutes to replicate in the right environment.

However, if conditions become unfavourable, the cells will stop dividing and eventually begin to die.

The changes in the size of a population of bacteria over time can be shown in a graph — see Figure 3.

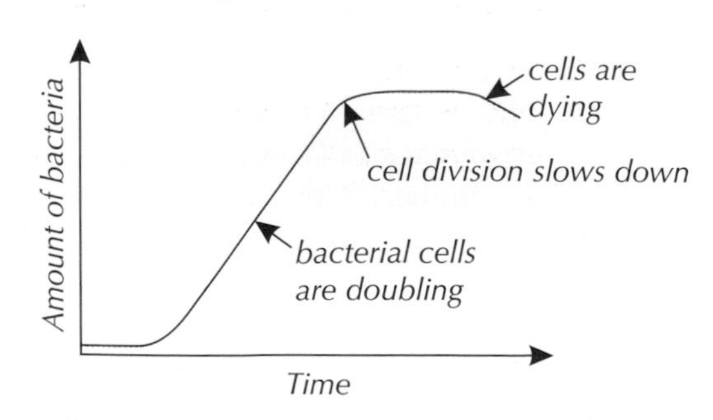

Figure 3: *A graph showing how a population of bacteria changes over time.*

Mean division time

If you want to estimate the number of bacteria in a population, you need to use the **mean division time**. The mean division time is just the average amount of time it takes for one bacterial cell to divide into two. If you know the mean division time of a cell, you can work out how many times it has divided in a certain amount of time, and so the number of cells it has produced in that time.

Example

A bacterial cell has a mean division time of 30 minutes. How many cells will it have produced after 2.5 hours?

1. Make sure both times are in the same units:

   2.5 hours × 60 = 150 minutes

2. Divide the total time that the bacteria are producing cells by the mean division time. This gives you the number of divisions:

   150 minutes ÷ 30 minutes = 5 divisions

3. Multiply 2 by itself for the number of divisions to find the number of cells:

   $2^5 = 2 \times 2 \times 2 \times 2 \times 2 =$ **32 cells**

Tip: 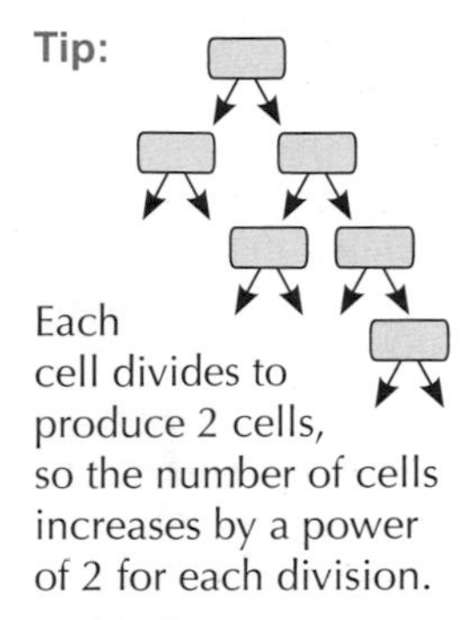

Each cell divides to produce 2 cells, so the number of cells increases by a power of 2 for each division.

Tip: H If you do a calculation like the one on the left and end up with a number of bacteria that's in the thousands, you can write the amount in standard form — have a look at pages 376-377 for some tips.

Practice Questions — Application

Q1 a) A bacterial cell divides by binary fission. It has a mean division time of 24 minutes. How many cells will it have produced after 6 hours? Give your answer to two significant figures.

b) Write your answer to part a) in standard form.

Q2 A scientist has found that under certain conditions, *Staphylococcus aureus* has a mean division time of 30 minutes.

a) How long will it take for one *Staphylococcus aureus* cell to produce a population of 128 cells? Give your answer in hours.

b) Under a different set of conditions, one *Staphylococcus aureus* cell produces 64 cells in 270 minutes. What is the mean division time of *S. aureus* under these conditions?

Tip: If you know the size of a population of bacteria that has been produced from one bacterial cell, you can work out how many divisions have taken place by finding the size of the population as a power of 2. E.g. to produce 16 bacterial cells, 4 divisions must have take place ($16 = 2 \times 2 \times 2 \times 2 = 2^4$).

7. Culturing Microorganisms

REQUIRED PRACTICAL 2

As well as understanding how bacteria multiply, you need to know how to grow them in the lab. That includes how to grow them safely.

Learning Objectives:
- Know that bacteria can be grown in the lab on an agar plate or in a nutrient broth.
- Understand why bacteria should not be grown at temperatures above 25 °C in a school laboratory.
- Be able to investigate the effect of antibiotics or antiseptics on the growth of bacteria (Required Practical 2).
- Know that uncontaminated cultures are required for bacterial growth investigations.
- Be able to describe how to prepare an uncontaminated bacterial culture and explain why the techniques involved are necessary.
- Be able to use πr^2 to calculate the area of an inhibition zone or bacterial colony.

Specification Reference 4.1.1.6

How are microorganisms cultured in the lab?

Bacteria (and some other microorganisms) are grown (cultured) in a "culture medium", which contains the carbohydrates, minerals, proteins and vitamins they need to grow. The culture medium used can be a nutrient broth solution or solid agar jelly. Bacteria grown on agar 'plates' will form visible colonies on the surface of the jelly, or will spread out to give an even covering of bacteria.

In the lab at school, cultures of microorganisms are not kept above 25 °C, because harmful pathogens (microorganisms that cause disease — see page 130) are more likely to grow above this temperature.

In industrial conditions, cultures are incubated at higher temperatures so that they can grow a lot faster.

Growing microorganisms on an agar plate

To make an agar plate, hot agar jelly is poured into shallow round plastic dishes called Petri dishes. When the jelly's cooled and set, inoculating loops (wire loops) can be used to transfer microorganisms to the culture medium. Alternatively, a sterile dropping pipette and spreader can be used to get an even covering of bacteria. The microorganisms then multiply.

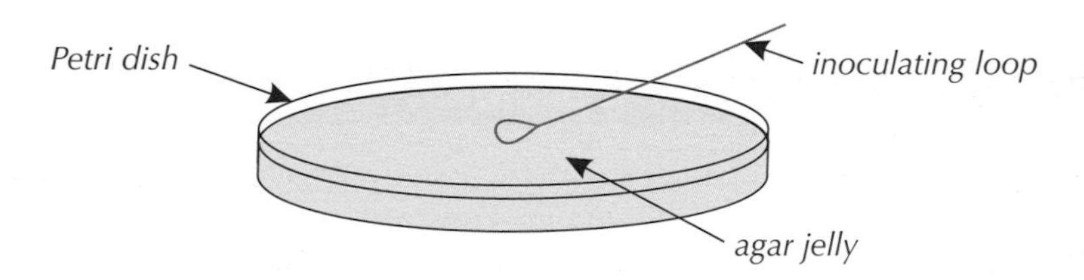

Figure 1: *An agar plate and inoculating loop.*

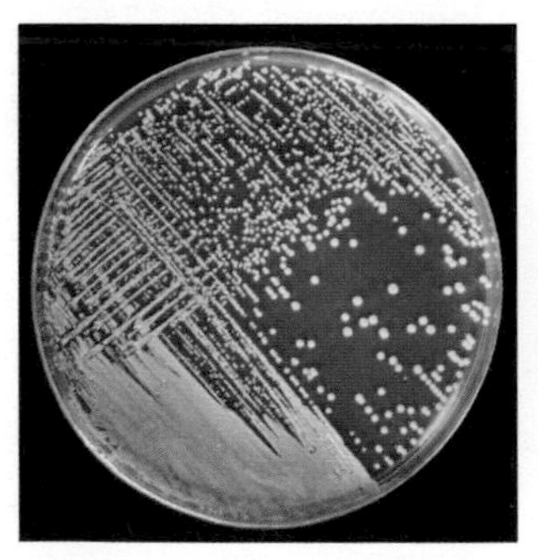

Figure 2: *Bacterial colonies on an agar plate.*

Tip: Working with microorganisms can be hazardous. Make sure you carry out a full risk assessment before you carry out this practical.

Investigating the effect of antibiotics on bacterial growth

You can test the action of antibiotics on cultures of bacteria by carrying out the following steps:

1. Place paper discs soaked in different types (or different concentrations) of antibiotics on an agar plate that has an even covering of bacteria. Leave some space between the discs.

2. The antibiotic should diffuse (soak) into the agar jelly. Antibiotic-resistant bacteria (i.e. bacteria that aren't affected by the antibiotic — see page 306) will continue to grow on the agar around the paper discs, but non-resistant strains will die. A clear area will be left where the bacteria have died — this is called an inhibition zone (see Figures 3 and 4).

3. Make sure you use a **control**. This is a paper disc that has not been soaked in an antibiotic. Instead, soak it in sterile water. You can then be sure that any difference between the growth of the bacteria around the control disc and around one of the antibiotic discs is due to the effect of the antibiotic alone (and not something weird in the paper, for example).

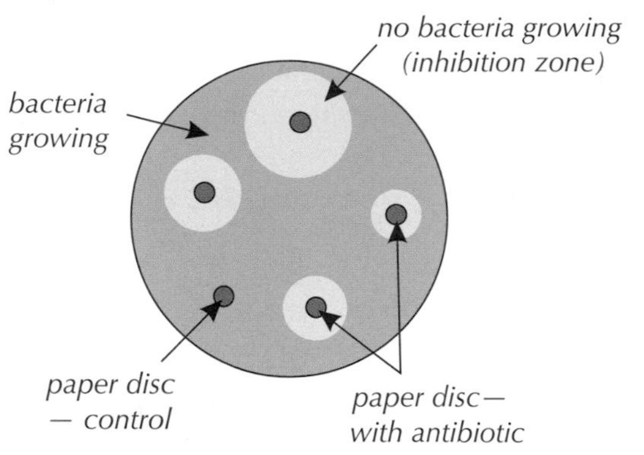

Figure 3: *An agar plate used to investigate the effect of antibiotics on bacteria.*

4. Leave the plate for 48 hours at 25 °C.
5. The more effective the antibiotic is against the bacteria, the larger the inhibition zone will be — see next page.

A similar method can be used to test the effects of antiseptics (or disinfectants) on bacterial growth — just replace the paper discs soaked in antibiotics with discs soaked in the solutions you're interested in. Alternatively, you could look at the effect of concentration of one antiseptic by soaking the discs in different concentrations of the same solution.

Tip: The control is not expected to have any effect on the bacteria.

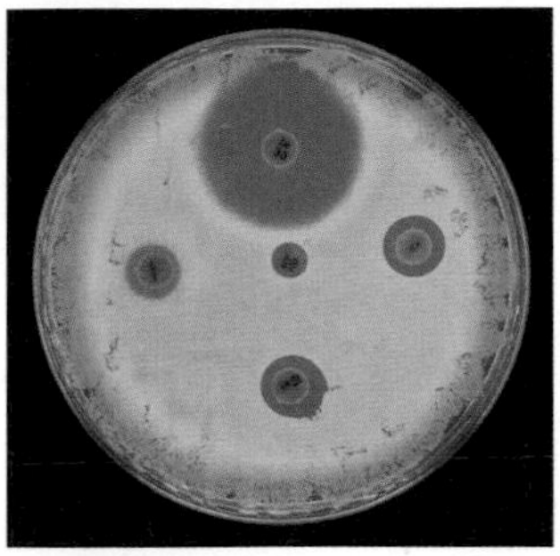

Figure 4: *Antibiotic discs on a bacterial culture. You can see clear zones around the discs where the bacteria can't grow.*

Preparing an uncontaminated culture

Contamination by unwanted microorganisms will affect your results and can potentially result in the growth of pathogens. To avoid this, follow these steps:

- The Petri dishes and culture medium must be sterilised before use (e.g. by heating to a high temperature), to kill any unwanted microorganisms that may be lurking on them.
- If an inoculating loop is used to transfer the bacteria to the culture medium, it should be sterilised first by passing it through a hot flame. 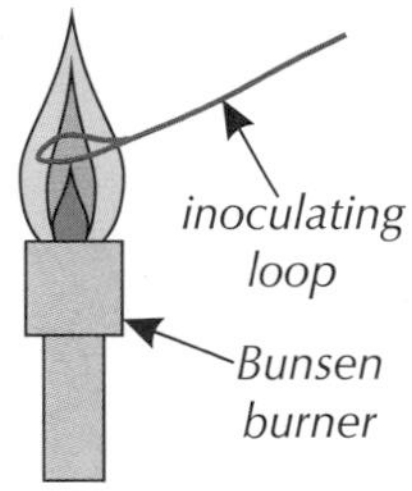
- After transferring the bacteria, the lid of the Petri dish should be lightly taped on — to stop microorganisms from the air getting in.
- The Petri dish should be stored upside down — to stop drops of condensation falling onto the agar surface.

inoculating loop

Bunsen burner

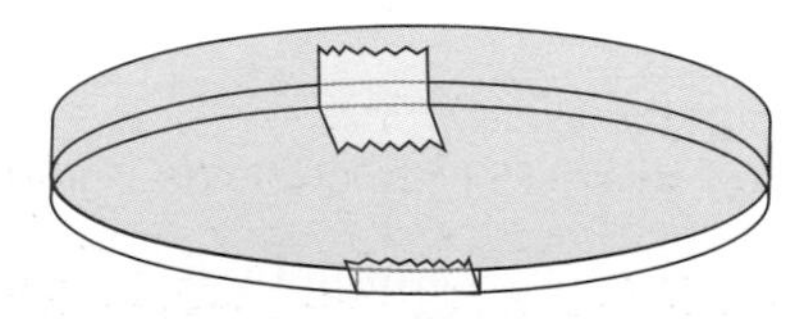

Figure 5: *The correct method of storing a Petri dish containing bacteria.*

Tip: You can sterilise equipment in a machine called an autoclave — it basically steams equipment at high pressure. Your school might have an autoclave.

Tip: Don't seal up the Petri dish completely before storing it — you need to let some oxygen get into the dish, otherwise you could end up growing some dangerous pathogens.

Measuring inhibition zones

You can compare the effectiveness of different antibiotics (or antiseptics) on bacteria by looking at the relative sizes of the inhibition zones. The larger the inhibition zone around a disc, the more effective the antibiotic is against the bacteria.

You can do this by eye if there are large differences in size. But to get more accurate results it's a good idea to calculate the area of the inhibition zones using their diameter (the distance across). You can measure this with a ruler — see Figure 6.

Tip: Don't open the Petri dish to measure the inhibition zones — they should be visible through the bottom of the dish.

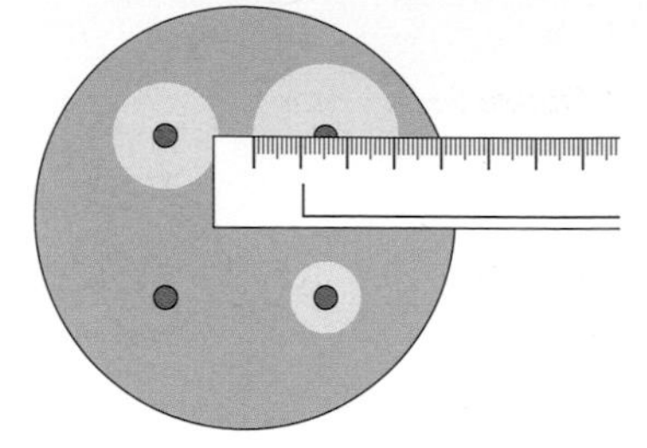

Figure 6: *Diagram showing a ruler being used to measure the diameter of an inhibition zone.*

To calculate the area of an inhibition zone, you need to use this equation:

$$\text{Area} = \pi r^2$$

This is the equation for the area of a circle.
You're likely to use the units cm^2 or mm^2 for the area.

Tip: Measuring the diameter and then halving it to get the radius is more accurate than measuring the radius straight off, as it's hard to judge by eye exactly where the centre of the inhibition zone is.

- r is the radius of the inhibition zone — it's equal to half the diameter.
- π is just a number. You should have a button for it on your calculator. If not, just use the value 3.14.

Example

The diagram below shows the inhibition zones produced by antibiotics A and B. Use the areas of the inhibition zones to compare the effectiveness of the antibiotics.

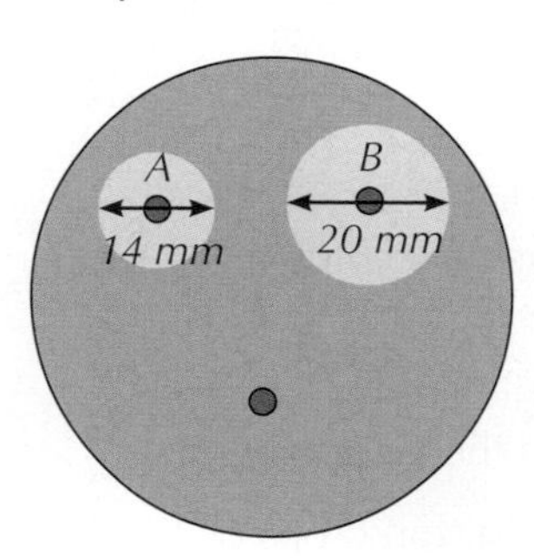

1. Divide the diameter of zone A by two to find the radius:
   Radius of A $= 14 \div 2 = 7$ mm
2. Stick the radius value into the equation, area $= \pi r^2$:
   Area of A $= \pi \times 7^2 =$ **154 mm²**
3. Repeat steps 1 and 2 for zone B:
   Radius of B $= 20 \div 2 = 10$ mm
   Area of B $= \pi \times 10^2 =$ **314 mm²**
4. Compare the sizes of the areas.
   314 mm^2 is just over twice 154 mm^2, so you could say that:

The inhibition zone of antibiotic B is roughly twice the size of the inhibition zone of antibiotic A.

Tip: When you're comparing sizes, make sure that the values you're using are all in the same units.

Finding the area of a colony

The equation on the previous page can also be used to calculate the area of a bacterial colony. You just need to measure the diameter of the colony you are interested in first.

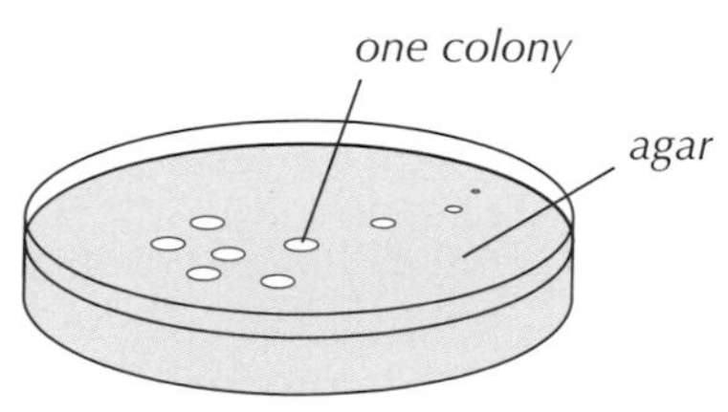

Figure 7: Diagram showing bacterial colonies on an agar plate.

> **Tip:** The size and shape of a bacterial colony on the surface of agar depends on the type of bacteria it's formed from.

Example

The diagram below shows bacterial colonies growing on an agar plate. A grid of 5 mm × 5 mm squares has been placed on the plate. Calculate the area covered by the colony labelled 'X'.

MATHS SKILLS

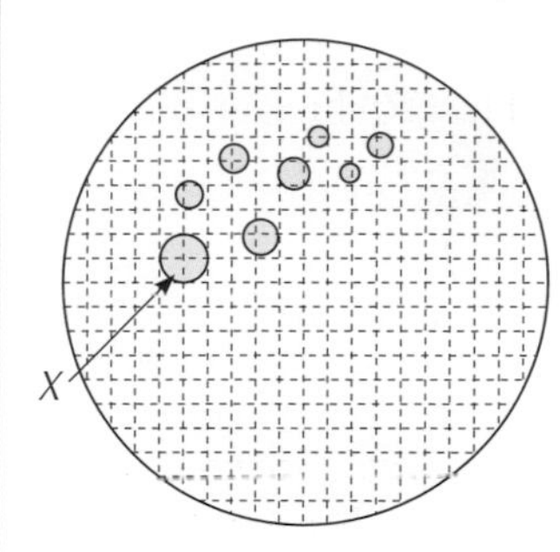

1. Work out the radius of X:

   The radius is the width of one square, which is 5 mm.

2. Stick the radius value into the equation, area = πr^2:

   Area of X = $\pi \times 5^2$ = **79 mm²**

Practice Questions — Application

Two different antibiotics were investigated for their effect on bacterial growth. The results of the investigation are shown below.

Q1 Copy and complete the table below by finding the diameter, radius and area of the inhibition zones for each of the antibiotics.

Give the value of the area to the nearest cm².

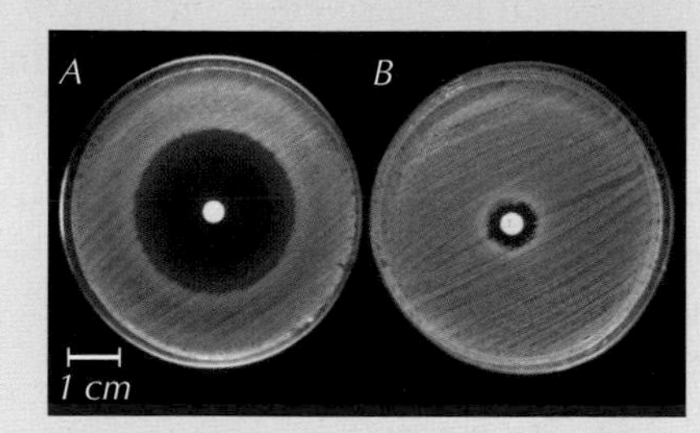

Antibiotic	Diameter (cm)	Radius (cm)	Area (cm²)
A			
B			

A control was used to show that the results of the investigation were due to the effect of the antibiotics on the bacteria.

Q2 What result would you expect to see if an antibiotic that was not effective against this strain of bacteria was used?

Q3 What can you conclude from the results of the investigation?

> **Tip:** If an image has a scale bar, you can use the bar to work out the size of the real object. E.g. if there is a 5 mm scale bar that represents 2 cm (so it looks like this: 2 cm) and the image is 14 mm wide, the real object is (14 ÷ 5) × 2 = 5.6 cm wide.

> **Tip:** When working out the area, remember to use 3.14 if you don't have a π button on your calculator.

8. Stem Cells

All the cells in an organism originate from stem cells. Scientists are attempting to use these cells to do some pretty amazing things...

Learning Objectives:

- Know that stem cells are undifferentiated cells.
- Know that stem cells are able to produce more stem cells and differentiate into specialised cells.
- Know that stem cells from adult bone marrow and embryonic stem cells can be cloned and made to differentiate into different cell types, e.g. blood cells.
- Know that stem cells could be used to treat some medical conditions, e.g. diabetes, paralysis.
- Know that embryonic stem cells can be produced by therapeutic cloning for use in medical treatments, and know the advantages of this technique.
- Know that the use of stem cells in medical treatments carries risks, such as the potential transfer of viruses to patients.
- Know that some people object to using stem cells in medical treatments and research on ethical or religious grounds.
- Know that plant stem cells can be found in meristems and can differentiate into any plant cell at any point in the plant's life.
- Know that plant stem cells can be used to grow clones of a plant and understand the benefits of this.

Specification Reference 4.1.2.3

What are stem cells?

Some cells are undifferentiated (i.e. they have not yet changed to become specialised for a particular job — see page 30). These undifferentiated cells are called **stem cells**. They can divide to produce lots more undifferentiated cells. They can also develop into different types of cell, depending on what instructions they're given.

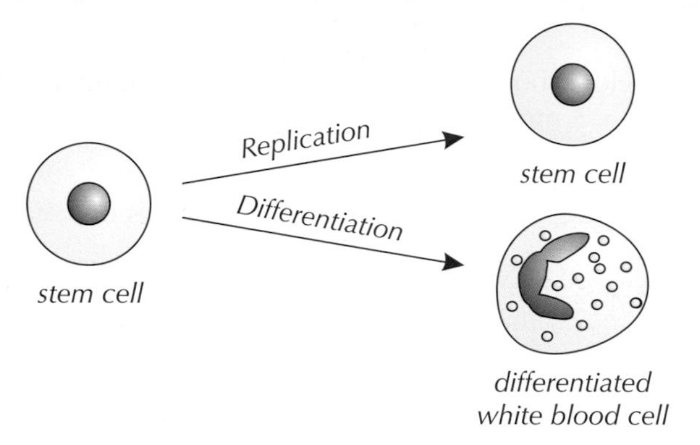

Figure 1: *Diagram showing stem cell replication and differentiation.*

Stem cells are found in early human embryos. They're exciting to doctors and medical researchers because they have the potential to turn into any kind of cell at all. This makes sense if you think about it — all the different types of cell found in a human being have to come from those few cells in the early embryo.

Adults also have stem cells, but they're only found in certain places, like bone marrow. These aren't as versatile as embryonic stem cells — they can't turn into any cell type at all, only certain ones, such as blood cells.

Stem cells from embryos and bone marrow can be grown in a lab to produce **clones** (genetically identical cells — there's more about cloning on pages 297-299) and made to differentiate into specialised cells to use in medicine or research.

Adult stem cells in medicine

Adult stem cells are already used to cure disease.

Example

People with some blood diseases (e.g. sickle cell anaemia — a disease which affects the shape of the red blood cells) can be treated by bone marrow transplants. Bone marrow is the tissue found inside bone. It contains stem cells that can turn into new blood cells to replace the faulty old ones.

Embryonic stem cells in medicine

Embryonic stem cells could also be used to replace faulty cells in sick people in the future (see Figure 3).

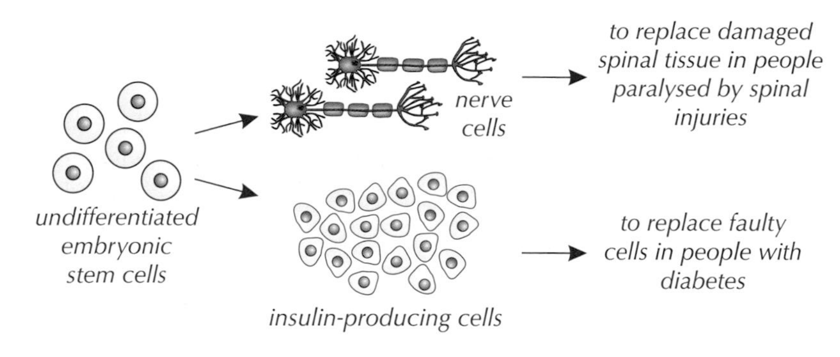

Figure 3: *Diagram showing some potential uses of embryonic stem cells in medicine.*

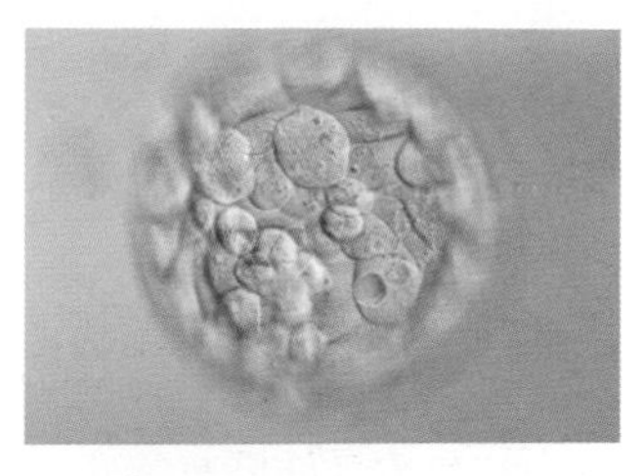

Figure 2: *Microscope image of a human embryo that's about 4 days old. Stem cells can be taken from the embryo at this stage.*

In a type of cloning called **therapeutic cloning**, an embryo could be made to have the same genetic information as the patient. This means that the stem cells produced from it would also contain the same genes as the patient and so wouldn't be rejected by the patient's body if used to replace faulty cells.

However, there are risks involved in using stem cells in medicine. For example, stem cells grown in the lab may become contaminated with a virus which could be passed on to the patient and so make them sicker. There are also ethical issues related to the use of embryonic stem cells — see below.

Tip: The advantage of using embryonic stem cells for therapies over adult stem cells is that they have the potential to differentiate into many different types of cell (unlike adult stem cells, which are limited to differentiating into certain types of cell). They are also easier to grow in culture than adult stem cells.

Issues involved in stem cell research

Embryonic stem cell research has exciting possibilities, but it's also pretty controversial.

Some people are against it, because they feel that human embryos shouldn't be used for experiments since each one is a potential human life.

However, others think that curing patients who already exist and who are suffering is more important than the rights of embryos. One fairly convincing argument in favour of this point of view is that the embryos used in the research are usually unwanted ones from fertility clinics which, if they weren't used for research, would probably just be destroyed. But of course, campaigners for the rights of embryos usually want this banned too.

Campaigners for the rights of embryos feel that scientists should concentrate more on finding and developing other sources of stem cells, so people could be helped without having to use embryos. Research is being done into getting stem cells from alternative sources. For example, it may be possible to produce stem cells from differentiated adult cells by reprogramming the adult cells back to an undifferentiated stage. However, further research has to be done to make sure that this technique is safe for use in medical treatments.

In some countries stem cell research is banned, but it's allowed in the UK as long as it follows strict guidelines.

Tip: Obtaining stem cells from an embryo destroys the embryo.

Exam Tip
Questions on the issues associated with the use of embryonic stem cells might crop up in the exam, so make sure you know arguments for and against the use of embryonic stem cells.

Stem cells in plants

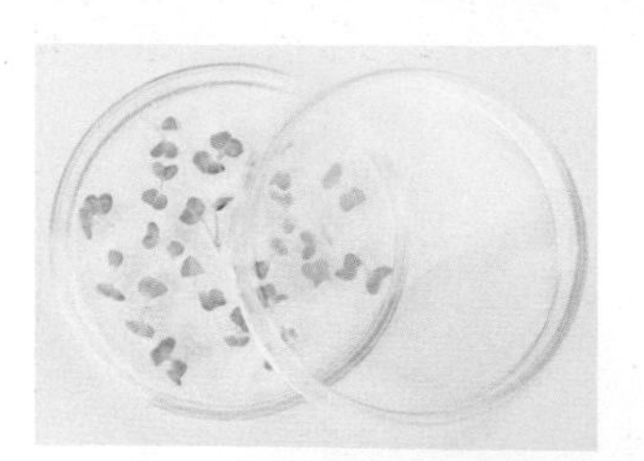

Figure 4: *Plants grown from meristem tissue.*

In plants, stem cells are found in the **meristems** (parts of the plant where growth occurs — see p. 89). Throughout the plant's entire life, cells in the meristem tissues can differentiate into any type of plant cell.

These stem cells can be used to produce clones (identical copies) of whole plants quickly and cheaply. This is useful for:

- growing crops of identical plants that have desired features for farmers.
- growing more plants of rare species (to prevent them being wiped out).

Example

Dutch elm disease is a fungal tree disease that has destroyed millions of elm trees worldwide. One way to protect elm trees from being wiped out by this disease could be to create and grow clones of naturally-resistant individuals.

Practice Questions — Fact Recall

Q1 What is a stem cell?

Q2 What type of cell can early embryonic stem cells turn into?

Q3 Give one place where stem cells can be found in an adult human.

Q4 Why could embryonic stem cells produced by therapeutic cloning be useful for replacing faulty cells in a sick person?

Q5 Name the tissue in plants where stem cells can be found.

Practice Questions — Application

Q1 Alzheimer's disease is a condition which damages the neurones (nerve cells) in the brain. These neurones die, but are not replaced, leading to a decrease in the amount of neurones in the brain. Symptoms of Alzheimer's disease include memory loss, confusion and changes in personality.

a) Suggest a way in which embryonic stem cells could potentially be used to treat people with Alzheimer's disease.

b) Give another possible use of embryonic stem cells in medicine.

Q2 Embryonic stem cells used for medical research are mostly taken from embryos left over from fertility clinics, which would otherwise be destroyed. Suggest a reason why some people are happier with using these embryos for stem cell research, rather than creating embryos purely with the purpose of being used for research.

Q3 Embryos begin to develop a nervous system 14 days after fertilisation. Some people think that it's morally acceptable to use an embryo for stem cell research before this point, as the embryo has no senses, so cannot be considered a life yet. Suggest a reason why some people may not agree with this view.

Exam Tip
It's important that you can look at something from the point of view of someone with different opinions to you.

Topic 1a Checklist — Make sure you know...

Cell Structure

- ❑ That plant and animal cells are eukaryotic cells, which make up eukaryotic organisms, and that bacterial cells are prokaryotic cells, which are much smaller than eukaryotic cells.
- ❑ That most animal cells have: a nucleus (which contains the genetic material that controls the activities of a cell), cytoplasm (where most of the chemical reactions happen), a cell membrane (which controls what goes in and out of the cell), mitochondria (where most of the reactions for aerobic respiration happen), and ribosomes (where proteins are made in the cell).
- ❑ That plant and algal cells have all the same parts as animal cells, plus a cell wall made of cellulose, and that most plant cells also have a permanent vacuole containing cell sap and chloroplasts which absorb the light needed for photosynthesis.
- ❑ That a bacterial cell has cytoplasm, a cell membrane and a cell wall, and that its genetic material is a single DNA loop found in the cytoplasm (rather than in a nucleus) and potentially one or more plasmids (small rings of DNA).

Microscopy

- ❑ That microscopy techniques have developed over time due to improving technology, and that electron microscopes can produce images of higher magnification and higher resolution than light microscopes.
- ❑ That smaller subcellular structures, such as ribosomes, can be seen under an electron microscope (but not a light microscope), which has led to a greater understanding of subcellular structures.
- ❑ How to prepare a slide to observe plant or animal cells, use a light microscope and draw your observations (Required Practical 1), and how to use the formula: magnification = image size ÷ real size, and how to write the answer to the calculation in standard form.
- ❑ About the size and scale of cells, and how to use estimations to work out the relative size or area of subcellular structures.

Cell Differentiation and Specialisation

- ❑ That many cells are specialised to carry out a specific function, and that cells in an organism become specialised as the organism develops. This involves developing specific subcellular structures, so that they can carry out their functions.
- ❑ That most animal cells lose the ability to differentiate (become specialised) early on, but lots of plant cells don't ever lose this ability. In mature animals, cell division is mainly used for growth and repair.
- ❑ How to relate a cell's structure (its shape and subcellular structures) to its function, including the structures of sperm cells, nerve cells, muscle cells, root hair cells, xylem cells and phloem cells.

Chromosomes

- ❑ That body cells contain two sets of chromosomes in the nucleus — these chromosomes found in pairs and are made up of long molecules of DNA
- ❑ That chromosomes carry genes, which are small sections of DNA.

cont...

Mitosis

- ☐ That body cells divide by mitosis, which is one stage of the cell cycle.
- ☐ That mitosis allows organisms to grow, develop and replace damaged cells.
- ☐ That during the cell cycle, the cell grows and increases its subcellular structures, copies its genetic material and then divides once (by mitosis), producing two cells which are genetically identical.
- ☐ How to recognise mitosis in cells and describe mitosis in a given context.

Binary Fission

- ☐ That a bacterial cell multiplies by binary fission, which produces two daughter cells, and that bacteria may divide as often as every 20 minutes if given the correct conditions (including enough nutrients and the correct temperature).
- ☐ How to calculate the number of bacteria in a population after a certain amount of time.
- ☐ H How to write the number of bacteria in standard form.

Culturing Microorganisms

- ☐ That bacteria can be grown in a culture medium, which can be a nutrient broth or agar jelly.
- ☐ That cultures of microorganisms in schools should not be grown above 25 °C, to make the growth of pathogens (disease-causing microorganisms) less likely.
- ☐ How to prepare an uncontaminated agar plate to grow colonies or an even covering of bacteria.
- ☐ How to investigate the effect of antibiotics (or antiseptics) on bacterial growth (Required Practical 2).
- ☐ How the contamination of bacterial cultures is prevented by sterilising inoculating loops, Petri dishes and culture media, by taping lids onto Petri dishes, and by storing Petri dishes upside down.
- ☐ How to calculate the area of an inhibition zone, or colony, using the equation: area = πr^2.

Stem Cells

- ☐ That stem cells are undifferentiated cells that can produce specialised cells and more stem cells.
- ☐ That adult stem cells can be taken from bone marrow and made to differentiate into certain types of cell, including blood cells. Adult stem cells already have some uses in medicine.
- ☐ That embryonic stem cells are found in early human embryos, and can be cloned and made to differentiate into any type of cell. This means they could be used to treat many medical conditions by replacing faulty cells, such as nerve cells in paralysis and insulin-producing cells in diabetes.
- ☐ That therapeutic cloning can be used to produce embryonic stem cells that are genetically identical to a patient, which means that the cells they produce will not be rejected by the patient's body.
- ☐ That the use of stem cells in medical treatments carries risks, e.g. viral infection, and that some people object to using stem cells for medical treatments and research on religious or ethical grounds.
- ☐ That plant stem cells, which can differentiate into any type of plant cell throughout the plant's life, are found in tissues called meristems, and that they can be used to grow clones of plants (including rare species and crop plants with desired features) quickly and cheaply.

Exam-style Questions

1 The genetic material of a multicellular organism is found in cells in the form of chromosomes.

1.1 Name the subcellular structure that contains chromosomes.

(1 mark)

1.2 There are twenty three different chromosomes in a female human body cell. How many copies of each chromosome does the cell have?

(1 mark)

1.3 A body cell doubles its genetic material as it prepares to divide. Name the process by which the body cell would divide to produce identical daughter cells.

(1 mark)

2 **Figure 1** shows a palisade cell.

Figure 1

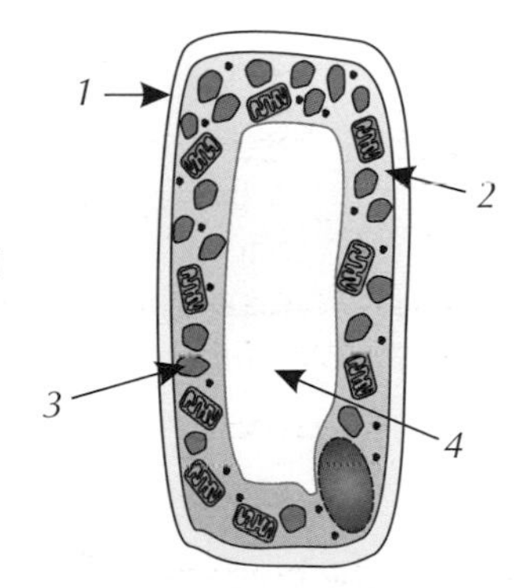

2.1 Look at the structures labelled 1-4 in **Figure 1**. Which of the following options gives the correct labels for these structures?

A 1 — permanent vacuole, 2 — cytoplasm, 3 — chloroplast, 4 — cell wall

B 1 — cytoplasm, 2 — chloroplast, 3 — cell wall, 4 — permanent vacuole

C 1 — cell wall, 2 — permanent vacuole, 3 — cytoplasm, 4 — chloroplast

D 1 — cell wall, 2 — cytoplasm, 3 — chloroplast, 4 — permanent vacuole

(1 mark)

2.2 Is a palisade cell an example of a eukaryotic cell or a prokaryotic cell? Give a reason for your answer that refers to the cell's structure.

(1 mark)

2.3 A palisade cell is a specialised cell. What is a specialised cell?

(1 mark)

2.4 What name is given to the unspecialised cells in an organism?

(1 mark)

2.5 Palisade cells are the main site of photosynthesis in a plant. They are found grouped together near the top of a leaf.

Suggest one way that the structure of a palisade cell makes it well adapted for its function.

(1 mark)

3 A bacterial cell divides once every 48 minutes.

3.1 Starting with one cell, calculate how many cells there will be after 4 hours.

(2 marks)

3.2 Will the offspring be genetically identical or genetically different to the original parent cell?

(1 mark)

3.3 Name the type of division used by bacterial cells.

(1 mark)

3.4 Describe two conditions that can lead to a high rate of bacterial replication.

(2 marks)

3.5 **Figure 2** shows a scale drawing of the bacterial cell. Calculate the magnification of the image.

(2 marks)

Figure 2

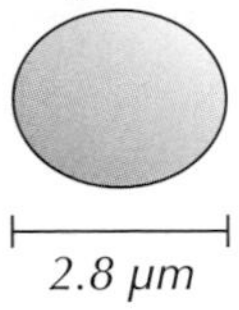

2.8 μm

4 A doctor needs to find out which antibiotics will treat a patient's infection, so she sends a sample of bacteria taken from the patient to be cultured in the lab.

A lab technician fills a Petri dish with agar jelly. When this has set, he transfers the bacterial sample to the dish using an inoculating loop.

4.1 Describe two steps the lab technician should take during this process to avoid contamination of the bacterial culture.

(2 marks)

Once the lab technician has transferred the bacteria, he places 5 different paper discs, each soaked in a different antibiotic, onto the agar. He also puts one paper disc that hasn't been soaked in antibiotic onto the agar. He leaves the Petri dish to incubate.

Figure 3 shows a diagram of the dish a few days later.

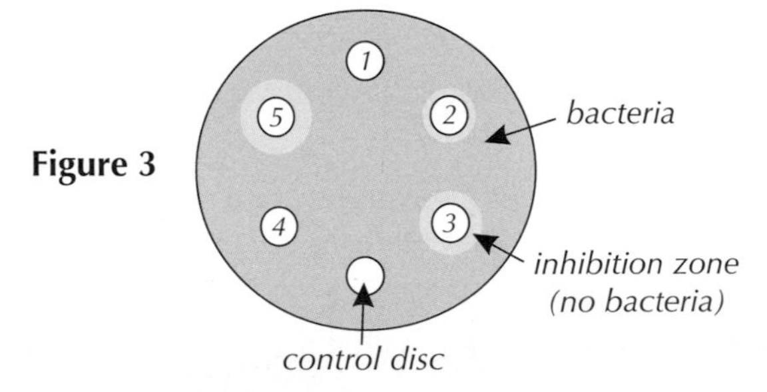

Figure 3

4.2 From these results, which antibiotic (1-5) is the best one to use to treat the patient's infection? Explain your answer.

(3 marks)

4.3 Give **two** variables that should have been controlled in order to make the results of this investigation valid.

(2 marks)

4.4 What is the purpose of the control disc?

(1 mark)

1. Diffusion

Particles tend to move about randomly and end up evenly spaced. This is important when it comes to getting substances in and out of cells.

Learning Objectives:

- Know the definition of 'diffusion'.
- Be able to describe how dissolved substances (such as oxygen) can move in and out of a cell, through the cell membrane, by diffusion.
- Understand that during diffusion, the net movement of substances will be from an area of higher concentration to an area of lower concentration.
- Be able to explain how the rate of diffusion is affected by the difference in the concentration of particles, the temperature and the surface area available for diffusion.

Specification Reference 4.1.3.1

What is diffusion?

"Diffusion" is simple. It's just the gradual movement of particles from places where there are lots of them to places where there are fewer of them. That's all it is — just the natural tendency for stuff to spread out.

Unfortunately you also have to learn the fancy way of saying the same thing, which is this:

> Diffusion is the spreading out of particles from an area of higher concentration to an area of lower concentration.

Diffusion happens in both solutions and gases — that's because the particles in these substances are free to move about randomly. The simplest type is when different gases diffuse through each other.

Example

When you spray perfume, the smell of perfume diffuses through the air in a room:

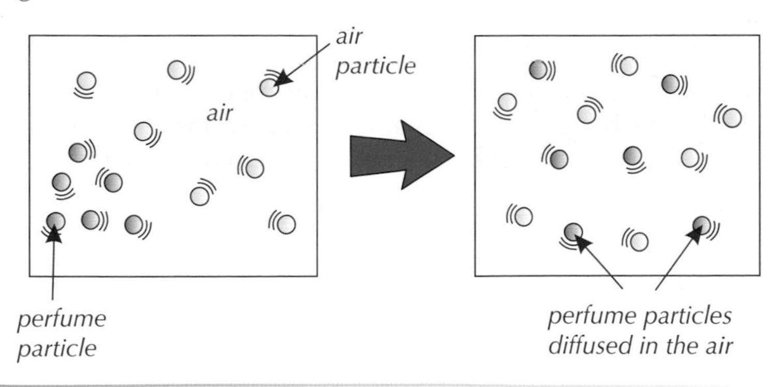

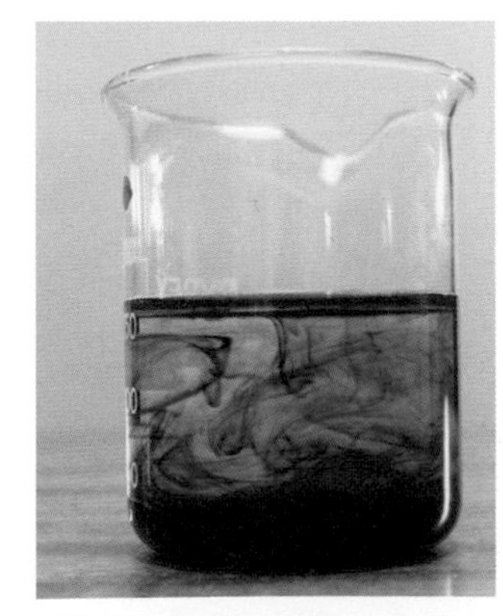

Figure 1: *The ink particles in this flask are diffusing into the water — they're moving from an area of high concentration (at the bottom of the flask) to an area of low concentration (higher up).*

Diffusion across cell membranes

Cell membranes are clever because they hold the cell together but they let stuff in and out as well. **Dissolved substances** can move in and out of cells by diffusion.

Only very small molecules can diffuse through cell membranes though — things like **oxygen** (needed for respiration — see page 173), glucose, amino acids and water. Big molecules like starch and proteins can't fit through the membrane (see Figure 2 on the next page).

Exam Tip
If you're struggling to remember which way the particles move in diffusion, think of it like this: if you were in a really crowded place, you'd probably want to get out of there to somewhere with a bit more room. It's the same with particles — they always diffuse from an area of higher concentration to an area of lower concentration.

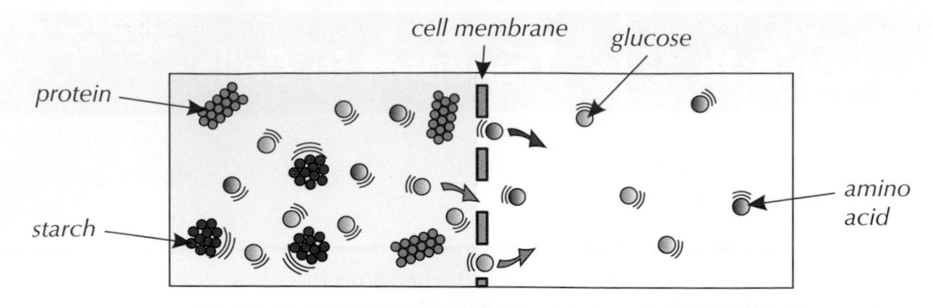

Figure 2: *Diagram to show diffusion across a cell membrane.*

Just like with diffusion in air, particles flow through the cell membrane from where there's a higher concentration (more of them) to where there's a lower concentration (fewer of them).

They're only moving about randomly of course, so they go both ways — but if there are a lot more particles on one side of the membrane, there's a **net** (overall) movement from that side.

Example

Particles are diffusing both in and out of this cell. However, the concentration of particles is higher inside the cell than outside, so the net movement of particles is out of the cell.

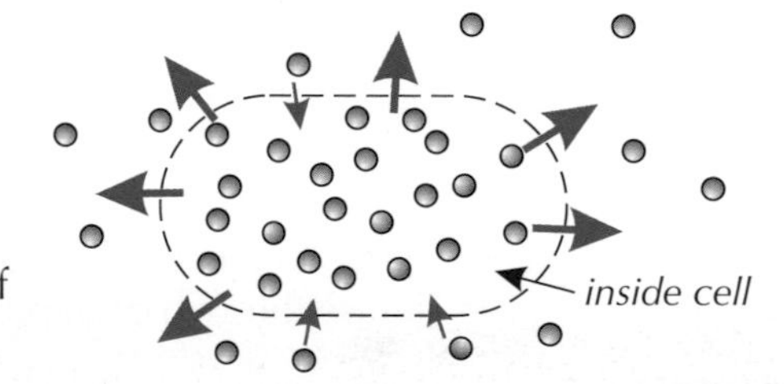

The rate of diffusion

The **rate** of diffusion can vary. It's affected by:

- The concentration gradient (the difference in concentration of the particles) — the bigger the concentration gradient, the faster the rate of diffusion. This is because the net movement from one side is greater.

Example

There will be a net movement of particles out of both Cell 1 and Cell 2 (shown below). However, the rate of diffusion will be faster out of Cell 1, because there's a bigger difference in the concentration of particles on either side of the cell membrane.

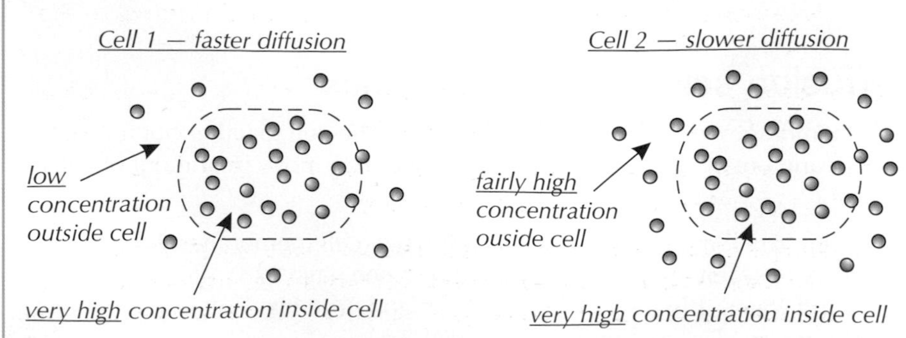

Tip: As diffusion takes place, the difference in concentration between the two sides of the membrane decreases, until the concentration on both sides is equal. This means that diffusion slows down over time.

- The temperature — the higher the temperature, the faster the rate of diffusion. This is because the particles have more energy, so move around faster.
- The surface area — the larger the surface area (e.g. of the cell membrane), the faster the rate of diffusion. This is because more particles can pass through at once.

Tip: The thickness of the cell membrane also affects the rate of diffusion — see p. 62.

Example

Some specialised cells (e.g. epithelial cells in the small intestine) have lots of small projections on their surface, formed from folds in the cell membrane. These projections give the cell a larger surface area, which means that more particles can diffuse across the membrane in the same amount of time — increasing the rate of diffusion.

Practice Questions — Fact Recall

Q1 Give the definition of diffusion.

Q2 Name four molecules that can diffuse through a cell membrane.

Q3 Explain how increasing the temperature affects the rate of diffusion.

Practice Questions — Application

Q1 The diagrams below show three cells in different glucose solutions. The concentration of glucose inside and outside the cell is shown in each case. Which diagram shows a situation where the net movement of glucose will be out of the cell — A, B or C?

A
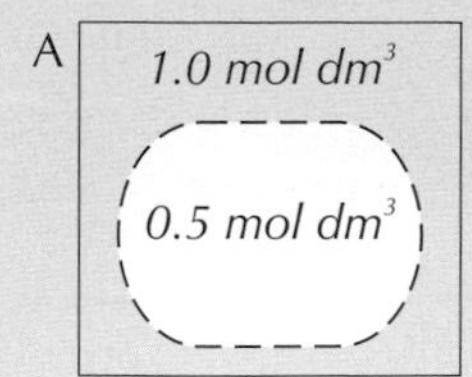

B
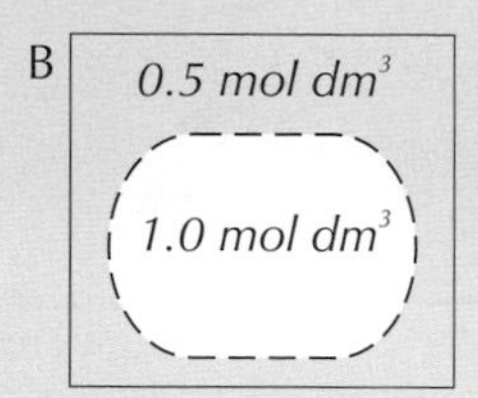

C
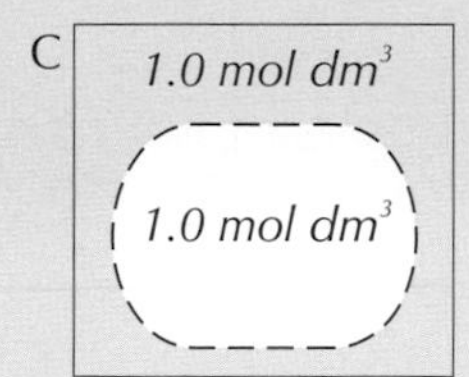

Q2 When cells respire they produce carbon dioxide as a waste product. The carbon dioxide diffuses from the cells into the bloodstream, so it can be removed from the body.
Is carbon dioxide concentration greater in the bloodstream or inside respiring cells? Explain your answer.

Q3 At a disco in a school hall, a DJ releases a short blast of smoke from a smoke machine at the front of the stage.

a) Explain how the smoke reaches the people standing at the opposite end of the hall from the stage.

b) Five minutes later the DJ sets the smoke machine off for a second time. Explain how the rate of diffusion of the smoke is now different from the first time the DJ set the smoke machine off.

2. Osmosis

Osmosis is an important process for life — it's how cells get the water they need to carry out chemical reactions.

Learning Objectives:
- Know the definition of osmosis.
- Be able to describe how water moves into or out of a cell by osmosis.
- Be able to investigate the effect of different concentrations of a sugar (or salt) solution on plant cells (Required Practical 3).
- Be able to work out the rate of osmosis.
- Be able to calculate percentage change in plant tissue mass.
- Be able to draw and interpret graphs showing the effect of concentration of a sugar or salt solution on the mass of plant tissue.

Specification Reference 4.1.3.2

What is osmosis?

Osmosis is the movement of water molecules across a partially permeable membrane from a region of higher water concentration to a region of lower water concentration.

A **partially permeable membrane** is just one with very small holes in it. So small, in fact, only tiny molecules (like water) can pass through them, and bigger molecules (e.g. sucrose, a sugar) can't. This is shown in Figure 1.

The water molecules actually pass both ways through the membrane during osmosis. This happens because water molecules move about randomly all the time. But because there are more water molecules on one side than on the other, there's a steady net flow of water into the region with fewer water molecules, i.e. into the stronger sugar solution. This means the strong sugar solution gets more dilute. The water acts like it's trying to "even up" the concentration either side of the membrane.

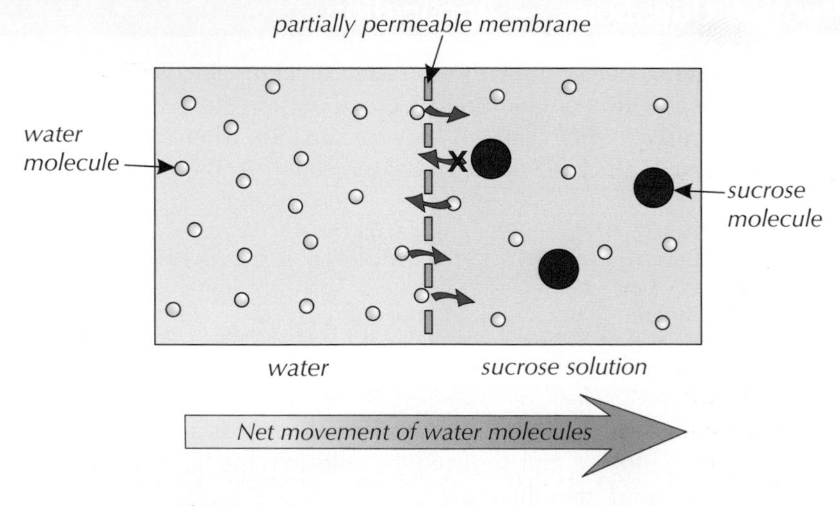

Figure 1: *Diagram to show how osmosis works.*

Tip: Remember, diffusion is the spreading out of particles from an area of higher concentration to an area of lower concentration. It's passive, which means it doesn't need energy.

Osmosis is a type of **diffusion** — passive movement of water particles from an area of higher water concentration to an area of lower water concentration.

The movement of water in and out of cells

Tip: Both plant and animal cells gain and lose water by osmosis.

The solution surrounding a cell will usually have a different concentration to the fluid inside the cell. This means that water will either move into the cell from the surrounding solution, or out of the cell, by osmosis.

If a cell is short of water, the solution inside it will become quite concentrated (i.e. there'll be a low concentration of water molecules). This usually means the solution outside the cell is more dilute (there's a higher concentration of water molecules), and so water will move into the cell by osmosis.
If a cell has lots of water, the solution inside it will be more dilute, and water will be drawn out of the cell and into the fluid outside by osmosis.

This is summarised in Figure 2.

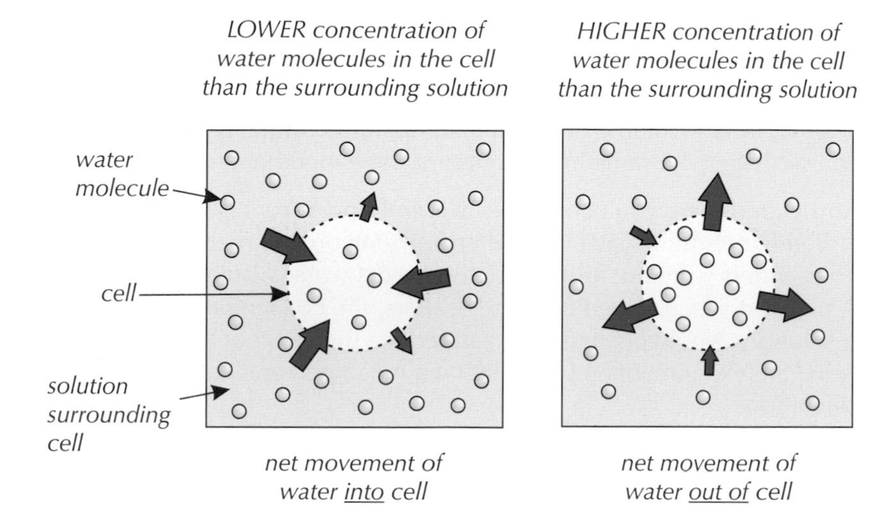

Figure 2: *Diagram to show the movement of water molecules into and out of a cell.*

Tip: When you exercise, you lose water and ions in your sweat. It's important to replace these because if the balance between ions and water in the body is wrong, too much or too little water is drawn into cells by osmosis, which can damage them. Sports drinks contain water and ions, so drinking them can replace those lost in sweat.

Tip: Cells that gain water by osmosis get a bit bigger (animal cells that gain too much water will eventually burst). Cells that lose water get a bit smaller.

Investigating the effect of sugar solutions on plant cells

REQUIRED PRACTICAL 3

There's an experiment you can do to show osmosis at work:

1. Cut up a potato into identical cylinders and measure their masses.
2. Get some beakers with different sugar solutions in them. One should be pure water and another should be a very concentrated sugar solution (e.g. 1.0 mol/dm^3). Then you can have a few others with concentrations in between (e.g. 0.2 mol/dm^3, 0.4 mol/dm^3, 0.6 mol/dm^3, etc.)
3. Place one potato cylinder in each beaker, as shown in Figure 3. Leave them in the beakers for twenty four hours.

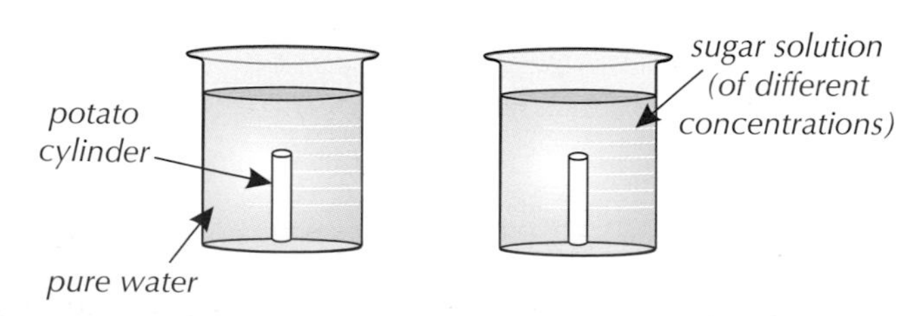

Figure 3: *A diagram to show the set-up of the experiment.*

Tip: Remember to be aware of any risks in this experiment. E.g. if you're using a knife, make sure that you cut away from yourself and that the blade is clean and sharp.

Tip: Make sure there's no skin left on your potato cylinders, as this could affect your results.

Tip: You could also carry out this experiment using different salt solutions and see what effect they have on potato cylinder mass.

4. Take the cylinders out, dry them with a paper towel and measure their masses again.

5. If the cylinders have drawn in water by osmosis, they'll have increased in mass. If water has been drawn out, they'll have decreased in mass. You can calculate the **percentage change** in mass (see p. 378), then plot a few graphs and things (see below).

Tip: By calculating the percentage change, you can compare the effect of sugar concentration on cylinders that didn't have the same initial mass. An increase in mass will give a positive percentage change and a decrease will give a negative percentage change.

Variables and errors

WORKING SCIENTIFICALLY

The **dependent variable** in this experiment is the cylinder mass and the **independent variable** is the concentration of the sugar solution. All other variables (volume of solution, temperature, time, type of sugar used, etc.) must be kept the same in each case or the experiment won't be a fair test.

Like any experiment, you need to be aware of how **errors** (see p. 13) may arise. Sometimes they may occur when carrying out the method, e.g. if some potato cylinders were not fully dried, the excess water would give a higher mass, or if water evaporated from the beakers, the concentrations of the sugar solutions would change. You can reduce the effect of these errors by repeating the experiment and calculating a mean percentage change at each concentration.

Tip: You can read all about how to calculate a mean on page 14.

Tip: You could do a similar experiment to work out the rate of osmosis by leaving the potato cylinders for 30 minutes, then dividing the change in mass (in grams) by the time taken (in minutes). So the units for the rate would be g/min.

Producing a graph of your results

You can create a graph of your results by plotting the percentage change in mass against the concentration of sugar solution. You can then draw a line of best fit, which you can use to determine the concentration of the solution in the potato cells:

Example

An experiment to investigate the effect of sugar concentration on potato cells gave the results in the table below.

Concentration of sugar solution (M)	*0.1*	*0.25*	*0.5*	*1.0*	*2.0*
Percentage change in mass (%)	*16.5*	*2.5*	*–8.0*	*–15.5*	*–20.0*

The results can be plotted as a graph with concentration of sugar solution on the *x*-axis and percentage change in mass on the *y*-axis, as shown in Figure 4.

The point at which the line of best fit crosses the *x*-axis (where the percentage change in mass is 0) is the point at which the concentration of the sugar solution is the same as the concentration of the solution in the potato cells (see Figure 4). You can estimate the concentration of the solution inside the potato cells by reading the value off the *x*-axis at this point.

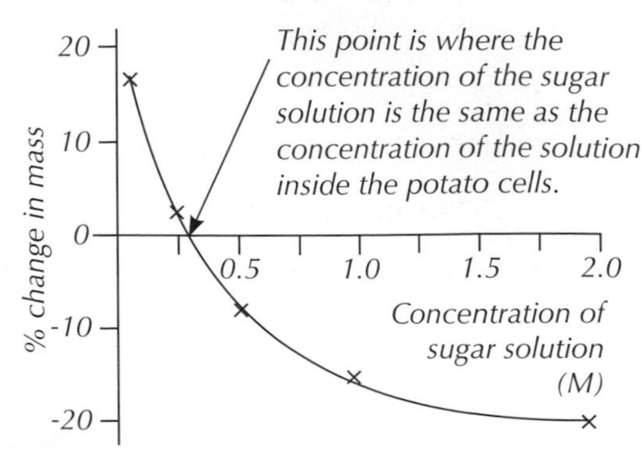

Figure 4: *A graph to find the concentration of the solution inside the potato cells.*

Tip: 'M' is a unit of concentration. The solution with a concentration of 0.00 M is pure water.

Tip: There's more about plotting graphs on page 18.

Practice Questions — Fact Recall

Q1 What is osmosis?

Q2 Explain how a cell that is short of water gains water from the surrounding solution.

Practice Question — Application

Q1 An experiment is carried out to investigate osmosis.

A potato is cut up into cylinders of the same known length and width.

Some of the potato cylinders are placed in a beaker of pure water. Others are placed in beakers with different concentrations of sugar solution.

The potato cylinders are left in the beakers for 24 hours. They are then removed and the length of each potato cylinder is recorded. The mean change in potato cylinder length for each beaker is then worked out.

The results of this experiment are shown in the table.

Concentration of solution (M)	Mean change in potato cylinder length (mm)
0.00	+ 3.1
0.25	+ 2.1
0.50	– 1.2
0.75	– 2.6
1.00	?

a) i) Describe what happened to the potato cylinders in the 0.00 M and 0.25 M solutions.

ii) Explain the changes you described in part a) i) in terms of osmosis.

b) Predict the mean change in length for the potato cylinders in the 1.00 M solution.

c) What was:

i) the dependent variable in this experiment?

ii) the independent variable in this experiment?

d) Give two variables that had to be controlled in this experiment to make it a fair test.

Exam Tip
You could get asked questions about any of the Required Practicals in the exam — so make sure you understand what's happening in each one.

Tip: There's more on variables on page 10.

3. Active Transport

Another way for molecules to move into and out of cells is via a process called active transport. But it needs energy...

Learning Objectives:

- Know the definition of active transport.
- Be able to describe how substances move in and out of a cell by active transport.
- Know how plant roots are adapted to maximise the exchange of substances.
- Understand how the structure of a root hair cell is related to its function.
- Know that plant root hair cells can absorb mineral ions from very dilute solutions using active transport.
- Know that plants need these ions for healthy growth.
- Know that sugar can be absorbed from the gut into the blood (against the concentration gradient) using active transport.
- Know that sugar is used for respiration in cells.
- Know the differences between diffusion, osmosis and active transport.

Specification References
4.1.3.1, 4.1.3.3

What is active transport?

Active transport is the movement of particles against a concentration gradient (i.e. from an area of lower concentration to an area of higher concentration) using energy transferred during respiration.

Active transport, like diffusion and osmosis, can be used to move substances in and out of cells. It allows cells to absorb ions from very dilute solutions.

Active transport in plant roots

The cells on the surface of plant roots grow into long "hairs" which stick out into the soil. Each branch of a root will be covered in millions of these microscopic hairs. This gives the plant a large surface area for absorbing water and mineral ions from the soil (see Figure 1).

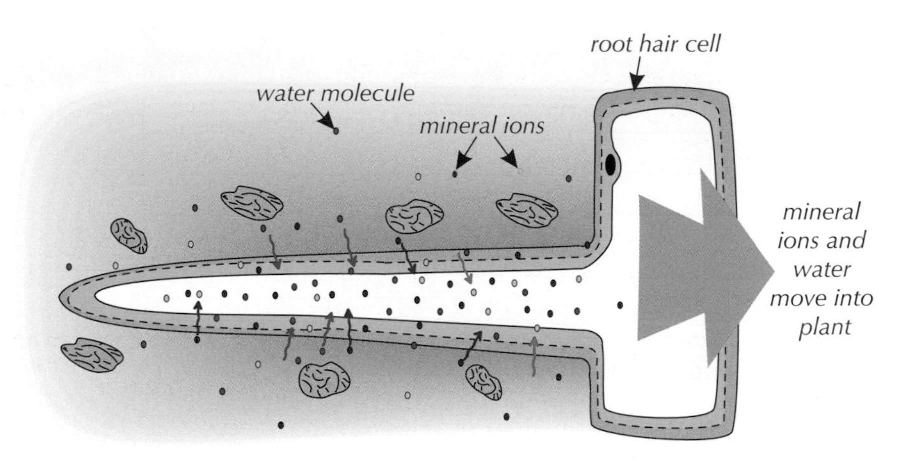

Figure 1: *Diagram to show the absorption of water and mineral ions by a root hair cell.*

Tip: Water is taken into root hair cells by osmosis. See page 54 for more on osmosis.

Plants need these mineral ions for healthy growth. However, the concentration of minerals is usually higher in the root hair cells than in the soil around them, so the root hair cells can't use diffusion to take up minerals from the soil. Instead they use active transport.

Tip: Minerals should move out of the root hairs if they followed the rules of diffusion.

Active transport allows the plant to absorb minerals from a very dilute solution, against a concentration gradient. This is essential for its growth. But active transport needs energy from respiration to make it work.

Active transport also happens in humans, for example in taking glucose from the gut (see next page), and from the kidney tubules.

Active transport in the gut

Active transport is used in the digestive system when there is a lower concentration of nutrients in the gut, but a higher concentration of nutrients in the blood. Here's how it works:

When there's a higher concentration of glucose and amino acids in the gut they diffuse naturally into the blood. BUT — sometimes there's a lower concentration of nutrients in the gut than there is in the blood (see Figure 2).

This means that the concentration gradient is the wrong way. The nutrients should go the other way if they followed the rules of diffusion. This is where active transport comes in.

Active transport allows nutrients to be taken into the blood, despite the fact that the concentration gradient is the wrong way. This is essential to stop us starving. It means that glucose can be taken into the bloodstream when its concentration in the blood is already higher than in the gut. The glucose can then be transported to cells, where it's used for respiration (see pages 173-175).

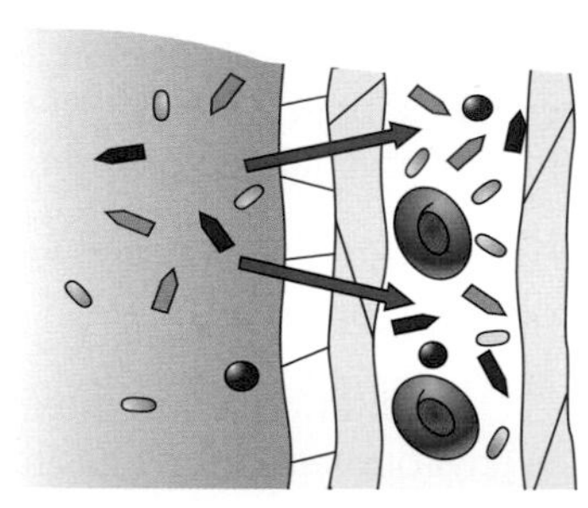

***Figure 2:** Diagram to show active transport in the gut.*

Tip: Active transport comes in useful for loads of processes in the body. E.g. it's used in the kidneys to reabsorb useful substances, like glucose and ions, when making urine.

Summary of transport methods

This topic covers three different ways that substances can get in and out of cells. Here's a handy table to help you remember their similarities and differences:

Type of transport:	Description
Diffusion (see p. 51)	▪ Movement of particles from an area of higher concentration to an area of lower concentration. ▪ Doesn't require energy.
Osmosis (see p. 54)	▪ Movement of water molecules across a partially permeable membrane from a region of higher water concentration to a region of lower water concentration. ▪ Doesn't require energy.
Active transport	▪ Movement of particles against a concentration gradient. ▪ Requires energy.

Exam Tip
Make sure you know the definition of each type of transport off by heart.

Practice Questions — Fact Recall

Q1 What is active transport?

Q2 True or false? Cells cannot absorb ions from very dilute solutions.

Q3 Why do plants need to take up mineral ions?

Q4 Why do root hair cells need to use active transport to take mineral ions up from the soil?

Q5 Name one molecule that is taken up by active transport in the gut.

4. Exchange Surfaces

Solutes are dissolved substances. Living organisms need to be able to exchange dissolved substances with their environment in order to survive. An organism's size and surface area affect how quickly this is done.

Learning Objectives:
- Be able to give examples of substances that move in and out of cells by diffusion.
- Be able to calculate surface area to volume ratios and know how to compare them.
- Understand that a single-celled organism has a larger surface area to volume ratio than a multicellular organism, which allows it to exchange enough substances across its membrane for its needs.
- Know that multicellular organisms have specialised exchange surfaces and transport systems for exchanging materials, and why they have these structures.
- Know how exchange surfaces in multicellular organisms are adapted for exchanging materials.

Specification Reference 4.1.3.1

Movement of substances

Life processes need gases or other dissolved substances (solutes) before they can happen. Dissolved substances move to where they need to be by diffusion and active transport. Water moves by osmosis.

Cells can use diffusion to take in substances they need and get rid of waste products. For example:

- Oxygen and carbon dioxide are transferred between cells and the environment during gas exchange.
- In humans, urea (a waste product produced from the breakdown of proteins) diffuses from cells into the blood plasma for removal from the body by the kidneys.

How easy it is for an organism to exchange substances with its environment depends on the organism's surface area to volume ratio (SA : V).

Surface area to volume ratios

A **ratio** shows how big one value is compared to another. The larger an organism is, the smaller its surface area is compared to its volume. You can show this by calculating surface area to volume ratios. To do that, you first need to know an organism's **volume** and **surface area**. The easiest way to find them is to estimate the size of the organism in the form of a block.

Calculating volume and surface area

The volume of a block (e.g. a cube or cuboid) is found by the equation:

volume = length × width × height

Tip: The units of volume are given in units cubed (e.g. μm^3).

Example

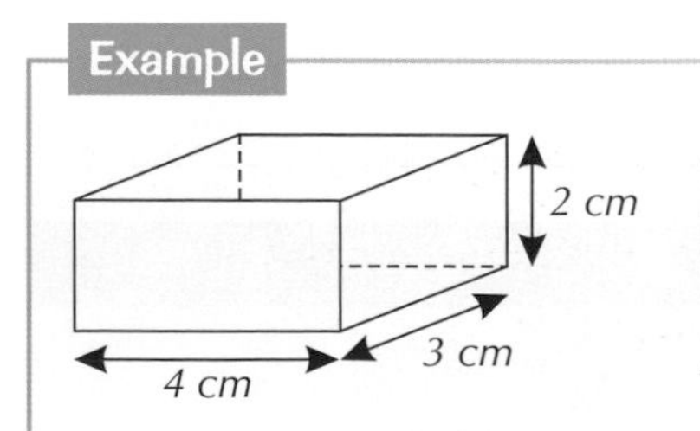

A block that measures 4 cm by 3 cm by 2 cm has the volume:

4 cm × 3 cm × 2 cm = **24 cm^3**

Tip: The area of a triangle is found by the equation: ½ × height × base.
E.g. height = 4 cm, base = 6 cm
Area = ½ × 4 cm × 6 cm = 12 cm^2

The area of a square or rectangular surface is found by the equation:

area = length × width

To calculate the surface area of an object, just calculate the area of each side and add them all together.

Example

A block that measures 4 cm by 3 cm by 2 cm has six surfaces:

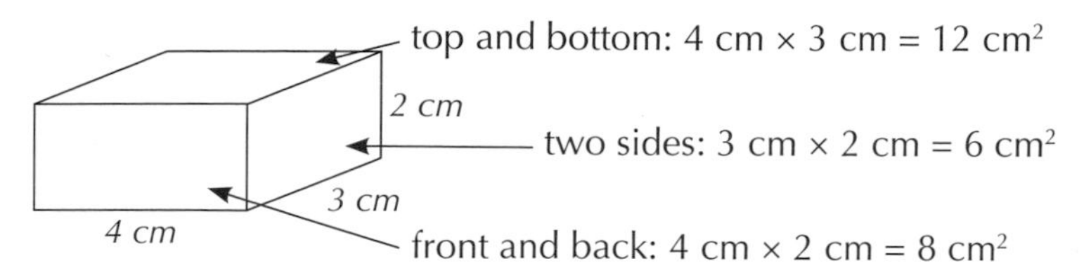

So the total surface area of the block is:
$12 + 12 + 6 + 6 + 8 + 8 =$ **$52\ cm^2$**

Tip: The units of area are given in units squared (e.g. μm^2).

Comparing surface area to volume ratios

Here's how to calculate surface area to volume ratios:

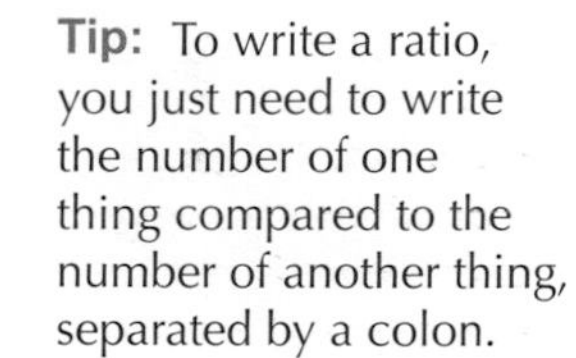

Tip: To write a ratio, you just need to write the number of one thing compared to the number of another thing, separated by a colon.

Example

MATHS SKILLS

A mouse can be represented by a cube measuring 1 cm × 1 cm × 1 cm.

Its volume is: $1 \times 1 \times 1 = 1\ cm^3$

Its surface area is: $6 \times (1 \times 1) = 6\ cm^2$

1 cm 1 cm 1 cm

"cube mouse"

So the mouse has a surface area : volume ratio of **6 : 1**.

Compare this to a cube hippo measuring 2 cm × 4 cm × 4 cm.

Its volume is: $2 \times 4 \times 4 = 32\ cm^3$

Its surface area is:

$2 \times (4 \times 4) = 32\ cm^2$
(top and bottom surfaces)

$+ (4 \times 2) \times 4 = 32\ cm^2$ (four sides)

Total surface area $= 64\ cm^2$

4 cm 4 cm 2 cm

"cube hippo"

The surface area to volume ratio of the hippo can be written as 64 : 32.

To simplify the ratio, divide both sides of the ratio by the volume.

So the hippo has a surface area : volume ratio of **2 : 1**.

Now you can compare the ratios:

The cube mouse's surface area is six times its volume, but the cube hippo's surface area is only twice its volume. So the mouse has a larger surface area compared to its volume.

Figure 1: *A hippo (top) has a small surface area:volume ratio. A mouse (bottom) has a large surface area:volume ratio.*

Tip: To compare two ratios, it's easier if you write both of them in the form X : 1. There's more about writing and simplifying ratios on page 377.

Why are exchange surfaces needed?

In single-celled organisms, gases and dissolved substances can diffuse directly into (or out of) the cell across the cell membrane. It's because they have a large surface area compared to their volume, so enough substances can be exchanged across the membrane to supply the volume of the cell.

Multicellular organisms have a smaller surface area compared to their volume — not enough substances can diffuse from their outside surface to supply their entire volume. This means they need some sort of **exchange surface** for efficient diffusion.

Tip: There's more on diffusion on pages 51-53, osmosis on pages 54-55 and active transport on pages 58-59.

Example

Oxygen is able to diffuse from the air into a bacterial cell across the cell surface. A bacterium can get all the oxygen it needs for respiration this way.

But in the human body, there are trillions of cells and they all need to get enough oxygen to respire. Oxygen can't just diffuse in through your skin — it would never reach the cells deep inside you.

So humans need a specialised gas exchange surface (the alveoli, see page 64) for the oxygen to diffuse across and a specialised breathing system to get it there. They also need a circulatory system (see page 79) to transport the oxygen to every cell.

Tip: One trillion is 10^{12} or 1 000 000 000 000. I wonder who counted all those cells...

The exchange surface structures have to allow enough of the necessary substances to pass through.

Exchange surfaces are adapted to maximise effectiveness:

- They have a thin membrane, so substances only have a short distance to diffuse. The posh way of saying this is that the substances have a 'short diffusion pathway'.
- They have a large surface area so lots of a substance can diffuse at once.
- Exchange surfaces in animals have lots of blood vessels, to get stuff into and out of the blood quickly.
- Gas exchange surfaces in animals (e.g. alveoli, page 64) are often ventilated too — air moves in and out.

Tip: Gas exchange in animals means taking in oxygen from the environment and releasing carbon dioxide.

Practice Questions — Fact Recall

Q1 a) Give three ways that substances can move into and out of a cell.

b) By which of these three ways can dissolved substances move into and out of a cell?

Q2 Give three ways in which exchange surfaces in animals may be adapted for effective exchange.

Practice Questions — Application

Q1 These photographs show an earthworm and an elephant. They are both animals.

An earthworm exchanges gases through its skin. An elephant has a specialised breathing system and exchanges gases in its lungs.

a) Suggest and explain two features that an earthworm's skin may have to make it effective as a gas exchange surface.

b) Explain why elephants need a specialised organ system for gas exchange while earthworms do not.

Q2 Three different cells can be represented by the shapes below:

A

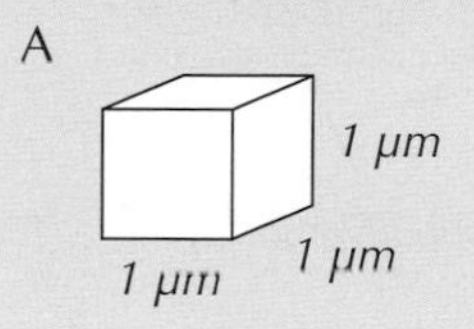

B

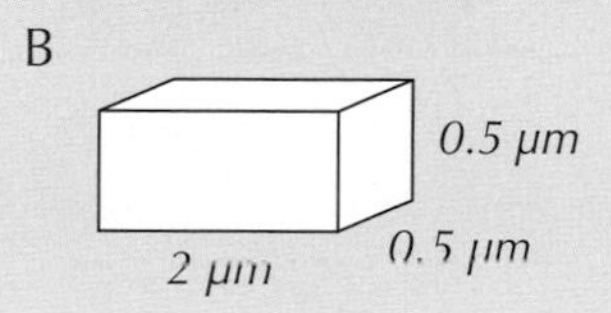

C

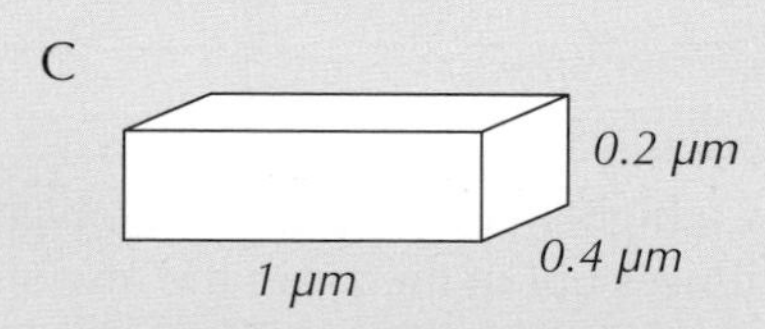

Tip: These cells are not drawn to scale.

a) Calculate the surface area of each cell.

b) Calculate the volume of each cell.

c) Calculate the surface area to volume ratio for each cell. Write each ratio in the form $n : 1$.

d) Predict which cell is most efficient at absorbing substances by diffusion. Explain your answer.

Learning Objectives:

- Know how alveoli are adapted to maximise the exchange of gases in the lungs in mammals.
- Know how villi are adapted to aid the absorption of nutrients in the gut.
- Know how leaves are adapted to maximise the exchange of gases in plants.
- Know how gills are adapted to maximise the exchange of gases in fish.

Specification Reference 4.1.3.1

Tip: It's one alveolus and two or more alveoli.

5. Exchanging Substances

There are a few examples of exchange surfaces that you need to know about. In humans (and other mammals), gases are exchanged in the lungs and nutrients are absorbed in the small intestine. In plants, gases are exchanged in the leaves. In fish, gases are exchanged in the gills.

Gas exchange in humans

The job of the **lungs** is to transfer oxygen to the blood and to remove waste carbon dioxide from it. To do this the lungs contain millions of little air sacs called **alveoli** (see Figure 1) where gas exchange takes place.

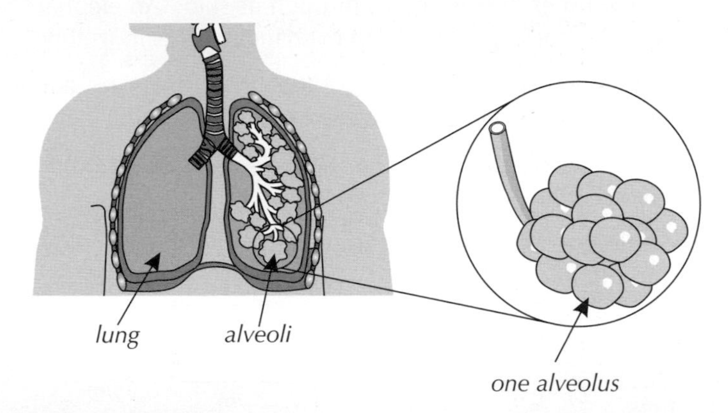

Figure 1: *Diagram to show the location and structure of the alveoli.*

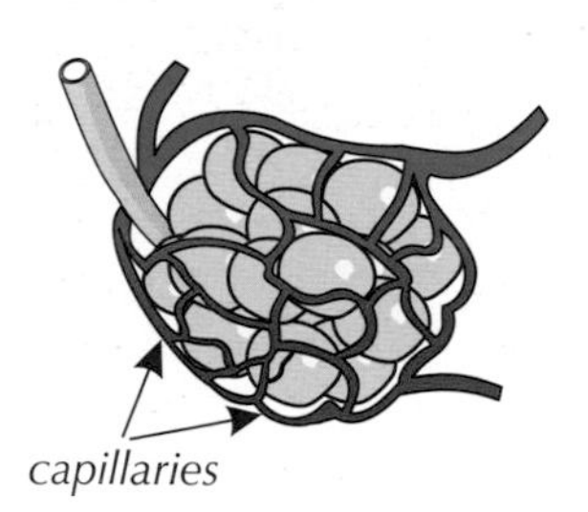

Figure 2: *The alveoli are surrounded by a network of capillaries.*

The alveoli are surrounded by a network of tiny blood vessels known as capillaries (see Figures 2 and 4). There is a higher concentration of oxygen in the air than in the blood, so oxygen diffuses out of the air in the alveoli and into the blood in the capillaries. Carbon dioxide (CO_2) diffuses in the opposite direction (see Figure 3). Air enters and leaves the alveoli via small tubes called bronchioles.

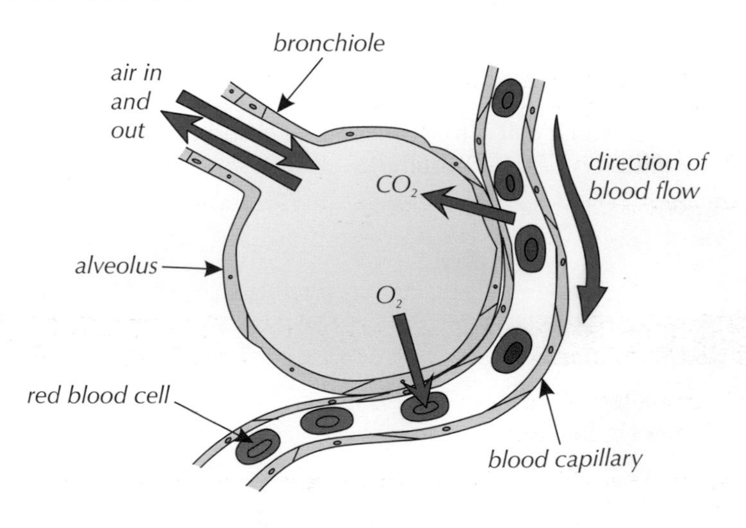

Figure 3: *Diagram to show gas exchange across an alveolus.*

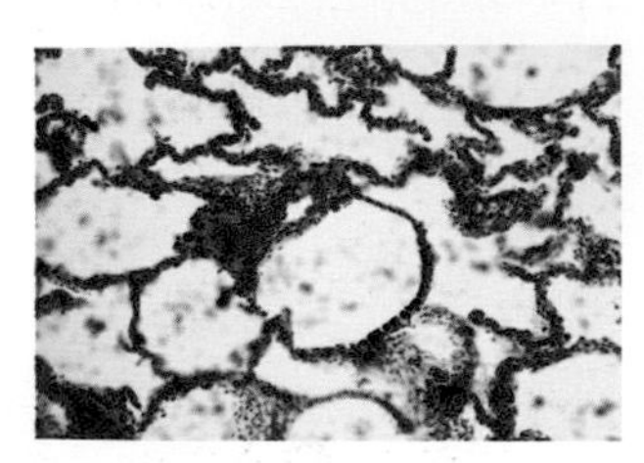

Figure 4: *Alveoli and surrounding capillaries as seen under a light microscope.*

Adaptations of the alveoli

The alveoli are specialised to maximise the diffusion of oxygen and CO_2. They have:

- An enormous surface area (about 75 m^2 in humans).
- A moist lining for dissolving gases.
- Very thin walls.
- A good blood supply.

Tip: Remember, having a large surface area means that lots of a substance (e.g. CO_2) can diffuse across the exchange surface at once. See page 62 for more.

Absorbing the products of digestion

Nutrients, e.g. glucose and amino acids, are absorbed into the bloodstream from the **small intestine** — either by diffusion or active transport (see page 59). To aid this absorption, the inside of the small intestine is covered in millions and millions of tiny little projections called **villi** (see Figure 5).

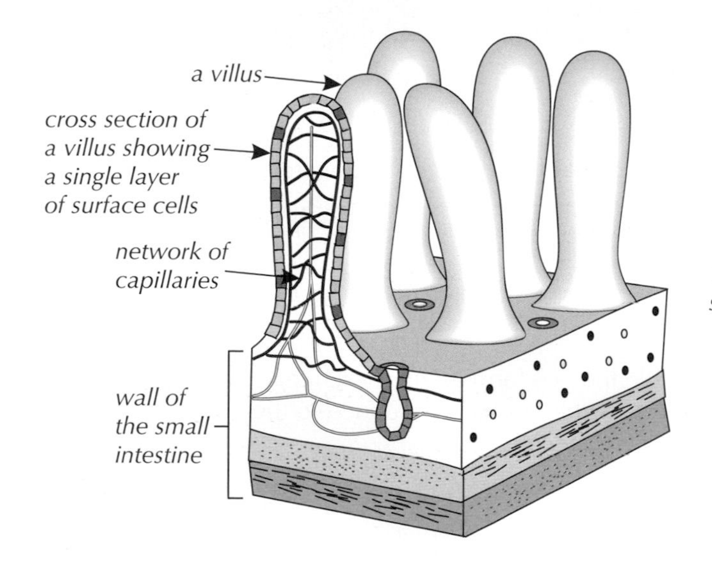

Figure 5: *Diagram showing the structure of villi in the small intestine.*

Tip: It's one vill<u>us</u> and two or more vill<u>i</u>.

Figure 6: *A 3D view of villi in the small intestine, as seen under a microscope.*

Villi increase the surface area in a big way so that digested food is absorbed much more quickly into the blood. They also have:

- a single layer of surface cells.
- a very good blood supply to assist quick absorption.

Tip: Notice how many features the alveoli and the villi have in common — they both have a large surface area, very thin walls and a good blood supply. Lots of exchange surfaces have these features as they make diffusion more effective — see p. 62.

Practice Questions — Fact Recall

Q1 Give one feature of the alveoli that helps them to maximise gas exchange in the lungs.

Q2 Explain how villi aid the absorption of nutrients in the small intestine.

Gas exchange in leaves

Plants need **carbon dioxide** for photosynthesis. Carbon dioxide **diffuses** into the air spaces within the leaf, then it diffuses into the cells where photosynthesis happens. The leaf's structure is adapted so that this can happen easily.

The underneath of the leaf is an **exchange surface**. It's covered in little holes called **stomata** which the carbon dioxide diffuses in through. Oxygen (produced in photosynthesis) and water vapour also diffuse out through the stomata. This is shown in Figure 7.

Tip: 'Stomata' is the plural of 'stoma'. So you get one stoma, but two or more stomata.

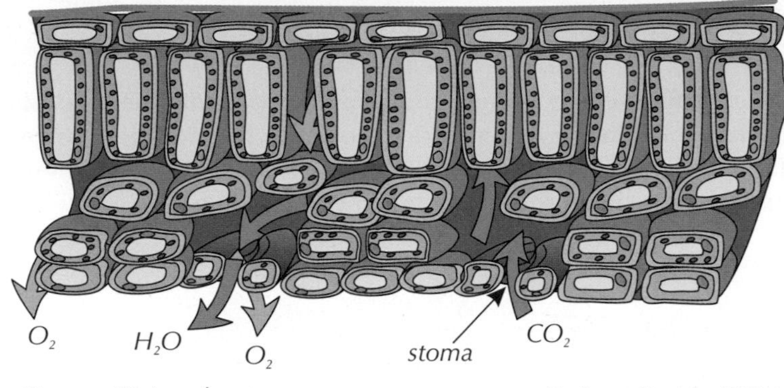

Figure 7: *Diagram to show movement of gases into and out of a leaf.*

Tip: Root hair cells are also adapted to maximise the surface area available for absorbing substances from the soil — see page 58.

The flattened shape of the leaf increases the area of this exchange surface so that it's more effective.

The walls of the cells inside the leaf form another exchange surface. The air spaces inside the leaf increase the area of this surface so there's more chance for carbon dioxide to get into the cells.

Water loss in plants

Water vapour is lost from all over the leaf surface, but most of it is lost through the stomata. If the plant is losing water faster through its leaves than it can be replaced by the roots the stomata can be closed by **guard cells**. Without these guard cells, the plant would soon lose so much water that it would **wilt** (droop). There's more about the guard cells on page 94.

Tip: The water vapour evaporates from the cells inside the leaf. Then it escapes by diffusion because there's a lot of it inside the leaf and less of it in the air outside.

Gas exchange in fish

There's a lower concentration of oxygen in water than in air. So fish have special adaptations to get enough of it. In a fish, the gas exchange surface is the **gills**.

Structure of gills

Water (containing oxygen) enters the fish through its mouth and passes out through the gills — see Figure 8 on the next page. As this happens, oxygen diffuses from the water into the blood in the gills and carbon dioxide diffuses from the blood into the water.

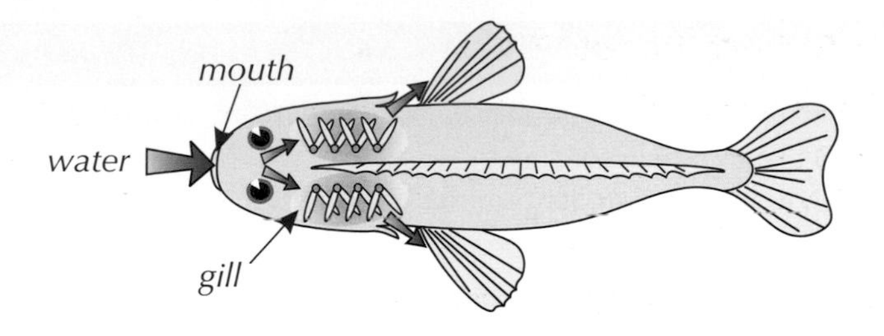

Figure 8: *Water flowing through a fish's gills.*

Tip: The gills are located inside a fish's head. Water flows in through the mouth, over the gills and out through openings behind the head.

Each gill is made of lots of thin plates called **gill filaments**, which give a large surface area for exchange of gases (and so increase the rate of diffusion). The gill filaments are covered in lots of tiny structures called **lamellae**, which increase the surface area even more — see Figure 10. The lamellae have lots of blood capillaries to speed up diffusion between the water and the blood. They also have a thin surface layer of cells to minimise the distance that the gases have to diffuse.

Figure 9: *The gills inside a catfish.*

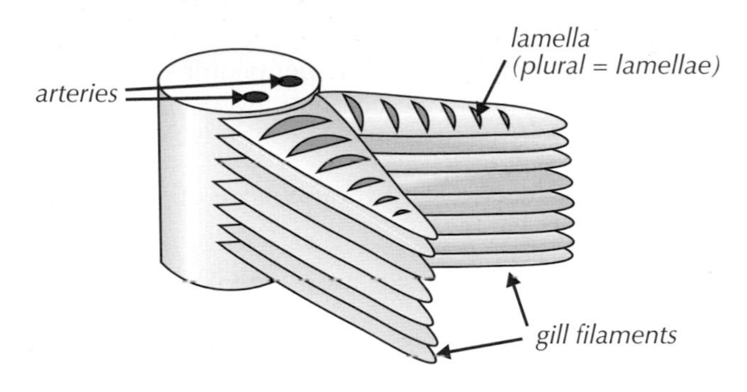

Figure 10: *A section of a fish's gill.*

Blood flows through the lamellae in one direction and water flows over in the opposite direction. This maintains a large concentration gradient between the water and the blood (see Figure 11). The concentration of oxygen in the water is always higher than that in the blood, so as much oxygen as possible diffuses from the water into the blood.

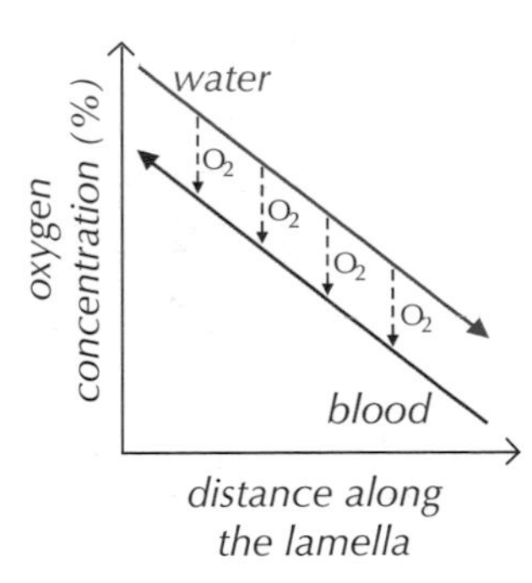

Figure 11: *The concentration of oxygen in the water and blood along the lamella. The concentration in the water is always higher than in the blood.*

Practice Questions — Fact Recall

Q1 How and where does carbon dioxide enter a leaf?

Q2 Name one other gas that is exchanged in a leaf.

Q3 How is the surface area of a leaf increased for gas exchange?

Q4 Explain how lamellae increase the efficiency of gas exchange across a gill.

Q5 How does the direction of the blood flow in the lamellae increase the efficiency of gas exchange in the gills?

Topic 1b Checklist — Make sure you know...

Diffusion

- ❑ That diffusion is the spreading out of particles from an area of higher concentration to an area of lower concentration.
- ❑ That dissolved substances such as oxygen (which is needed for respiration) move in and out of a cell by diffusion, and that the net (overall) movement will be to an area of lower concentration.
- ❑ That the bigger the difference in concentration, the faster the diffusion rate is, because the net movement from one side is greater.
- ❑ That the higher the temperature, the faster the diffusion rate is, because the particles have more energy so move faster.
- ❑ That the larger the surface area, the faster the diffusion rate is, because more particles can pass through at once.

Osmosis

- ❑ That osmosis is the movement of water molecules across a partially permeable membrane from a region of higher water concentration (a dilute solution) to a region of lower water concentration (a concentrated solution).
- ❑ How differences in the concentration of water molecules inside and outside of a cell will cause water molecules to move into or out of the cell by osmosis.
- ❑ How to investigate the effect of different concentrations of a sugar or salt solution on potato cells (Required Practical 3).
- ❑ How to work out the rate of osmosis and the units for it (g/min).
- ❑ How to calculate percentage change in mass for plant tissue (e.g. a piece of potato).
- ❑ How to draw and interpret graphs showing the effect of different concentrations of sugar or salt solution on potato cells.

Active Transport

- ❑ That active transport is the movement of particles against a concentration gradient (i.e. from an area of lower concentration to an area of higher concentration) using energy transferred by respiration.
- ❑ That the structure of a root hair cell gives it a large surface area to absorb mineral ions and water.
- ❑ That root hair cells use active transport to absorb mineral ions (which plants need for healthy growth) against a concentration gradient.
- ❑ That sugar is absorbed from the gut into the blood by active transport, for use in cellular respiration.
- ❑ How to describe the transport of substances in and out of cells by active transport, diffusion and osmosis.
- ❑ The differences between active transport, diffusion and osmosis.

cont...

Exchange Surfaces

- ☐ That oxygen and carbon dioxide move in and out of cells by diffusion in gas exchange, and urea diffuses out of cells into the blood plasma for excretion.
- ☐ How to calculate and compare surface area to volume ratios.
- ☐ That exchanging materials with the environment is more difficult in multicellular organisms than in single-celled organisms, due to their smaller surface area to volume ratios — so multicellular organisms need specialised organ systems for exchange, and transport systems to carry materials around the body.
- ☐ How exchange surfaces are adapted for exchanging materials — they're thin (so substances only have a short distance to diffuse), they have a large surface area (so lots of a substance can diffuse at once), they have lots of blood vessels (in animals, to get stuff in and out of the blood quickly), and they're often ventilated (for gas exchange in animals).

Exchanging Substances

- ☐ That alveoli (tiny air sacs) increase the surface area of the lungs in humans, maximising the diffusion of oxygen into the blood and carbon dioxide out of the blood.
- ☐ That alveoli also have a thin layer of cells, a moist lining and an efficient blood supply to aid the diffusion of gases.
- ☐ That villi increase the surface area of the small intestine in humans, contain a network of capillaries and have a single layer of surface cells, aiding the absorption of nutrients (by active transport and diffusion) into the blood.
- ☐ That carbon dioxide diffuses into a plant leaf from the air through the stomata (tiny holes on the underside of the leaf), and that oxygen and water vapour diffuse out of the leaf.
- ☐ That the flattened shape of the leaf and the internal air spaces between cells increase the surface area of the leaf for gas exchange.
- ☐ How the gills of a fish are adapted for efficient gas exchange — the gill filaments and lamellae increase the surface area, and the lamellae have a thin surface layer and an efficient blood supply to maximise the diffusion between water and the blood.

Exam-style Questions

1 Methods of exchanging substances with the environment vary between organisms. Some organisms contain specialised exchange surfaces for this purpose.

Figure 1 shows a rod-shaped bacterium, with its approximate dimensions.

Figure 1

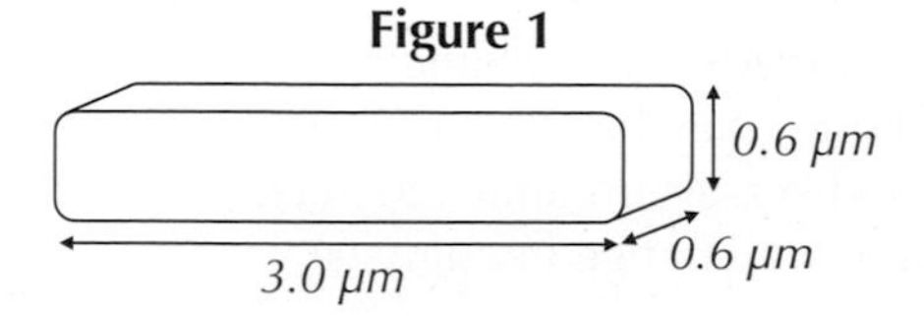

1.1 Calculate the surface area to volume ratio of this bacterium.

(3 marks)

1.2 Explain why this bacterium doesn't need any specialised surfaces for exchanging substances with the environment.

(3 marks)

Plants are organisms that have specialised exchange surfaces.
Most plants exchange gases through their stomata.

1.3 What are stomata?

(1 mark)

1.4 Carbon dioxide diffuses out through the stomata into the air outside the leaf. What does this tell you about the concentration of carbon dioxide in the air?

(1 mark)

1.5 Other than stomata, describe **one** leaf adaptation for efficient gas exchange.

(1 mark)

Figure 2 shows a section of an organ with a specialised gas exchange surface. This organ is found in fish.

Figure 2

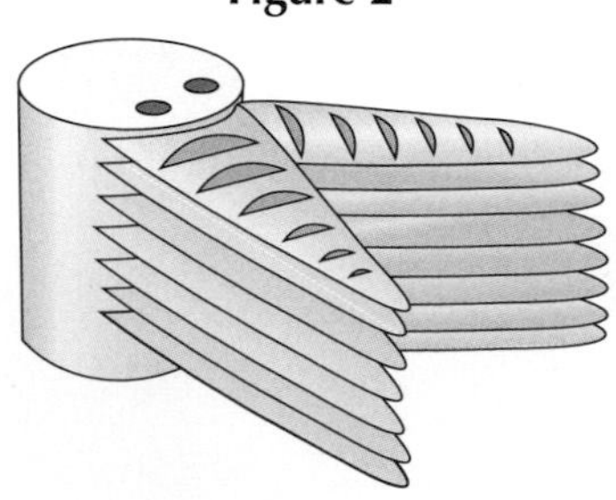

1.6 Name the organ specialised for gas exchange in fish.

(1 mark)

1.7 If the water in a lake becomes polluted, the amount of oxygen dissolved in the water is reduced. What effect would this have on the rate of diffusion of oxygen into the gas exchange organs of fish that live in the lake? Explain your answer.

(2 marks)

2 In humans, glucose is absorbed into the bloodstream in the small intestine. This is often done using active transport, which requires respiration.

2.1 Explain why glucose may be absorbed using active transport, and why this requires respiration.

(2 marks)

2.2 *Giardia* is a parasite that attaches itself to the walls of the small intestine. Infection with *giardia* can cause the villi in the small intestine to become damaged.

Suggest how this would affect the absorption of glucose and other nutrients in the small intestine. Explain your answer.

(2 marks)

2.3 Water from food is absorbed into the bloodstream in the large intestine.

State by what process water is absorbed in the large intestine and explain how it works.

(2 marks)

3 Eleanor is investigating osmosis using potatoes.

She cuts small wells into two pieces of potato, and sets up two experiments (A and B). She then leaves the potatoes for five hours.

The diagrams show how the appearance of each experiment changes after five hours.

Experiment A

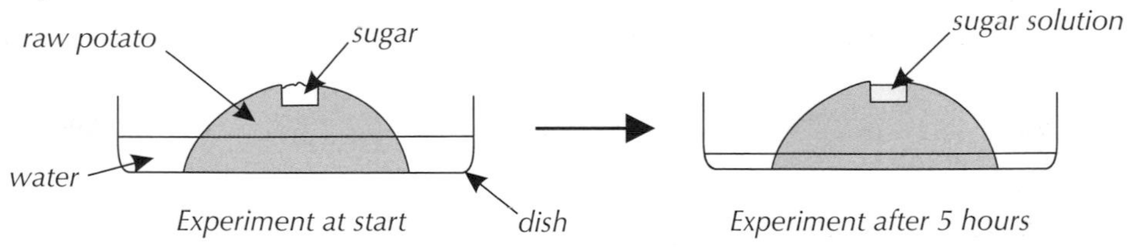

Experiment B

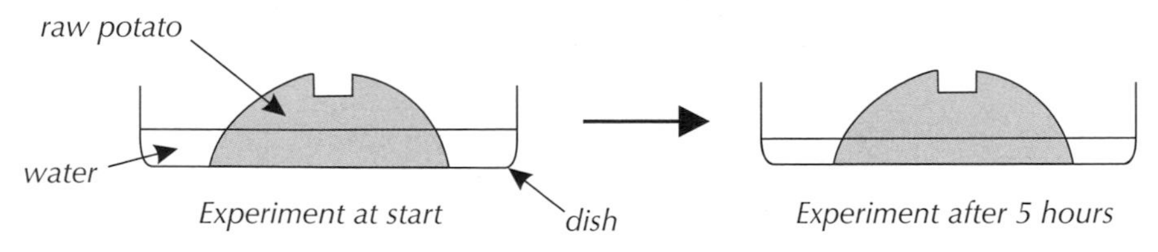

3.1 Explain the changes that Eleanor saw in experiment A.

(3 marks)

3.2 Suggest an explanation for why the water level in the dish in experiment B didn't decrease as much as much as that in experiment A.

(3 marks)

1. Cell Organisation

Multicellular organisms (like humans) can have trillions of cells. To keep the organism going, those cells have to work together — which needs organisation.

Learning Objectives:

- Know that living things are made up of cells, which can be thought of as the basic building blocks of all organisms.
- Know that a tissue is a group of similar cells that work together to carry out a particular function.
- Know that organs are groups of different tissues that work together to perform a certain function.
- Know that organ systems are groups of organs that work together to perform a particular function.
- Know that the digestive system is an organ system and that its function is to digest and absorb food.
- Know that organ systems work together to form whole organisms.

Specification References 4.2.1, 4.2.2.1

How are cells organised?

Cells are the basic building blocks that make up all living organisms. As you know from page 30, **specialised cells** carry out a particular function. The process by which cells become specialised for a particular job is called **differentiation**. Differentiation occurs during the development of a multicellular organism.

Specialised cells are organised to form tissues, which form organs, which form organ systems — there's more about each of these over the next few pages.

Large **multicellular organisms** have different systems inside them for exchanging and transporting materials.

Example

Mammals have a breathing system — this includes the airways and lungs. The breathing system is needed to take air into and out of the lungs, so that oxygen and carbon dioxide can be exchanged between the body and the environment.

Tissues

You need to know what a tissue is:

> A tissue is a group of similar cells that work together to carry out a particular function.

A tissue can include more than one type of cell. Mammals (like humans), have several different types of tissue:

Examples

- Muscular tissue, which contracts (shortens) to move whatever it's attached to.
- Glandular tissue, which makes and secretes substances like enzymes (proteins that control chemical reactions, see p. 115) and hormones (chemical messengers, see p. 207).
- Epithelial tissue, which covers some parts of the body, e.g. the inside of the gut.

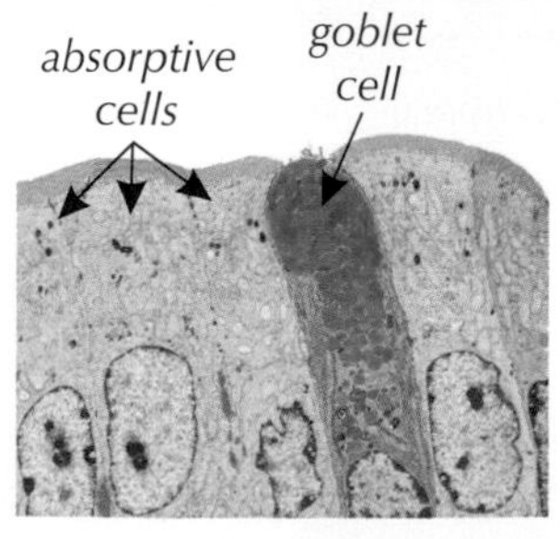

Figure 1: *Epithelial tissue in the small intestine. It contains absorptive cells and goblet cells.*

Organs

Tissues are organised into organs:

An organ is a group of different tissues that work together to perform a certain function.

Mammals have many different organs, which are made up of different tissues.

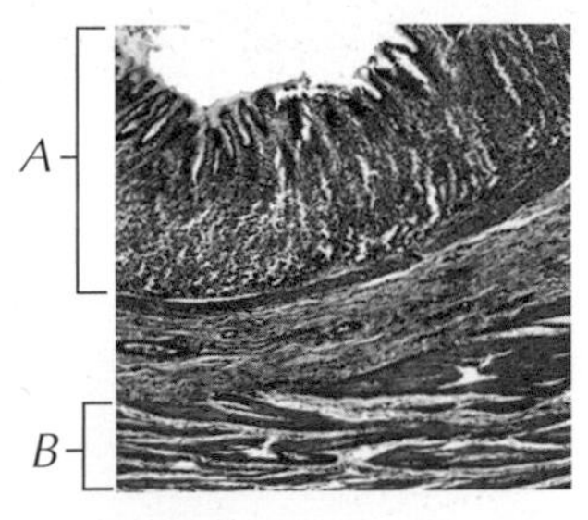

Figure 2: *A cross-section through part of the stomach. There's glandular tissue at the top (A) and muscular tissue at the bottom (B).*

Example

The **stomach** is an organ made of these tissues:

- Muscular tissue, which moves the stomach wall to churn up the food.
- Glandular tissue, which makes digestive juices to digest food.
- Epithelial tissue, which covers the outside and inside of the stomach.

Tip: Digestive juices are secretions from the digestive system that help to break down food. They contain enzymes (see pages 120-121).

Organ systems

Organs are organised into organ systems:

An organ system is a group of organs working together to perform a particular function.

Example

The **digestive system** is the organ system that breaks down food in humans and other mammals. It's also an exchange system — it exchanges materials with the environment by taking in nutrients and releasing substances such as bile (see page 122).

The digestive system (see Figure 3) is made up of these organs:

- **Glands** (e.g. the **pancreas** and **salivary glands**), which produce digestive juices.
- The **stomach**, where food is digested.
- The **liver**, which produces bile.
- The **small intestine**, where food is digested and soluble food molecules are absorbed.
- The **large intestine**, where water is absorbed from undigested food, leaving faeces.

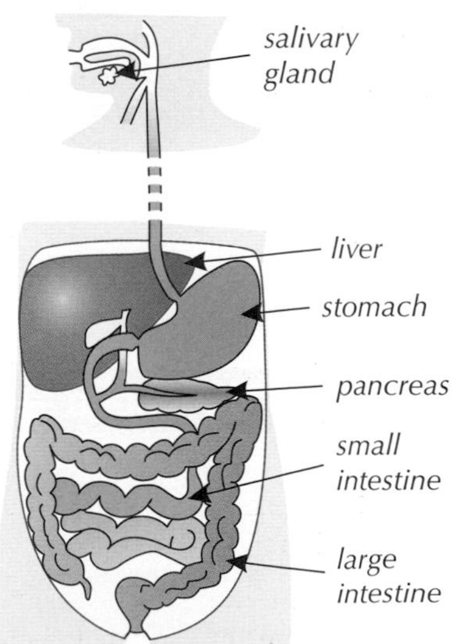

Figure 3: *A diagram of the human digestive system.*

Tip: There's more on the digestive system on pages 120-122.

Organ systems work together to make entire organisms.

Size and scale

It will help to have an understanding of the size and scale of all the structures that make up an organ system — from the tiny individual specialised cells to the organ system as a whole.

Exam Tip
You don't need to learn the lengths given in this example — they're just to give you an idea of size.

Tip: The digestive tract is the big long 'tube' which food passes through, including the small and large intestines. The intestines are all folded up, which is part of the reason why the digestive tract can be so long and yet still fit inside you.

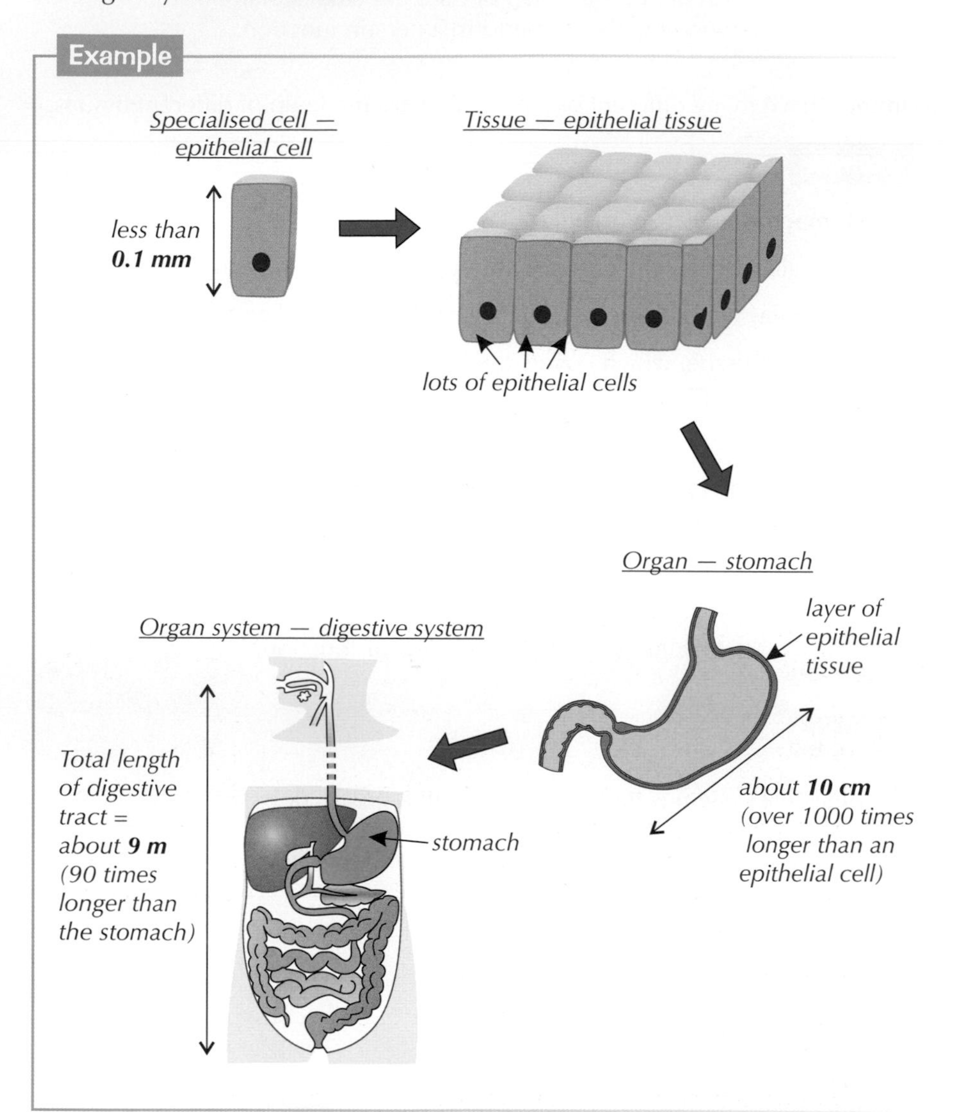

Practice Questions — Fact Recall

Q1 What can be thought of as the basic building blocks that make up all living organisms?

Q2 What is a tissue?

Q3 What is an organ?

Q4 What term describes a group of organs which work together to perform a particular function?

Q5 Look at the diagram of the digestive system.

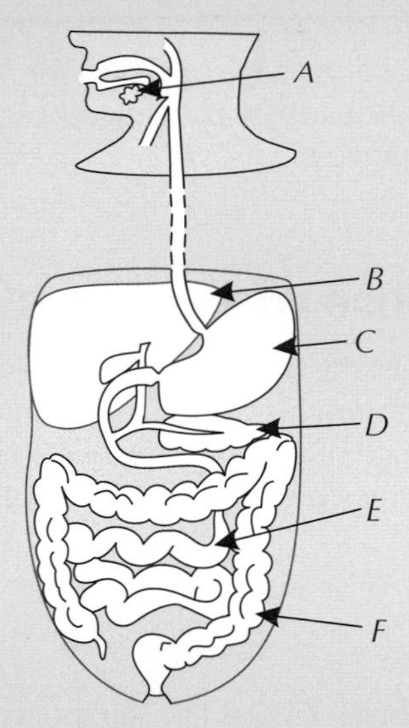

Write down the names of the organs labelled A-F and briefly say what each one does.

Exam Tip
You need to know where the organs of the digestive system are on a diagram for your exam, so make sure you can get them all right in Q5.

Practice Questions — Application

Q1 Blood is made up of specialised red blood cells and several different types of white blood cells. Is blood an example of a cell, tissue, organ or organ system? Explain your answer.

Q2 The fallopian tubes and uterus are part of the female reproductive system. A fallopian tube contains muscular and epithelial tissue. These work together to move a fertilised egg cell along the fallopian tube to the uterus (an organ).

a) Suggest why a fallopian tube has muscular tissue.

b) Is a fallopian tube a cell, tissue, organ or organ system? Explain your answer.

c) Rewrite the following list of structures in order of size. Start with the smallest:

uterus reproductive system egg cell muscular tissue

Tip: The definitions of a tissue, an organ and an organ system should be really clear in your head. Take a look back at pages 72-73 if you're not sure of them.

2. The Lungs

Learning Objectives:
- Know how the trachea, bronchi, alveoli and the capillary network surrounding the alveoli make up the structure of the human lungs.
- Understand the function of human lungs.
- Know how human lungs are adapted for exchanging gases.

Specification Reference 4.2.2.2

Exchanging substances is a complicated process for large, complex organisms. The lungs are essential for humans (and many other animals), because they play a vital role in the exchange of gases.

What is the purpose of the lungs?

You need to get oxygen from the air into your bloodstream so that it can get to your cells for respiration. You also need to get rid of the carbon dioxide in your blood. This exchange of gases all happens inside your **lungs**. Air is forced in and out of your lungs by the action of breathing.

The thorax

The lungs are in the **thorax** (see Figure 1). The thorax is the top part of your body. It's separated from the lower part of the body (the **abdomen**) by a muscle called the **diaphragm**. The lungs are like big pink sponges and are protected by the **ribcage**. They're surrounded by the **pleural membranes**.

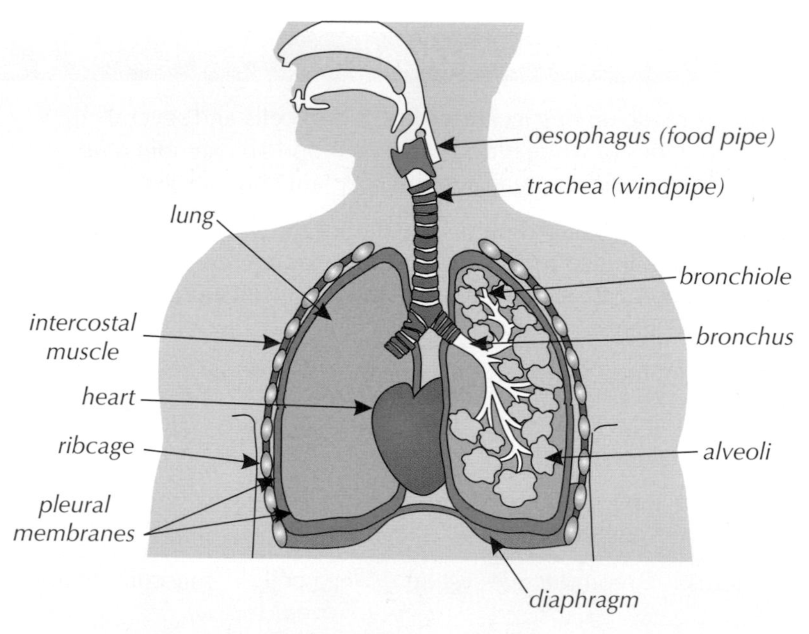

Figure 1: *Diagram showing the main structures in the thorax.*

Exam Tip
Make sure you know exactly where the trachea, bronchi and alveoli are in relation to each other in the lungs.

The air that you breathe in goes through the **trachea**. This splits into two tubes called '**bronchi**' (each one is 'a bronchus'), one going to each lung. The bronchi split into progressively smaller tubes called **bronchioles**.

The bronchioles finally end at small bags called **alveoli** where the gas exchange takes place by diffusion (see page 64).

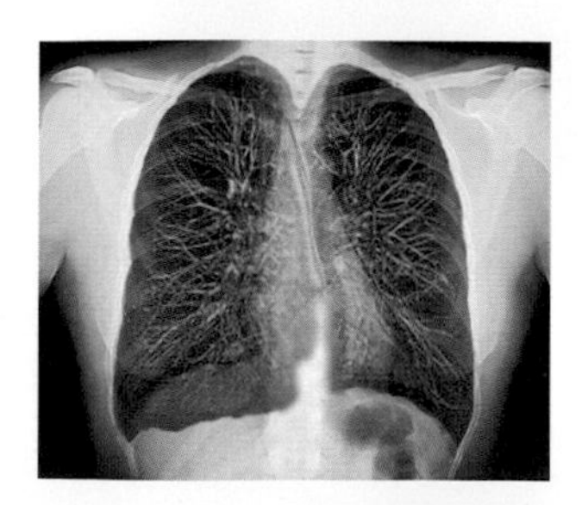

Figure 2: *A 3D MRI scan of the thorax. The lungs appear black and the bronchioles orange.*

Alveoli

The lungs contain millions and millions of little air sacs called alveoli, surrounded by a network of blood capillaries (see Figure 3). This is where **gas exchange** happens.

The blood passing next to the alveoli has just returned to the lungs from the rest of the body, so it contains lots of carbon dioxide and very little oxygen. Oxygen diffuses out of the alveolus (high concentration) into the blood (low concentration). Carbon dioxide diffuses out of the blood (high concentration) into the alveolus (low concentration) to be breathed out (see Figure 3).

Tip: The alveoli have an enormous surface area, a moist lining, very thin walls and a good blood supply to maximise gas exchange. This has already been covered back on page 65, so head back there if you need a recap.

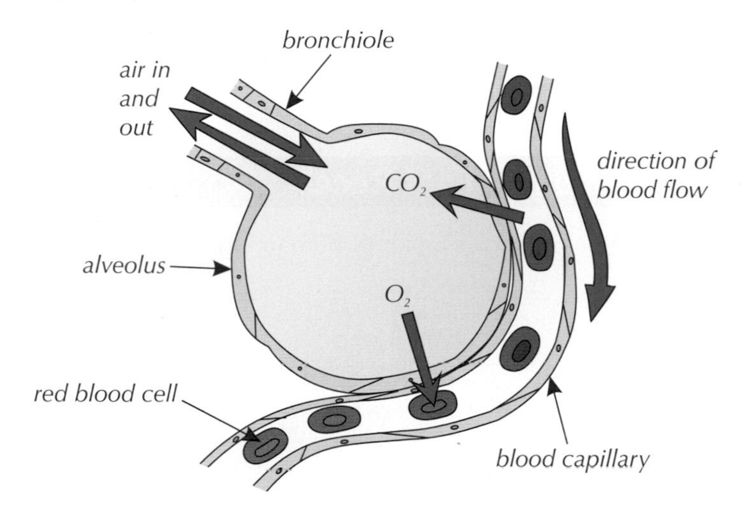

Figure 3: *Diagram showing gas exchange across an alveolus.*

When the blood reaches body cells, oxygen is released from the red blood cells (where there's a high concentration) and diffuses into the body cells (where the concentration is low). At the same time, carbon dioxide diffuses out of the body cells (where there's a high concentration) into the blood (where there's a low concentration). It's then carried back to the lungs (see Figure 4).

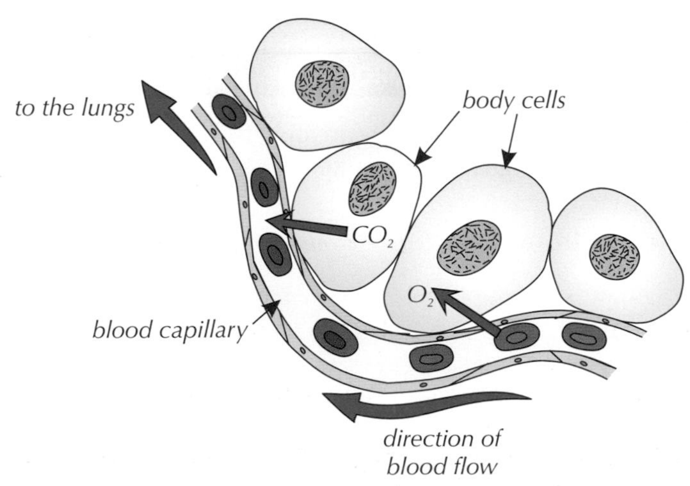

Figure 4: *Diagram showing gas exchange between a capillary and body cells.*

Calculating breathing rate

Rate calculations pop up all the time in biology, and you're expected to know how to do them — thankfully, they're pretty easy. Breathing rate is the sort of thing that you could get asked to work out in your exam. You can calculate someone's breathing rate in breaths per minute.

Tip: You can work out someone's heart rate in beats per minute in exactly the same way as this. Just replace number of breaths with number of beats.

Example

Chris takes 112 breaths in 8 minutes.
Calculate his average breathing rate in breaths per minute.

breaths per minute = number of breaths ÷ number of minutes
= 112 ÷ 8
= 14 breaths per minute

Practice Question — Fact Recall

Q1 The diagram below shows a pair of human lungs.

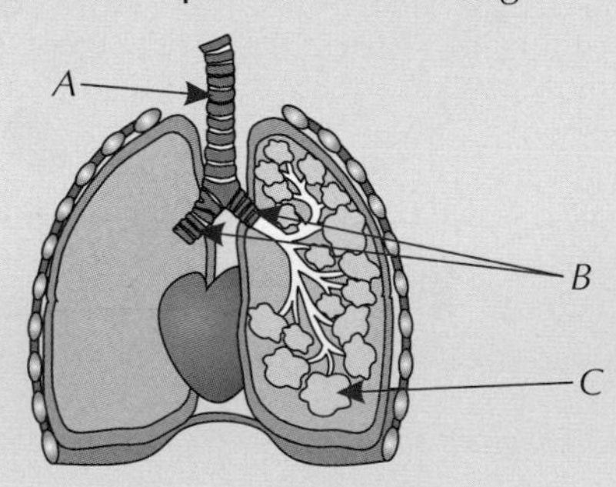

a) Name the tube labelled A.

b) Name the tubes labelled B.

c) i) What is the name given to the part of the lungs labelled C?

ii) Explain why part C is surrounded by a network of capillaries.

Exam Tip
Quick calculations like this can be a really easy way to pick up marks in the exam. It'd be a good idea to make sure that you're happy with how to do them.

Practice Question — Application

Q1 While taking part in a race, Gibson took 366 breaths in 9 minutes. Calculate his average breathing rate in breaths per minute.

3. Circulatory System — The Heart

Multicellular organisms need a way to transport materials between their cells. Humans (like many animals) have a transport system called the circulatory system to do just that, and the heart has a major role to play in it...

Learning Objectives:

- Know that humans have a double circulatory system to transport blood around the body, in which the heart acts as a pump.
- Know that one of the 'circuits' in the human circulatory system is powered by the right ventricle and is responsible for pumping blood to the lungs, where gas exchange can take place.
- Know that the other 'circuit' in the human circulatory system is powered by the left ventricle and pumps blood around all the other parts of the body.
- Know the structure of the human heart, including its blood vessels: the aorta, vena cava, pulmonary artery, pulmonary vein and coronary arteries.
- Understand how the human heart functions.
- Know that the heart contains a group of cells located in the right atrium that act as a pacemaker to control the natural resting heart rate.
- Know that patients with an irregular heart beat can be fitted with an electrical artificial pacemaker to correct the issue.

Specification Reference 4.2.2.2

Function of the circulatory system

The circulatory system's main function is to get food and oxygen to every cell in the body. As well as being a delivery service, it's also a waste collection service — it carries waste products like carbon dioxide and urea to where they can be removed from the body. The circulatory system includes the heart, blood vessels and the blood.

A double circulatory system

Humans have a double circulatory system — two circuits joined together:

- In the first circuit, the right ventricle (see next page) pumps deoxygenated blood (blood without oxygen) to the lungs to take in oxygen. The blood then returns to the heart (see Figure 1a).
- In the second circuit, the left ventricle (see next page) pumps oxygenated blood around all the other organs of the body. The blood gives up its oxygen at the body cells and the deoxygenated blood returns to the heart to be pumped out to the lungs again (see Figure 1b).

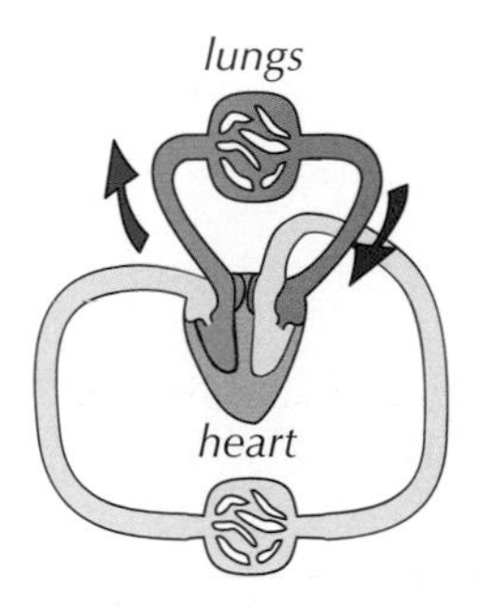

Figure 1a: *Blood circulation between the heart and the lungs.*

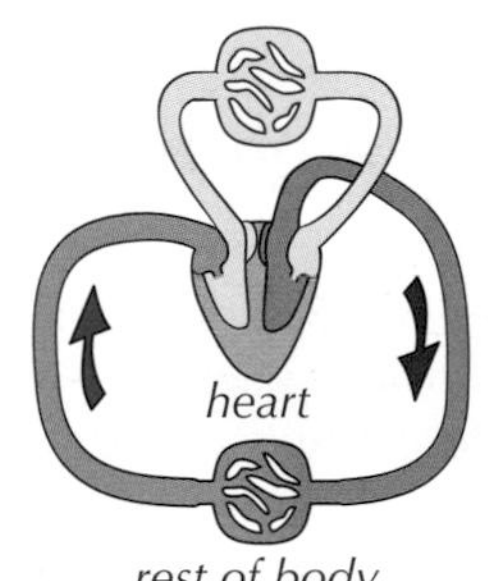

Figure 1b: *Blood circulation between the heart and the rest of the body.*

Structure of the heart

The heart is a pumping organ that keeps the blood flowing around the body. The walls of the heart are mostly made of **muscle tissue**, which contracts to pump the blood.

The heart has four chambers (the **right atrium**, **right ventricle**, **left atrium** and **left ventricle**) which it uses to pump blood around (see Figure 2).
The main blood vessels leading into and out of these chambers are the **vena cava**, **pulmonary artery**, **aorta** and **pulmonary vein**.

Exam Tip
You need to learn the names of these chambers and the main blood vessels leading into and out of the heart for your exam.

Tip: Don't be confused about the way the right and left side of the heart are labelled in this diagram — this is the right and left side of the person whose heart it is.

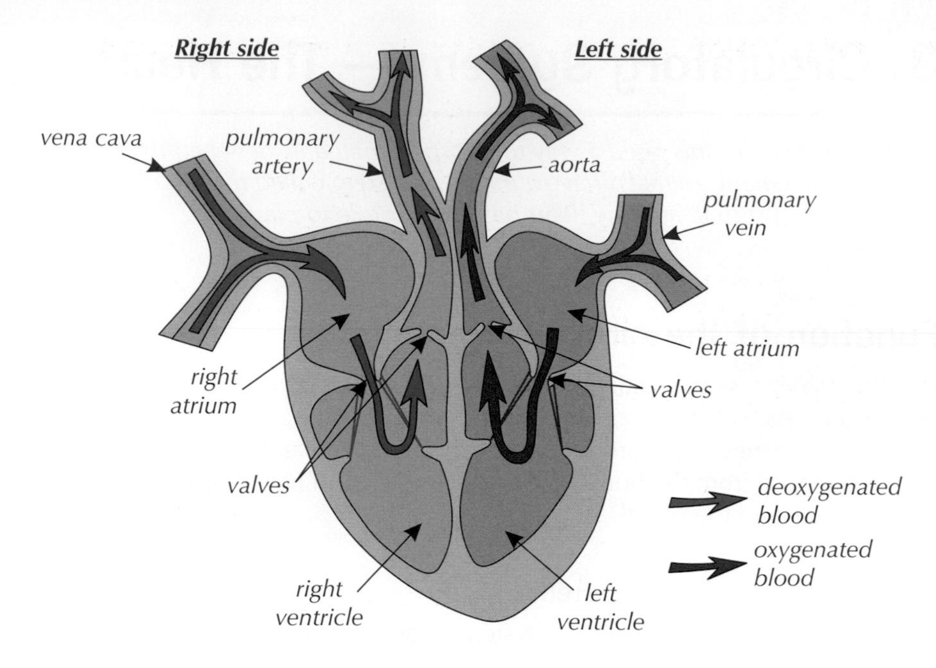

Figure 2: *Diagram showing the structure of the heart and the direction of blood flow through the heart.*

The heart has **valves** to make sure that blood goes in the right direction — they prevent it flowing backwards.

The heart also needs its own supply of oxygenated blood. Arteries called **coronary arteries** branch off the aorta and surround the heart, making sure that it gets all the oxygenated blood it needs.

Tip: The coronary arteries are really important. If they become blocked, the heart can become starved of oxygen, potentially leading to a heart attack and heart failure.

Blood flow through the heart

You need to know how the heart uses its four chambers to pump blood around the body. Here's what happens:

1. Blood flows into the two atria from the vena cava and the pulmonary vein (see Figure 3).

Tip: 'Atria' is the plural of 'atrium'. So you get one atrium, but two atria.

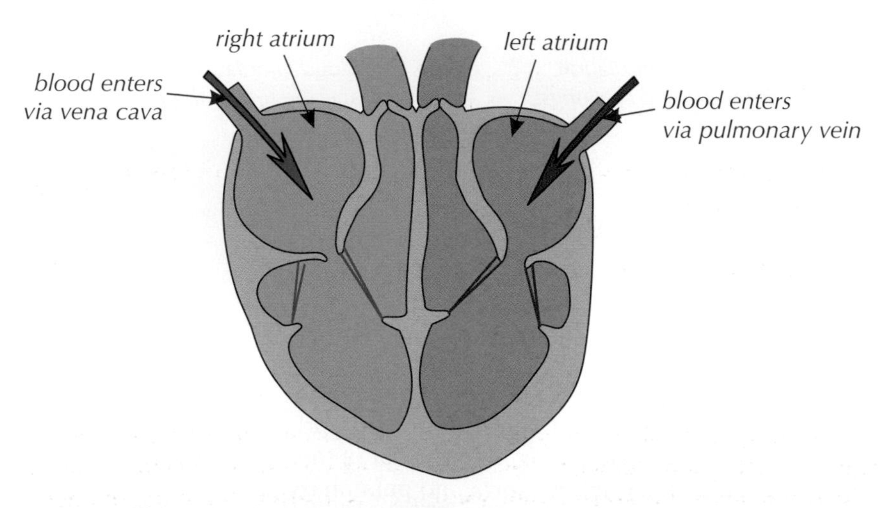

Figure 3: *Diagram showing blood flowing into the heart.*

2. The atria contract, pushing the blood into the ventricles (see Figure 4).

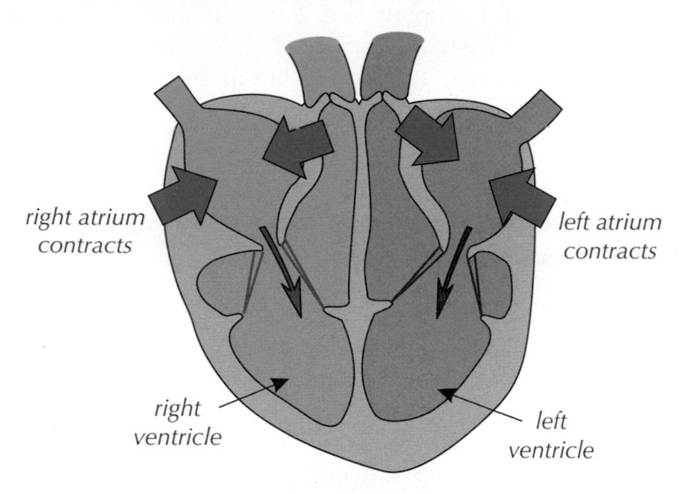

Figure 4: *Diagram showing blood being forced into the ventricles.*

3. The ventricles contract, forcing the blood into the pulmonary artery and the aorta, and out of the heart (see Figure 5).

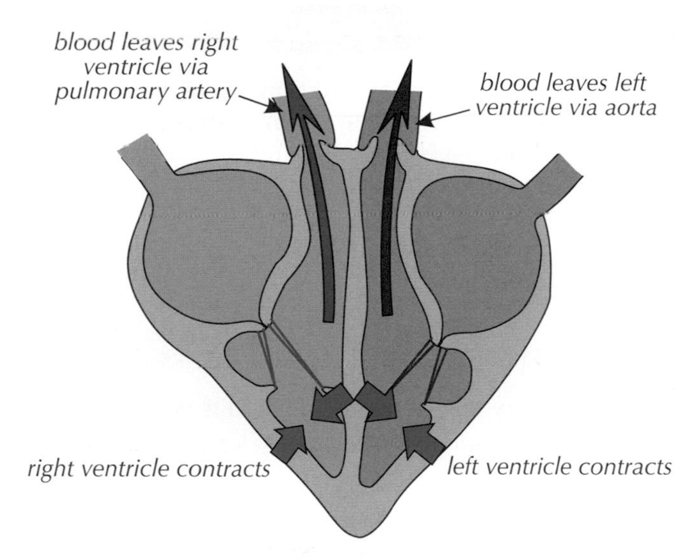

Figure 5: *Diagram showing blood being forced out of the heart.*

4. The blood then flows to the organs through arteries, and returns through veins.

5. The atria fill again and the whole cycle starts over.

Tip: There's more about the different types of blood vessels, including arteries and veins, on pages 83-84.

The heart's pacemaker

Your resting heart rate is controlled by a group of cells in the right atrium wall that act as a pacemaker. These cells produce a small electric impulse which spreads to the surrounding muscle cells, causing them to contract.

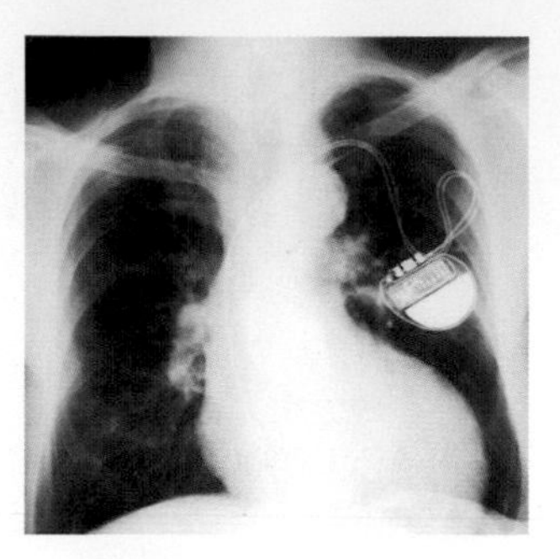

Figure 6: *An X-ray of the chest of a patient fitted with an artificial pacemaker.*

An artificial pacemaker is often used to control heartbeat if the natural pacemaker cells don't work properly (e.g. if the patient has an irregular heartbeat). It's a little device that's implanted under the skin and has a wire going to the heart. It produces an electric current to keep the heart beating regularly.

Practice Questions — Fact Recall

Q1 What is the function of the circulatory system?

Q2 The human circulatory system is made up of two separate circuits. What is the function of each circuit?

Q3 Name the four chambers of the heart.

Q4 Name the two blood vessels which:

a) carry blood into the heart, b) carry blood out of the heart.

Q5 Explain how blood flows through the heart.

Practice Questions — Application

Q1 Coronary heart disease is a condition where the coronary arteries become blocked. Suggest why coronary heart disease can cause heart attacks and heart failure.

Q2 Some heart conditions are caused by an irregular heart beat.

a) Describe how the resting heart rate is controlled in a healthy heart.

b) Pacemakers can be used to treat an irregular heartbeat. Describe a pacemaker and explain how it works.

Q3 In the circulatory system of a fish, blood is pumped from the heart, across the gills (to pick up oxygen), to the body tissues and then back to the heart in a single circuit.

Give one similarity and one difference between the circulatory system of a fish and that of a human.

4. Circulatory System — The Blood Vessels

Our blood makes its way around our body in blood vessels. There are three different types of blood vessels you need to know about...

Learning Objectives:
- Know that arteries, veins and capillaries are the three different types of blood vessel found in the body.
- Be able to explain how the structures of arteries, veins and capillaries are adapted to allow them to perform their functions.
- Be able to calculate the rate of blood flow.

Specification Reference 4.2.2.2

Arteries

Arteries are blood vessels which carry blood away from the heart, towards the organs. The heart pumps the blood out at high pressure, so the artery walls are strong and elastic. They contain thick layers of muscle to make them strong, and elastic fibres to allow them to stretch and spring back. The walls are thick compared to the size of the hole down the middle (the lumen) — see Figure 1.

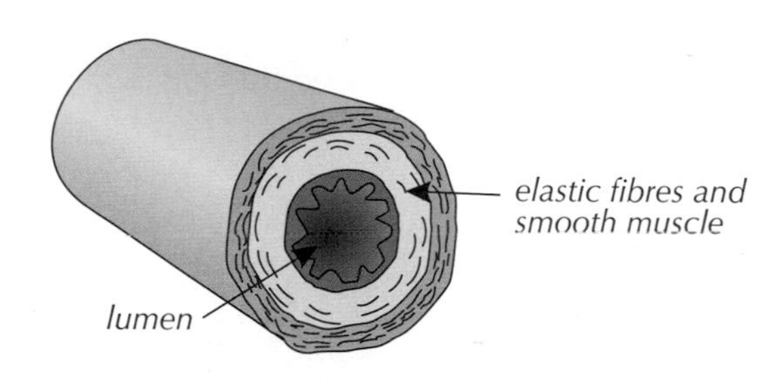

Figure 1: *Diagram to show the structure of an artery.*

Capillaries

Arteries branch into **capillaries**. Capillaries are involved in the exchange of materials at the tissues — they carry the blood really close to every cell in the body to exchange substances with them. They supply food and oxygen to the cells, and take away waste products like carbon dioxide.

Capillaries are really tiny — too small to see. They have permeable walls, so the substances being exchanged with the cells can diffuse in and out. Their walls are usually only one cell thick (see Figure 2). This increases the rate of diffusion by decreasing the distance over which it occurs. Capillaries are also very narrow. This gives them a large surface area compared to their volume, which also increases the rate of diffusion.

Tip: Remember, diffusion is the process by which substances move from an area of higher concentration to an area of lower concentration.

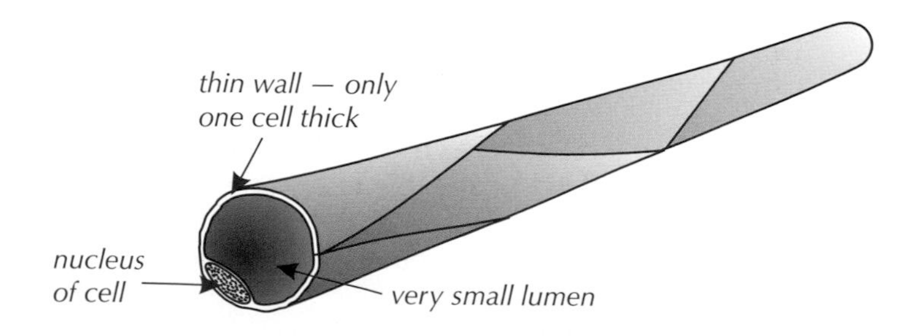

Figure 2: *Diagram to show the structure of a capillary.*

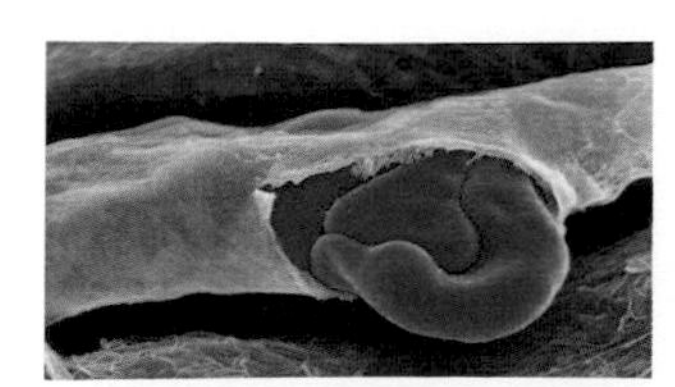

Figure 3: *A torn capillary with blood cells leaking out — you can see how thin the walls of the capillary are.*

Veins

Capillaries eventually join up to form **veins**. Veins carry blood to the heart.

The blood is at lower pressure in the veins, so the walls don't need to be as thick as artery walls. Veins have a bigger lumen than arteries to help the blood flow despite the lower pressure (see Figure 4).

Figure 4: *Diagram to show the structure of a vein.*

They also have valves to help keep the blood flowing in the right direction (see Figure 6).

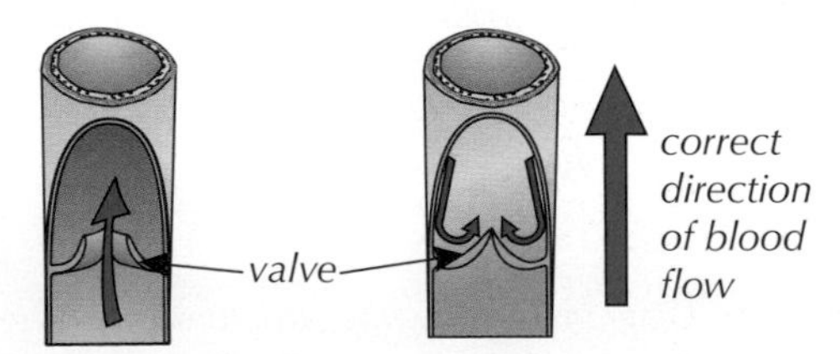

Figure 6: *Diagram to show how valves prevent the back flow of blood.*

Exam Tip

If you're struggling to remember which way arteries and veins carry blood in the exam, think: arteries are the 'way art' (way out) of the heart, and veins are the 'vey in' (way in).

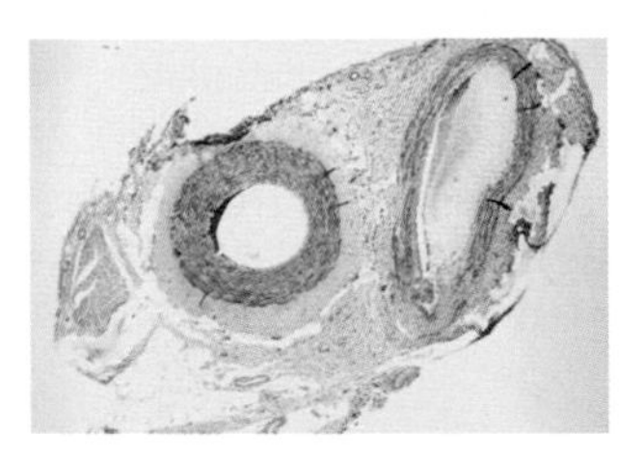

Figure 5: *A cross-section through an artery and a vein — the walls of the blood vessels are pink. The walls of the artery (left) are much thicker than the walls of the vein (right).*

Calculating the rate of blood flow

You might get asked to calculate the rate of blood flow in your exam. Thankfully, it's not too tricky. You just need to use this formula:

rate of blood flow = volume of blood ÷ number of minutes

Take a look at this example:

Example

1464 ml of blood passed through an artery in 4.5 minutes. Calculate the rate of blood flow through the artery in ml/min.

rate of blood flow = volume of blood ÷ number of minutes
= 1464 ÷ 4.5
= **325 ml/min**

Practice Questions — Fact Recall

Q1 Describe the structure of an artery.

Q2 What type of blood vessel has walls that are only one cell thick?

Q3 Describe briefly how substances in the blood pass into body cells.

Q4 Give two differences between the structure of a vein and the structure of an artery.

Practice Questions — Application

Q1 Alan has a condition known as SVCS. This condition is very serious as it obstructs blood flow going into the heart. Which type of blood vessel (artery, capillary or vein) do you think SVCS affects? Give a reason for your answer.

Q2 The graph below shows the relative pressure inside two different blood vessels — one is a vein and one is an artery.

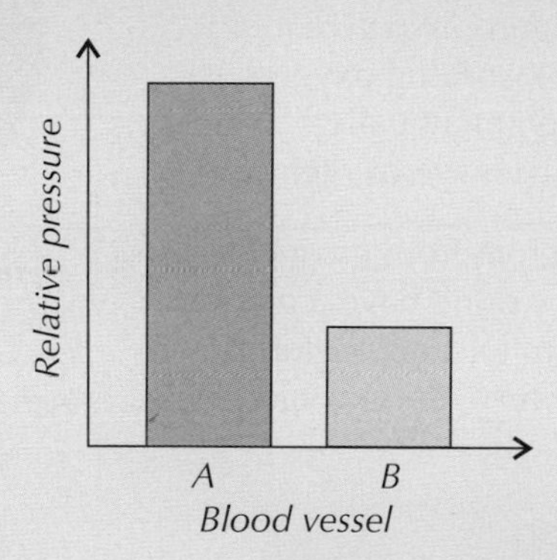

Which blood vessel (A or B) would you expect to contain a higher proportion of muscle tissue in its walls? Explain your answer.

Q3 In a particular artery in the body, the average rate of blood flow should be between 270 and 315 ml/min. A doctor measured the blood flow through this artery of one of his patients. In five minutes, 1075 ml of blood flowed through the patient's artery. The doctor thought that the patient may have a blockage in this artery.

Calculate the rate of blood flow through the patient's artery to determine whether you agree with the doctor. Explain your answer.

5. Circulatory System — The Blood

Learning Objectives:
- Know that blood is a tissue that contains red blood cells, white blood cells and platelets, all suspended in a liquid called plasma.
- Know the functions of the components of blood.
- Understand how blood cells are adapted to their functions.
- Be able to identify the different blood cells in photographs and diagrams.

Specification Reference 4.2.2.3

Blood is a vital part of the circulatory system — it's constantly being pumped out by the heart, travelling through blood vessels to other organs of the body and then returning to the heart. Time to find out just what it's made of...

What is blood?

Blood is a tissue — it's a group of similar cells which work together to perform a specific function (see page 72). The function of the blood is to transport substances around the body. It's made up of **red blood cells**, **white blood cells** and **platelets**, which are all suspended in a liquid called **plasma**.

Red blood cells

The job of red blood cells is to transport oxygen around the body. They have a biconcave shape (see Figure 1) to give them a large surface area for absorbing oxygen and they contain a red pigment called **haemoglobin**, which carries the oxygen.

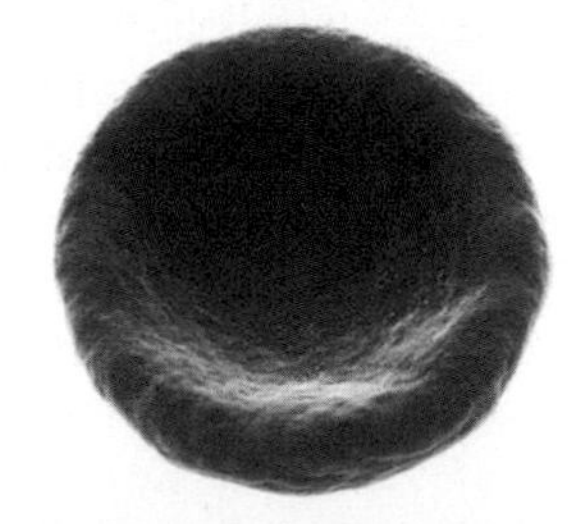

Figure 1: *A red blood cell has a biconcave shape — it looks like it's been squashed in the middle on both sides.*

Red blood cells are different from most types of animal cell because they don't have a nucleus — this allows more room for haemoglobin, which means they can carry more oxygen.

Tip: The more red blood cells you've got, the more oxygen can get to your cells. At high altitudes there's less oxygen in the air — so people who live there produce more red blood cells to compensate.

Transporting oxygen

All body cells need oxygen for respiration — a process which releases energy. Oxygen enters the lungs when you breathe in, then red blood cells transport the oxygen from the lungs to all the cells in the body:

- In the lungs, oxygen diffuses into the blood. The oxygen combines with haemoglobin in red blood cells to become oxyhaemoglobin (see Figure 2).
- In body tissues, the reverse happens — oxyhaemoglobin splits up into haemoglobin and oxygen, to release oxygen to the cells.

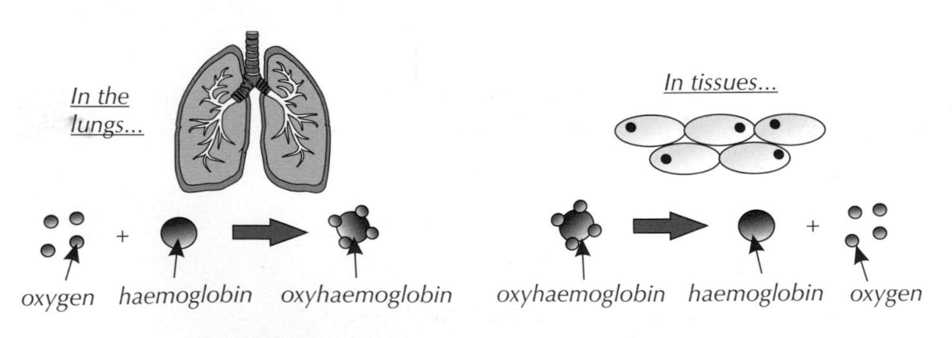

Figure 2: *Simplified diagram illustrating the formation and breakdown of oxyhaemoglobin.*

White blood cells

There are different types of white blood cell but they all have the same function — they defend against microorganisms that cause disease. They can do this in different ways:

- They can engulf unwelcome microorganisms (in a process called phagocytosis) and digest them.
- They can produce antibodies to fight microorganisms.
- They can produce antitoxins to neutralise any toxins produced by the microorganisms.

Figure 3: *A white blood cell.*

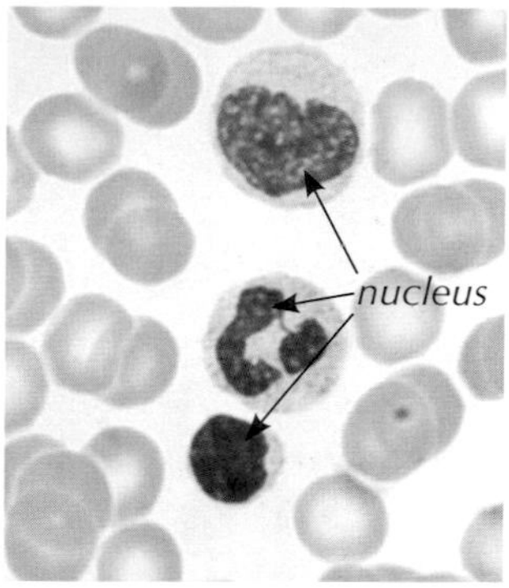

Unlike red blood cells, white blood cells do have a nucleus.

Figure 4: *A picture taken down a light microscope showing three different types of white blood cell (stained purple) surrounded by red blood cells (pale red).*

Tip: There's more on white blood cells on page 136.

Platelets

Platelets are small fragments (pieces) of cells. They have no nucleus. They help the blood to **clot** (clump together) at a wound — this seals the wound and stops you from losing too much blood (see Figure 5). It also stops microorganisms from getting in at the wound. A lack of platelets can cause excessive bleeding and bruising.

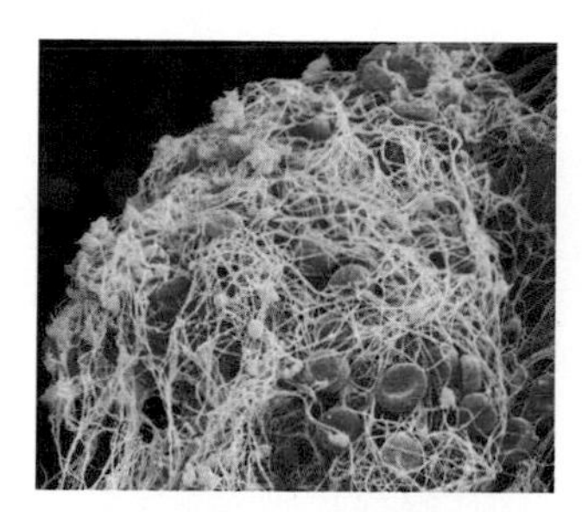

Figure 5: *A blood clot formed by the action of platelets.*

Plasma

Plasma is a pale straw-coloured liquid which carries just about everything in blood. It carries:

- Red and white blood cells and platelets.
- Nutrients like glucose and amino acids. These are the soluble products of digestion, which are absorbed from the small intestine and taken to the cells of the body.
- Carbon dioxide from the organs to the lungs.
- Urea from the liver to the kidneys.
- Hormones.
- Proteins.
- Antibodies and antitoxins produced by the white blood cells.

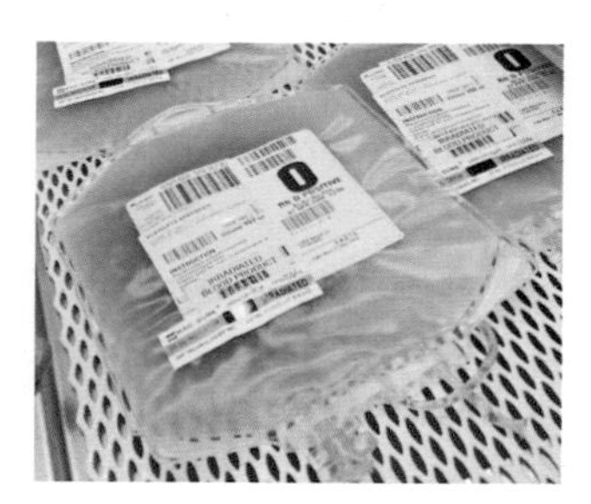

Figure 6: *A pouch of plasma, which has been extracted from donated blood.*

Tip: Urea is a waste product produced from the breakdown of amino acids in the liver.

Practice Questions — Fact Recall

Q1 a) Name the red pigment found in red blood cells.

b) Describe the role of this red pigment.

Q2 What is the function of white blood cells?

Q3 True or false? White blood cells don't have a nucleus.

Q4 What are platelets?

Practice Questions — Application

Q1 Sufferers of Bernard-Soulier syndrome often bleed for a long time after an injury, even when they only have a very small wound. Suggest what component of the blood is abnormal in someone with Bernard-Soulier syndrome. Explain your answer.

Q2 Thalassaemia is a blood disorder in which the body does not make enough haemoglobin. Explain how this disorder can lead to organs not functioning properly because of a lack of oxygen.

Q3 Dr McKenna is looking at blood test results for two of her patients, Fay and Imogen. Their results, along with the normal range for each of the blood components tested, are shown in the following table:

	Red Blood Cells (10^{12}/L)	White Blood Cells (10^{9}/L)	Platelets (10^{9}/L)	Urea (mmol/L)	Blood Glucose (mmol/L)
Normal Range	3.9-5.6	4.0-11.0	150-400	3.0-7.0	3.3-5.6
Fay	4.2	3.2	250	3.2	3.6
Imogen	3.7	10.2	315	5.4	5.2

Which patient, Fay or Imogen, do you think is the most at risk of getting an infection? Explain your answer.

Tip: Don't worry about the unfamiliar units in this table. You don't need to know what they mean to answer the question.

6. Plant Cell Organisation

Plant cells are organised in the same way as the cells of other multicellular organisms (see page 72). Cells are grouped into tissues, tissues are grouped into organs, and organs are grouped into organ systems.

Learning Objectives:

- Know that the roots, stem and leaves work together in an organ system that transports substances around the plant.
- Know these examples of plant tissues and their functions:
  - epidermal tissue
  - palisade mesophyll tissue
  - spongy mesophyll tissue
  - xylem
  - phloem
  - meristem tissue
- Know that leaves are organs.
- Understand that the structures of the tissues in leaves are adapted to allow them to perform their functions.

Specification References
4.2.3.1, 4.2.3.2

How are plant cells organised?

Plants are made of **organs** like stems, roots and leaves. Plant organs work together to make **organ systems**. These can perform the various tasks that a plant needs to carry out to survive and grow — for example, transporting substances around the plant. Plant organs are made of **tissues** — see below for some examples:

Examples

- Epidermal tissue — this covers the whole plant.
- Palisade mesophyll tissue — this is the part of the leaf where most photosynthesis happens.
- Spongy mesophyll tissue — this is also in the leaf, and contains big air spaces to allow gases to diffuse in and out of cells.
- Xylem and phloem — they transport things like water, mineral ions and food around the plant (through the roots, stems and leaves — see page 91 for more).
- Meristem tissue — this is found at the growing tips of shoots and roots and is able to differentiate (change) into lots of different types of plant cell, allowing the plant to grow.

Leaf structure

The leaf is an organ made up of several types of tissue. Leaves contain epidermal, mesophyll, xylem and phloem tissues (see Figures 1 and 2).

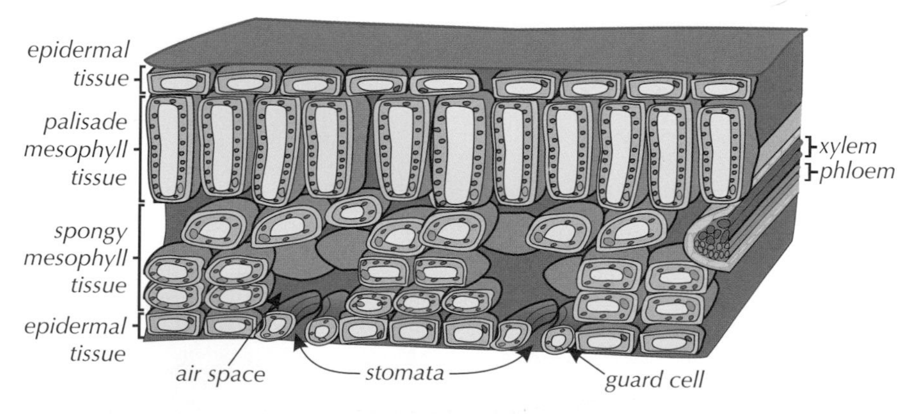

***Figure 2:** Diagram showing a cross-section of a typical plant leaf.*

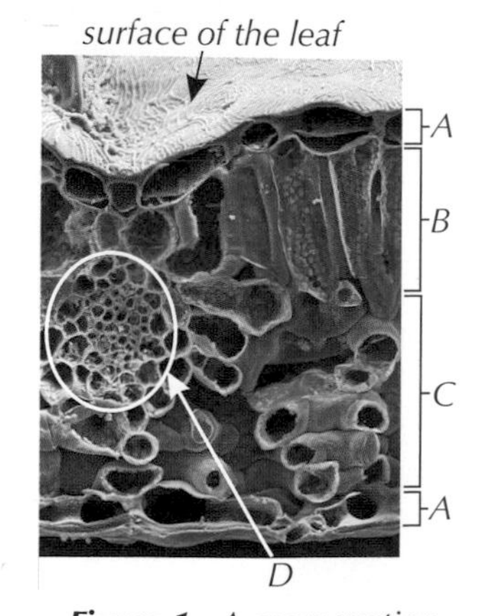

***Figure 1:** A cross-section through a leaf, seen under a microscope. It shows the epidermal tissue (A), palisade mesophyll tissue (B), spongy mesophyll tissue (C) and xylem and phloem tissue (D).*

You need to know how the structures of the tissues that make up the leaf are related to their function:

- The epidermal tissues are covered with a waxy cuticle, which helps to reduce water loss by evaporation.
- The upper epidermis is transparent so that light can pass through it to the palisade layer.
- The palisade layer has lots of chloroplasts (the little structures where photosynthesis takes place). This means that they're near the top of the leaf where they can get the most light.
- The xylem and phloem form a network of vascular bundles, which deliver water and other nutrients to the entire leaf and take away the glucose produced by photosynthesis. They also help support the structure.
- The tissues of leaves are also adapted for efficient gas exchange (see page 66). E.g. the lower epidermis is full of little holes called stomata, which let CO_2 diffuse directly into the leaf. The opening and closing of stomata is controlled by guard cells in response to environmental conditions. The air spaces in the spongy mesophyll tissue increase the rate of diffusion of gases.

Exam Tip
If you're struggling to remember where the different tissues are in a leaf, think of it as epidermis is external and mesophyll is in the middle.

Tip: See page 94 for more on stomata and guard cells.

Practice Questions — Fact Recall

Q1 True or False? Leaves are organ systems.

Q2 Name the plant tissue where photosynthesis occurs.

Q3 What is the function of the air spaces in spongy mesophyll tissue?

Q4 Where in a plant is meristem tissue found?

Q5 Explain why the upper epidermis of a leaf is transparent.

7. Transpiration and Translocation

Just like animals, plants also have transport systems to move substances around...

Transport tissues in plants

Flowering plants have two separate types of tissue — phloem and xylem — for transporting substances around. Both types of tissue form 'tubes', which go to every part of the plant, but they are totally separate.

Phloem

Phloem tubes are made of columns of elongated living cells with small pores in the end walls to allow cell sap to flow through (see Figures 1 and 2). Cell sap is a liquid that's made up of the substances being transported and water.

Phloem tubes transport food substances (mainly dissolved sugars) made in the leaves to the rest of the plant for immediate use (e.g. in growing regions) or for storage. The transport goes in both directions — from the leaves down to the roots, and from the roots up to the leaves. This process is called **translocation**.

Figure 1: *Diagram showing the inside of a phloem tube.*

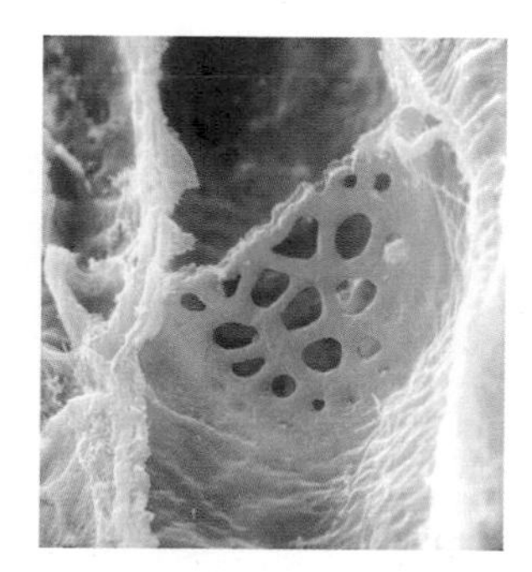

Figure 2: *Microscope image of the inside of a cut phloem tube — the small holes between the cells can be seen clearly.*

Xylem

Xylem tubes are made of dead cells joined end to end with no end walls between them and a hole down the middle (see Figures 3 and 4). They're strengthened with a material called lignin.

Xylem tubes carry water and mineral ions up the plant — from the roots to the stem and leaves. The movement of water from the roots, through the xylem and out of the leaves is called the **transpiration stream** (see next page).

Figure 3: *Diagram showing the inside of a xylem tube.*

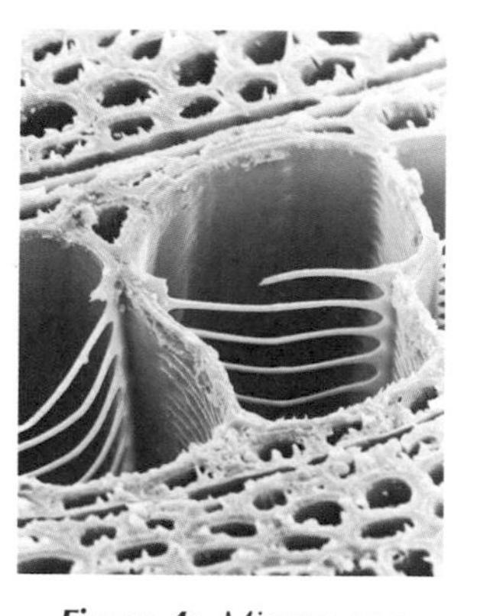

Figure 4: *Microscope image showing a cut xylem tube (dark green).*

Learning Objectives:

- Know that phloem tubes are made up of columns of elongated cells with pores in the end walls through which cell sap can move, which makes them adapted for their function.
- Understand the role of phloem tissue in the transport of food (usually dissolved sugars) from the leaves to the rest of the plant and know that this process is called translocation.
- Know that xylem is made up of hollow tubes strengthened by lignin.
- Understand the role of xylem tissue in the transport of water and mineral ions to the stems and leaves from the roots.
- Understand that the structure of xylem is adapted for transporting water in the transpiration stream.
- Know how root hair cells are adapted for taking in water and mineral ions efficiently.

Specification Reference 4.2.3.2

Tip: The root hair cells take up water from the soil by osmosis, and take up mineral ions by active transport. There's more information about how they're adapted for this on page 58.

The transpiration stream

Transpiration is the loss of water from a plant. The transpiration stream is the movement of water through a plant from the roots to the leaves.
It happens like this:

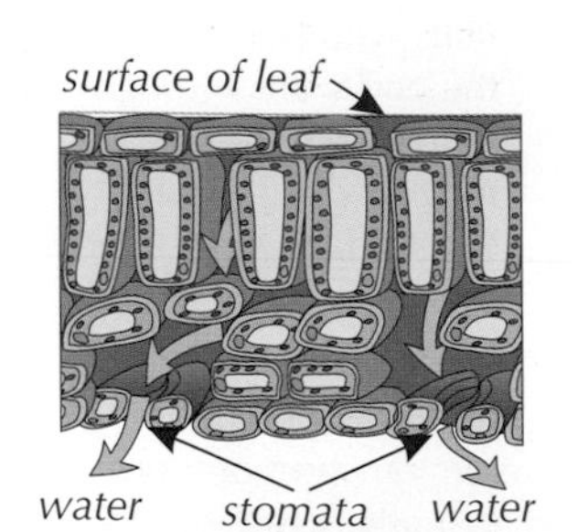

Figure 5: *Cross-section of a leaf showing how water moves out during transpiration.*

1. Water from inside a leaf evaporates and diffuses out of the leaf, mainly through the stomata (tiny holes found mainly on the lower surface of the leaf) — see Figure 5.
2. This creates a slight shortage of water in the leaf, and so more water is drawn up from the rest of the plant through the xylem vessels to replace it.
3. This in turn means more water is drawn up from the roots, and so there's a constant transpiration stream of water through the plant.

This is shown in Figure 6.

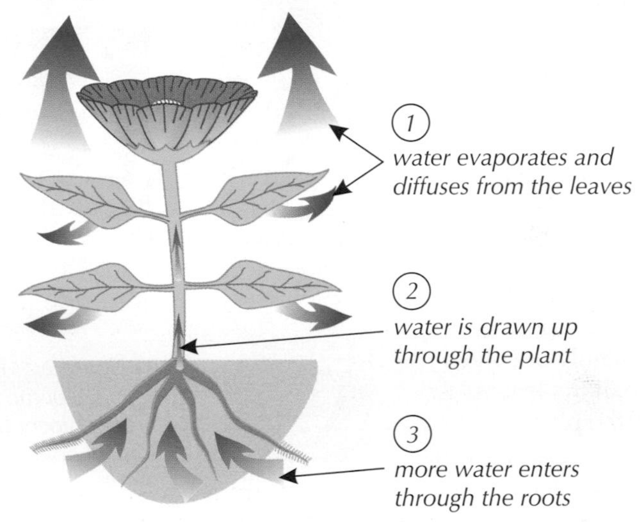

Figure 6: *Diagram showing how water moves through a plant in the transpiration stream.*

Tip: The rate of transpiration depends on the conditions in the plant's environment. Transpiration happens fastest in bright, hot and dry conditions where the air flow around the leaves is good, because this is when evaporation happens fastest — see next page for more.

Transpiration is just a side-effect of the way leaves are adapted for photosynthesis. They have to have stomata in them so that gases can be exchanged easily (see page 66). Because there's more water inside the plant than in the air outside, the water escapes from the leaves through the stomata by diffusion.

Practice Questions — Fact Recall

Q1 a) What is transported in phloem tubes?

b) Which area of a plant do the phloem tubes transport substances from?

Q2 Give two substances that are transported in xylem tubes.

Q3 Explain how water is transported from the roots of a plant to its leaves.

8. Transpiration and Stomata

How quickly transpiration takes place is affected by several factors...

Factors affecting transpiration rate

Transpiration rate is affected by four main things:

Light intensity

The brighter the light, the greater the transpiration rate. Stomata begin to close as it gets darker. Photosynthesis can't happen in the dark, so they don't need to be open to let CO_2 in. When the stomata are closed, very little water can escape.

Temperature

The warmer it is, the faster transpiration happens. When it's warm, the water particles have more energy to evaporate and diffuse out of the stomata.

Air flow

The better the air flow around a leaf (e.g. stronger wind), the greater the transpiration rate. If air flow around a leaf is poor, the water vapour just surrounds the leaf and doesn't move away. This means there's a high concentration of water particles outside the leaf as well as inside it, so diffusion doesn't happen as quickly. If there's good air flow, the water vapour is swept away, maintaining a low concentration of water in the air outside the leaf. Diffusion then happens quickly, from an area of higher concentration to an area of lower concentration.

Humidity

The drier the air around a leaf, the faster transpiration happens. This is like what happens with air flow. If the air is humid there's a lot of water in it already, so there's not much of a difference between the inside and the outside of the leaf. Diffusion happens fastest if there's a really high concentration in one place, and a really low concentration in the other.

Investigating transpiration rate

You can estimate the rate of transpiration by measuring the uptake of water by a plant. This is because you can assume that water uptake by the plant is directly related to water loss by the leaves (transpiration). Figure 1 shows the apparatus you'll need.

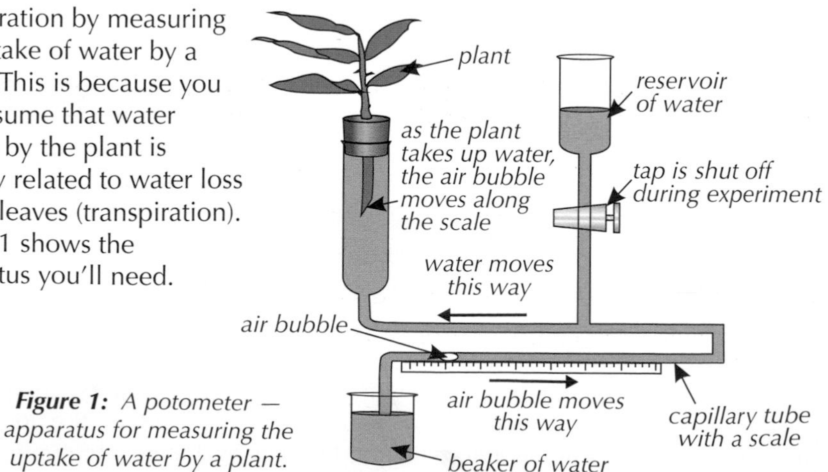

Figure 1: *A potometer — apparatus for measuring the uptake of water by a plant.*

Learning Objectives:
- Understand how light intensity, temperature, air flow and humidity affect the rate of transpiration.
- Be able to read off a graph, draw graphs and interpret graphs, charts and tables relating to transpiration.
- Be able to calculate the rate of transpiration of a plant.
- Understand the role of stomata and guard cells in controlling gas exchange and water loss from leaves.

Specification Reference 4.2.3.2

Tip: Wet clothes dry as water evaporates from the fabric. You wouldn't expect clothes hung out in the garden to dry very quickly on a cold, wet day — so don't expect evaporation from a plant to be quick under those conditions either.

Exam Tip
In your exam you might need to read off data related to transpiration from a graph, chart or table and explain what the data is showing. You also need to be able to create your own graphs with suitable scales for the axes. See pages 16-18 and page 21 for more on these skills.

Tip: It's quite tough to set up a potometer. See page 371 for some tips.

To investigate the rate of transpiration, set up the apparatus as in the diagram, and then record the starting position of the air bubble. Start a stopwatch and record the distance moved by the bubble per unit of time, e.g. per hour. Keep the conditions constant throughout the experiment, e.g. the temperature and air humidity.

Calculating the rate of transpiration

You can use the data you collect in your experiment to calculate the rate of transpiration. Take a look at this example:

Tip: Rates of transpiration can also be given in mm^3/min, but this requires a bit more maths to find the actual volume of water that the plant has taken up.

Example

A potometer was set up to measure the rate of transpiration of a plant. In 60 minutes, the bubble in the potometer moved 6.6 cm. Calculate the rate of transpiration. Give your answer in mm/min.

1. First, convert 6.6 cm into mm:

   $6.6 \times 10 = 66$ mm

2. Then divide the distance moved by the time taken:

   $66 \div 60 =$ **1.1 mm/min**

Tip: Stoma is the singular word, stomata is the plural.

Guard cells

Two guard cells surround each stoma. They have a kidney shape which opens and closes the stomata in a leaf.

When the plant has lots of water, the guard cells fill with it and go plump and turgid. This makes the stomata open so gases can be exchanged for photosynthesis. When the plant is short of water, the guard cells lose water and become flaccid, making the stomata close (see Figure 3). This helps stop too much water vapour escaping.

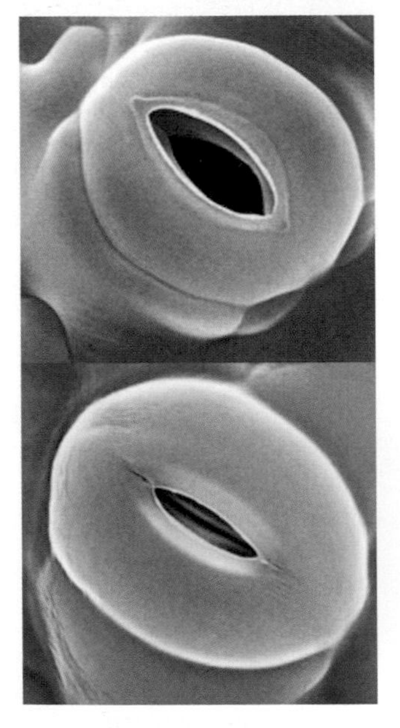

Figure 2: *An open stoma (top) and a closed stoma (bottom).*

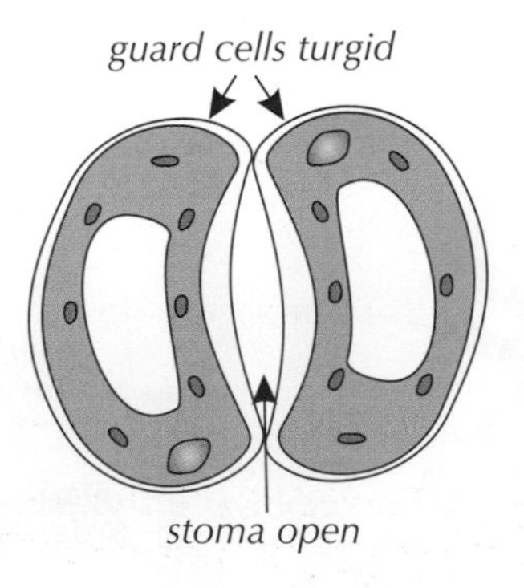

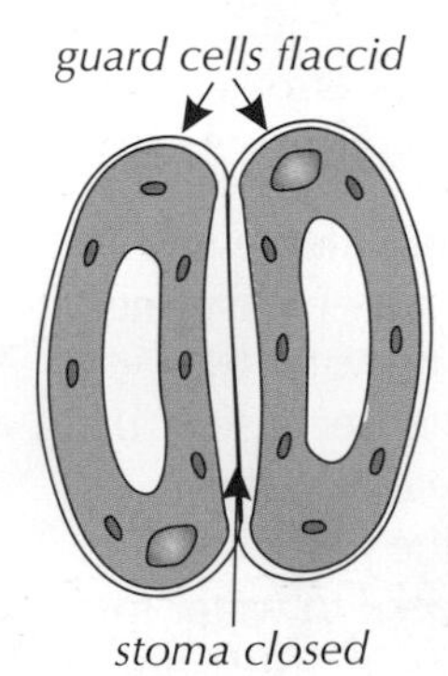

Figure 3: *Diagram showing how guard cells open and close a stoma.*

Thin outer walls and thickened inner walls make the opening and closing work. They're also sensitive to light and close at night to save water without losing out on photosynthesis. You usually find more stomata on the undersides of leaves than on the top. The lower surface is shaded and cooler — so less water is lost through the stomata than if they were on the upper surface. Guard cells are therefore adapted for gas exchange and controlling water loss within a leaf.

Investigating stomata and guard cells

Using a light microscope, you can investigate the distribution of stomata and guard cells on a leaf.

1. For some plants (e.g. *Kniphofia*), you can peel the epidermal tissue straight off the leaf and mount a piece on a microscope slide (see page 25). Using a pipette, put a large drop of water onto the slide, then place the tissue into the water using tweezers. Finally put a cover slip over the top.
2. For a leaf where peeling the top layer off isn't possible, you need to make an 'imprint' of its surface. You can do this using clear nail varnish. Apply a thin layer of varnish to the underside of the leaf and allow it to dry. Then use a piece of sticky tape to remove the varnish from the leaf by gently peeling it off and stick the sticky tape (with attached nail varnish imprint) to a microscope slide.
3. Now you're ready to view your slide. Put it on the microscope's stage and focus the microscope. Start with the least powerful objective lens and work your way up until you can see the stomata and guard cells clearly. You could draw a labelled diagram of what you can see down the microscope.
4. Now you can estimate the total number of stomata on the surface you're investigating. It would take ages to count every stoma on the surface, so it makes sense to just count a 'sample'.
5. Count the stomata that you can see in your 'field of vision' (the bit of the specimen that you can see when you look down the microscope). When you've finished, move the slide so you're looking at a new field of vision and count the stomata again. Do this at least three times and calculate the mean to determine the average number of stomata per field of vision.
6. To estimate the total number of stomata on the surface that you're investigating, you need to work out how many times the field of vision would fit into it. Find out the area of the field of vision and divide the total area of the surface by it. Finally, multiply your answer by the average number of stomata per field of vision. This gives you an estimate of the total number of stomata on the surface.

You can easily compare the distribution of stomata in different areas of a leaf (e.g. the upper and lower surfaces) or between different leaves. To do this, carry out steps 1-5 for each of the leaf surfaces you're investigating and compare the mean number of stomata per field of vision on each of the surfaces.

Tip: There is lots more about using microscopes, including how to make slides, on pages 25-27.

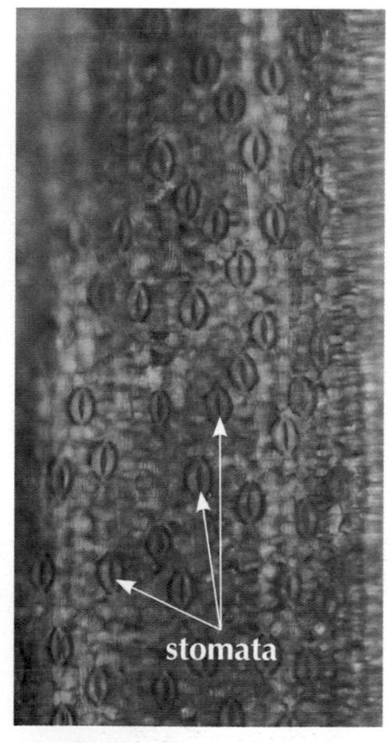

Figure 4: *A view of a lily leaf taken with a light microscope, showing many stomata and guard cells.*

Practice Questions — Fact Recall

Q1 What is the effect of increasing light intensity on the rate of transpiration?

Q2 Potometers can be used to estimate the rate of transpiration. What does a potometer actually measure?

Q3 a) Name the cells that control the size of the stomata.
b) Explain how these cells close the stomata.

Practice Questions — Application

Q1 Ben is investigating water loss in plants. He weighs two of the same type of plant, then leaves both plants in a room for 24 hours. One of the plants is placed next to a fan, the other one isn't. All other variables are controlled. After 24 hours, Ben weighs the plants again.

a) Both plants have lost mass. Explain why.

b) Which plant do you think will have lost the most mass? Explain your answer.

c) Ben wants to find out how temperature affects water loss in plants. Suggest how he could alter his experiment to investigate this.

Q2 A group of students set up a potometer as shown in the diagram.

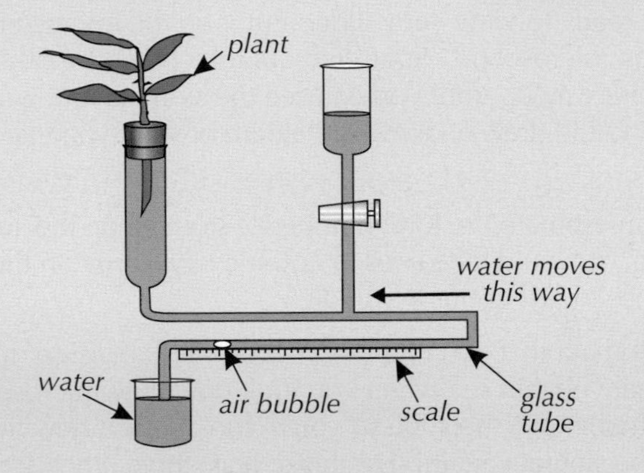

The students wanted to find out how temperature affected transpiration rate, so they measured how far the bubble moved in the potometer when they did the experiment at three different temperatures. Their results are shown in the table:

Temperature (°C)	Distance moved by the bubble in 20 minutes (mm)
10	26
15	31
20	37

a) Explain the trend shown in the students' results.

b) Calculate the rate of transpiration of the plant at 10 °C. Give your answer in mm/min.

c) Suggest one way in which the students could have improved the validity of their investigation.

d) The students coated the underside of half of the plant's leaves with nail varnish, then did the experiment again at 10 °C. Would you expect the bubble to have moved more or less than 26 mm in this experiment? Explain your answer.

Tip: Head back to page 9 for information on what makes an experiment valid if you're struggling to answer Q2c).

Topic 2a Checklist — Make sure you know...

Cell Organisation

- ❑ That all living things are made up of basic building blocks called cells.
- ❑ That similar cells sometimes work together to form tissues, which have specific functions.
- ❑ That organs are made up of different types of tissue grouped together to perform a certain function.
- ❑ That organs work together in organ systems to perform a particular function in the body (e.g. the digestive system consists of glands, the stomach, the liver, the small intestine and the large intestine, which all work together to break down food and exchange materials with the environment).
- ❑ That organ systems work together to form whole organisms.

The Lungs

- ❑ That the human lungs consist of the trachea, the bronchi and lots of little air sacs called alveoli, which are surrounded by a network of capillaries.
- ❑ That when you breathe in, the air travels down the trachea, through the bronchi and into the alveoli, where gas exchange takes place. Oxygen diffuses from the alveoli into the blood in the capillaries and carbon dioxide diffuses the other way, into the alveoli, where it can be breathed out.
- ❑ That the alveoli and capillary network mean that the lungs are adapted for efficient gas exchange.

Circulatory System — The Heart

- ❑ That humans have a double circulatory system to get food and oxygen to every cell in the body via the blood, which is pumped around the body by the heart. One circuit, powered by the right ventricle, pumps deoxygenated blood to the lungs to take in oxygen, and the other circuit, powered by the left ventricle, pumps oxygenated blood to the rest of the body.
- ❑ That the four chambers of the human heart are the right atrium, right ventricle, left atrium and left ventricle, and that the chambers have valves to prevent backflow.
- ❑ That the vena cava, pulmonary artery, aorta and pulmonary vein are the four major blood vessels connected to the heart.
- ❑ That the coronary arteries branch off the aorta and supply the heart muscle with oxygenated blood.
- ❑ How the human heart functions to pump blood around the body.
- ❑ That a group of cells in the right atrium control the rate that the heart beats at, and that an artificial pacemaker (a small electrical device) can be fitted to treat patients with an irregular heartbeat.

Circulatory System — The Blood Vessels

- ❑ That there are three types of blood vessel found in the body — arteries, which carry blood away from the heart, capillaries, which carry blood close to the cells to allow substances to be exchanged, and veins, which carry blood back to the heart.
- ❑ How the structures of the different blood vessels are adapted to allow them to carry out their roles in the circulatory system.
- ❑ How to calculate the rate of blood flow in a blood vessel.

cont...

Circulatory System — The Blood

- ❑ That blood is a tissue and its main components are red and white blood cells, platelets and plasma.
- ❑ That red blood cells have no nucleus, have a biconcave shape and contain haemoglobin (a red pigment) that binds to oxygen, allowing them to carry oxygen from the lungs to the rest of the body.
- ❑ That white blood cells have a nucleus and help to defend against disease caused by microorganisms.
- ❑ That platelets are fragments of cells, don't have a nucleus and help blood to clot at a wound.
- ❑ That plasma carries just about everything in the blood including blood cells and platelets, nutrients released from digestion, carbon dioxide, urea, hormones, proteins, antibodies and antitoxins.
- ❑ How to recognise the different types of blood cell in diagrams and photographs.

Plant Cell Organisation

- ❑ That the roots, stem and leaves work together in an organ system that transports substances.
- ❑ That leaves are organs consisting of upper and lower epidermal tissue, palisade and spongy mesophyll tissue, xylem and phloem tissues, stomata and guard cells.
- ❑ That epidermal tissue covers the entire plant.
- ❑ That palisade mesophyll tissue is the location of photosynthesis and spongy mesophyll tissue contains air spaces for gas exchange.
- ❑ That xylem and phloem are the major transport tissues in plants.
- ❑ That meristem tissue is found at the tips of the growing regions of plants and can differentiate into different types of plant cell.
- ❑ How the tissues in leaves are adapted to carry out their roles in the organ.

Transpiration and Translocation

- ❑ That phloem is the plant tissue responsible for transporting food from the leaves to the rest of the plant in a process called translocation.
- ❑ That phloem is made up of columns of elongated living cells that have pores in the end walls to allow the cell sap containing the food (usually dissolved sugars) to pass through.
- ❑ That xylem transports water and mineral ions from the roots to the stems and leaves and is made up of hollow tubes of dead cells strengthened by lignin that have no end walls between them.
- ❑ That the structure of xylem is suited to carrying water in the transpiration stream, which is the movement of water through the plant from the roots to the leaves.

Transpiration and Stomata

- ❑ How high light intensity, high temperature, increased air flow and low humidity (drier air) increase the rate of transpiration and vice versa.
- ❑ How to read from and interpret graphs, charts and tables containing data on transpiration, and how to produce your own transpiration-related graphs.
- ❑ How to calculate the rate of transpiration.
- ❑ How stomata allow water to diffuse out of leaves and gas exchange to take place, and how guard cells control the opening of the stomata to regulate these processes.

Exam-style Questions

1 Humans have a specialised system for exchanging gases with their environment.

1.1 Which of the following options gives the correct route taken by oxygen into the bloodstream?

A bronchi, trachea, capillaries, alveoli
B capillaries, alveoli, bronchi, trachea
C trachea, bronchi, alveoli, capillaries
D trachea, alveoli, capillaries, bronchi

(1 mark)

1.2 Which of the following sentences best describes alveoli?

A Small blood vessels that carry deoxygenated blood to the lungs.
B Tubes that split from the trachea, one into each lung.
C Small air sacs found in the lungs that play a major role in gas exchange.
D Specialised cells found only in the lungs that regulate the breathing rate.

(1 mark)

2 **Figure 1** represents the heart.

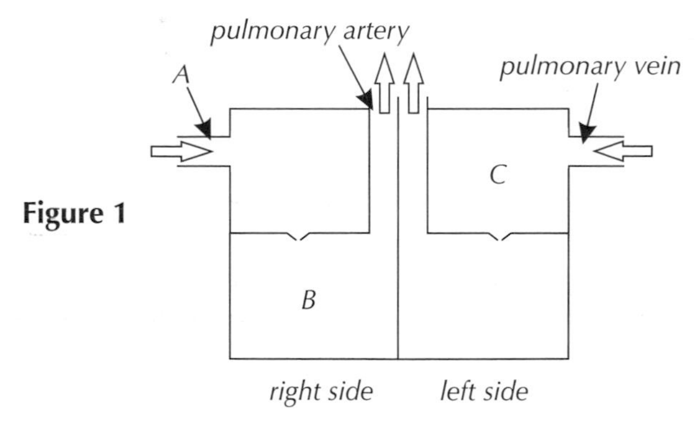

Figure 1

2.1 Name each of the structures labelled A-C on the diagram.

(3 marks)

Figure 2 shows cross-sections through the pulmonary artery and the pulmonary vein.

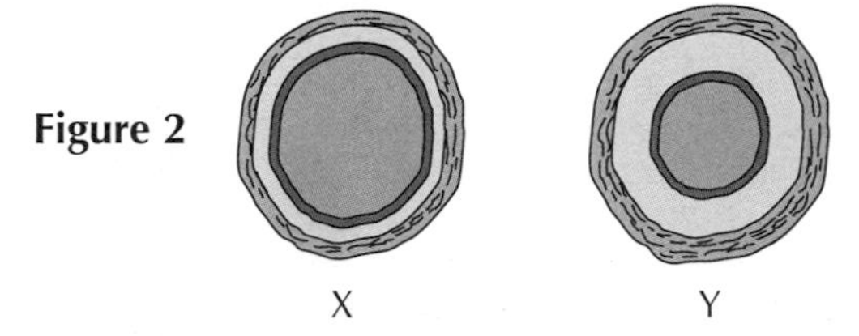

Figure 2

2.2 Which diagram (X or Y) shows the pulmonary vein? Explain your answer.

(1 mark)

3 **Figure 3** shows a group of blood cells.

Figure 3

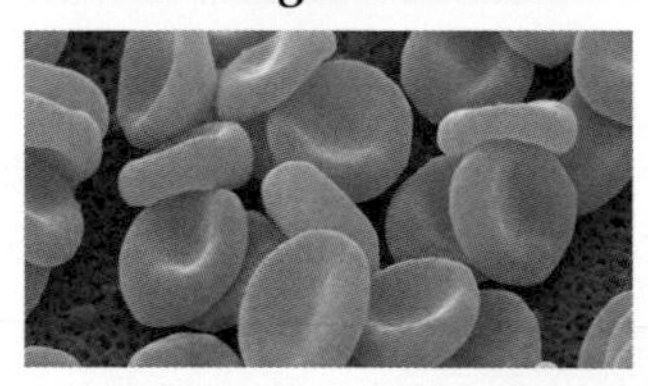

Identify the blood cells shown in **Figure 3** and explain **three** ways in which they are adapted to their function.

(4 marks)

4 **Figure 4** shows a cross-section through a leaf.

Figure 4

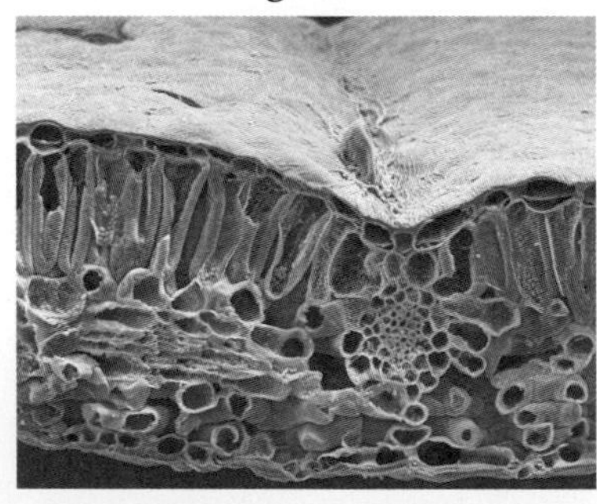

4.1 What name is given to the tissue that covers the leaf?

(1 mark)

4.2 Some of the cells in the photograph contain lots of chloroplasts. What plant tissue are these cells part of? Explain your answer.

(3 marks)

4.3 Throughout the day, the rate at which carbon dioxide diffuses into the leaf cells changes. Explain how a change in the concentration of carbon dioxide in the air spaces outside the cells could slow down the rate of diffusion.

(2 marks)

5 Desert plants are adapted for life in hot, dry conditions. Like most plants, they exchange gases through their stomata.

Suggest and explain how the following adaptations help plants to survive in desert conditions:

5.1 Stomata are kept closed during the day, when it is hot and dry, and only opened at night, when it is cool.

(3 marks)

5.2 Stomata are located in sunken pits. The pits help to trap water vapour close to the surface of the leaf.

(2 marks)

1. Introduction to Health and Disease

Diseases are one cause of poor health. Sometimes, different diseases interact, leading to further health complications that don't appear related at first.

Learning Objectives:
- Know the definition of health, and know that diseases are often the cause of ill health.
- Know that a disease can be communicable (infectious) or non-communicable (not infectious).
- Understand that diseases can interact with each other.
- Know these examples of interacting diseases: immune system problems and communicable diseases; viruses and cancer; immune reactions and allergies; physical health problems and mental illness.
- Know that factors other than disease, such as diet, stress and life situations can also lead to ill health.
- Be able to read off a graph, and draw and interpret tables, charts and graphs relating to the incidence of disease.
- Be able to identify a correlation on a graph relating to the incidence of disease.
- Understand why sampling is used when collecting epidemiological data.

Specification Reference 4.2.2.5

Diseases

Diseases are often responsible for causing ill health.

Health is the state of physical and mental wellbeing.

Diseases can be communicable or non-communicable.

Communicable Diseases

Communicable diseases are those that can spread from person to person or between animals and people. They can be caused by things like bacteria, viruses, parasites and fungi. They're sometimes described as contagious or infectious diseases. Measles and malaria are examples of communicable diseases. There's more about them on pages 132-133.

Non-Communicable Diseases

Non-communicable diseases are those that cannot spread between people or between animals and people. They generally last for a long time and get worse slowly. Asthma, cancer and coronary heart disease (see page 103) are examples of non-communicable diseases.

Interaction of diseases

Sometimes diseases can interact and cause other physical and mental health issues that don't immediately seem related. Here are a few examples:

Examples

- People who have problems with their immune system (the system that your body uses to help fight off infection — see p. 136) have an increased chance of suffering from communicable diseases such as influenza (flu), because their body is less likely to be able to defend itself against the pathogen that causes the disease.
- Some types of cancer can be triggered by infection by certain viruses. For example, infection with some types of hepatitis virus can cause long-term infections in the liver where the virus lives in the cells. This can lead to an increased chance of developing liver cancer. Another example is infection with HPV (human papilloma virus), which can cause cervical cancer in women.
- Immune system reactions in the body caused by infection by a pathogen can sometimes trigger allergic reactions such as skin rashes or worsen the symptoms of asthma for asthma sufferers.

- Mental health issues such as depression can be triggered when someone is suffering from severe physical health problems, particularly if they have an impact on the person's ability to carry out everyday activities or if they affect the person's life expectancy.

Tip: Stress can cause all sorts of physical health issues (e.g. high blood pressure, headaches and issues with the immune system) as well as mental health issues (e.g. depression).

Exam Tip
In the exam, you may need to read off a graph, or draw or explain tables, graphs and charts showing data on the incidence of disease. The incidence of disease is the number of new cases of the disease within a given time. It may be given as, e.g. the number of new cases per 100 000 persons per year, so make sure you understand the units the data is shown in before answering the question.

Exam Tip
In the exam, you might need to identify a correlation between variables relating to disease incidence from a graph. To recap what correlation is, look at page 18.

Tip: Sampling is often used when collecting data related to health and patterns in diseases (epidemiological data). This is because it would take too long and cost too much to collect data from the entire population. See p. 373 for more on sampling.

Other factors affecting health

There are plenty of factors other than diseases that can also affect your health. Take a look at these examples:

Examples

- Whether or not you have a good, balanced diet that provides your body with everything it needs, and in the right amounts. A poor diet can affect your physical and mental health.
- The stress you are under — being constantly under lots of stress can lead to health issues.
- Your life situation — for example, whether you have easy access to medicines to treat illness, or whether you have access to things that can prevent you from getting ill in the first place, e.g. being able to buy healthy food or access condoms to prevent the transmission of some sexually transmitted diseases.

Practice Questions — Fact Recall

Q1 What is meant by the term 'health'?

Q2 What is the main difference between communicable and non-communicable diseases?

Q3 Give three factors other than disease that could affect health.

Practice Question — Application

Q1 The graph below shows the results of a study into the average number of colds suffered by employees in high-stress jobs and low-stress jobs in the same company per year.

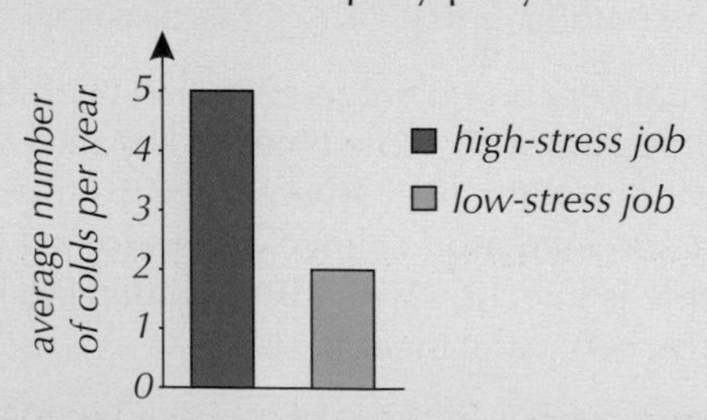

Describe and suggest an explanation for the results.

2. Cardiovascular Disease

Lots of things can go wrong with the heart and blood vessels.
We have ways to effectively treat many of these issues. Read on for more...

Learning Objectives:

- Understand the causes of coronary heart disease and the complications it can lead to.
- Know that stents can be used to treat coronary heart disease by keeping the coronary arteries open.
- Know that statins can be used to reduce the amount of cholesterol in the blood and slow the formation of fatty deposits in arteries.
- Know that surgeons can transplant the heart, or heart and lungs, of an organ donor into a patient with heart failure.
- Know that in certain circumstances, artificial hearts can be fitted to a patient, usually on a temporary basis.
- Understand that heart valves can become damaged and faulty, leading to them being leaky or not opening fully when required.
- Know that replacement biological or mechanical heart valves are available.
- Be able to assess the pros and cons of different treatments for cardiovascular disease.

Specification Reference 4.2.2.4

What is cardiovascular disease?

Cardiovascular disease is a term used to describe diseases of the heart or blood vessels. A common example is **coronary heart disease**. Coronary heart disease is when the arteries that supply the blood to the heart muscle (the **coronary arteries**, see Figure 1) get blocked by layers of fatty material building up. This causes the arteries to become narrow, so blood flow is restricted and there's a lack of oxygen to the heart muscle — this can result in a heart attack.

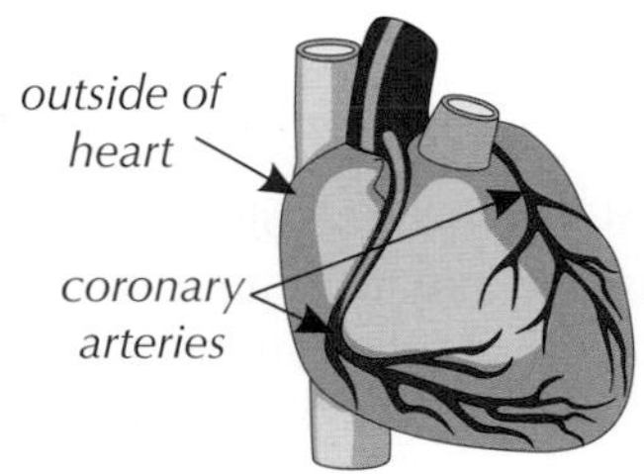

Figure 1: *The location of the coronary arteries.*

Cardiovascular disease can be treated in a few different ways:

Stents

Stents are wire mesh tubes that can be inserted inside arteries to widen them and keep them open (see Figure 1). They are particularly useful in people with coronary heart disease.

Stents keep the coronary arteries open, making sure blood can pass through to the heart muscles (see Figure 2). This keeps the person's heart beating (and the person alive).

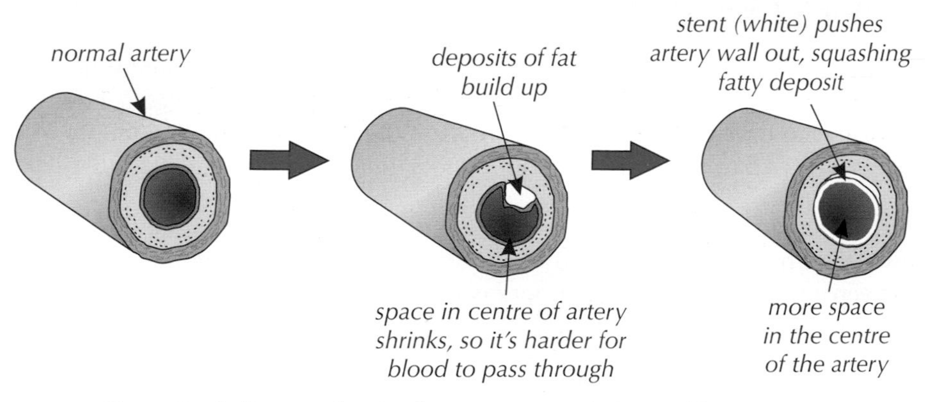

Figure 2: *A diagram showing how a stent can help to widen an artery.*

Advantages of stents

Stents are a way of lowering the risk of a heart attack in people with coronary heart disease. They are effective for a long time and the recovery time from the surgery is relatively quick.

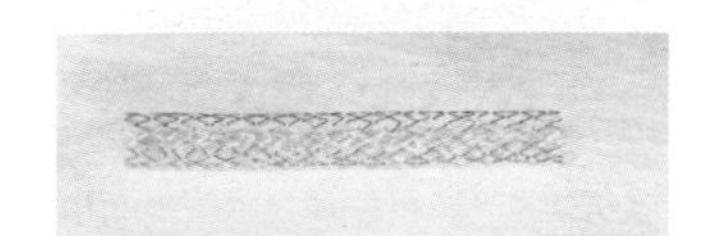

Figure 3: *A stent that can be used to widen an artery.*

Disadvantages of stents

There is a risk of complications during the operation (e.g. heart attack) and a risk of infection from surgery. There's also the risk of patients developing a blood clot near the stent. This is called thrombosis.

Statins

Statins are drugs that can reduce the amount of 'bad' **cholesterol** present in the bloodstream. Cholesterol is an essential lipid that your body produces and needs to function properly. However, too much of a certain type of cholesterol (known as 'bad' or LDL cholesterol) can cause health problems. Having too much of this 'bad' cholesterol in the bloodstream can cause fatty deposits to form inside arteries, which can lead to coronary heart disease. Taking statins slows down the rate of fatty deposits forming.

Advantages of statins

By reducing the amount of 'bad' cholesterol in the blood, statins can reduce the risk of strokes, coronary heart disease and heart attacks. As well as reducing the amount of 'bad' cholesterol, statins can increase the amount of a beneficial type of cholesterol (known as 'good' or HDL cholesterol) in your bloodstream. This type can remove 'bad' cholesterol from the blood. Some studies suggest that statins may also help prevent some other diseases.

Disadvantages of statins

Statins are a long-term drug that must be taken regularly. There's the risk that someone could forget to take them. Also, statins can sometimes cause negative side effects, e.g. headaches. Some of these side effects can be serious, e.g. kidney failure, liver damage and memory loss. Another disadvantage is that the effect of statins isn't instant. It takes time for their effect to kick in.

Artificial hearts

If a patient has heart failure, doctors may perform a heart transplant (or heart and lungs transplant if the lungs are also diseased) using donor organs from people who have recently died. However, if donor organs aren't available right away or they're not the best option, doctors may fit an artificial heart.

Artificial hearts are mechanical devices that pump blood for a person whose own heart has failed (see Figure 4). They're usually used as a temporary fix, to keep a person alive until a donor heart (a replacement heart from another person) can be found or to help a person recover by allowing the heart to rest and heal. In some cases they're used as a permanent fix, which reduces the need for a donor heart.

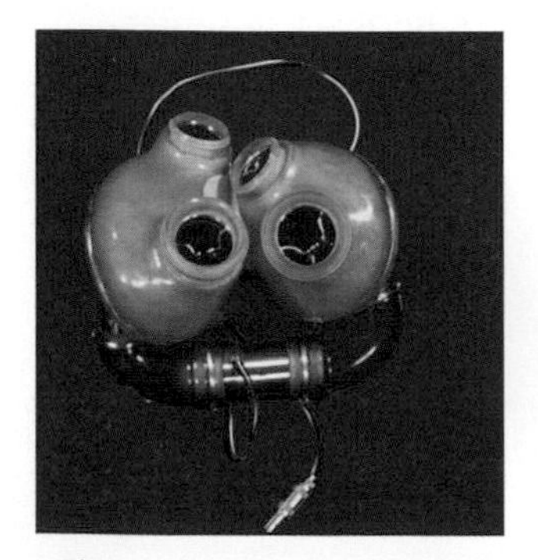

Figure 4: *The first artificial heart to be inserted into a patient, in 1982.*

Example

The SynCardia Total Artificial Heart is a modern artificial heart with a high success rate. The device kept one patient alive for 1374 days (which is nearly 4 years) before they were able to have a heart transplant.

Advantages of artificial hearts

The main advantage of artificial hearts is that they're less likely to be rejected by the body's immune system than a donor heart. This is because they're made from metals or plastics, so the body doesn't recognise them as 'foreign' and attack in the same way as it does with living tissue.

Disadvantages of artificial hearts

Surgery to fit an artificial heart (as with transplant surgery) can lead to bleeding and infection. Also, artificial hearts don't work as well as healthy natural ones — parts of the heart could wear out or the electrical motor could fail.

Blood doesn't flow through artificial hearts as smoothly as through natural hearts either, which can cause blood clots and lead to strokes (a problem caused when the blood supply to part of the brain is cut off). A patient receiving an artificial heart has to take drugs to thin their blood and make sure clots don't occur. This can cause problems with bleeding if they're hurt in an accident because their blood can't clot normally to heal the wounds.

Also, having an artificial heart in the body (and any battery and controller that might be inserted as well) may be uncomfortable for the patient.

Replacement heart valves

The valves in the heart can be damaged or weakened by heart attacks, infection or old age. The damage may cause the valve tissue to stiffen, so it won't open properly. Or a valve may become leaky, allowing blood to flow in both directions rather than just forward. This means that blood doesn't circulate as effectively as normal.

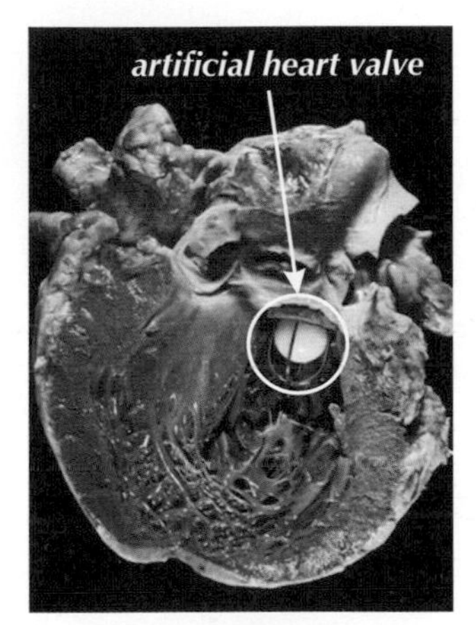

Figure 5: *An artificial valve in a human heart (cut open).*

Severe valve damage can be treated by replacing the valve. Replacement valves can be ones taken from humans or other mammals (e.g. cows or pigs) — these are **biological valves**. Or they can be man-made — these are **mechanical valves**.

Replacing a valve is a much less drastic procedure than a whole heart transplant. But fitting artificial valves is still major surgery and there can still be problems with blood clots.

Artificial blood

When someone loses a lot of blood, e.g. in an accident, their heart can still pump the remaining red blood cells around (to get oxygen to their organs), as long as the volume of their blood can be topped up.

Artificial blood is a blood substitute, e.g. a salt solution ("saline"), which is used to replace the lost volume of blood. It's safe (if no air bubbles get into the blood) and can keep people alive even if they lose 2/3 of their red blood cells. This may give the patient enough time to produce new blood cells. If not, the patient will need a blood transfusion.

Ideally, an artificial blood product would replace the function of the lost red blood cells, so that there's no need for a blood transfusion. Scientists are currently working on products that can do this.

Practice Questions — Fact Recall

Q1 What is cardiovascular disease?

Q2 What causes coronary heart disease?

Q3 a) What is a stent?

b) Describe how stents can be used to treat people with coronary heart disease.

Practice Question — Application

Q1 A patient with heart failure requires a heart transplant. However, doctors cannot immediately find a suitable donor organ. They decide to fit an artificial heart instead.

a) Other than better availability, suggest another advantage of artificial hearts over donor hearts.

b) Suggest two disadvantages of artificial hearts.

c) The doctor who will carry out the surgery reads an article in a medical journal about artificial hearts. He makes notes about two of the artificial hearts he's read about:

Artificial heart one:

- Has been implanted in more than 1100 patients.
- Is connected to a battery pack outside the body with wires that pass through a small hole in the skin.

Artificial heart two:

- Has been implanted into 14 patients.
- Is powered by a battery which is inserted into the body and can be charged through the skin.

The doctor decides to recommend artificial heart two to the patient. Based on the information given, evaluate the decision to treat a patient with artificial heart two.

Exam Tip
For questions where you're asked to evaluate something, make sure you take into account and write about both the advantages and the disadvantages.

3. Risk Factors for Non-Communicable Diseases

Certain things can make developing a non-communicable disease more likely...

What are risk factors?

Risk factors are things that are linked to an increase in the likelihood that a person will develop a certain disease during their lifetime. They don't guarantee that someone will get the disease.

Risk factors are often aspects of a person's lifestyle (e.g. how much exercise they do). They can also be the presence of certain substances in the environment (e.g. air pollution) or substances in your body (e.g. asbestos fibres — asbestos was a material used in buildings until it was realised that the fibres could build up in your airways and cause diseases such as cancer later in life).

Many non-communicable diseases are caused by several different risk factors interacting with each other rather than one factor alone.

Lifestyle factors can have different impacts locally, nationally and globally. E.g. in developed countries, non-communicable diseases are more common as people generally have a higher income and can buy high-fat food. Nationally, people from deprived areas are more likely to smoke, have a poor diet and not exercise. This means the incidence of cardiovascular disease, obesity and Type 2 diabetes is higher in those areas. Your individual choices affect the local incidence of disease.

Risk factors that cause disease

Some risk factors are able to directly cause a disease.

Examples

- Smoking has been proven to directly cause cardiovascular disease, lung disease and lung cancer. It damages the walls of arteries and the cells in the lining of the lungs.
- It's thought that obesity can directly cause Type 2 diabetes by making the body less sensitive or resistant to insulin, meaning that it struggles to control the concentration of glucose in the blood.
- Drinking too much alcohol has been shown to cause liver disease. The liver breaks down alcohol, but the reaction can damage its cells. Liver cells may also be damaged when toxic chemicals leak from the gut due to damage to the intestines caused by alcohol. Too much alcohol can affect brain function too. It can damage the nerve cells in the brain, causing the brain to lose volume.
- Smoking when pregnant reduces the amount of oxygen the baby receives in the womb and can cause lots of health problems for the unborn baby. Drinking alcohol has similar effects. Alcohol can damage the baby's cells, affecting its development and causing a wide range of health issues.

Learning Objectives:

- Know that risk factors are things that are linked to an increase in the likelihood that a person will develop a certain disease.
- Know that risk factors may be related to lifestyle, or the presence of substances in someone's body or in the environment.
- Know that many non-communicable diseases are caused by multiple factors interacting.
- Understand how lifestyle factors can affect how common a disease is at a local, national and global level.
- Understand that some risk factors have been proven to cause a disease, but others have not, and be able to give examples of risk factors that have been proven to cause a disease.
- Be able to identify a correlation on a graph relating to disease risk factors.
- Be able to read from and interpret charts, graphs and tables containing data related to risk factors for diseases.
- Know the human and financial costs of non-communicable diseases for individuals, local communities, nations and globally.
- Know why sampling is used when collecting disease risk factor data.

Specification Reference 4.2.2.6

Tip: There's more about the risk factors of cancer on pages 109-110.

- Cancer can be directly caused by exposure to certain substances or radiation. Things that cause cancer are known as carcinogens. Carcinogens work in different ways. For example, some damage a cell's DNA in a way that makes the cell more likely to divide uncontrollably. Ionising radiation (e.g. from X-rays) is an example of a carcinogen.

Correlation and cause

Risk factors are identified by scientists looking for correlations in data, and correlation doesn't always equal cause. Some risk factors aren't capable of directly causing a disease, but are related to another risk factor that is.

Exam Tip
In the exam, you might need to identify a correlation between risk factors and diseases from a graph. To recap what correlation is, look at page 18.

Example

A lack of exercise and a high fat diet are heavily linked to an increased chance of cardiovascular disease, but they can't cause the disease directly. It's the resulting high blood pressure and high 'bad' cholesterol levels (see p. 104) that can actually cause it.

The cost of non-communicable diseases

Non-communicable diseases can be costly on different levels.

Human cost

The human cost of non-communicable diseases is obvious. Tens of millions of people around the world die from non-communicable diseases per year. People with these diseases may have a lower quality of life or a shorter lifespan. This not only affects the sufferers themselves, but their loved ones too.

Exam Tip
In the exam, you might need to read off data related to disease risk factors from a table, graph or chart, or explain what the data is showing. Remember to always check any units or graph scales carefully before answering any questions.

Financial cost

It's also important to think about the financial cost. The cost to the NHS of researching and treating these diseases is huge — and it's the same for other health services and organisations around the world. Families may have to move or adapt their home to help a family member with a disease, which can be costly. Also, if the family member with the disease has to give up work or dies, the family's income will be reduced. A reduction in the number of people able to work can also affect a country's economy.

Tip: To save time and money, scientists often use sampling when collecting data on risk factors for diseases rather than collecting data from the entire population. Turn to page 373 for more on sampling.

Practice Questions — Fact Recall

Q1 What is a risk factor?

Q2 Other than the presence of substances in the body, give two other things that risk factors can be.

Q3 Give an example of how lifestyle risk factors can affect the incidence of non-communicable diseases on a national level.

Q4 Give two diseases that smoking has been proven to cause.

Q5 Give a risk factor that is thought to cause Type 2 diabetes.

Q6 What is a carcinogen?

4. Cancer

There are risk factors that make it more likely that someone will develop cancer. Here's all you need to know about the disease and its risk factors.

Learning Objectives:

- Know that cancer is caused when changes occur in cells that cause them to grow and divide out of control.
- Know what benign tumours are and that they stay in one place in the body and don't spread to other body tissues.
- Know that malignant tumours are cancerous and that the cells are able to travel in the blood to spread to other parts of the body and form secondary tumours.
- Know that risk factors linked to lifestyle have been identified for cancer by scientists.
- Know that genetic risk factors have been identified for certain types of cancer.

Specification Reference 4.2.2.7

Tumours

Cancer is caused by uncontrolled cell growth and division. This uncontrolled growth and division is a result of changes that occur to the cells and results in the formation of a tumour (a growth of abnormal cells). Not all tumours are cancerous. They can be benign or malignant.

Benign tumours

This is where the tumour grows until there's no more room. The tumour stays in one place (usually within a membrane) rather than invading other tissues in the body. This type isn't normally dangerous, and the tumour isn't cancerous.

Malignant tumours

This is where the tumour grows and spreads to neighbouring healthy tissues. Cells can break off and spread to other parts of the body by travelling in the bloodstream. The malignant cells then invade healthy tissues elsewhere in the body and form secondary tumours. Malignant tumours are dangerous and can be fatal — they are cancers.

Risk factors for cancer

Anyone can develop cancer. Cancer survival rates have increased due to medical advances such as improved treatment, being able to diagnose cancer earlier and increased screening for the disease. However, risk factors can increase the chance of developing some cancers. Having risk factors doesn't mean that you'll definitely get cancer. It just means that you're at an increased risk of developing the disease compared to people who have fewer risk factors, or don't have any risk factors at all. Some risk factors are associated with aspects of a person's lifestyle, and some are associated with genetics.

Lifestyle risk factors

Scientists have identified lots of lifestyle risk factors for various types of cancer. Here are a few examples:

Examples

- Smoking — It's a well known fact that smoking is linked to lung cancer, but research has also linked it to other types of cancer too, including mouth, bowel, stomach and cervical cancer.
- Obesity — Obesity has been linked to many different cancers, including bowel, liver and kidney cancer. It's the second biggest preventable cause of cancer after smoking.
- UV exposure — People who are often exposed to UV radiation from the Sun have an increased chance of developing skin cancer. People who live in sunny climates and people who spend a lot of time outside are at higher risk of the disease. People who frequently use sun beds are also putting themselves at higher risk of developing skin cancer.

Figure 1: *The lung of a smoker, showing blackening by tar and a cancerous tumour near the top (white).*

Figure 2: A cross section of a diseased liver with multiple cancerous tumours (white). These tumours may have been caused by infection with hepatitis B or C viruses.

- Viral infection — Infection with some viruses has been shown to increase the chances of developing certain types of cancer. For example, infection with hepatitis B and hepatitis C viruses can increase the risk of developing liver cancer. The likelihood of becoming infected with these viruses sometimes depends on lifestyle — e.g. they can be spread between people through unprotected sex or sharing needles.

Genetic risk factors

Genes control the activities of your cells and your characteristics — see page 265. The genes you have are inherited from your parents. Sometimes you can inherit faulty genes that make you more susceptible to cancer.

Example

Mutations (see page 249) in the BRCA genes have been linked to an increased likelihood of developing breast and ovarian cancer.

Tip: A mutation is a random change in an organism's DNA.

Practice Questions — Fact Recall

Q1 Are benign tumours cancerous?

Q2 Describe the characteristics of malignant tumours.

Practice Question — Application

Q1 The formula below shows how to calculate a person's BMI, which is one way to determine whether someone is a healthy weight.

$$\text{BMI} = \frac{\text{weight in kg}}{(\text{height in metres})^2}$$

Tip: $(\text{height in metres})^2$ just means that you multiply height in metres by itself.

The table below shows how BMI relates to a person's weight.

BMI	Weight Classification
less than 18.5	underweight
18.5 – 24.9	ideal weight
25 – 29.9	overweight
30 – 39.9	obese
40 or higher	very obese

Robyn weighs 77 kg and is 178 cm tall.
Lauren weighs 75 kg and is 155 cm tall.

a) Calculate the BMI of each of the women.

b) Based on this information alone, who has the higher risk of developing cancer? Explain your answer.

Topic 2b Checklist — Make sure you know...

Introduction to Health and Disease

- ❑ That health is the state of physical and mental wellbeing and that diseases are a large cause of ill health.
- ❑ That diseases can be communicable (can spread from person to person or between animals and people) or non-communicable (cannot be spread).
- ❑ That diseases can interact with each other to cause other physical and/or mental health issues. For example, people with immune system problems are more likely to develop communicable diseases, some cancers can be triggered by infection with certain viruses, certain immune reactions brought on by infection with a pathogen can lead to skin rashes and make asthma symptoms worse, and depression and other mental illness can develop in patients suffering from physical health problems.
- ❑ That as well as disease, a poor and unbalanced diet, high levels of stress, and disadvantaged life situations (e.g. no access to drugs) can contribute to ill health.
- ❑ How to draw, read from and interpret tables, charts and graphs (for example, frequency tables, frequency diagrams, bar charts and histograms) relating to the incidence of disease.
- ❑ How to identify a correlation between two variables in data relating to the incidence of disease by looking at data plotted on a graph.
- ❑ That sampling is often used when collecting epidemiological data (data relating to health and patterns in diseases) to save time and money.

Cardiovascular Disease

- ❑ That cardiovascular diseases are diseases of the heart or blood vessels.
- ❑ That coronary heart disease is a common example of a cardiovascular disease.
- ❑ That coronary heart disease occurs when the arteries that supply blood to the heart tissue (coronary arteries) are blocked by deposits of fatty material, starving the heart muscle of the oxygen it needs.
- ❑ That fitting stents to narrowed arteries to keep them open can help reduce the risk of a heart attack in patients with coronary heart disease.
- ❑ That statins are drugs used to reduce the amount of cholesterol in the bloodstream, which slows down the rate of fatty deposits forming inside arteries.
- ❑ That surgeons can treat a patient with heart failure by transplanting a healthy donor heart, or heart and lungs, into the patient.
- ❑ That as an alternative to a donor heart, a mechanical artificial heart can be fitted to a patient while they are waiting for a donor heart to become available, or to allow their own heart to rest and heal.
- ❑ That heart valves can become damaged or weakened, causing them to not open properly or to leak, meaning that the blood can't circulate as effectively as normal.
- ❑ That faulty heart valves can be replaced, and that the replacement valves can either be biological (from humans or other mammals) or mechanical (man-made devices).
- ❑ How to assess the pros and cons of using stents, statins, artificial hearts, transplants and heart valve replacements to treat cardiovascular disease.

cont...

Risk Factors for Non-Communicable Diseases

- [ ] That risk factors are things that increase the likelihood that someone will develop a certain disease, but having a risk factor doesn't mean that someone will definitely get the disease.
- [ ] That risk factors could be related to aspects of a person's lifestyle (e.g. how much alcohol they drink), the presence of certain substances in the environment (e.g. air pollution) or the presence of substances in someone's body (e.g. asbestos fibres).
- [ ] That many non-communicable diseases are caused by multiple risk factors interacting with each other — for example, Type 2 diabetes, cardiovascular disease, and some lung and liver diseases.
- [ ] How the incidence of non-communicable diseases can be affected by lifestyle-related risk factors (e.g. smoking and diet) at a global, national and local level.
- [ ] That some risk factors have been proven to directly cause a disease (e.g. smoking has been shown to directly cause lung disease and lung cancer), whereas some risk factors have not.
- [ ] That smoking can cause cardiovascular disease, that obesity can cause Type 2 diabetes, that alcohol can affect liver and brain function, that smoking can cause lung disease and lung cancer, that smoking and drinking alcohol while pregnant can cause issues for unborn babies, and that exposure to carcinogens (such as ionising radiation) can cause cancer.
- [ ] How to use graphs to identify correlation between risk factors and diseases.
- [ ] How to read off and interpret data on disease risk factors presented in charts, graphs and tables.
- [ ] The human and financial costs that non-communicable diseases can have on a local, national and global level.
- [ ] That scientists often use sampling to collect data about the risk factors for different diseases because collecting data from the entire population would take too long and cost too much.

Cancer

- [ ] That cancer is caused by uncontrolled cell growth and division triggered by certain changes in cells, which leads to the formation of growths of abnormal cells called tumours.
- [ ] That benign tumours are tumours that stay in one place in the body and aren't able to spread to and invade other tissues.
- [ ] That malignant tumours are cancerous and are able to grow and spread into neighbouring tissues.
- [ ] That cells can break off malignant tumours and travel in the bloodstream to cause secondary tumours in other parts of the body.
- [ ] That there are lots of different lifestyle-related risk factors for developing cancer that have been identified by scientists, such as smoking, obesity, exposure to too much UV radiation (e.g. from the Sun or sun beds) or infection with certain viruses.
- [ ] That there are also risk factors for cancer related to genetics — inheriting certain genes could increase a person's chances of developing certain cancers.

Exam-style Questions

1 A doctor was viewing the X-ray of a patient's chest shown in **Figure 1**. She noticed that a large tumour had formed in the patient's left lung.

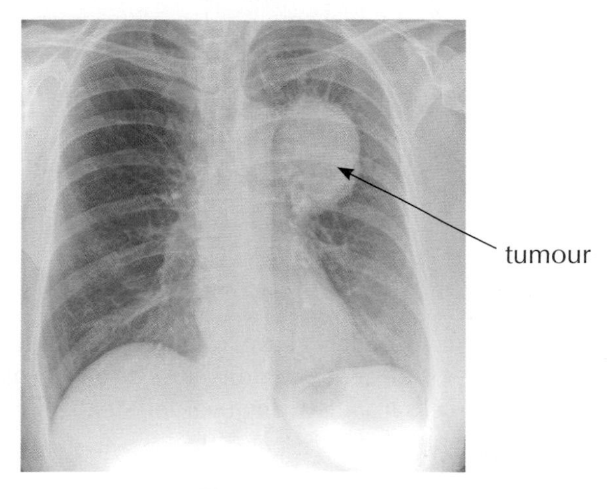

Figure 1

1.1 Which of these sentences best describes how tumours form?

- **A** Tumours are formed when body cells are unable to replicate.
- **B** Tumours are formed when body cells die.
- **C** Tumours are formed when body cells grow too slowly.
- **D** Tumours are formed when body cells grow and divide out of control.

(1 mark)

After further investigation, the doctor found another tumour in the patient's liver made up of similar cells to the tumour in the lungs. This led her to suspect that the tumours were cancerous.

1.2 What is the name for a cancerous tumour?

(1 mark)

1.3 Suggest why the doctor suspected the tumours to be cancerous.

(3 marks)

1.4 Which of these statements is true?

- **A** Neither genetic factors nor aspects of a person's lifestyle can be cancer risk factors.
- **B** Cancer risk factors are always related to genetics.
- **C** Both genetic factors and aspects of a person's lifestyle can be cancer risk factors.
- **D** All cancer risk factors are related to aspects of a person's lifestyle.

(1 mark)

1.5 The doctor asks if the patient takes any medication.
The patient takes statins. How do statins prevent coronary heart disease?

(2 marks)

2 A doctor is treating a patient who has damage to a heart valve. The doctor has decided that the best treatment for the patient would be to replace the damaged valve.

2.1 Describe the complications that can be caused by a damaged heart valve.

(3 marks)

2.2 The doctor decides to use a mechanical replacement valve.
Describe how mechanical replacement valves differ from biological replacement valves.

(2 marks)

2.3 Another patient requires a heart transplant. A donor has not yet been found.
Suggest what might be done to keep the patient alive until a donor has been found.

(1 mark)

3 After a heart attack, many patients have a stent fitted into the affected coronary artery. This could be a bare-metal stent or a drug-eluting stent, which gradually releases a drug to prevent new cells from growing over the stent.

In a trial, patients needing a stent were randomly given either a bare-metal stent or a drug-eluting stent. Following their surgery, the patients were monitored and any health problems recorded.

The results for 154 patients who received the drug-eluting stent and 153 patients who received the bare-metal stent are shown in **Table 1**:

Table 1

	Number of patients who died following treatment	Number of patients who had another heart attack	% of patients whose artery renarrowed
Drug-eluting stent	3	3	9
Bare-metal stent	7	3	21

3.1 Calculate the percentage of patients that died after receiving the drug-eluting stent.

(1 mark)

3.2 Explain why stents are often used in people who have had a heart attack.

(3 marks)

3.3 Based on the information given, some doctors may conclude that drug-eluting stents are more beneficial than bare-metal stents. Do you think this is a valid conclusion? Explain your answer.

(3 marks)

1. Enzymes

Enzymes are essential. There are loads of different types in the body. They play an important role in speeding up chemical reactions in your insides.

Enzymes

Living things have thousands of different chemical reactions going on inside them all the time, including the reactions involved in metabolism (see page 179). These reactions need to be carefully controlled — to get the right amounts of substances.

You can usually make a reaction happen more quickly by raising the temperature. This would speed up the useful reactions but also the unwanted ones too... not good. There's also a limit to how far you can raise the temperature inside a living creature before its cells start getting damaged.

So... living things produce enzymes that act as **biological catalysts**. Enzymes reduce the need for high temperatures and we only have enzymes to speed up the useful chemical reactions in the body.

A catalyst is a substance which increases the speed of a reaction, without being changed or used up in the reaction.

Enzymes are all large proteins and all proteins are made up of chains of amino acids. These chains are folded into unique shapes, which enzymes need to do their jobs (see below).

Active sites

Chemical reactions usually involve things either being split apart or joined together. Every enzyme has an **active site** with a unique shape that fits onto the substance involved in a reaction. Enzymes are really picky — they usually only catalyse one specific reaction. This is because, for the enzyme to work, the substrate has to fit into its active site. If the substrate doesn't match the enzyme's active site, then the reaction won't be catalysed.

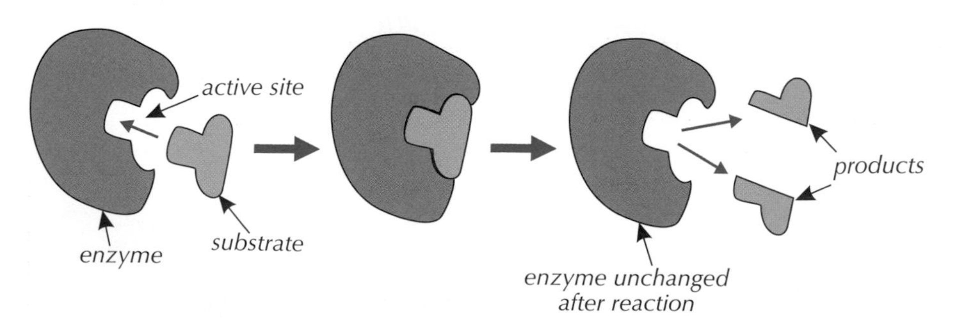

Figure 1: *Diagram to show how an enzyme works.*

Learning Objectives:
- Know that enzymes play a role in metabolism.
- Understand the function and characteristics of enzymes.
- Know that each enzyme has an active site with a unique shape that will only fit onto the substance involved in a specific reaction.
- Know that the 'lock and key' model of enzyme action is a simpler version of what actually takes place when an enzyme binds to its substrate.
- Understand how the activity of enzymes is affected by pH and temperature.

Specification Reference 4.2.2.1

Tip: The substance that an enzyme acts on is called the substrate.

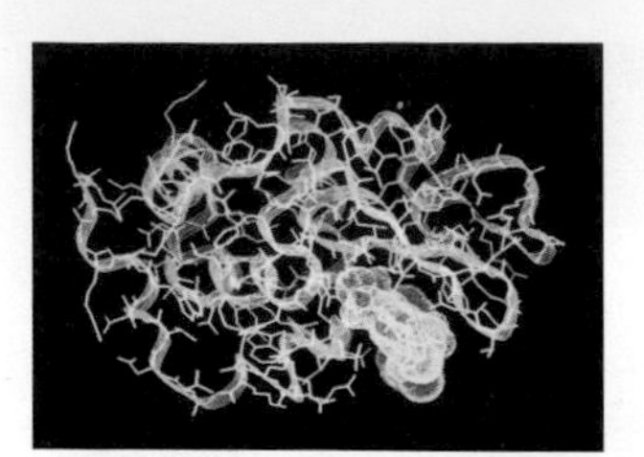

Figure 2: Computer model of an enzyme bound to a substance (yellow).

Tip: Increasing the temperature increases the rate of a reaction because the reactants have more energy, so they move about more, and collide with each other more often.

Tip: If an enzyme is involved in speeding up a reaction, then the reaction is known as an 'enzyme-catalysed reaction', an 'enzymatic reaction' or an 'enzyme-controlled reaction'.

Tip: A low pH means that an environment is acidic. A high pH means that it is alkaline.

Figure 1 (on the previous page) shows the 'lock and key' model of enzyme action. This is simpler than how enzymes actually work. In reality, the active site changes shape a little as the substrate binds to it to get a tighter fit. This is called the 'induced fit' model of enzyme action.

Optimum conditions for enzymes

Enzymes need the right conditions, such as the right temperature and pH, for them to work best — these are called **optimum conditions**.

Temperature

Changing the temperature changes the rate of an enzyme-catalysed reaction. Like with any reaction, a higher temperature increases the rate at first. But if it gets too hot, some of the bonds holding the enzyme together break. This changes the shape of the enzyme's active site, so the substrate won't fit any more. The enzyme is said to be **denatured**.

Therefore enzymes have a temperature at which they are most active — this is called the optimum temperature (see Figure 3). Enzymes in the human body normally work best at around 37 °C.

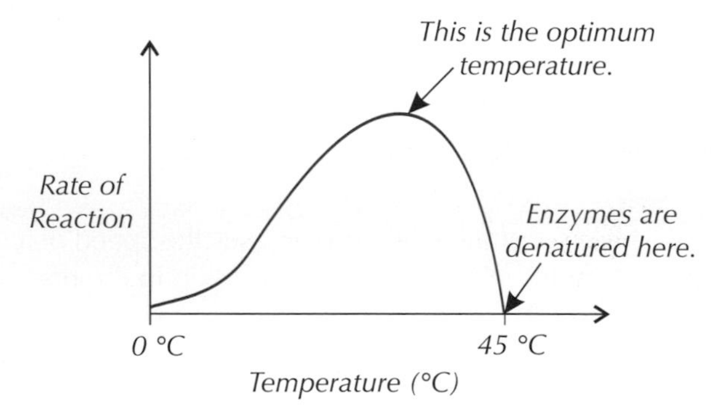

Figure 3: *Graph to show how temperature affects enzyme action.*

pH

The pH also affects enzymes. If it's too high or too low, the pH interferes with the bonds holding the enzyme together. This changes the shape of the active site and denatures the enzyme.

All enzymes have an optimum pH that they work best at — this is called the optimum pH (see Figure 4).

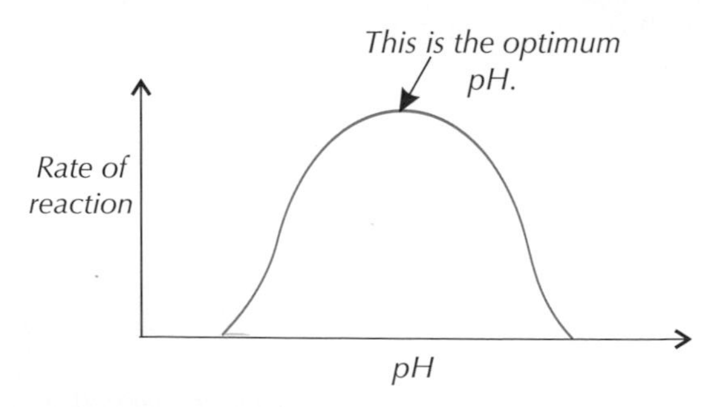

Figure 4: *Graph to show how pH affects enzyme action.*

The optimum pH is different for different enzymes depending on where they work. For many enzymes it's neutral pH 7, but not always.

Example

Pepsin is an enzyme used to break down proteins in the stomach. It works best at pH 2, which means it's well-suited to the acidic conditions in the stomach.

Practice Questions — Fact Recall

Q1 Why can enzymes be described as biological catalysts?

Q2 What is the name of the part of an enzyme that binds to the substrate?

Q3 True or false? All enzymes work best at pH 7.

Practice Question — Application

Q1 Hexokinase is an enzyme found in the human body that's involved in respiration. It catalyses this reaction:

$$\text{glucose} \xrightarrow{\text{hexokinase}} \text{substance A}$$

Enzymes in the human body work best at 37 °C.

A scientist heats up hexokinase to 50 °C and adds it to some glucose. Suggest the effect this will have on the rate of the reaction compared to if the reaction was at 37 °C. Explain your answer.

Tip: Respiration is the process of breaking down glucose, which transfers energy. There's loads more about it on pages 173-176.

2. Investigating Enzymatic Reactions

You need to know how to investigate the effect of pH on the rate of the reaction catalysed by the enzyme amylase. Read on for more.

Learning Objective:
- Be able to investigate the effect of pH on the rate of starch breakdown by the enzyme amylase (Required Practical 5).
- Be able to calculate the rate of chemical reactions.

Specification Reference
4.2.2.1

Investigating the effect of pH on amylase

REQUIRED PRACTICAL 5

The enzyme amylase catalyses the breakdown of starch to maltose. It's easy to detect starch using iodine solution — if starch is present, the iodine solution will change from browny-orange to blue-black. This is how you can investigate how pH affects amylase activity:

1. Put a drop of iodine solution into every well of a spotting tile.
2. Place a Bunsen burner on a heat-proof mat, and a tripod and gauze over the Bunsen burner. Put a beaker of water on top of the tripod and heat the water until it is 35 °C (use a thermometer to measure the temperature). Try to keep the temperature of the water constant throughout the experiment.
3. Use a syringe to add 1 cm^3 of amylase solution and 1 cm^3 of a buffer solution with a pH of 5 to a boiling tube. Using test tube holders, put the boiling tube into the beaker of water and wait for five minutes.
4. Next, use a different syringe to add 5 cm^3 of a starch solution to the boiling tube.
5. Immediately mix the contents of the boiling tube and start a stop clock.
6. Use continuous sampling to record how long it takes for the amylase to break down all of the starch. To do this, use a dropping pipette to take a fresh sample from the boiling tube every thirty seconds and put a drop into a well. When the iodine solution remains browny-orange, starch is no longer present.
7. Repeat the whole experiment with buffer solutions of different pH values to see how pH affects the time taken for the starch to be broken down.
8. Remember to control any variables each time (e.g. concentration and volume of amylase solution) to make it a fair test.

Tip: Don't forget to do a risk assessment before you do this experiment. You should always take basic safety precautions like wearing goggles and a lab coat.

Tip: Instead of using the Bunsen burner, tripod and gauze, you could set an electric water bath to 35 °C and place the boiling tube in that. See page 371 for more on water baths.

Tip: Remember to repeat the experiment at least three times at each pH and work out the mean time taken for the starch to break down.

Tip: See page 125 for more on testing for starch using iodine.

Tip: You could use a pH meter to accurately measure the pH of your solutions.

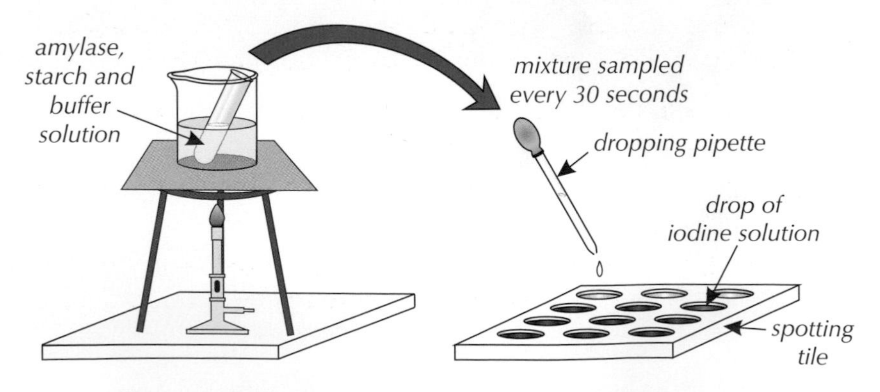

Figure 1: *Apparatus for investigating the breakdown of starch.*

Calculating the rate of a reaction

It's often useful to calculate the rate of reaction after an experiment. Rate is a measure of how much something changes over time.

For the experiment on the previous page, you can calculate the rate of reaction using this formula:

$$\text{Rate} = \frac{1000}{\text{time}}$$

The units are s^{-1} since rate is given per unit of time.

Example

At pH 6, the time taken for amylase to break down all of the starch in a solution was 90 seconds. So the rate of the reaction $= 1000 \div 90 =$ **11 s^{-1}**

If an experiment measures how much something changes over time, you calculate the rate of reaction by dividing the amount that it has changed by the time taken.

Example

The enzyme catalase catalyses the breakdown of hydrogen peroxide into water and oxygen. During an investigation into the activity of catalase, 24 cm^3 of oxygen was released in 50 seconds (s). Calculate the rate of the reaction. Write your answer in cm^3/s.

Amount of product formed = change = 24 cm^3

Rate of reaction = change ÷ time
= 24 cm^3 ÷ 50 s
= **0.48 cm^3/s**

Tip: Look back at page 94 for how to work out the rate of transpiration. Also look at page 168 for how to work out the rate of photosynthesis.

Practice Question — Application

Q1 A group of students were investigating the effect of pH on the rate of starch breakdown by the enzyme amylase. They mixed the enzyme and starch in a boiling tube, then every 10 seconds added a drop of the mixture to a well on a spotting tile containing iodine solution. They timed how long it took for all of the starch to be broken down.

a) Explain how the pH level of the enzyme and starch mixture could be controlled.

b) Explain how the students would know when all of the starch had been broken down.

c) It took 120 seconds for all of the starch to be broken down. Calculate the rate of the reaction. Give your answer in s^{-1}.

3. Enzymes and Digestion

Digestion involves loads of reactions where large food molecules are broken down into smaller ones. Lots of enzymes are needed to catalyse all these reactions, so the body creates the ideal conditions for these enzymes to work in.

Learning Objectives:

- Know that digestive enzymes break down big food molecules to smaller, soluble food molecules, which are able to be absorbed into the bloodstream.
- Know that the products of digestion are used to make new carbohydrates, lipids and proteins, and that some of the glucose produced is used for respiration.
- Know that carbohydrases convert carbohydrates to simple sugars.
- Know amylase (which breaks down starch) is a carbohydrase.
- Know that proteases convert proteins into amino acids.
- Know that lipases convert lipids into glycerol and fatty acids.
- Be able to understand word equations for digestion reactions.
- Know where in the digestive system amylase, proteases and lipases are made and where they act.
- Know that bile is an alkaline substance made by the liver and stored in the gall bladder.
- Know that bile neutralises the stomach acid and emulsifies fats into smaller droplets with increased surface area to speed up the action of lipase.

Specification Reference 4.2.2.1

The digestive system

As you know from page 73, the digestive system is the organ system that breaks down food, so that nutrients can be absorbed into the body from the gut. (There's a reminder of what it looks like in Figure 1.)

Food is broken down in two ways:

- By mechanical digestion — this includes our teeth grinding down food and our stomach churning up food.
- By chemical digestion — where enzymes help to break down food.

You need to know all about how enzymes work in digestion.

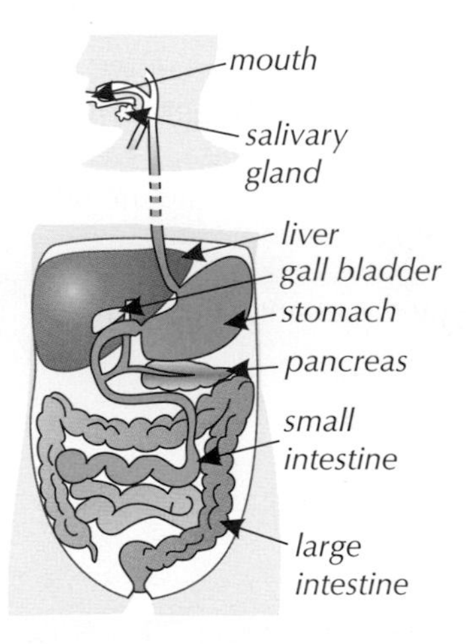

Figure 1: *The digestive system.*

Digestive enzymes

The enzymes involved in digestion work outside body cells. They're produced by specialised cells in glands and in the gut lining, and then released into the gut to mix with food molecules, such as carbohydrates, proteins and lipids (fats or oils).

Starch, proteins and lipids are big molecules. They're too big to pass through the walls of the digestive system, so digestive enzymes break these big molecules down into smaller ones like sugars (e.g. glucose and maltose), amino acids, glycerol and fatty acids. These smaller, soluble molecules can pass easily through the walls of the digestive system, allowing them to be absorbed into the bloodstream. The body makes good use of the products of digestion. They can be used to make new carbohydrates, proteins and lipids. Some of the glucose (a sugar) that's made is used in respiration (see p. 173).

Different digestive enzymes help to break down different types of food.

Carbohydrases

Carbohydrases convert carbohydrates into simple sugars. Amylase is an example of a carbohydrase. It breaks down starch — see Figure 2.

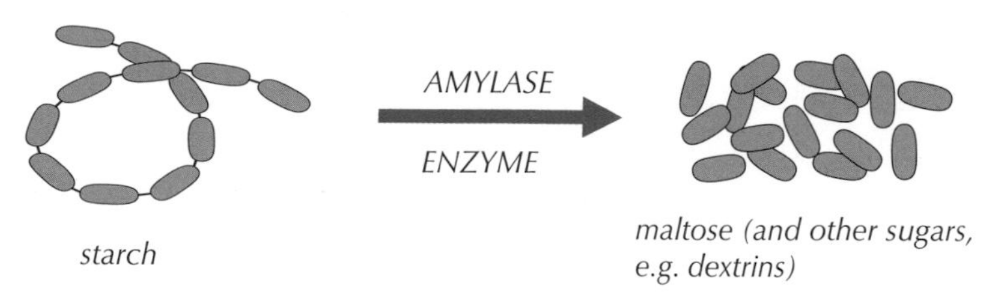

Figure 2: *Diagram to show the breakdown of starch by amylase.*

Amylase is made in the salivary glands, the pancreas and the small intestine. It works in the mouth and the small intestine.

Tip: Most absorption of food molecules happens in the small intestine.

Proteases

Protease enzymes are digestive enzymes that catalyse the conversion of proteins into amino acids — see Figure 3.

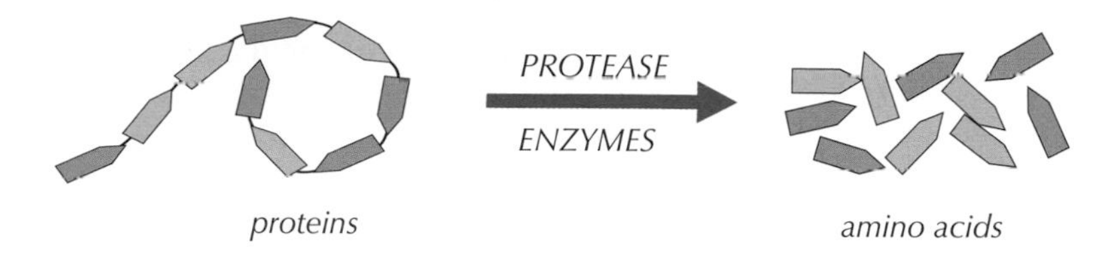

Figure 3: *Diagram to show the breakdown of proteins by proteases.*

Proteases are made in the stomach, the pancreas and the small intestine. They work in the stomach and the small intestine.

Lipases

Lipase enzymes are digestive enzymes that catalyse the conversion of lipids into glycerol and fatty acids — see Figure 4.

Tip: Protease enzymes break down proteins. Lipase enzymes break down lipids.

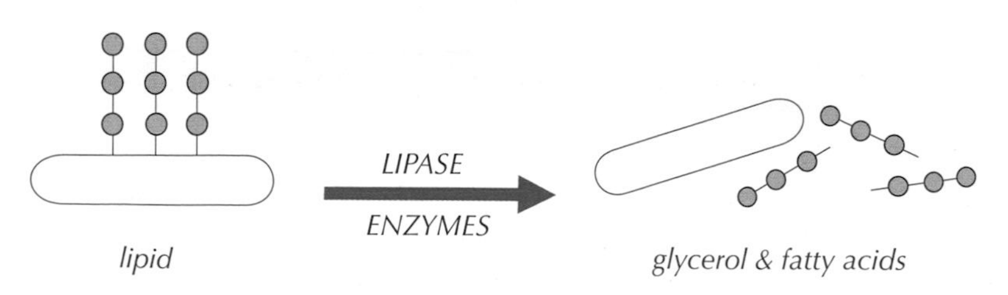

Figure 4: *Diagram to show the breakdown of lipids by lipases.*

Lipases are made in the pancreas and the small intestine. They work in the small intestine.

Enzymes in the digestive system

The table in Figure 5 summarises all the information you need to know about the digestive enzymes for your exam. Use Figure 5 to make sure you know exactly where each enzyme is made and where it works.

Enzyme(s) →	Amylase	Proteases	Lipases
Help(s) to break down...	Starch	Proteins	Lipids
...into...	Sugars	Amino acids	Glycerol and fatty acids
Made in the...	Salivary glands, pancreas and small intestine	Stomach, pancreas and small intestine	Pancreas and small intestine
Work(s) in the...	Mouth and small intestine	Stomach and small intestine	Small intestine

Figure 5: *Table summarising the key facts about amylase, proteases and lipases.*

Bile

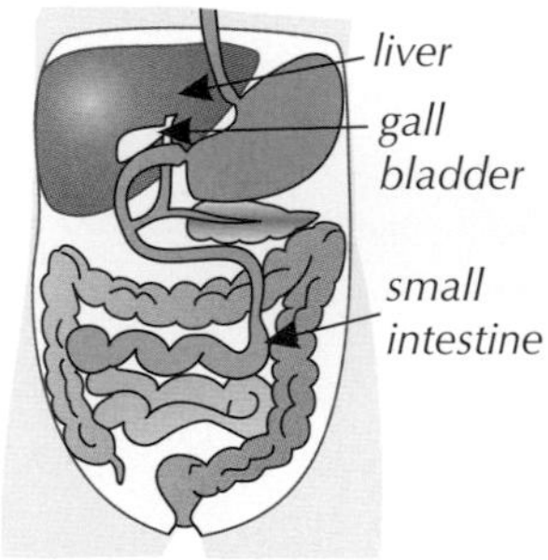

Bile is produced in the liver. It's stored in the gall bladder before it's released into the small intestine (see Figure 6).

Figure 6: *Bile is produced in the liver, stored in the gall bladder and released into the small intestine.*

The hydrochloric acid in the stomach makes the pH too acidic for enzymes in the small intestine to work properly. Bile is alkaline — it neutralises the acid and makes conditions alkaline. The enzymes in the small intestine work best in these alkaline conditions.

Bile also emulsifies fats. In other words it breaks the fat into tiny droplets (see Figure 7). This gives a much bigger surface area of fat for the enzyme lipase to work on — which makes its digestion faster.

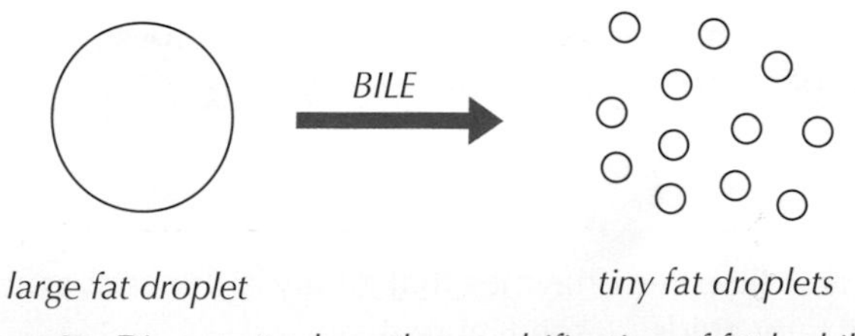

Figure 7: *Diagram to show the emulsification of fat by bile.*

Practice Questions — Fact Recall

Q1 True or false? Digestive enzymes build small molecules up into larger molecules.

Q2 Name the digestive enzyme produced in the salivary glands.

Q3 Where are protease enzymes involved in digestion produced?

Q4 Name the type of enzyme that catalyses the breakdown of lipids.

Q5 Describe how bile produces ideal conditions for enzymes in the small intestine to work in.

Practice Question — Application

Q1 Here is a labelled diagram of the digestive system.

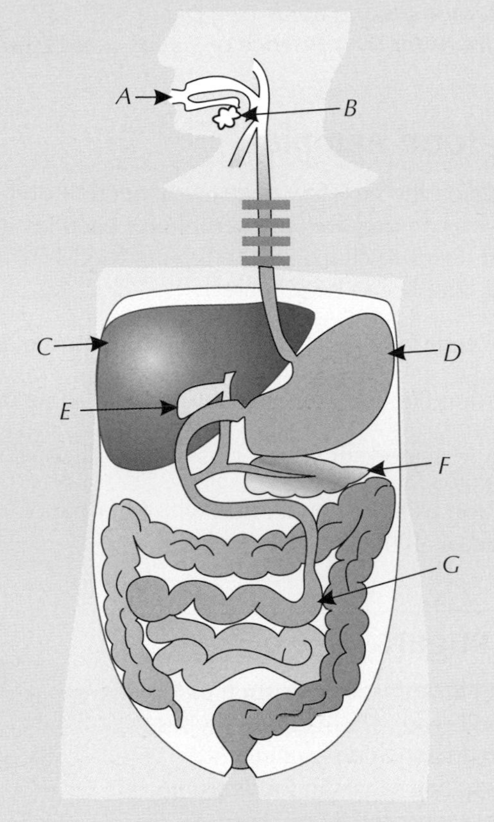

Complete this table using the appropriate letters from the diagram. (If there's more than one answer to the question, make sure you write down all the answers in the table.)

	Amylase	Proteases	Lipases	Bile
Made where?	B, F			
Work(s) where?				

4. Food Tests

Learning Objective:

- Know how to carry out qualitative tests to detect the presence of sugars (Benedict's test), starch (iodine test), proteins (biuret test) and lipids (Sudan III test) in foods (Required Practical 4).

Specification Reference 4.2.2.1

Your digestive system breaks down all sorts of substances from food, e.g. carbohydrates, proteins and lipids (see pages 120-122). You need to know how to test foods for the presence of sugars, starch, proteins and lipids.

Preparing a food sample

For each of the tests on the next few pages, you need to prepare a food sample. Luckily, you can prepare your sample for each test in the same way. The only thing that needs to change is the type of food you're using. Here's how a food sample can be prepared:

1. Get a piece of your food and break it up using a pestle and mortar.
2. Transfer the ground up food to a beaker and add some distilled water.
3. Give the mixture a good stir with a glass rod to dissolve some of the food.
4. Filter the solution using a funnel lined with filter paper. This will get rid of all the solid bits of food.

Tip: Make sure you're aware of the risks of the Benedict's test before you start. Wear eye protection and be very careful around the hot water in the water bath.

Testing for sugars

REQUIRED PRACTICAL 4

Sugars are found in all sorts of foods such as biscuits, cereal and bread. There are two types of sugars — non-reducing and reducing. You can test for reducing sugars in foods using the Benedict's test. Here's what you'd do:

1. Prepare a food sample using the method described above and transfer 5 cm^3 to a test tube.
2. Prepare a water bath so that it's set to 75 °C.
3. Add some Benedict's solution to the test tube (about 10 drops) using a pipette.
4. Place the test tube in the water bath using a test tube holder and leave it in there for 5 minutes. Make sure the tube is pointing away from you.
5. During this time, if the food sample contains a reducing sugar, the solution in the test tube will change from the normal blue colour to green, yellow or brick-red (see Figure 2) — it all depends on how much sugar is in the food.

Figure 1: A negative (left) and a positive (right) Benedict's test.

Tip: A positive result is one where the test has identified the presence of what's being tested for (in this case a reducing sugar). A negative result is one where the test has not identified the presence of what's being tested for.

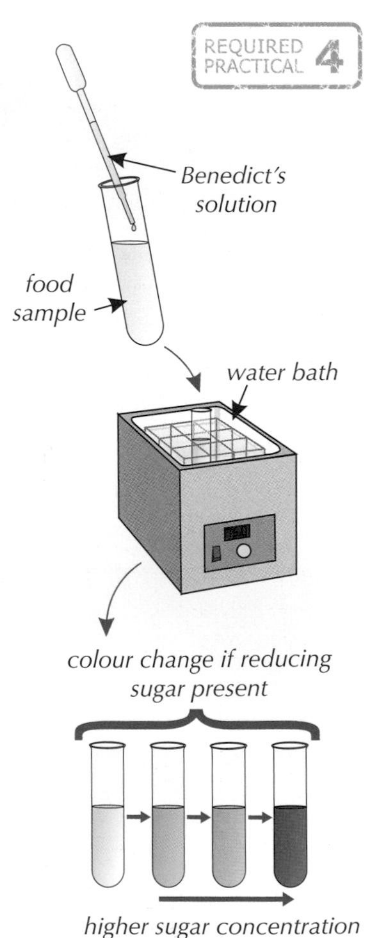

Figure 2: Steps involved in the Benedict's test.

Testing for starch

REQUIRED PRACTICAL 4

You can check food samples for the presence of starch using the iodine test. Foods like pasta, rice and potatoes contain a lot of starch. Here's how to do the test:

1. Make a food sample using the method on the previous page and transfer 5 cm^3 of your sample to a test tube.
2. Then add a few drops of iodine solution and gently shake the tube to mix the contents. If the sample contains starch, the colour of the solution will change from browny-orange to black or blue-black (see Figure 4).

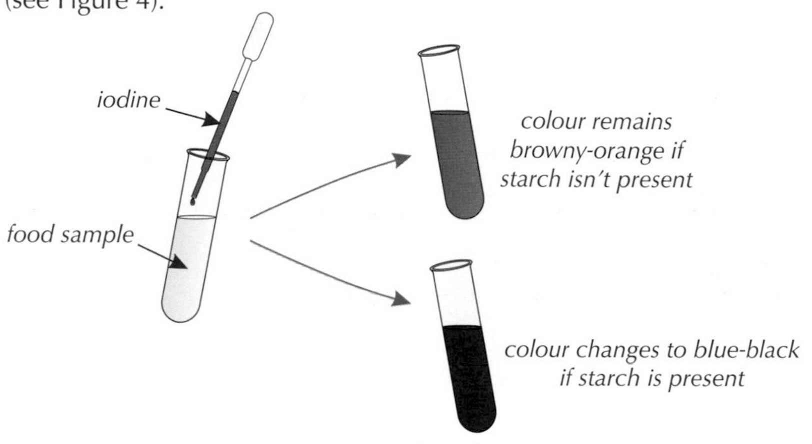

Figure 3: The iodine test for starch.

Tip: Make sure you assess and address the risks of the iodine starch test before you start. Wear eye protection, as iodine is an irritant to the eyes.

Figure 4: A dark blue-black colour indicates the presence of starch in an iodine test.

Testing for proteins

You can use the biuret test to see if a type of food contains protein. Meat and cheese are protein rich and good foods to use in this test. Here's how it's done:

1. Use the method on the previous page to prepare a sample of your food and transfer 2 cm^3 of your sample to a test tube.
2. Add 2 cm^3 of biuret solution to the sample and mix the contents of the tube by gently shaking it.
3. If the food sample contains protein, the solution will change from blue to pink or purple. If no protein is present, the solution will stay blue (see Figure 5).

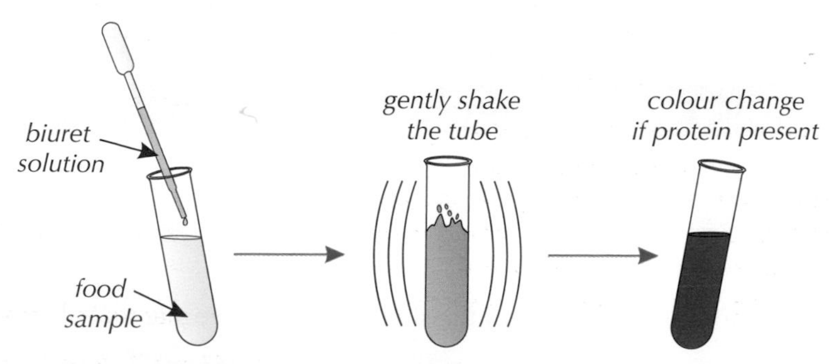

Figure 5: Steps involved in the biuret test.

Tip: Make sure you carry out a risk assessment before you perform the biuret test. The chemicals in the biuret solution are dangerous, so you'll need to wear safety goggles and make sure that you wash it off straight away if you get any on your skin.

Figure 6: Diagram showing a negative (left) and positive (right) result for the biuret test for the presence of proteins.

Testing for lipids

REQUIRED PRACTICAL 4

Lipids are found in foods such as olive oil, margarine and milk. You can test for the presence of lipids in a food using Sudan III stain solution.

1. Prepare a sample of the food you're testing using the method on page 124 (but you don't need to filter it). Transfer about 5 cm^3 into a test tube.
2. Use a pipette to add 3 drops of Sudan III stain solution to the test tube and gently shake the tube.
3. Sudan III stain solution stains lipids. If the food sample contains lipids, the mixture will separate out into two layers. The top layer will be bright red (see Figure 7). If no lipids are present, no separate red layer will form at the top of the liquid.

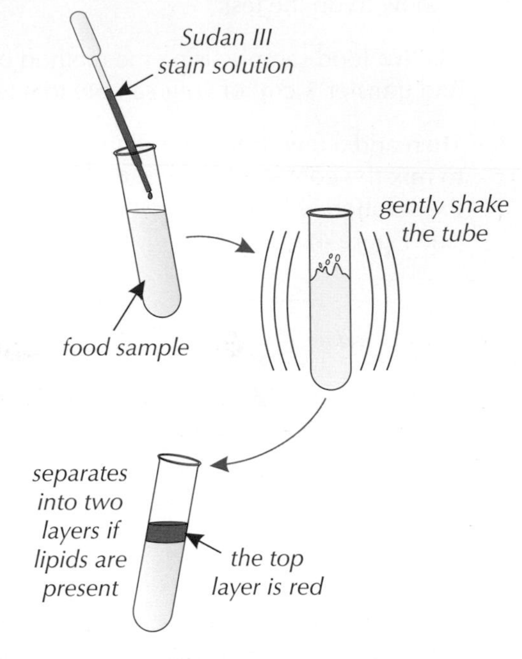

Figure 7: *Steps involved in the Sudan III test.*

Tip: Make sure you do a risk assessment before you start the test for lipids. Make sure you wear safety goggles. Also, Sudan III stain solution is flammable, so make sure you keep it away from any lit Bunsen burners.

Practice Questions — Fact Recall

Q1 Name the reagent used to test for the presence of reducing sugars.

Q2 Describe the colour change that would occur in the iodine test if starch was present in the food sample.

Q3 What colour would biuret solution be if it was added to a food sample that didn't contain protein?

Practice Question — Application

Q1 A student has a piece of food and wants to know if it contains lipids. Outline a test he could carry out to find out.

Topic 2c Checklist — Make sure you know...

Enzymes

- ☐ That enzymes are large proteins that act as biological catalysts, speeding up the rate of chemical reactions involved in metabolism.
- ☐ That every enzyme has its own unique active site, and it's the shape of this active site that allows the enzyme to bind to its substrate.

cont...

- [ ] That enzymes will only bind to one specific substrate, so if the substrate doesn't fit the active site, the reaction won't be catalysed. This is the 'lock and key' model of enzyme action.
- [ ] That increasing the temperature increases the rate of an enzyme-controlled reaction until it gets so hot that some of the bonds holding the enzyme together break and the enzyme becomes denatured, so the active site no longer fits the substrate.
- [ ] That enzymes all have an optimum temperature at which they work best, and this is different for different enzymes.
- [ ] That extreme pH levels can denature enzymes.
- [ ] That enzymes all have an optimum pH level that they work best at, and this optimum level varies between different enzymes.

Investigating Enzymatic Reactions

- [ ] How to investigate the effects of pH on the rate of the breakdown of starch by the enzyme amylase (Required Practical 5).
- [ ] How to calculate the rate of a chemical reaction such as an enzyme controlled reaction.

Enzymes and Digestion

- [ ] That digestive enzymes break down large food molecules (which can't be absorbed into the bloodstream) into smaller, soluble food molecules that can be absorbed into the bloodstream.
- [ ] That the body makes new carbohydrates, proteins and lipids from the products of digestion and also uses some of the glucose produced in the process of respiration.
- [ ] That carbohydrases convert carbohydrates into simple sugars.
- [ ] That amylase is a carbohydrase, produced in the salivary glands, pancreas and small intestine, and that it works in the mouth and small intestine to break down starch.
- [ ] That protease enzymes catalyse the conversion of proteins into amino acids.
- [ ] That proteases are made in the stomach, pancreas and small intestine, and act in the stomach and small intestine.
- [ ] That lipases catalyse the conversion of lipids (fats and oils) into glycerol and fatty acids.
- [ ] That lipases are made in the pancreas and small intestine, and act in the small intestine.
- [ ] That bile is produced by the liver and stored in the gall bladder.
- [ ] That bile helps to speed up the action of lipases by neutralising the hydrochloric acid from the stomach (providing more suitable pH conditions for the lipases to work in) and emulsifying fats (breaking fats down into smaller droplets with increased surface area for the lipases to work on).

Food Tests

- [ ] How to test for the presence of sugars, starch, proteins and lipids in different foods using the Benedict's test, iodine test, biuret test and Sudan III test (Required Practical 4).

Exam-style Questions

1 Enzymes are needed for many processes in the body.

1.1 What type of biological molecule is an enzyme?

A a lipid **B** a hormone **C** a protein **D** an amino acid

(1 mark)

1.2 What is the function of an enzyme?

(1 mark)

1.3* One of the processes in the body that requires enzymes is digestion.
Describe the action, production and release of different types of enzymes in digestion.

(6 marks)

2 A student was measuring the rate that carbon dioxide (CO_2) was produced in a reaction controlled by enzyme X. He performed the experiment at three different temperatures — 10 °C, 25 °C and 50 °C. The experiment at 50 °C produced no carbon dioxide at all.

His results are shown in **Figure 1**.

Figure 1

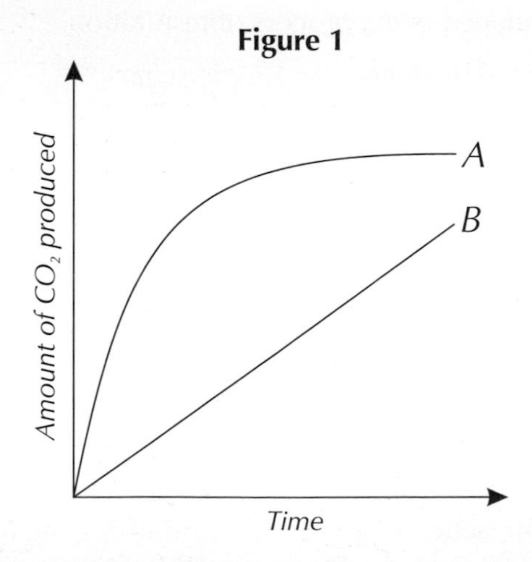

2.1 Enzyme X is only able to catalyse this reaction.
Explain why most enzymes can only catalyse a single reaction.

(2 marks)

2.2 Suggest which line on **Figure 1** represents the experiment carried out at 25 °C.
Explain your answer.

(1 mark)

2.3 Suggest why the experiment at 50 °C failed to produce any CO_2.

(3 marks)

3 Scientists can use a range of simple tests to detect the presence of certain substances in different types of food. A scientist was testing a food sample for the presence of glucose, a reducing sugar.

3.1 Which of the following reagents should she use?

A Benedict's solution

B Biuret solution

C Iodine solution

D Sudan III stain solution

(1 mark)

3.2 The result of the test was no colour change.
What does this suggest about the presence of glucose in the sample?

(1 mark)

3.3 Describe how the scientist should have prepared the food sample before the reagent was added.

(4 marks)

3.4 Following the test for reducing sugars, she decided to test a fresh sample of the food for starch. Name the reagent that she would have used and describe what she would have seen if the result was positive.

(3 marks)

4 The conditions inside the stomach are acidic due to the presence of hydrochloric acid. An experiment was done to show how digestion in the stomach works. The enzyme pepsin is a protease found in the stomach. In the experiment, three equally sized pieces of meat were added to each of three test tubes. Then the test tubes were filled with equal volumes of the following substances:

Test tube 1 — hydrochloric acid (HCl acid)
Test tube 2 — pepsin
Test tube 3 — pepsin and hydrochloric acid (HCl acid)

The results are shown in **Figure 2**.

Figure 2

Describe the results and explain what they suggest about pepsin.

(3 marks)

Learning Objectives:

- Know that pathogens are microorganisms that cause infectious disease in plants and animals.
- Know that a pathogen can be a bacterium, virus, protist or fungus.
- Know that bacteria and viruses can reproduce quickly.
- Understand that bacteria can make people feel ill by producing poisons that damage body tissues.
- Know that viruses make people feel ill by causing cell damage when they reproduce.
- Know how pathogens can be spread (water, air and direct contact).

Specification Reference 4.3.1.1

1. Communicable Disease

Disease isn't a particularly pleasant thing, but you've still got to learn about it. First up in the section is meeting the microorganisms that cause disease.

Pathogens

Pathogens are microorganisms that enter the body and cause disease. They cause **communicable (infectious) diseases** — diseases that can spread (see p. 101). Both plants and animals can be infected by pathogens. There are several types of pathogen.

Bacteria

Bacteria are very small cells (about 1/100th the size of your body cells), which can reproduce rapidly inside your body. They can make you feel ill by producing **toxins** (poisons) that damage your cells and tissues.

Example

Salmonella bacteria — these bacteria cause food poisoning (see page 134).

Viruses

Viruses are not cells. They're tiny, about 1/100th the size of a bacterium. Like bacteria, they can reproduce rapidly inside your body. They live inside your cells and replicate themselves using the cells' machinery to produce many copies of themselves. The cell will usually then burst, releasing all the new viruses. This cell damage is what makes you feel ill.

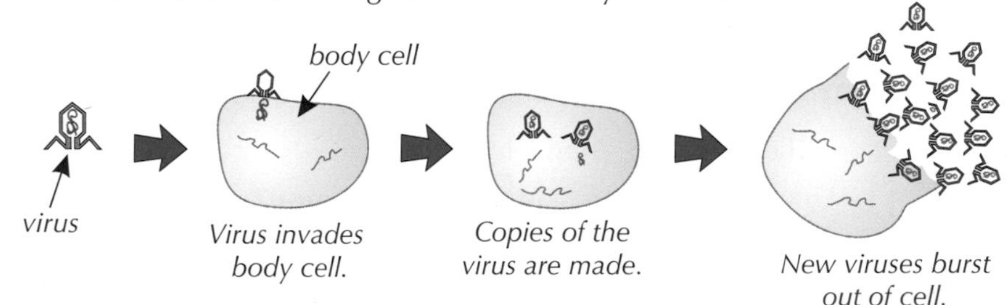

Figure 1: *A virus replicating in a human cell.*

Example

HIV (see page 132) — this virus infects and destroys cells that normally help to defend the body against disease (like white blood cells, see page 87). This makes HIV sufferers more likely to get ill from infection by other pathogens.

Protists

There are lots of different types of protists. But they're all eukaryotes (see page 23) and most of them are single-celled. Some protists are parasites. Parasites live on or inside other organisms and can cause them damage. They are often transferred to the organism by a **vector**, which doesn't get the disease itself — e.g. an insect that carries the protist.

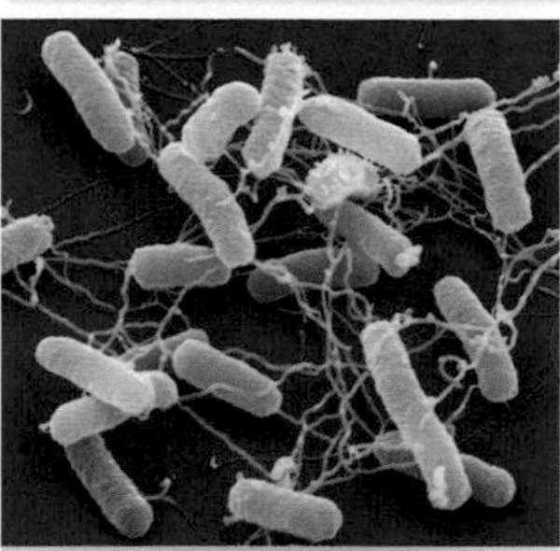

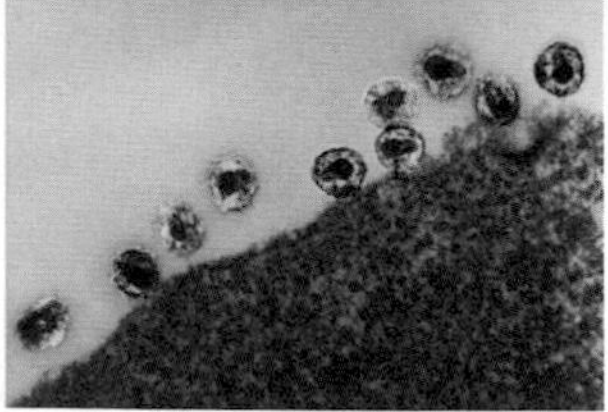

Figure 2: *A Salmonella typhimurium bacterium (top). HIV particles (blue) infecting a white blood cell (bottom).*

Example

Malaria is caused by a protist that lives inside a mosquito (see page 133).

Fungi

Some fungi are single-celled. Others have a body which is made up of hyphae (thread-like structures). These hyphae can grow and penetrate human skin and the surface of plants, causing diseases. The hyphae can produce spores, which can be spread to other plants and animals.

Example

The black spot fungus infects rose plants — see page 133 for more.

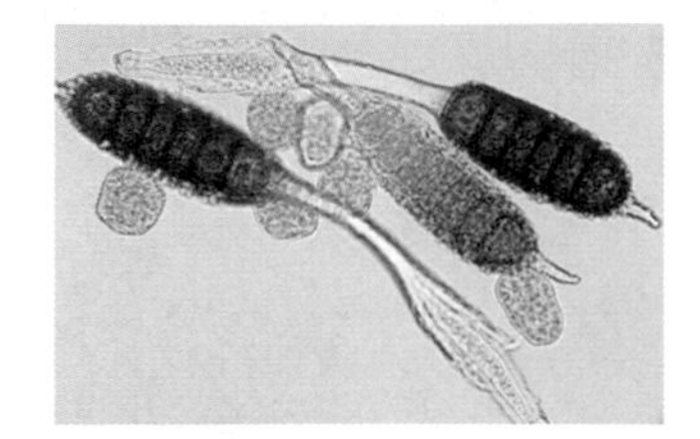

Figure 3: *Black spot fungus spores.*

The spread of pathogens

Pathogens can be spread in many ways. Here are a few that you need to know about.

Water

Some pathogens can be picked up by drinking or bathing in dirty water.

Example

Cholera is a bacterial infection that's spread by drinking water contaminated with the diarrhoea of other sufferers.

Exam Tip

If you're asked in the exam to suggest how someone may have been infected with a disease, just use the information you've been given to work out whether it was likely to be via water, air or by direct contact.

Air

Pathogens can be carried in the air and can then be breathed in.

Example

The influenza virus that causes flu is spread in the air in droplets produced when you cough or sneeze.

Direct contact

Some pathogens can be picked up by touching contaminated surfaces, including the skin.

Example

Athlete's foot is a fungus which makes skin itch and flake off. It's most commonly spread by touching the same things as an infected person, e.g. shower floors and towels.

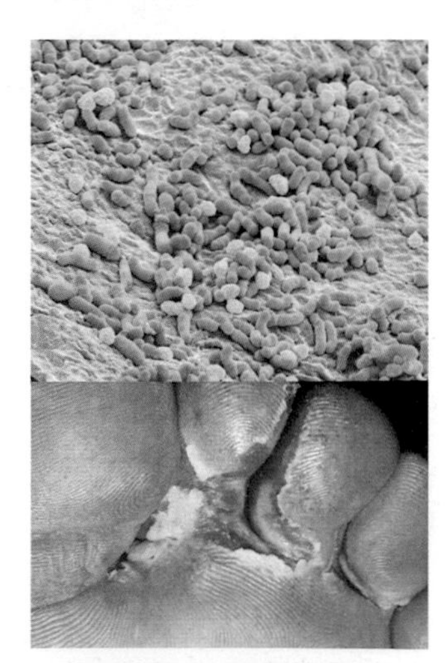

Figure 4: *Athlete's foot fungus on human skin (top). Toes infected with athlete's foot (bottom).*

Practice Questions — Application

Q1 *Clostridium tetani* is the bacterium that causes tetanus. It usually enters the body through deep cuts. Suggest how *Clostridium tetani* bacteria may make you feel ill.

Q2 The chickenpox virus can be carried in droplets in the air. Suggest how an infected student could infect someone else who is sitting in another part of the same classroom.

2. Viral Diseases

So, you know that diseases can be caused by viruses, fungi, protists and bacteria. It's time to look at some in a bit more detail. First up, viral diseases...

Learning Objectives:

- Know how the measles virus is spread.
- Know the symptoms of the measles virus and understand that it can lead to complications and even death in some circumstances.
- Know the symptoms of HIV and that they can be controlled with antiretroviral drugs.
- Understand how the virus attacks the body and how the disease progresses to late stage HIV infection, or AIDS.
- Know how HIV is spread.
- Know that the tobacco mosaic virus affects some plants.
- Know the symptoms of tobacco mosaic virus, and how the growth of the plant is affected.

Specification References
4.3.1.2, 4.3.3.1

Examples of viral diseases

You need to know about these three viral diseases.

Measles

Measles is a viral disease. It is spread by through droplets from an infected person's sneeze or cough. People with measles develop a red skin rash, and they'll show signs of a fever (a high temperature).

Measles can be very serious or even fatal if there are complications. For example, measles can sometimes lead to pneumonia (a lung infection) or a brain infection called encephalitis. Most people are vaccinated against measles when they're young.

HIV

HIV is a virus spread by sexual contact, or by exchanging bodily fluids such as blood. This can happen when people share needles when taking drugs.

HIV initially causes flu-like symptoms for a few weeks. Usually, the person doesn't then experience any symptoms for several years. During this time, HIV can be controlled with antiretroviral drugs. These stop the virus replicating in the body. The virus attacks the immune cells (cells of the immune system — see page 136). If the body's immune system is badly damaged, it can't cope with other infections or cancers. At this stage, the virus is known as late stage HIV infection, or AIDS.

Tobacco mosaic virus

Tobacco mosaic virus (TMV) is a virus that affects many species of plants, e.g. tomatoes. It causes a mosaic pattern on the leaves of the plants — parts of the leaves become discoloured (see Figure 1). The discolouration means the plant can't carry out photosynthesis as well, so the virus affects growth.

Figure 1: *A plant infected with tobacco mosaic virus.*

Practice Questions — Fact Recall

Q1 How is the measles virus spread?

Q2 Give two symptoms of measles.

Q3 Outline how the HIV virus can be spread.

Q4 What type of drugs are used in the early stages of HIV?

Q5 Name one type of plant that can be infected by the tobacco mosaic virus.

Q6 Explain why growth is affected in a plant infected with the tobacco mosaic virus.

3. Fungal and Protist Diseases

Two types of pathogen to contend with on this page — fungi and protists.

Learning Objectives:
- Know that rose black spot is an example of a fungal disease in plants.
- Know the symptoms of rose black spot.
- Understand why being infected with rose black spot affects the growth of a plant.
- Know how rose black spot is spread, and how it can be treated.
- Know that malaria is caused by a protist.
- Know that part of the life cycle of the malarial protist takes place inside a mosquito.
- Be able to give the symptoms of malaria.
- Know how the spread of malaria can be controlled.

Specification References
4.3.1.4, 4.3.1.5, 4.3.3.1

Fungal diseases

Rose black spot

Rose black spot is a fungus that causes purple or black spots to develop on the leaves of rose plants (see Figure 1). The leaves can then turn yellow and drop off. This means that less photosynthesis can happen, so the plant doesn't grow very well. It spreads through the environment in water or by the wind.

Gardeners can treat the disease using fungicides and by stripping the plant of its affected leaves. These leaves then need to be destroyed so that the fungus can't spread to other rose plants.

Figure 1: *A rose plant infected with rose black spot fungus.*

Protist diseases

Malaria

Malaria is caused by a protist (see page 130). Part of the malarial protist's life cycle takes place inside the mosquito. The mosquitoes are vectors — they pick up the malarial protist when they feed on an infected animal. Every time the mosquito feeds on another animal, it infects it by inserting the protist into the animal's blood vessels.

Malaria causes repeating episodes of fever. It can be fatal. The spread of malaria can be reduced by stopping the mosquitoes from breeding.

Example

Mosquitoes can breed in standing water, so mosquitoes can be prevented from breeding by removing these water sources. For example, people who live in areas with a risk of malaria should make sure that they empty or get rid of any containers on their land that fill with rainwater.

People can also be protected from mosquito bites to stop the spread of the disease.

Examples

- Spraying any exposed skin with an insect repellant can prevent people from being bitten by mosquitoes.
- Sleeping under a mosquito net can prevent people from being bitten when they're asleep.

Practice Questions — Fact Recall

Q1 Describe the signs that suggest that a plant has been infected with black spot fungus.

Q2 Why are mosquito nets used in areas with a high risk of malaria?

4. Bacterial Diseases and Preventing Disease

Learning Objectives:
- Know that *Salmonella* is a type of bacteria that causes food poisoning.
- Know the symptoms of *Salmonella* food poisoning.
- Know how *Salmonella* food poisoning is spread and controlled.
- Know that gonorrhoea is a sexually transmitted disease caused by bacteria.
- Know how gonorrhoea is spread.
- Know the symptoms of gonorrhoea.
- Understand why gonorrhoea is not easily treated with penicillin anymore.
- Know how the spread of gonorrhoea is controlled.
- Understand ways that the spread of disease can be reduced or prevented.

Specification References
4.3.1.1, 4.3.1.3

Last but not least, you need to know about bacterial diseases. Then it's on to how we can reduce (or even prevent) the spread of diseases.

Bacterial diseases

Salmonella food poisoning

Salmonella is a type of bacteria that causes food poisoning. Infected people can suffer from fever, stomach cramps, vomiting and diarrhoea. These symptoms are caused by the toxins that the bacteria produce (see page 130).

You can get *Salmonella* food poisoning by eating food that's been contaminated with *Salmonella* bacteria.

Examples
- Eating chicken that caught the disease whilst it was alive.
- Eating food that has been contaminated by being prepared in unhygienic conditions.

In the UK, most poultry (e.g. chickens and turkeys) is given a vaccination against *Salmonella*. This is to control the spread of the disease. The vaccination prevents the poultry from catching the disease, which stops it from being passed on to humans.

Gonorrhoea

Gonorrhoea is a sexually transmitted disease (STD). STDs are passed on by sexual contact, e.g. having unprotected sex. Gonorrhoea is caused by bacteria. A person with gonorrhoea will get pain when they urinate. Another symptom is a thick yellow or green discharge from the vagina or the penis.

Gonorrhoea was originally treated with an antibiotic called penicillin, but this has become trickier now because strains of the bacteria have become resistant to it (see page 140). To prevent the spread of gonorrhoea, people can be treated with other antibiotics and should use barrier methods of contraception (see page 230), such as condoms.

Tip: Being resistant to an antibiotic means that the bacteria won't be killed by it.

Reducing and preventing the spread of disease

There are things that we can do to reduce, and even prevent, the spread of disease. For example:

Being hygienic

Using simple hygiene measures can prevent the spread of disease.

Example

Doing things like washing your hands thoroughly before preparing food or after you've sneezed can stop you infecting another person.

Destroying vectors

Some protist diseases are carried by vectors (see page 130). These spread the disease to humans. By getting rid of these organisms that spread disease, you can prevent the disease from being passed on.

Example

Vectors that are insects can be killed using insecticides or by destroying their habitat so that they can no longer breed.

Exam Tip
How to prevent the spread of a disease is linked to how it spreads. For example, if a disease is spread by direct contact, the way to prevent it will be to prevent infected people coming into contact with people who aren't infected.

Isolating infected individuals

If you isolate someone who has a communicable disease, it prevents them from passing it on to anyone else.

Example

Ebola is a virus spread by direct contact with the bodily fluids of a person with the disease. In an Ebola outbreak, patients with the disease are isolated to prevent the spread of the disease, and medical staff wear protective clothing (see Figure 1).

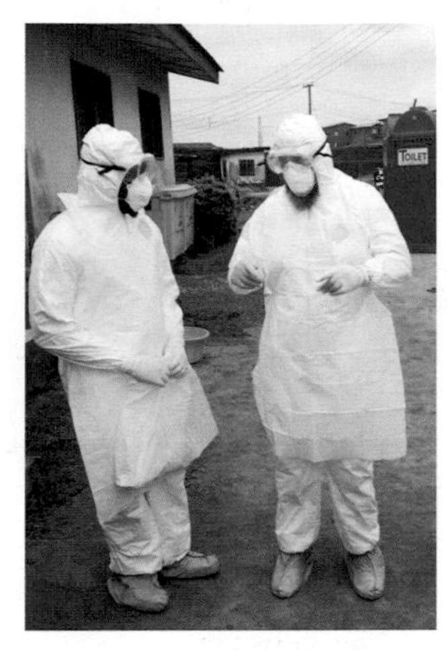

Figure 1: *Medical staff dressed in protective clothing whilst treating Ebola patients.*

Vaccination

Vaccinating people and animals against communicable diseases means that they can't develop the infection and then pass it on to someone else. There's more about how vaccination works on page 138.

Example

Children often get vaccinated against common diseases — for example, they are often given the meningitis C vaccine at around 3 months of age. The meningitis B vaccine is usually given at 2 months, 4 months, and 12 months of age.

Tip: Remember, only communicable diseases can be spread from person-to-person — non-communicable diseases can't.

Practice Questions — Fact Recall

Q1 List three symptoms of *Salmonella* food poisoning.

Q2 Suggest one way that *Salmonella* food poisoning is spread.

Q3 How is gonorrhoea treated?

Q4 How does destroying vectors prevent the spread of disease?

Practice Questions — Application

Q1 A patient in a hospital has a communicable disease where they need to be put into isolation. Suggest why they have been put into isolation.

Q2 A nurse working in a GP surgery washes his hands carefully after seeing patients. Suggest how this will prevent the spread of disease.

5. Fighting Disease

Learning Objectives:
- Be able to explain the ways that the body defends itself against the entry of pathogens using non-specific defence systems.
- Understand the role of the immune system and how the white blood cells help to defend the body against pathogens and destroy them.

Specification Reference 4.3.1.6

Part of being healthy means being free from disease. But that's easier said than done...

The body's defences

In order to cause disease, pathogens first need to enter the body. The human body has got features that stop a lot of nasties getting inside in the first place. These are the body's non-specific defence systems.

Examples

- The skin acts as a barrier to pathogens. It also secretes antimicrobial substances which kill pathogens.
- Hairs and mucus in your nose trap particles that could contain pathogens.
- The trachea and bronchi (breathing pipework — see page 76) secrete mucus to trap pathogens.
- The trachea and bronchi are lined with cilia. These are hair-like structures, which waft the mucus up to the back of the throat where it can be swallowed.
- The stomach produces hydrochloric acid. This kills pathogens that make it that far from the mouth.

Figure 1: *Bacteria (yellow) sticking to mucus (blue) on microscopic hairs in the nose.*

The immune system

If pathogens do make it into your body, your immune system kicks in to destroy them. The most important part of your immune system is the **white blood cells**. They travel around in your blood and crawl into every part of you, constantly patrolling for microbes. When they come across an invading microbe they have three lines of attack.

1. Consuming them

White blood cells can engulf foreign cells and digest them. This is called phagocytosis.

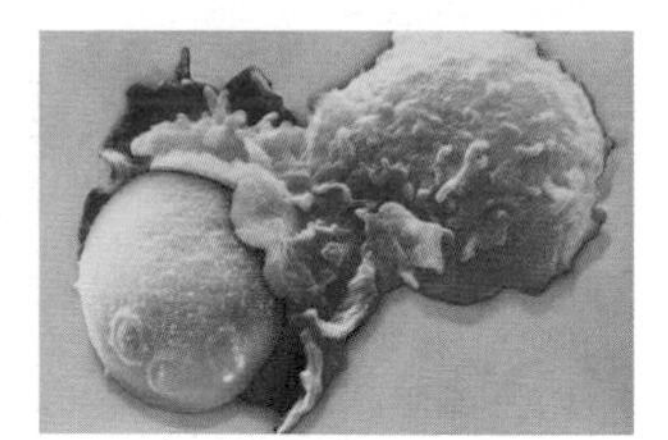

Figure 2: *A white blood cell (blue/pink) engulfing a pathogen (yellow/green). Seen under a microscope.*

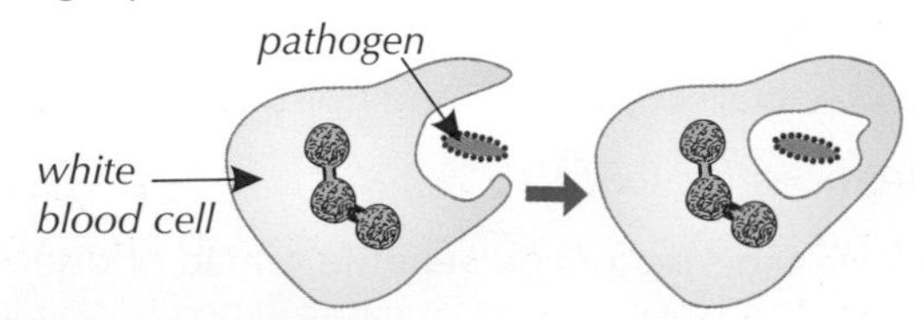

Figure 3: *A white blood cell engulfing a pathogen.*

2. Producing antitoxins

These counteract toxins produced by the invading bacteria.

3. Producing antibodies

Every invading pathogen has unique molecules (called **antigens**) on its surface. When your white blood cells come across a foreign antigen (i.e. one they don't recognise), they will start to produce proteins called **antibodies** to lock onto the invading cells so that they can be found and destroyed by other white blood cells.

Exam Tip
Don't be tempted to write 'white blood cells engulf disease' in the exam — white blood cells engulf pathogens (which cause disease).

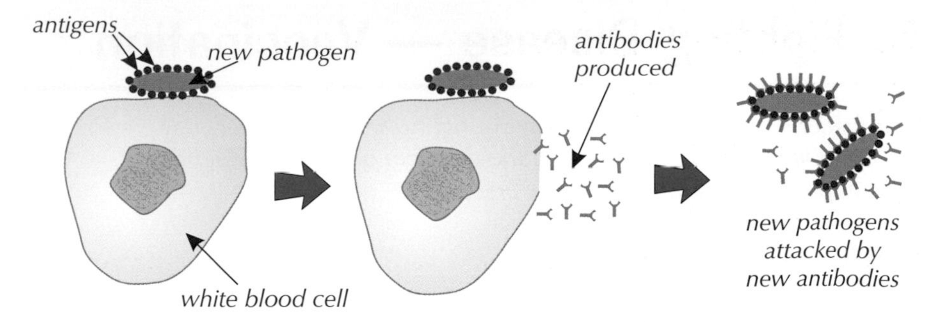

Figure 4: *Diagram showing the production of antibodies.*

The antibodies produced are specific to that type of antigen — they won't lock on to any others. Antibodies are then produced rapidly and carried around the body to find all similar bacteria or viruses.

If the person is infected with the same pathogen again the white blood cells will rapidly produce the antibodies to kill it — the person is naturally immune to that pathogen and won't get ill.

Exam Tip
Don't get antigens and antibodies mixed up in the exam. Remember, the body makes antibodies against the antigens on pathogens.

Tip: The white blood cells that produce antibodies are also known as B-lymphocytes.

Practice Questions — Fact Recall

Q1 Give two ways in which the skin helps to defend the body against disease.

Q2 Give three ways in which white blood cells help to defend the body against disease.

Practice Questions — Application

Q1 The SARS coronavirus (SARS-CoV) causes SARS, a respiratory disease. It can enter the body by being inhaled in droplets. Describe one feature of the body that helps to prevent SARS-CoV from entering the body.

Q2 As a child, John suffered from measles. Measles are caused by a virus.

a) Explain why John is unlikely to get ill with measles again.

b) German measles are caused by a different virus, which John has never had before. Explain why John could still become ill from German measles.

Exam Tip
In the exam, you might get asked questions about pathogens or diseases you've never heard of before. Don't let that put you off though — you should be able to work out the answers by applying what you already know.

6. Fighting Disease — Vaccination

Learning Objectives:
- Understand how vaccinations work.
- Understand that if enough people are immune to a particular pathogen, the spread of the pathogen is reduced.

Specification Reference 4.3.1.7

Vaccinations have changed the way we fight disease. We don't always have to deal with the problem once it's happened — we can prevent it happening in the first place.

What are vaccinations?

When you're infected with a new pathogen, it takes your white blood cells a few days to learn how to deal with it. But by that time, you can be pretty ill. Vaccinations can stop you feeling ill in the first place.

Vaccinations involve injecting small amounts of dead or inactive pathogens. These carry antigens, which cause your white blood cells to produce antibodies to attack them — even though the pathogen is harmless (since it's dead or inactive).

Example

The MMR vaccine is given to children. It contains weakened versions of the viruses that cause measles, mumps and rubella (German measles) all in one vaccine.

Tip: The 'weakened' viruses in the MMR vaccine are still alive, but aren't able to cause disease.

If live pathogens of the same type then appear at a later date, the white blood cells can rapidly mass-produce antibodies to kill off the pathogen. The vaccinated person is now immune to that pathogen and won't get ill (see Figure 1).

Tip: Remember, antibodies are specific to a particular type of pathogen (see p. 137). So a vaccination against the typhoid virus will only protect you against typhoid — it won't make you immune to anything else.

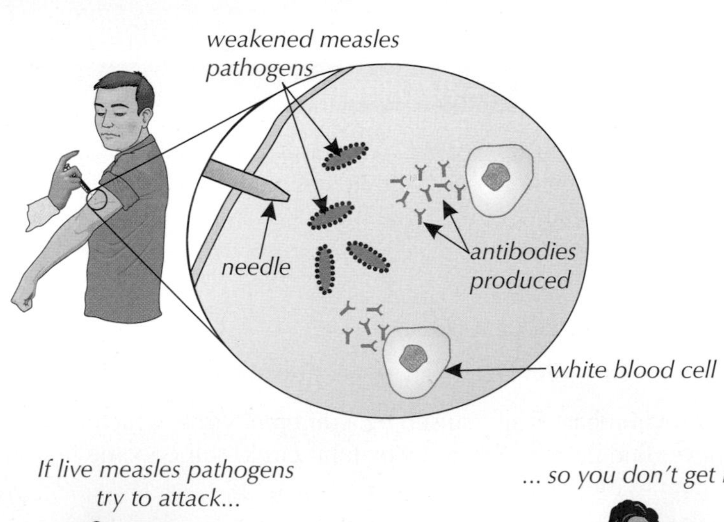

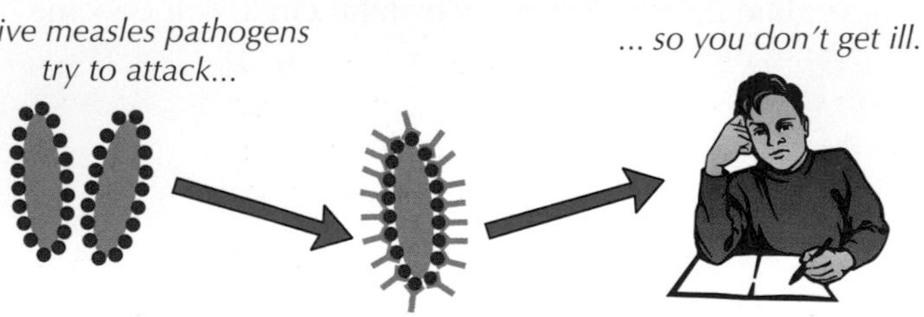

Figure 1: *A diagram to show how vaccination works.*

Tip: Some vaccinations "wear off" over time. So booster injections may need to be given to increase levels of antibodies again.

Vaccinations — pros and cons

There are both pros and cons of vaccinations.

Pros

- Vaccines have helped control lots of infectious diseases that were once common in the UK (e.g. polio, measles, whooping cough, rubella, mumps, tetanus...). Smallpox no longer occurs at all, and polio infections have fallen by 99%.
- Big outbreaks of disease — called **epidemics** — can be prevented if a large percentage of the population is vaccinated. That way, even the people who aren't vaccinated are unlikely to catch the disease because there are fewer people able to pass it on. But if a significant number of people aren't vaccinated, the disease can spread quickly through them and lots of people will be ill at the same time.

Exam Tip
In the exam, you might get asked to 'evaluate' the pros and cons of vaccinations. This means giving a balanced account of both the advantages and the disadvantages, before trying to come up with an overall judgement.

Cons

- Vaccines don't always work — sometimes they don't give you immunity.
- You can sometimes have a bad reaction to a vaccine (e.g. swelling, or maybe something more serious like a fever or seizures). But bad reactions are very rare.

Balancing the risks

Deciding whether to have a vaccination means balancing risks — the risk of catching the disease if you don't have a vaccine, against the risk of having a bad reaction if you do. As always, you need to look at the evidence.

Example

If you get measles (the disease), there's about a 1 in 15 chance that you'll get complications (e.g. pneumonia) — and about 1 in 500 people who get measles actually die. However, the number of people who have a problem with the vaccine is more like 1 in 1 000 000.

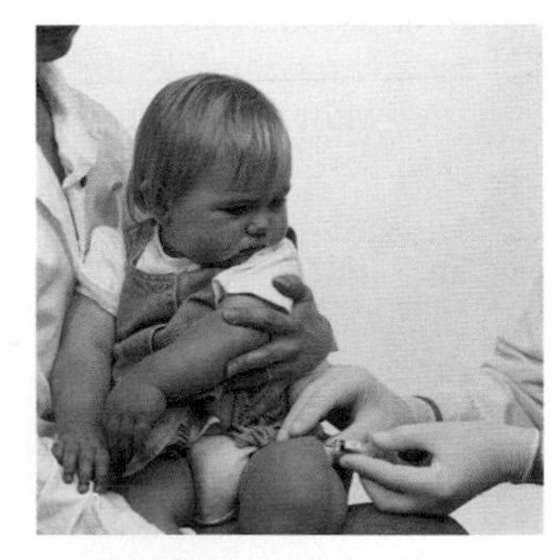

Figure 2: *A baby being given the MMR vaccine.*

Practice Question — Application

Q1 Whooping cough is a severe cough that can cause serious breathing difficulties in young babies.

In 2012, the US state of Washington declared a whooping cough epidemic. The Department of Health encouraged all adults to get booster vaccinations against the disease, to help protect babies who were too young to receive the vaccination.

a) Suggest how a whooping cough vaccination might work.

b) Explain how a large-scale vaccination of adults could help to protect babies who haven't been vaccinated.

7. Fighting Disease — Drugs

There aren't vaccinations against every pathogen (not just yet, anyway) so you're still going to get ill sometimes. And that's when drugs come in...

Learning Objectives:

- Know that painkillers and some other drugs can relieve the symptoms of a disease, but don't kill the pathogens.
- Understand how antibiotics can be used to treat disease.
- Understand that antibiotics (e.g. penicillin) are drugs used to kill bacteria.
- Understand that different antibiotics kill different bacteria and that antibiotics have no effect on viruses.
- Understand why it's hard to produce drugs to kill viruses.
- Know that the use of antibiotics has greatly reduced deaths from communicable diseases.
- Understand that antibiotic-resistant strains of bacteria can develop, and that this is a problem.
- Know that drugs were once extracted from plants and from microorganisms, including digitalis, aspirin and penicillin.
- Know that many new drugs are now synthesised by chemists working in pharmaceutical companies, but they may still begin with a plant chemical.

Specification References
4.3.1.8, 4.3.1.9

Different types of drugs

Painkillers (e.g. aspirin) are drugs that relieve pain. However, they don't actually tackle the cause of the disease, they just help to reduce the symptoms. Other drugs do a similar kind of thing — reduce the symptoms without tackling the underlying cause. For example, lots of "cold remedies" don't actually cure colds — they have no effect on the cold virus itself.

Antibiotics (e.g. **penicillin**) work differently — they actually kill (or prevent the growth of) the bacteria causing the problem without killing your own body cells. Different antibiotics kill different types of bacteria, so it's important to be treated with the right one.

But antibiotics don't destroy viruses (e.g. flu or cold viruses). Viruses reproduce using your own body cells, which makes it very difficult to develop drugs that destroy just the virus without killing the body's cells.

The use of antibiotics has greatly reduced the number of deaths from communicable diseases caused by bacteria.

Example

During the First World War, nearly 20% of US soldiers who caught pneumonia (a bacterial infection) died as a result. During World War II, this figure dropped to less than 1% thanks to treatment with penicillin.

Antibiotic resistance

Bacteria can **mutate** — sometimes the mutations cause them to be resistant to (not killed by) an antibiotic.

Examples

- **MRSA** (methicillin-resistant *Staphylococcus aureus*) causes serious wound infections and is resistant to the powerful antibiotic, methicillin.
- Some strains of the bacteria that cause TB (a lung disease) are now resistant to several of the antibiotics that would normally be used to fight them.

If you have an infection, some of the bacteria might be resistant to antibiotics. This means that when you treat the infection, only the non-resistant strains of bacteria will be killed. The individual resistant bacteria will survive and reproduce, and the population of the resistant strain will increase. This is an example of **natural selection** (see page 287). This resistant strain could cause a serious infection that can't be treated by antibiotics.

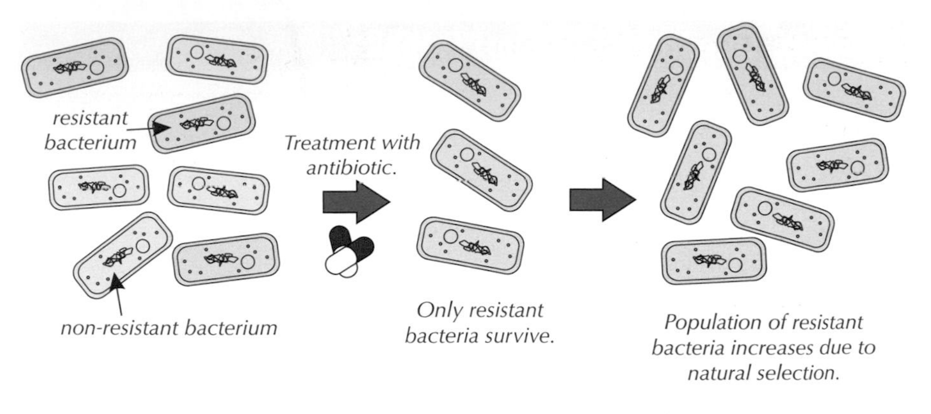

Figure 1: *How antibiotic resistance develops.*

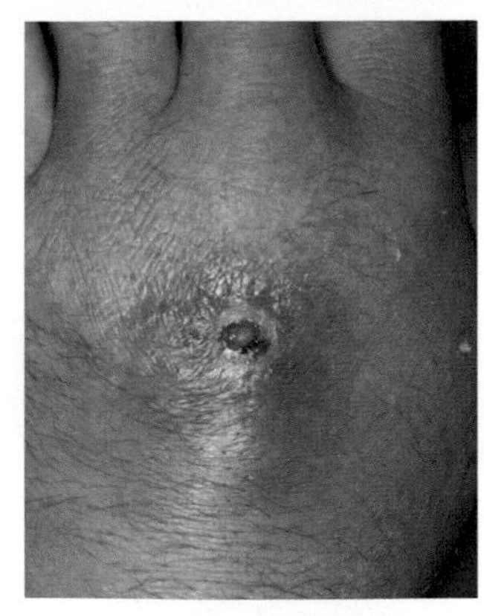

Figure 2: *A wound infected with MRSA.*

Slowing down the development of antibiotic resistance

To slow down the rate of development of resistant strains, it's important for doctors to avoid over-prescribing antibiotics. So you won't get them for a sore throat, only for something more serious. It's also important that you finish the whole course of antibiotics and don't just stop once you feel better.

The origins of drugs

Plants produce a variety of chemicals to defend themselves against pests and pathogens (see page 152). Some of these chemicals can be used as drugs to treat human diseases or relieve symptoms. A lot of our current medicines were discovered by studying plants used in traditional cures.

Examples

- Aspirin is used as a painkiller and to lower fever. It was developed from a chemical found in willow.
- Digitalis is used to treat heart conditions. It was developed from a chemical found in foxgloves.

Some drugs were extracted from microorganisms.

Example

Alexander Fleming was clearing out some Petri dishes containing bacteria. He noticed that one of the dishes of bacteria also had mould on it and the area around the mould was free of the bacteria. He found that the mould (called *Penicillium notatum*) on the Petri dish was producing a substance that killed the bacteria — this substance was penicillin.

Tip: A Petri dish is a circular dish with a lid. Samples of bacteria can be grown in them on a layer of agar jelly — see Required Practical 2 on pages 40-41.

These days, drugs are made on a large scale in the pharmaceutical industry — they're synthesised by chemists in labs. However, the process still might start with a chemical extracted from a plant.

Example

The World Health Organisation has a list of drugs that it considers to be essential for a basic health care system. In 2000, 11% of the 252 drugs were plant-based.

Tip: There's more about how drugs are developed on pages 143-144.

Practice Questions — Fact Recall

Q1 True or false?
Painkillers can be used to tackle the cause of an infection.

Q2 What are antibiotics? Give one example.

Q3 Why is it important to be treated with the right antibiotic for a particular infection?

Q4 Explain why it can be difficult to develop drugs that kill viruses.

Q5 Why is it important not to over-use antibiotics?

Q6 Which drug originates from willow?

Q7 Who discovered penicillin, and how was it discovered?

Practice Question — Application

Q1 Chloe is suffering from the flu, which is caused by a virus.

a) Explain why Chloe's doctor will not prescribe her antibiotics.

b) Chloe's flu is causing a headache. Chloe's doctor recommends taking a painkiller. Suggest why the painkiller will not help to clear Chloe's infection any quicker.

Tip: To answer part b), think about what the flu remedy will and won't do.

8. Developing Drugs

Before drugs make it onto the pharmacy shelves, they have to be tested...

Learning Objectives:
- Understand why new medicinal drugs have to be tested before becoming available to the public (to check safety and effectiveness).
- Know that during drug trials, new drugs are tested for efficacy and toxicity, and to find the optimum dose.
- Be able to describe the process of preclinical testing of new drugs.
- Be able to describe what happens when drugs are tested in clinical trials.
- Know what the terms 'optimum dose', 'placebo' and 'double-blind trials' mean.

Specification Reference 4.3.1.9

Drug testing

New drugs are constantly being developed. But before they can be given to the general public, they have to go through a thorough testing procedure to make sure they are safe and effective. There are three main stages in drug testing — preclinical testing on human cells and tissues, preclinical testing on live animals, and clinical testing on human volunteers.

Preclinical trials

Cells and tissues

In preclinical testing, drugs are tested on human cells and tissues in the lab. However, you can't use human cells and tissues to test drugs that affect whole or multiple body systems, e.g. testing a drug for blood pressure must be done on a whole animal because it has an intact circulatory system.

Live animals

The next step in preclinical testing is to test the drug on live animals. This is to test **efficacy** (whether the drug works and produces the effect you're looking for), to find out about its **toxicity** (how harmful it is) and to find the best dosage (the concentration that should be given, and how often it should be given).

The law in Britain states that any new drug must be tested on two different live mammals. Some people think it's cruel to test on animals, but others believe this is the safest way to make sure a drug isn't dangerous before it's given to humans. Other people think that animals are so different from humans that testing on animals is pointless.

Figure 1: A piece of tissue from a human intestine being used in drug testing.

Clinical trials

If the drug passes the tests on animals then it's tested on human volunteers in a clinical trial.

First, the drug is tested on healthy volunteers. This is to make sure that it doesn't have any harmful side effects when the body is working normally. At the start of the trial, a very low dose of the drug is given and this is gradually increased.

If the results of the tests on healthy volunteers are good, the drugs can be tested on people suffering from the illness. The **optimum dose** is found — this is the dose of drug that is the most effective and has few side effects.

Placebos

To test how well the drug works, patients are randomly put into two groups. One is given the new drug, the other is given a **placebo** (a substance that's like the drug being tested but doesn't do anything). This is so the doctor can see the actual difference the drug makes — it allows for the placebo effect (when the patient expects the treatment to work and so feels better, even though the treatment isn't doing anything).

Tip: A placebo is usually a tablet or an injection with no drug added.

Tip: People in the control group of a clinical trial should be similar to people in the treatment group, e.g. of a similar age, gender, etc. This helps to make sure the study is a fair test — see page 10 for more on making a study a fair test.

WORKING SCIENTIFICALLY

Example

A pharmaceutical company has developed a drug, Drug X, to help people with high blood cholesterol to lower their cholesterol level. The drug was tested on 1500 patients with high cholesterol. The patients were divided into two groups and over six months each patient had one tablet per day. One group had a tablet containing Drug X and the other group had a tablet without Drug X — the placebo.

Blind and double-blind trials

Clinical trials are **blind** — the patient in the study doesn't know whether they're getting the drug or the placebo. In fact, they're often **double-blind** — neither the patient nor the doctor knows until all the results have been gathered (see Figure 2). This is so the doctors monitoring the patients and analysing the results aren't subconsciously influenced by their knowledge.

	Blind trial	Double-blind trial
Does the **patient** know whether they're getting the drug or the placebo?	no	no
Does the **doctor** know whether the patient is getting the drug or the placebo?	yes	no

Figure 2: *Table summarising blind and double-blind trials.*

Tip: Peer review is when other scientists check that the work is valid and has been carried out rigorously — see page 2.

WORKING SCIENTIFICALLY

The results of drug testing and drug trials aren't published until they've been through peer review. This helps to prevent false claims.

The importance of drug testing

It's really important that drugs are tested thoroughly before being used to make sure they're safe. An example of what can happen when drugs are not thoroughly tested is the case of thalidomide...

Example

Thalidomide is a drug that was developed in the 1950s. It was intended as a sleeping pill, and was tested for that use. But later it was also found to be effective in relieving morning sickness in pregnant women.

Unfortunately, thalidomide hadn't been tested as a drug for morning sickness, and so it wasn't known that it could pass through the placenta and affect the fetus, causing abnormal limb development. In some cases, babies were born with no arms or legs at all. About 10 000 babies were affected by thalidomide, and only about half of them survived.

The drug was banned, and more rigorous testing procedures were introduced. More recently thalidomide has been used in the treatment of leprosy and other diseases, e.g. some cancers.

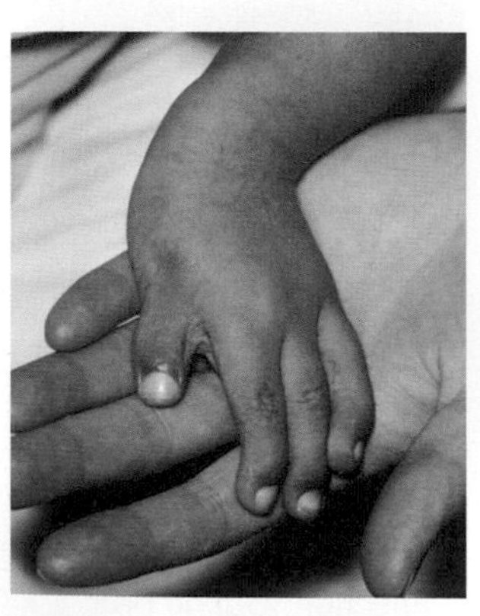

Figure 3: *This baby was born with a deformed hand due to thalidomide.*

Practice Questions — Fact Recall

Q1 What are drugs tested on in preclinical testing?

Q2 Give two things that drug testing aims to find out.

Q3 a) In a clinical trial, what type of volunteer is the drug tested on first — healthy volunteers or people suffering from the illness?

b) Describe the dosage of drug that is first used in a clinical trial.

Q4 What is a placebo?

Q5 Briefly explain what is involved in a double-blind trial.

Exam Tip
If you're asked what a placebo is in the exam, remember that it's not a drug — this is because it doesn't actually do anything.

Practice Questions — Application

Q1 For each of the drugs below, suggest what would have been given as a placebo when that drug went through clinical trials.

a) a paracetamol capsule

b) a steroid inhaler

c) a cortisone injection

Q2 A new weight-loss pill has been developed called Drug X. The pill was tested in a double-blind clinical trial. It involved three groups of obese volunteers (50 per group), all of whom wanted to lose weight.
Each group member took a pill three times a day for a year, alongside diet and exercise. The first group took Drug X, the second group took a similar weight-loss pill already available (Drug Y) and the third group took a placebo. The average weight loss per person after one year is shown on the graph.

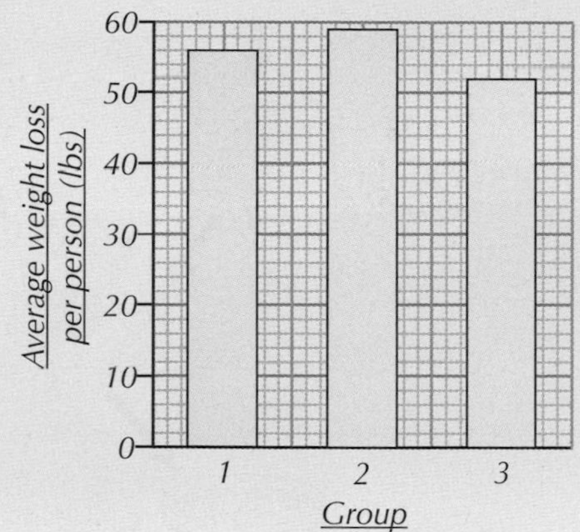

a) i) Would the doctors involved in this trial have known which patients were being given the placebo? Explain your answer.

ii) Suggest what could have been used as a placebo in this trial.

b) Why do you think Group 2 was included in the trial?

c) From these results, what can you conclude about Drug X? Use data from the graph to support your answer.

Exam Tip
When answering a question about a clinical trial in the exam, make sure you read all the information you're given. Here you need to have spotted that it's a double-blind trial — this will help you answer Q2 a) i).

9. Monoclonal Antibodies Higher

Learning Objectives:

- H Know that monoclonal antibodies are made by cloning cells, and that they will only bind to one specific protein antigen and so can be used to target specific cells or chemicals.
- H Be able to describe the process for producing monoclonal antibodies from mouse lymphocytes and a tumour cell to make a hybridoma cell which is then cloned.

Specification Reference 4.3.2.1

Scientists can mass-produce antibodies for all sorts of uses. They're called monoclonal antibodies, and can be engineered to bind to anything you want.

Producing monoclonal antibodies

Antibodies are produced by B-lymphocytes — a type of white blood cell (see page 87). **Monoclonal antibodies** are produced from lots of clones of a single white blood cell (a lymphocyte). This means all the antibodies are identical and will only target one specific protein antigen. However, you can't just grab the lymphocyte that made the antibody and grow more — lymphocytes don't divide very easily. Tumour cells, on the other hand, don't produce antibodies but divide lots — so they can be grown really easily. It's possible to fuse a mouse B-lymphocyte with a tumour cell to create a cell called a **hybridoma**. Hybridoma cells can be cloned to get lots of identical cells. These cells all produce the same antibodies (monoclonal antibodies). The antibodies can be collected and purified.

Tip: H A B-lymphocyte is just one type of lymphocyte.

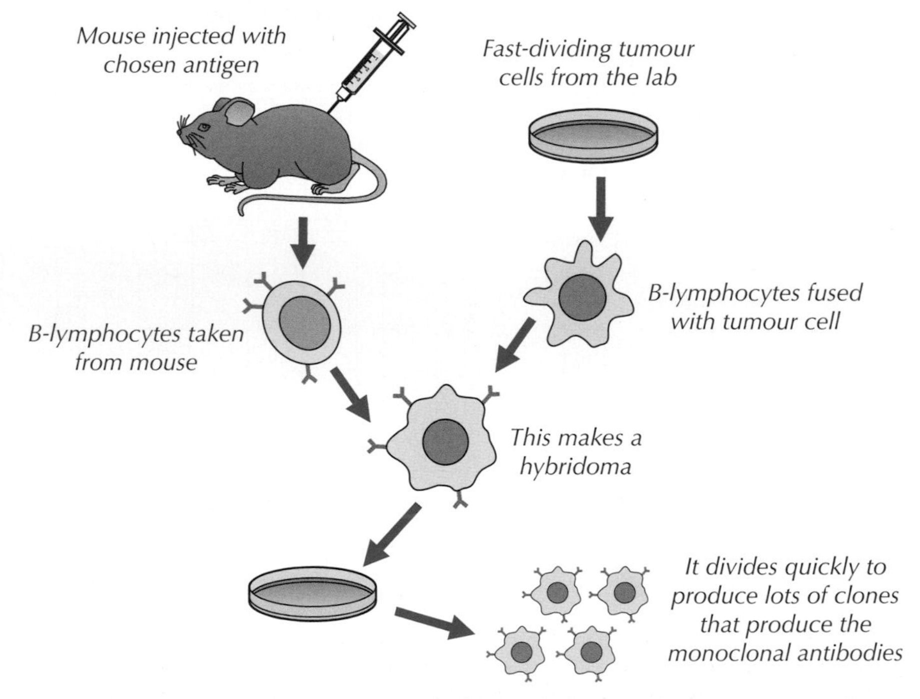

Figure 1: *The process of producing monoclonal antibodies.*

Tip: H White blood cells are involved in the body's immune response. They attack pathogens in many ways — by engulfing them, by producing antitoxins and by producing antibodies. (see page 136 for more).

You can make monoclonal antibodies that bind to anything you want, e.g. an antigen that's only found on the surface of one type of cell. Monoclonal antibodies are really useful because they will only bind to (target) this molecule — this means you can use them to target a specific cell or chemical in the body.

Exam Tip H
Make sure you know the process of producing monoclonal antibodies inside-out — you could be asked to describe it in the exam.

Practice Questions — Fact Recall

Q1 What two things are fused to make a hybridoma cell?

Q2 What happens in monoclonal antibody production after a hybridoma cell has been made?

10. Monoclonal Antibody Uses Higher

With their ability to bind to a specific chemical or pathogen, monoclonal antibodies are useful for lots of things. These pages just cover a few of the things they're used for — read on to find out...

Pregnancy tests

Monoclonal antibodies can be used to diagnose diseases and other health conditions. For example, monoclonal antibodies are used in pregnancy tests. A hormone called HCG is found in the urine of women only when they are pregnant. Pregnancy testing sticks use monoclonal antibodies to detect this hormone.

Here's how they work:

The test stick

The bit of the stick you wee on has some antibodies to the hormone, with blue beads attached. The test strip (the bit of the stick that turns blue if you're pregnant) has some more antibodies to the hormone stuck onto it (so that they can't move).

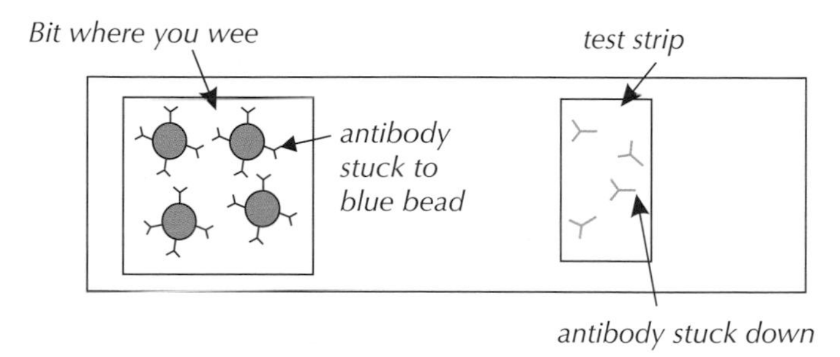

Figure 1: *Diagram showing a pregnancy test before being used.*

If you're pregnant

If you're pregnant and you wee on the stick, the hormone binds to the antibodies on the blue beads. The urine moves up the stick, carrying the hormone and the beads. The beads and hormone bind to the antibodies on the strip. So the blue beads get stuck on the strip, turning it blue.

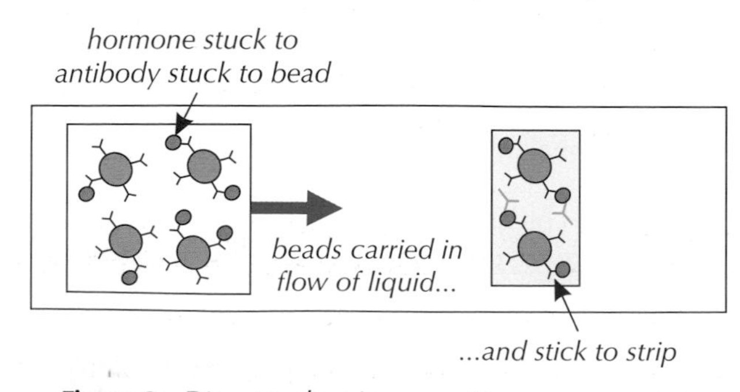

Figure 2: *Diagram showing a positive pregnancy test.*

Learning Objectives:

- H Be able to describe how monoclonal antibodies can be used for diagnosis, e.g. in pregnancy tests.
- H Be able to describe how monoclonal antibodies can be used to treat disease, and understand how they are used to treat cancer.
- H Be able to describe how monoclonal antibodies can be used to measure the levels of certain hormones and chemicals in blood samples.
- H Be able to describe how monoclonal antibodies can be used to detect pathogens.
- H Be able to describe how monoclonal antibodies can be used to locate specific molecules using a fluorescent dye.
- H Understand why monoclonal antibodies are not yet used as widely as expected.

Specification Reference 4.3.2.2

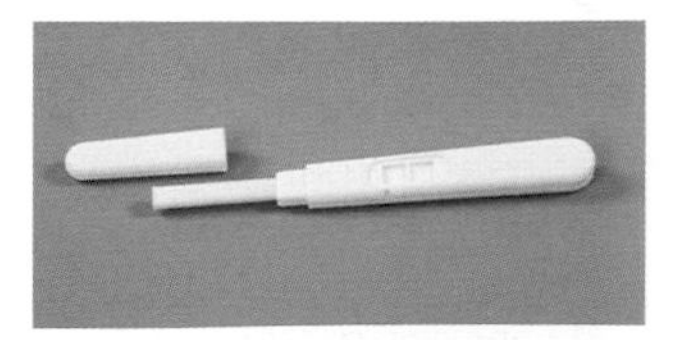

Figure 3: *A pregnancy test. The left end is where urine is applied. The window halfway along shows the test strip and the control strip (see next page).*

Tip: **H** A 'positive' test is one where the woman is pregnant. A 'negative' test means that the woman is not pregnant.

Tip: **H** Remember, monoclonal antibodies will only attach to one specific protein antigen or chemical.

If you're not pregnant

If you're not pregnant and you wee on the stick, the urine still moves up the stick, carrying the blue beads. But there's nothing to stick the blue beads onto the test strip, so it doesn't go blue.

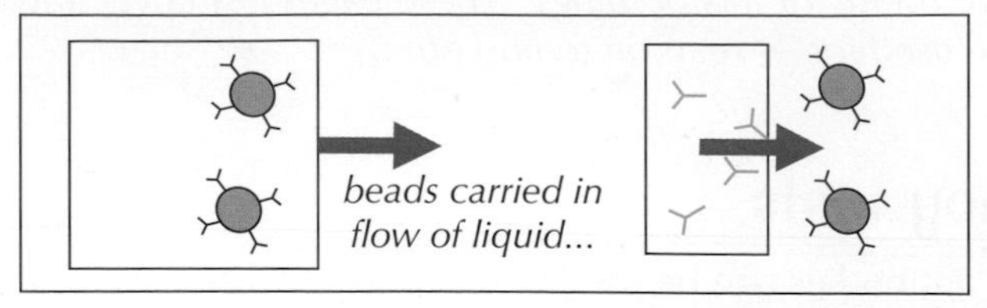

Figure 4: *Diagram showing a negative pregnancy test.*

Control window

Pregnancy tests also have a control window. This shows the person using the test whether the test has worked correctly. The control window contains antibodies to the antibodies that are attached to the blue beads. This means that once the blue beads reach the control window, the antibodies attached to them will bind to the antibodies stuck to the control window. The control window should go blue both when a pregnancy test is positive as well as negative. If it doesn't, it means the test has not worked.

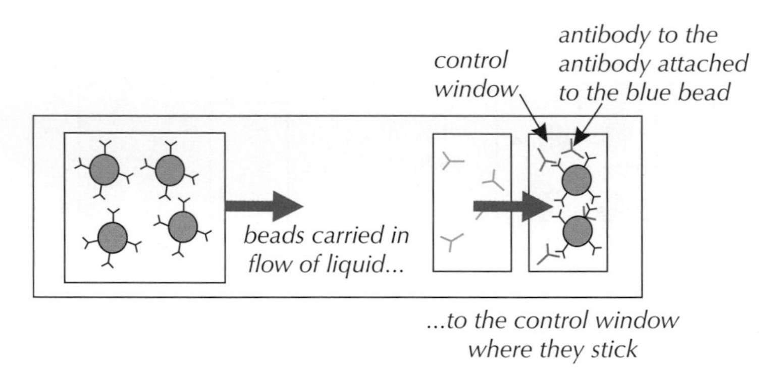

Figure 5: *Diagram showing the control window in a negative test.*

Treating diseases

Different cells in the body have different antigens on their cell surface. So you can make monoclonal antibodies that will bind to specific cells in the body (e.g. just liver cells). Cancer cells have antigens on their cell membranes that aren't found on normal body cells. They're called tumour markers. In the lab, you can make monoclonal antibodies that will bind to these tumour markers.

An anti-cancer drug can be attached to these monoclonal antibodies. This might be a radioactive substance, a toxic drug or a chemical which stops cancer cells growing and dividing. The antibodies are given to the patient through a drip. The antibodies target specific cells (the cancer cells) because they only bind to the tumour markers. The drug kills the cancer cells but doesn't kill any normal body cells near the tumour.

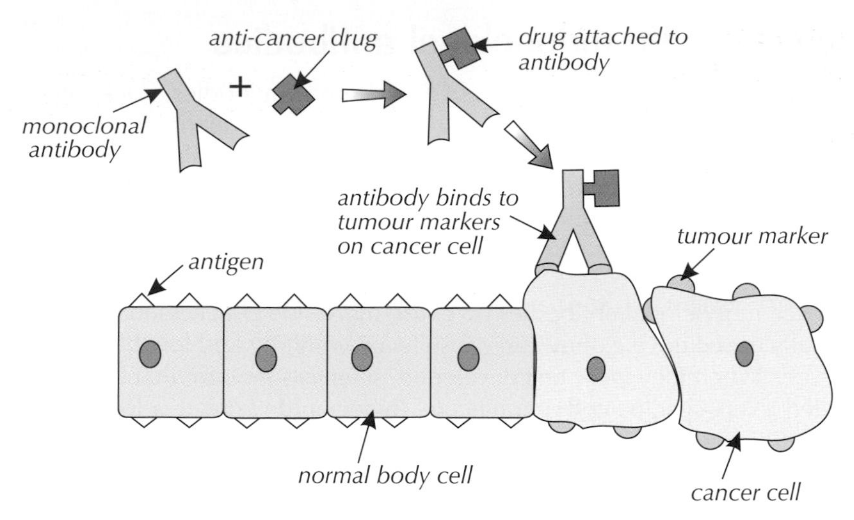

Figure 6: *Monoclonal antibodies being used in cancer treatment.*

Using monoclonal antibodies in laboratories

Monoclonal antibodies can be used to bind to hormones and other chemicals in blood to measure their levels.

Example **Higher**

A technique known as the ELISA test can be used to detect banned drugs in athletes' urine.

They're also used to test blood samples in laboratories for certain pathogens.

Example **Higher**

Different strains of the pathogen that causes gonorrhoea (page 134) can be identified using monoclonal antibodies.

Exam Tip **H**

In the exam, you won't be expected to explain any specific uses of monoclonal antibodies on your own. Instead, you'll be given information about a test or treatment — just use this to figure out how the monoclonal antibodies are being used.

Using monoclonal antibodies in research

Monoclonal antibodies can be used to locate specific molecules on a cell or in a tissue. First, monoclonal antibodies are made that will bind to the specific molecules you're looking for. The antibodies are then bound to a fluorescent dye. If the molecules are present in the sample you're analysing, the monoclonal antibodies will attach to them, and they can be detected using the dye.

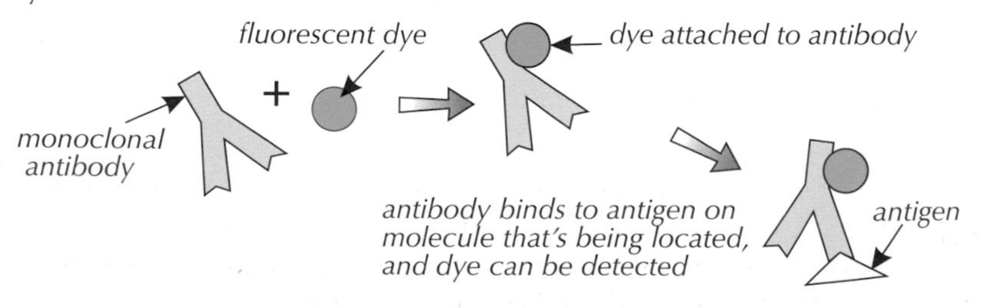

Figure 8: *Using monoclonal antibodies with fluorescent dye to detect molecules.*

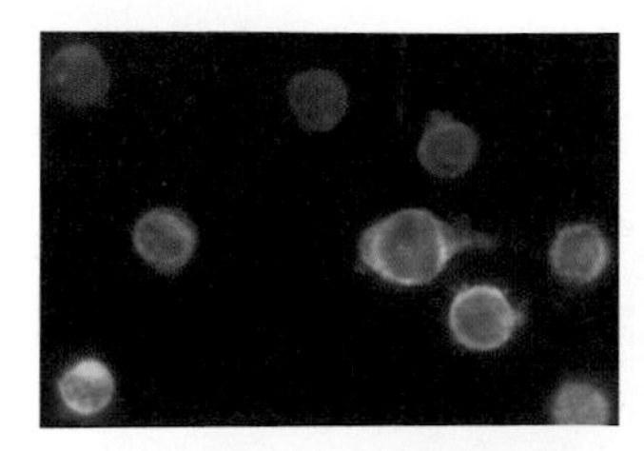

Figure 7: *Cancer cells that have been targeted by monoclonal antibodies that have fluorescent dye attached.*

Problems with monoclonal antibodies

There are some obvious advantages of monoclonal antibodies. One big one is in cancer treatment. Other cancer treatments (like standard chemotherapy and radiotherapy) can affect normal body cells as well as killing cancer cells, whereas monoclonal antibodies target specific cells. This means the side effects of an antibody-based drug are lower than for standard chemotherapy or radiotherapy.

Tip: H Suffering from really unpleasant side-effects can outweigh the benefits of a treatment.

However, monoclonal antibodies do cause more side effects than were originally expected, e.g. they can cause fever, vomiting and low blood pressure. When they were first developed, scientists thought that because they targeted a very specific cell or molecule, they wouldn't create a lot of side effects. This means that they are not as widely used as treatments as scientists had originally thought they might be.

Practice Questions — Fact Recall

Q1 In a positive pregnancy test, what does the monoclonal antibody bind to?

Q2 Name a disease that can be treated using monoclonal antibodies.

Q3 Give one problem associated with monoclonal antibodies.

Practice Questions — Application

Q1 An athletics committee is concerned that some athletes are using banned performance-enhancing drugs during a competition.

How could monoclonal antibodies with fluorescent dye attached to them be used to identify the athletes using banned drugs?

Q2 In a pregnancy test, urine is applied to an area that contains antibodies to a hormone that is present in urine during pregnancy. These antibodies are attached to blue beads.

A control window on the test contains antibodies to the antibodies attached to the blue beads.

Suggest how the control window shows whether or not the pregnancy test has worked.

11. Plant Diseases and Defences

Plants aren't just affected by diseases — they can also be up against pests who either want to eat them or live on them. Luckily, they have some defences.

Learning Objectives:

- Know that plants can be damaged by ion deficiency diseases, including nitrate deficiency and magnesium deficiency, and know the symptoms of these deficiency diseases.
- Know that plant diseases can be caused by infection with pathogens.
- Know that plants can be infested with pests like aphids.
- H Know how plant diseases can be detected and identified.
- Be able to describe physical and chemical defence responses of plants, along with mechanical adaptations.

Specification References
4.3.3.1, 4.3.3.2

Types of plant disease

Plant diseases can be caused by a range of different things.

Deficiency diseases

Plants need mineral ions from the soil. If there aren't enough, plants suffer deficiency symptoms.

Examples

- Nitrates are needed to make proteins and therefore for growth. A lack of nitrates causes stunted growth.
- Magnesium ions are needed for making chlorophyll, which is needed for photosynthesis. Plants without enough magnesium suffer from chlorosis (when not enough chlorophyll is made) and have yellow leaves.

Diseases caused by pathogens

Plants can be infected by viral, bacterial and fungal pathogens — see pages 132-133 for some examples.

Infestation with pests

Plants can also be infested and damaged by insects. For example, aphids are an insect that can cause huge damage to plants. Infestations of pests are easy to spot — you should be able to see them on the plants.

Detecting plant diseases **Higher**

It's usually pretty clear that a plant has a disease. The common signs are:

- Stunted growth
- Spots on the leaves
- Patches of decay (rot)
- Abnormal growths, e.g. lumps
- Malformed stems or leaves
- Discolouration

Example **Higher**

- A fungal disease called brown rot affects the fruit on plum trees, causing patches of the plums to go brown and rot.
- Crown gall is a bacterial disease that causes growths on plants — see Figure 1.

Figure 1: *A tomato plant affected by crown gall disease. It has a growth at the bottom of the stem.*

Identifying plant diseases Higher

Different plant diseases have different signs. They can be identified by:

- Looking up the signs in a gardening manual or on a gardening website.
- Taking the infected plant to a laboratory, where scientists can identify the pathogen.
- Using testing kits that identify the pathogen using monoclonal antibodies (see page 146).

Plant defences

Plants have physical, chemical and mechanical defences against pests and pathogens.

Physical defences

- Most plant leaves and stems have a waxy cuticle, which provides a barrier to stop pathogens entering.
- Plant cells themselves are surrounded by cell walls made from cellulose. These form a physical barrier against pathogens that make it past the waxy cuticle.
- Plants have layers of dead cells around their stems, for example, the outer part of the bark on trees. These act as a barrier to stop pathogens entering.

Chemical defences

- Some can produce antibacterial chemicals which kill bacteria — e.g. the mint plant and witch hazel.
- Other plants produce poisons which can deter herbivores (organisms that eat plants) — e.g. tobacco plants, foxgloves and deadly nightshade.

Mechanical defences

- Some plants have adapted to have thorns and hairs. These stop animals from touching and eating them.
- Other plants have leaves that droop or curl when something touches them. This means that they can prevent themselves from being eaten by knocking insects off themselves and moving away from things.
- Some plants can cleverly mimic other organisms. E.g. the passion flower has bright yellow spots on its leaves which look like butterfly eggs. This stops other butterflies laying their eggs there. Several species of plant in the 'ice plant family' in southern Africa look like stones and pebbles. This tricks other organisms into not eating them.

Figure 2: *Passion flower leaves with yellow spots that mimic butterfly eggs (top). Plants that mimic pebbles (bottom).*

Tip: Adaptations are features that help an organism survive in its habitat — see page 321 for more on adaptations.

Practice Questions — Fact Recall

Q1 What happens to plants that are lacking in magnesium ions?

Q2 Give one use that a gardener might have for monoclonal antibodies.

Q3 Describe two mechanical adaptations that plants can have to prevent animals from eating them.

Topic 3 Checklist — Make sure you know...

Communicable Disease

- ☐ That pathogens are microorganisms that cause communicable disease in plants or animals, and include bacteria, viruses, protists and fungi.
- ☐ That bacteria and viruses reproduce rapidly — bacteria can make you feel ill by producing toxins that damage cells, and viruses make you feel ill by reproducing inside cells and damaging them.
- ☐ That pathogens can be spread between people by water, air or by direct contact.

Viral Diseases

- ☐ That measles is caused by a virus, and that the symptoms are fever and a red skin rash — it is spread in droplets when people with the virus sneeze or cough.
- ☐ That if complications occur, measles can be fatal, and so most children have the measles vaccine.
- ☐ That HIV is spread by sexual contact or exchange of bodily fluids, and the first symptoms are flu-like.
- ☐ That at the early stage, HIV can be controlled with antiretroviral drugs.
- ☐ That HIV attacks the cells of the immune system — when the immune system is so damaged that it can no longer fight off other infections or cancers, HIV is called late stage HIV infection, or AIDS.
- ☐ That tobacco mosaic virus (TMV) affects many types of plants, including tomatoes, and causes leaves to become discoloured — this affects growth by preventing photosynthesis.

Fungal and Protist Diseases

- ☐ That rose black spot is a fungal plant disease that causes purple or black spots to develop on the leaves — the leaves sometimes then turn yellow and fall off.
- ☐ That changes to the leaves in rose black spot mean that photosynthesis is reduced, affecting growth.
- ☐ That rose black spot is spread in water or by the wind, and that it can be treated using fungicides and by removing and destroying the affected leaves.
- ☐ That malaria is caused by a protist that spends part of its life cycle in the mosquito host.
- ☐ That malaria causes recurring fever and can be fatal, but that its spread can be reduced by preventing the mosquitoes from breeding and by taking precautions to avoid being bitten (e.g. mosquito nets).

Bacteria Diseases and Preventing Disease

- ☐ That *Salmonella* bacteria cause food poisoning — fever, abdominal cramps, vomiting and diarrhoea.
- ☐ That *Salmonella* can be spread by eating contaminated food, and know how it can be controlled.
- ☐ That gonorrhoea is a bacterial sexually transmitted disease that causes pain when urinating and a yellow or green discharge from the genitals.
- ☐ That gonorrhoea can often no longer be treated with penicillin because of antibiotic resistance — it is now treated with other antibiotics and can be prevented by using condoms.
- ☐ That the spread of diseases can be reduced or prevented by being hygienic, destroying vectors, isolating infectious people, and by using vaccinations.

cont...

Fighting Disease

- ☐ That the skin, nose, trachea and bronchi, and the stomach have features to stop pathogens entering the body.
- ☐ That if pathogens do manage to enter the body, the immune system can attack them — white blood cells produce antibodies and antitoxins, and engulf pathogens by phagocytosis.

Fighting Disease — Vaccination

- ☐ How vaccinations work — dead or inactive pathogens are used to trigger the production of antibodies by white blood cells. If the same, live pathogen then enters the body, the white blood cells will rapidly mass-produce the right antibodies to kill it, so that the person doesn't get ill.
- ☐ That if enough people are immune to a certain pathogen, it can reduce the spread of the pathogen.

Fighting Disease — Drugs

- ☐ That many drugs (such as painkillers) reduce the symptoms of a disease but don't kill pathogens.
- ☐ That antibiotics (such as penicillin) are drugs that kill bacteria but don't destroy viruses.
- ☐ That antibiotics kill specific bacteria, so it's important to be treated with the right one.
- ☐ That viruses reproduce inside body cells, making it difficult to develop drugs against them.
- ☐ That the use of antibiotics has greatly reduced the number of deaths from communicable diseases caused by bacteria.
- ☐ That bacteria can mutate, leading to the development of antibiotic-resistant strains.
- ☐ That originally, drugs were extracted from plants and microorganisms (e.g. aspirin from willow, digitalis from foxgloves, and penicillin from mould).
- ☐ That drugs are now synthesised in labs, but that the process may still involve a plant chemical.

Developing Drugs

- ☐ That drugs have to be tested to make sure they are safe and effective.
- ☐ That during drug trials, new drugs are tested for efficacy and toxicity, and also to find the best dose.
- ☐ That new drugs are first tested in preclinical trials on human cells and tissues as well as on animals.
- ☐ That after preclinical testing, drugs are tested in clinical trials on healthy volunteers at low doses. They're then tested on ill volunteers to find the optimum dose of a drug (the dose the drug works best at).
- ☐ What a placebo is (a substance that's similar to the drug being tested but that doesn't do anything) and why they are used in clinical trials (to make sure the drug works).
- ☐ That in a double-blind trial, neither the doctors nor the volunteers know which volunteers have been given the drug and who has been given the placebo.

cont...

Monoclonal Antibodies

- ❑ [H] That monoclonal antibodies are produced from lots of clones of a single white blood cell, and that they will only bind to one specific protein antigen, so can target a particular cell or chemical.
- ❑ [H] How monoclonal antibodies are made — a lymphocyte and tumour are fused to create a hybridoma, which is cloned to produce lots of identical cells which all produce the same antibodies.

Monoclonal Antibody Uses

- ❑ [H] That monoclonal antibodies are used for diagnosis, such as in pregnancy tests — the monoclonal antibody used targets a hormone found in the urine of pregnant women.
- ❑ [H] That monoclonal antibodies can be used to treat cancer by being attached to a radioactive substance, a toxic drug or a chemical which stops cancer cells growing and dividing.
- ❑ [H] How monoclonal antibodies can be used to measure the levels of hormones and chemicals in blood and to test blood samples for particular pathogens.
- ❑ [H] How monoclonal antibodies with a fluorescent dye attached to them can be used to detect and locate specific molecules on a cell or in a tissue sample.
- ❑ [H] That monoclonal antibodies are not yet as widely used as scientists expected because they cause more side effects than was originally expected.

Plant Diseases and Defences

- ❑ What happens to plants that suffer from a deficiency of nitrate ions (stunted growth) and magnesium ions (chlorosis).
- ❑ That plants can be infected by viral, bacterial and fungal pathogens, and by insects, e.g. aphids.
- ❑ [H] What the common signs of plant disease are — stunted growth, spots on the leaves, patches of decay, abnormal growths, malformed stems or leaves and discolouration.
- ❑ [H] That plant diseases can be identified using gardening manuals or websites, by taking the plants to a laboratory, or by using testing kits that work using monoclonal antibodies.
- ❑ That plants have physical, chemical and mechanical defences against pathogens and pests, and be able to give examples of each.

Exam-style Questions

1 A patient in a hospital has been diagnosed with malaria after returning from Kenya.

1.1 Give **one** symptom of malaria.

(1 mark)

1.2 Name the type of pathogen that causes malaria.

(1 mark)

1.3 Another disease that can be caught in Kenya is typhoid.
The patient was vaccinated against typhoid before his trip to Kenya.
Outline how the typhoid vaccine can prevent someone from developing the disease.

(4 marks)

2 A garden centre has to keep their plants healthy and free from disease.

2.1 Some of the plants are suffering from stunted growth.
Suggest which ion deficiency condition they might be suffering from.

(1 mark)

2.2 Stunted growth can be a sign of other plant diseases too.
Suggest **three** more signs of disease in plants.

(3 marks)

The staff discover that a rose plant has been infected with rose black spot.

2.3 What type of pathogen causes rose black spot?

(1 mark)

2.4 Suggest how the disease may have been spread to the plant.

(2 marks)

2.5 Suggest how the garden centre should treat the disease.

(3 marks)

3 Pathogens can enter the body in several ways.

3.1 Outline how the stomach helps to defend the body against the entry of pathogens.

(1 mark)

3.2 The human airways include the trachea and bronchi.
Outline how the trachea and bronchi help to defend the body against pathogens.

(3 marks)

Sometimes pathogens get past the body's defences.

3.3 How can bacteria make people feel ill?

(1 mark)

3.4 Name **one** disease in humans caused by bacteria.

(1 mark)

3.5 The immune system responds to pathogens getting inside the body.
Outline the immune system's response to bacteria inside the body.

(4 marks)

4 The graph below shows the number of measles cases reported each year worldwide, along with the estimated vaccine coverage, from 1980 to 2014.

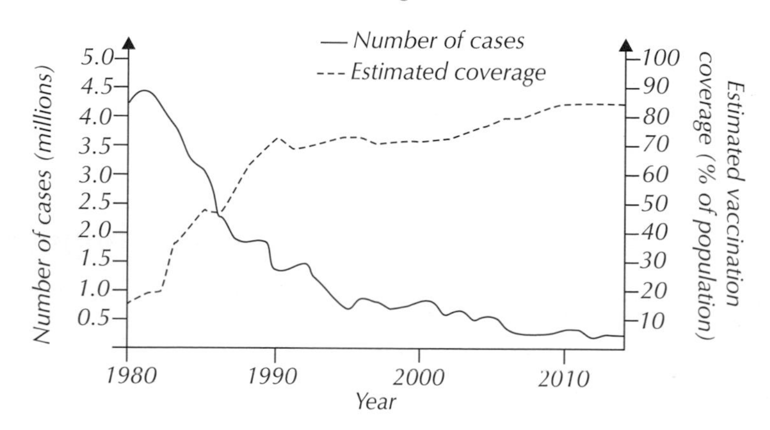

4.1 Measles is caused by a pathogen. What is a pathogen?

(1 mark)

4.2 Describe the trends shown in the graph.
Use data from the graph to support your answer.

(4 marks)

4.3 Evaluate how far this data supports the case for vaccinating people against measles.

(3 marks)

5 A new drug is being developed.

5.1 Describe the difference between how drugs were made traditionally and how drugs are made these days.

(2 marks)

5.2 The new drug enters preclinical testing. Preclinical testing happens on cells and tissues, and it is sometimes done using live animals too.
Suggest why a drug designed to affect the absorption of nutrients from the small intestine into the bloodstream might be tested on live animals rather than just on cells and tissues.

(2 marks)

The drug then enters clinical trials.

5.3* Describe what happens at this stage.
Refer to placebos and double-blind testing in your answer.

(6 marks)

1. The Basics of Photosynthesis

Plants make their own food using energy from the Sun, a substance called chlorophyll, carbon dioxide from the air and water from the soil.

Learning Objectives:
- Know that energy is transferred to the chloroplasts of plants when they absorb light.
- Know that photosynthesis is endothermic.
- Be able to write the word equation for photosynthesis.
- Know the chemical symbols for carbon dioxide, water, oxygen and glucose.

Specification Reference 4.4.1.1

What is photosynthesis?

Photosynthesis is the process that produces 'food' in plants. The 'food' it produces is **glucose** (a sugar). Photosynthesis uses energy to change carbon dioxide and water into glucose and oxygen. Plants use this glucose for a number of things, such as making cell walls and proteins (see pages 160-161).

Where does photosynthesis happen?

Photosynthesis takes place in **chloroplasts** in green plant cells — they contain pigments like **chlorophyll** that absorb light. Energy is transferred to the chloroplasts from the environment by light. Photosynthesis is **endothermic** — this means energy is transferred from the environment in the process.

Tip: Exothermic reactions transfer energy to the surroundings and endothermic reactions transfer energy from the surroundings.

Photosynthesis happens in the leaves of all green plants — this is largely what the leaves are for. Figure 1 shows a cross-section of a leaf showing the four raw materials (carbon dioxide, water, light and chlorophyll) needed for photosynthesis.

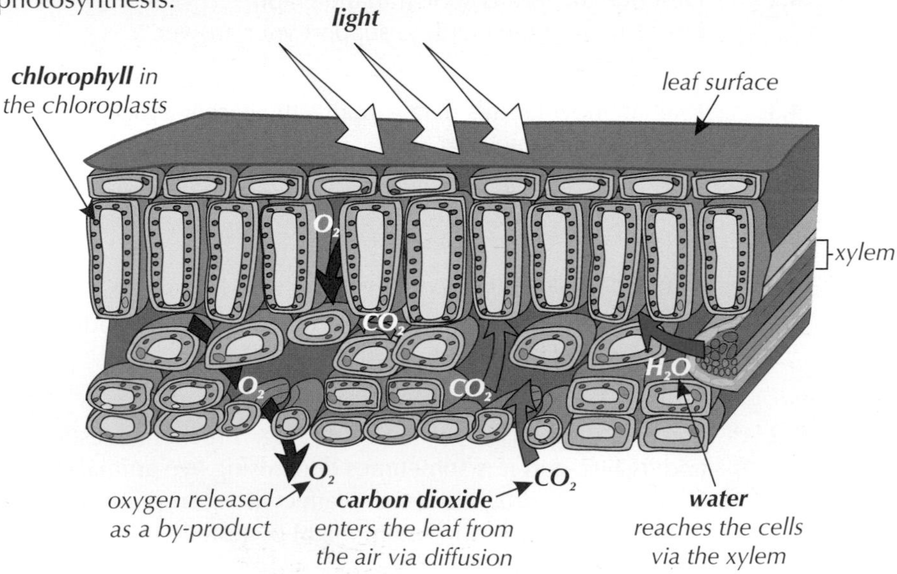

Figure 1: *Diagram of a cross-section through a leaf, showing the four raw materials required for photosynthesis.*

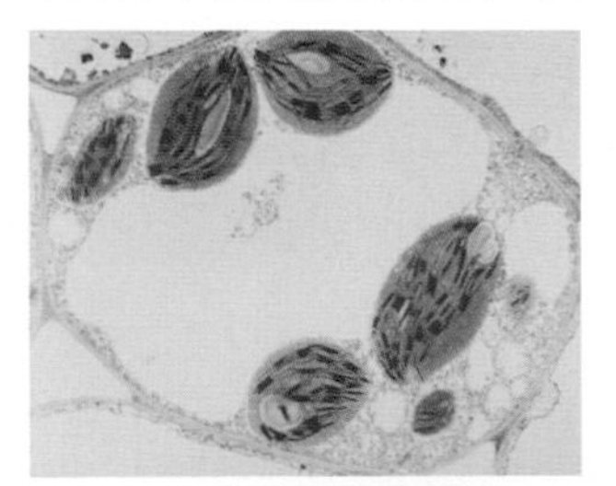

Figure 2: *Microscope image of a plant cell. The green structures are chloroplasts.*

Tip: Xylem is the tissue that transports water around a plant (see page 91).

Word equation

The word equation for photosynthesis is:

$$\text{carbon dioxide} + \text{water} \xrightarrow{\text{light}} \text{glucose} + \text{oxygen}$$

Exam Tip
You should learn this photosynthesis equation — it could get you easy marks in the exam.

Symbol equation

There's a symbol equation for photosynthesis too:

$$6CO_2 + 6H_2O \xrightarrow{\text{light}} C_6H_{12}O_6 + 6O_2$$

Exam Tip
You don't need to know this equation for the exam, but you do need to know the individual symbols for carbon dioxide (CO_2), water (H_2O), glucose ($C_6H_{12}O_6$) and oxygen (O_2).

Practice Questions — Fact Recall

Q1 Which product produced by photosynthesis is 'food' for the plant?

Q2 Which gas is used in photosynthesis?

Q3 Which gas is produced by photosynthesis?

Q4 a) Name the parts of a plant cell that contain chlorophyll.

b) What role does chlorophyll play in photosynthesis?

Q5 Write out the word equation for photosynthesis.

Q6 Name the chemical substance with the symbol $C_6H_{12}O_6$.

Practice Questions — Application

Q1 Three identical plants were grown for a week. They were treated in exactly the same way, except they each received different amounts of sunlight per day, as shown in this table:

Plant	Hours of sunlight received per day
A	3
B	7
C	10

The plants didn't have any source of light other than sunlight.

Which plant do you think will have produced the most glucose after one week? Explain your answer.

Q2 A scientist studied three different types of plant cell (taken from the same plant) under a microscope. She recorded the average number of chloroplasts she found in each type of cell, as shown in this table:

Type of plant cell	Average number of chloroplasts
1	20
2	0
3	51

Suggest which type of plant cell (1-3) is likely to have the highest rate of photosynthesis. Explain your answer.

Tip: Think about what the job of the chloroplasts is.

Learning Objectives:

- Know the ways that plants use the glucose produced during photosynthesis.
- Know that plants need nitrate ions from the soil, as well as glucose, to make proteins.

Specification Reference 4.4.1.3

2. How Plants Use Glucose

As you know, plants photosynthesise to make 'food' in the form of glucose. They make use of this glucose in a number of ways...

What is glucose used for?

During photosynthesis, plants convert carbon dioxide and water into glucose (see page 158). They use the glucose they produce for the following things:

1. Respiration

Plants use some of the glucose they make for respiration. Respiration is a process which occurs in all living organisms — it's where energy is released from the breakdown of glucose (see page 173).
Plants use this energy to convert the rest of the glucose into various other useful substances, which they can use to build new cells and grow. To produce some of these substances plants also need to gather a few minerals from the soil (e.g. nitrates — see below).

Figure 1: Microscope image of a plant cell wall made up of strands of cellulose.

2. Making cellulose

Glucose is converted into **cellulose** for making strong cell walls (see page 24), especially in a rapidly growing plant. These cell walls support and strengthen the cells.

3. Making amino acids

In plant cells, glucose is combined with **nitrate ions** (which are absorbed from the soil) to make **amino acids**. Amino acids are the building blocks which make up proteins. When amino acids are joined together in a particular sequence they make up a particular protein.

Tip: Plants use proteins for growth. If a plant has a nitrate ion deficiency it can't make as much protein, and so it will have stunted growth.

Storage of glucose

Glucose can be converted into other substances for storage:

Lipids

Plants can convert some of the glucose they produce into **lipids** (fats and oils). Plants store lipids in seeds.

Example

Sunflower seeds contain a lot of oil — we get cooking oil and margarine from them.

Starch

Plants can convert glucose into starch. Glucose is stored as starch so it's ready for use when photosynthesis isn't happening as much, like in the winter. In plants, starch is stored in roots, stems, seeds and leaves.

Example

Potato and parsnip plants store a lot of starch underground over the winter so a new plant can grow from it the following spring. We eat the swollen storage organs (see Figure 2).

Figure 2: *Potato plants that have just been dug up — the potatoes that we eat are starch stores for the plant.*

Starch is **insoluble** which makes it much better for storing than glucose — a cell with lots of glucose in would draw in loads of water and swell up.

Tip: Water moves from regions of higher concentration of water to regions of lower concentration — this is known as osmosis (see page 54). A cell with lots of glucose would have a lower concentration of water than its surroundings, and so water would move in from outside by osmosis.

Practice Questions — Fact Recall

Q1 By what process do plants break down glucose in order to transfer energy to their cells?

Q2 What do plants use cellulose for?

Q3 What do plants absorb from the soil to make proteins with?

Q4 Glucose can be stored as starch in plants. How else is glucose stored?

Q5 Why is starch better for storage than glucose?

3. The Rate of Photosynthesis

Photosynthesis doesn't always happen at the same rate — factors such as light intensity, carbon dioxide level and temperature can all affect the rate.

Learning Objectives:

- Be able to explain how the rate of photosynthesis can be affected by temperature, the level of carbon dioxide, the light intensity, and also by the amount of chlorophyll in the plant.
- Be able to draw, read from and interpret graphs showing the effect of one limiting factor on the rate of photosynthesis.
- H Know that the factors that affect the rate of photosynthesis can interact.
- H Be able to interpret rate graphs that show two or three factors to identify which is the limiting factor.

Specification Reference 4.4.1.2

Limiting factors in photosynthesis

The rate of photosynthesis is affected by the intensity of light, the volume of carbon dioxide (CO_2), and the temperature. All three of these things need to be at the right level to allow a plant to photosynthesise as quickly as possible. If any one of these factors is too high or too low, it will become the **limiting factor**. This just means it's the factor which is stopping photosynthesis from happening any faster.

Which factor is limiting at a particular time depends on the environmental conditions:

Examples

- At night there's much less light than there is during the day, so light intensity is usually the limiting factor at night.
- In winter it's usually cold, so a low temperature is often the limiting factor.
- If it's warm enough and bright enough, the amount of CO_2 is usually limiting.

The amount of chlorophyll can also be a limiting factor of photosynthesis.

Example

The amount of chlorophyll in a plant can be affected by disease (e.g. infection with the tobacco mosaic virus) or environmental stress, such as a lack of nutrients. These factors can cause chloroplasts to become damaged or to not make enough chlorophyll. This means the rate of photosynthesis is reduced because they can't absorb as much light.

Tip: Plants also need water for photosynthesis, but when a plant is so short of water that it becomes the limiting factor in photosynthesis, it's already in such trouble that this is the least of its worries.

Tip: There's more about tobacco mosaic virus on page 132.

Interpreting data on the rate of photosynthesis

You need to be able to interpret data showing how different factors affect the rate of photosynthesis.

1. Effect of light intensity

Light provides the energy needed for photosynthesis. As you can see from Figure 1, as the light level is raised, the rate of photosynthesis increases steadily — but only up to a certain point. Beyond that, it won't make any difference — as light intensity increases, the rate will no longer increase. This is because it'll be either the temperature or the carbon dioxide level which is now the limiting factor, not light.

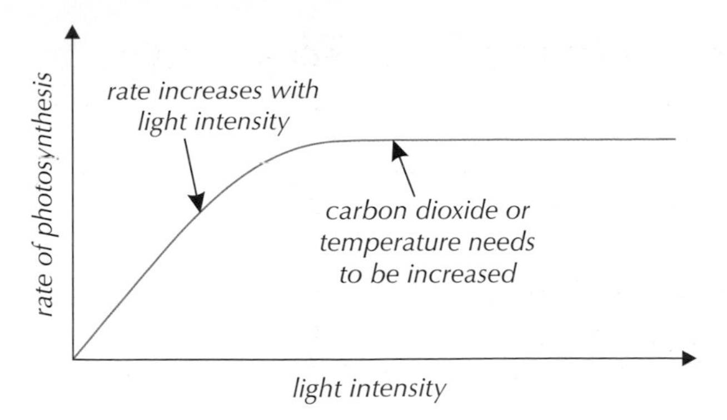

Figure 1: *Graph showing how light intensity affects the rate of photosynthesis.*

Exam Tip
You need to know how to plot a graph like this when given appropriate data — see p.18 if you need some tips. You also need to be able to read off values from graphs like this and explain what the data is showing, so make sure you really understand what these pages are talking about.

Remember, photosynthesis hasn't stopped when the graph levels off — it's just not increasing anymore.

In the lab you can change the light intensity by moving a lamp closer to or further away from your plant (see page 167 for this experiment). But if you just plot the rate of photosynthesis against "distance of lamp from the plant", you get a weird-shaped graph. To get a graph like the one above you either need to measure the light intensity at the plant using a light meter or do a bit of nifty maths with your results.

Figure 2: *The number of bubbles produced in a set time can be counted to estimate the rate of photosynthesis, but a more accurate set-up can be seen on page 167.*

2. Effect of carbon dioxide level

CO_2 is one of the raw materials needed for photosynthesis.
As with light intensity, the amount of CO_2 will only increase the rate of photosynthesis up to a point. After this the graph flattens out — as the amount of CO_2 increases, the rate no longer increases. This shows that CO_2 is no longer the limiting factor.

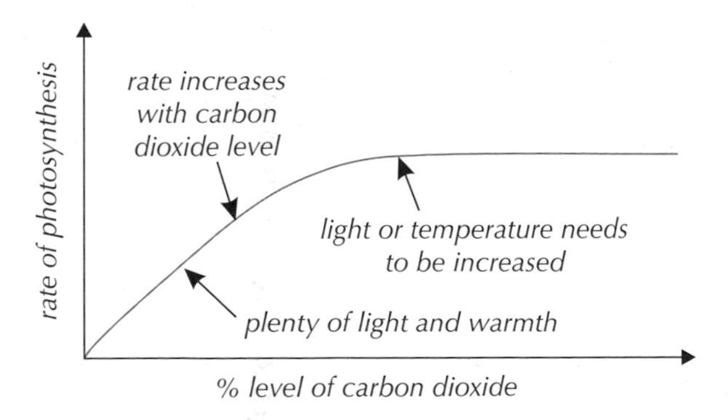

Figure 3: *Graph showing how carbon dioxide level affects the rate of photosynthesis.*

As long as light and CO_2 are in plentiful supply then the factor limiting photosynthesis must be temperature.

Tip: If you're investigating the rate of photosynthesis using a plant in water, you can increase the CO_2 level by dissolving sodium hydrogencarbonate in the water. This gives off carbon dioxide.

3. Effect of temperature

Enzymes are proteins which increase the speed of chemical reactions in living things — so enzymes increase the rate of photosynthesis in plant cells. The speed at which enzymes work is affected by temperature.

Tip: You can read more about enzymes on pages 115-117.

Usually, if the temperature is the limiting factor in photosynthesis it's because it's too low — the enzymes needed for photosynthesis work more slowly at low temperatures. But if the plant gets too hot, the enzymes it needs for photosynthesis and its other reactions will be damaged. This happens at about 45 °C (which is pretty hot for outdoors, although greenhouses can get that hot if you're not careful). Figure 4 shows the effect of temperature on the rate of photosynthesis.

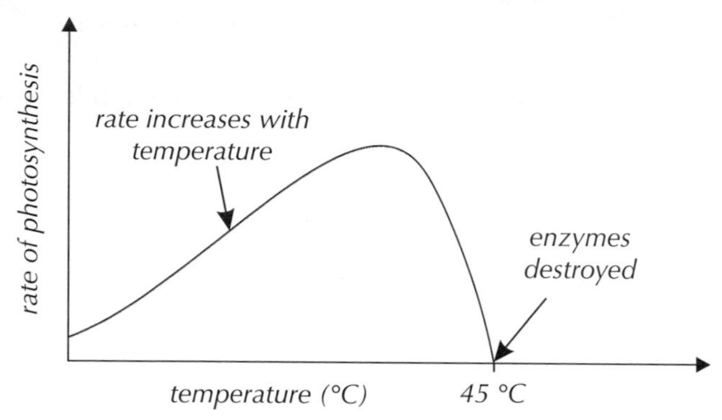

Figure 4: *Graph showing how temperature affects the rate of photosynthesis.*

4. Graphs with more than one limiting factor Higher

You could get a graph that shows more than one limiting factor on the rate of photosynthesis.

Exam Tip H
Make sure you get to grips with interpreting graphs like this — they could come up in the exam. Just look at what factors the graph shows, and look at how they're affecting the line that shows the rate of photosynthesis. There's a question on page 166 to practise with.

Example 1 Higher

Figure 5 shows how the rate of photosynthesis is affected by light intensity and temperature. At the start, both of the lines show that as the light intensity increases, the rate of photosynthesis increases steadily. But the lines level off when light is no longer the limiting factor. The line at 25 °C levels off at a higher point than the one at 15 °C, showing that temperature must have been a limiting factor at 15 °C.

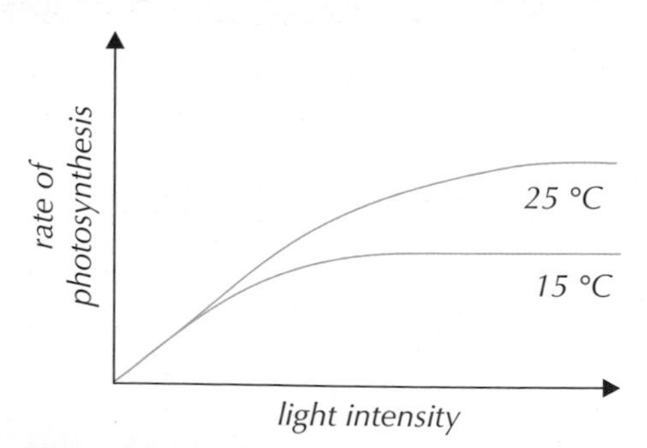

Figure 5: *Graph showing how photosynthesis is affected by light intensity and temperature.*

Example 2 | Higher

Figure 6 shows how the rate of photosynthesis is affected by light intensity and CO_2 concentration. Again, both the lines level off when light is no longer the limiting factor. The line at the higher CO_2 concentration of 0.4% levels off at a higher point than the one at 0.04%. This means CO_2 concentration must have been a limiting factor at 0.04% CO_2. The limiting factor here isn't temperature because it's the same for both lines (25 °C).

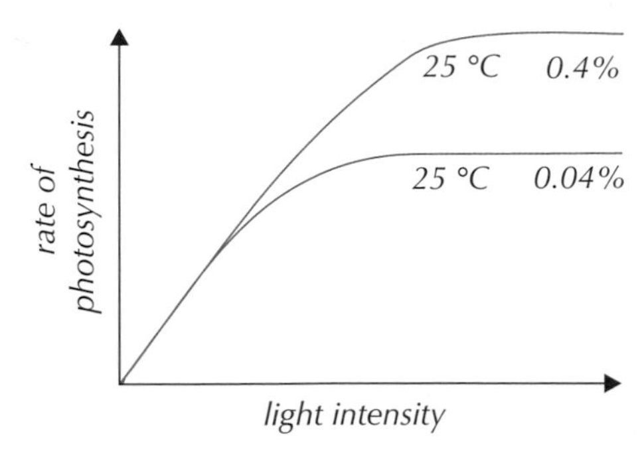

Figure 6: *Graph showing how photosynthesis is affected by light intensity and CO_2 concentration.*

Practice Questions — Fact Recall

Q1 The rate of photosynthesis is affected by limiting factors.

a) What is meant by a 'limiting factor' of photosynthesis?

b) Give three environmental conditions that can affect the rate of photosynthesis.

Q2 True or false? The rate of photosynthesis can be affected by the amount of chlorophyll in the leaves.

Practice Questions — Application

Q1 Complete the table to show what is most likely to be the limiting factor of photosynthesis in the environmental conditions listed.

Environmental conditions	Most likely limiting factor
Outside on a cold winter's day.	
In an unlit garden at 1:30 am, in the UK, in summer.	
On a windowsill on a warm, bright day.	

Q2 Peter was investigating the rate of photosynthesis in pondweed. He put equally sized samples of pondweed from the same plant in two separate flasks containing different solutions. The solution in flask A had a lower carbon dioxide concentration than the solution in flask B. Peter put the flasks an equal distance away from a light source and gradually increased the intensity of the light throughout the experiment. He kept both flasks at a constant temperature of 25 °C. During the experiment Peter measured the amount of gas produced in each flask using a gas syringe. His results are shown on the graph:

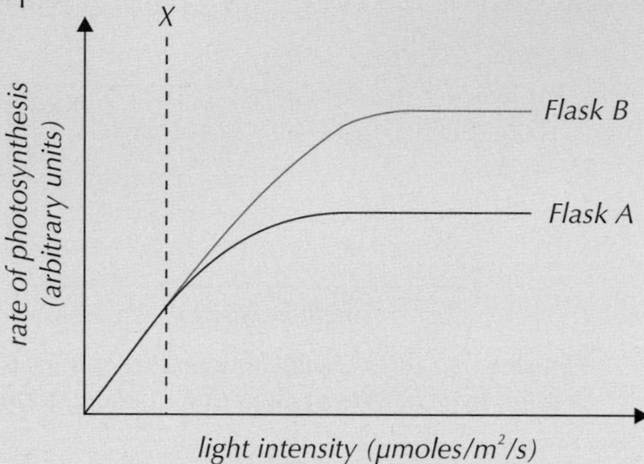

a) Peter thinks that the limiting factor before point X in his experiment is light. Why does he think this?

b) Both graphs level off. Suggest why Flask A levels off at a lower point than Flask B. Explain your answer.

Exam Tip H
μmoles/m^2/s is a unit used to measure light intensity. Don't panic if you see unfamiliar units in the exam — just focus on what the axis is showing you, e.g. here it's light intensity.

4. Investigating Photosynthesis Rate

So you've seen a graph showing how light intensity affects the rate of photosynthesis back on page 163, but you also need to know how you'd actually get the data for that graph.

Investigating the rate of photosynthesis

To investigate the effect of different light intensities on the rate of photosynthesis, you need to be able to measure the rate. One way to do this involves an aquatic plant, such as Canadian pondweed. The rate at which the pondweed produces oxygen corresponds to the rate at which it's photosynthesising — the faster the rate of oxygen production, the faster the rate of photosynthesis.

Here's how the experiment works:

A source of white light is placed at a specific distance from the pondweed. You should then leave the pondweed for a couple of minutes to adjust to the new light intensity before starting the experiment. When you're ready to start, the pondweed is left to photosynthesise for a set amount of time. As it photosynthesises, the oxygen released will collect in the capillary tube.

At the end of the experiment, the syringe is used to draw the gas bubble in the tube up alongside a ruler and the length of the gas bubble is measured. This is proportional to the volume of O_2 produced.

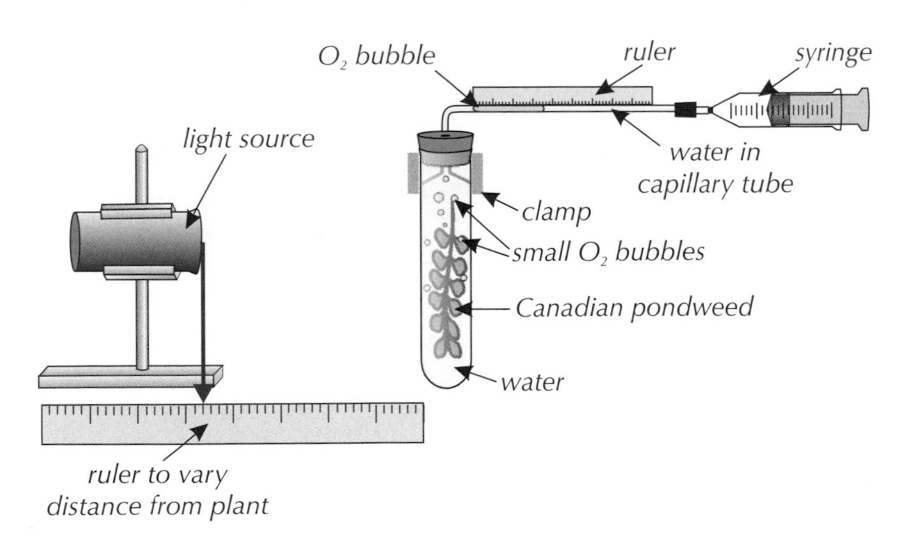

Figure 1: *Diagram showing the experimental set up used to measure the rate of photosynthesis in water plants.*

The experiment is repeated twice with the light source at the same distance and the mean volume of O_2 produced is calculated. Then the whole experiment is repeated with the light source at different distances from the pondweed.

Learning Objectives:
- Be able to investigate how light intensity affects the rate of photosynthesis, using an aquatic plant (Required Practical 6).
- Be able to calculate the rate of photosynthesis.

Specification Reference 4.4.1.2

Tip: Make sure you do a risk assessment before you start this investigation, so that you carry it out safely. E.g. there's a slight risk of infection from the pond water, so make sure you wash your hands after handling the pondweed and cover up any cuts.

Figure 2: *Pondweed giving off oxygen bubbles as a by-product of photosynthesis.*

Tip: Canadian pondweed is not native to the UK, so it must be disposed of carefully after the experiment — it shouldn't be released into the environment.

Tip: Taking repeat readings and calculating a mean reduces the effect of random errors, and will make your results more precise — see page 13.

WORKING SCIENTIFICALLY

Controlling variables

In this experiment, you're investigating the effect of different light intensities on the rate of photosynthesis — the light intensity is the **independent variable**. The **dependent variable** (the variable you're measuring) is the amount of oxygen produced, which is represented by the length of the bubble.
You have to try to keep all the other variables constant, so that it's a fair test.

Tip: The independent variable is the variable that you change in an experiment — see page 10.

WORKING SCIENTIFICALLY

Examples

- If your plant's in a flask, keep the flask in a water bath to help keep the temperature constant.
- There's not much you can do to keep the carbon dioxide level constant — you may just have to use a large container for your plant, and do the experiments as quickly as you can, so that the plant doesn't use up too much of the carbon dioxide in the container.

Tip: The variables in these experiments include all of the limiting factors of photosynthesis.

Tip: For more on designing investigations and controlling variables see pages 9-11.

Calculating the rate of photosynthesis

You can compare the results at different light intensities by giving the rate of photosynthesis as the length of the bubble per unit time, e.g. cm/min.

To calculate the rate, just divide the total length of the bubble by the time taken to produce it.

Example

MATHS SKILLS

A student measured the amount of gas produced by a piece of pondweed by collecting it in a capillary tube.
After 5 minutes, the gas bubble was 25 mm long.
Calculate the rate of photosynthesis. Give your answer in cm/min.

- First, convert 25 mm into cm:

$$25 \div 10 = 2.5 \text{ cm}$$

- Then divide the length by the time taken:

$$\text{Rate} = 2.5 \text{ cm} \div 5 \text{ min} = \mathbf{0.5 \text{ cm/min}}$$

Tip: Alternatively, you could work out the rate as the volume of oxygen per unit time (e.g. cm^3/min), but this requires a bit more maths to find the volume of the bubble first.

Tip: When you're working out a rate, you should always be dividing by time.

Practice Question — Application

Q1 A student set up an experiment to investigate the rate of photosynthesis using a piece of Canadian pondweed.

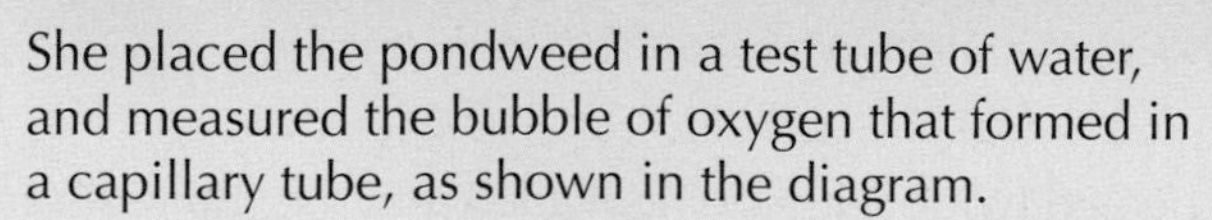

She placed the pondweed in a test tube of water, and measured the bubble of oxygen that formed in a capillary tube, as shown in the diagram.

In 3 minutes, the pondweed produced a bubble of oxygen in the capillary tube that was 25 mm long.
Calculate the rate of photosynthesis for the pondweed.
Give your answer in cm/min, to 1 significant figure.

5. The Inverse Square Law Higher

This can be a bit tricky to get your head around, but stick with it...

What is the inverse square law?

In the experiment on page 167, when the lamp is moved away from the pondweed, the amount of light that reaches the pondweed decreases. You can say that as the distance increases, the light intensity decreases. In other words, distance and light intensity are **inversely proportional** to each other.

However, it's not quite as simple as that. It turns out that light intensity decreases in proportion to the square of the distance. This is called the **inverse square law** and is written out like this:

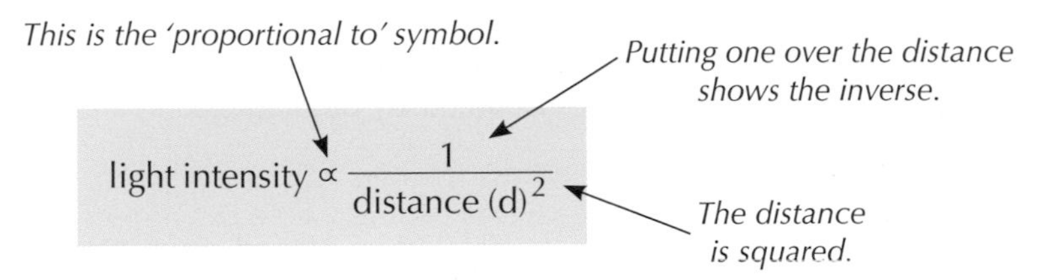

The inverse square law means that if you halve the distance, the light intensity will be four times greater and if you third the distance, the light intensity will be nine times greater. Likewise, if you double the distance, the light intensity will be four times smaller and if you treble the distance, the light intensity will be nine times smaller.

You can use $1/d^2$ as a measure of light intensity.

Example **Higher**

Use the inverse square law to calculate the light intensity when the lamp is 10 cm from the pondweed.

- Use the formula $\frac{1}{d^2}$: light intensity $= \frac{1}{d^2}$
- Fill in the values you know — you're given the distance, so put that in. light intensity $= \frac{1}{10^2}$
- Calculate the answer. $\frac{1}{10^2} = 0.01$ a.u.

Practice Question — Application

Q1 A student is carrying out an experiment to study the effect of light intensity on the rate of photosynthesis. He positions a lamp at different distances from samples of pondweed.
Use the inverse square law to calculate the light intensity when the lamp is positioned:

a) 15 cm from the pondweed, b) 30 cm from the pondweed.

Learning Objectives:

- H Know what inverse proportion is.
- H Know that the inverse square law links light intensity and distance.
- H Understand how to use the inverse square law.

Specification Reference 4.4.1.2

Tip: H If two variables are proportional to each other, when one variable changes, the other will change at the same rate. Inverse proportion just means that as one variable increases, the other decreases at the same rate.

Tip: H a.u. stands for arbitrary units.

Learning Objective:

- H Understand that the temperature, light intensity and carbon dioxide concentration in a plant's environment can be managed artificially in order to increase the rate of photosynthesis and therefore increase the profit made from the plants.

Specification Reference 4.4.1.2

6. Artificially Controlling Plant Growth Higher

As you've read on the last few pages, different factors affect the rate of photosynthesis. Farmers and other plant growers can use this knowledge to try to increase the rate of growth in their plants.

Greenhouses

If you know the ideal conditions for photosynthesis, then you can create an environment which maximises the rate of photosynthesis, which in turn maximises the rate of plant growth (plants use some of the glucose produced by photosynthesis for growth — see page 160). The most common way to artificially create the ideal environment for plants is to grow them in a greenhouse. Commercial growers (people who grow plants to make money, such as farmers) often grow large quantities of plants in commercial greenhouses. The following conditions can easily be managed in a greenhouse:

***Figure 1:** Lettuces growing in a commercial greenhouse.*

1. Temperature

Greenhouses help to trap the Sun's heat (see Figure 2), and make sure that the temperature doesn't become limiting.

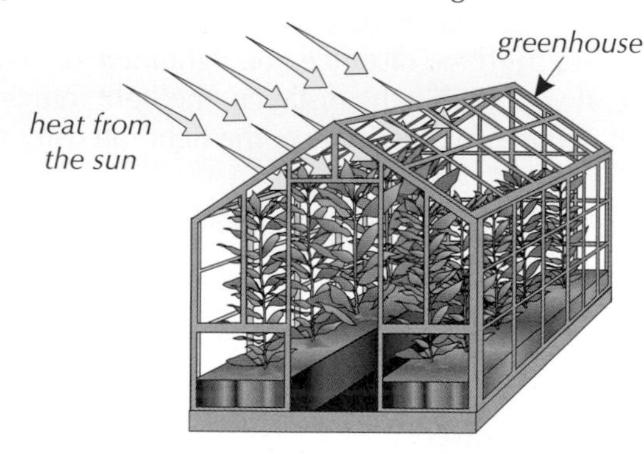

***Figure 2:** Greenhouses trap the Sun's heat.*

Temperature can be controlled in other ways too:

Example Higher

- In winter, a farmer might use a heater in their greenhouse to keep the temperature at the ideal level.
- In summer, greenhouses could get too hot, so farmers might use shades and ventilation to cool things down.

***Figure 3:** Lamps in greenhouses supply light so plants can continue to photosynthesise at night.*

2. Light

Light is always needed for photosynthesis, so farmers often supply artificial light after the Sun goes down to give their plants more quality photosynthesis time.

3. Carbon dioxide concentration

Farmers can also increase the level of carbon dioxide in the greenhouse. A fairly common way is to use a paraffin heater to heat the greenhouse. As the paraffin burns, it makes carbon dioxide as a by-product.

4. General health of plants

Keeping plants enclosed in a greenhouse also makes it easier to keep them free from pests and diseases. The farmer can add fertilisers to the soil as well, to provide all the minerals needed for healthy growth.

Greenhouse costs

Controlling the conditions in a greenhouse costs money — but if the farmer can keep the conditions just right for photosynthesis, the plants will grow much faster and a decent crop can be harvested much more often, which can then be sold.

It's important that a farmer supplies just the right amount of heat, light, etc. — enough to make the plants grow well, but not more than the plants need, as this would just be wasting money.

Interpreting data on artificial environments

In the exam you could be given some data about controlling photosynthesis in artificial environments (like greenhouses) and asked questions about it.

Example **Higher**

A farmer normally grows his plants outside when the average air temperature is between 9 and 11 °C. The farmer wants to increase the growth rate of his plants, so he is considering getting a greenhouse. He finds some data on how temperature affects the rate of photosynthesis for the type of plants he grows. This is shown in the graph:

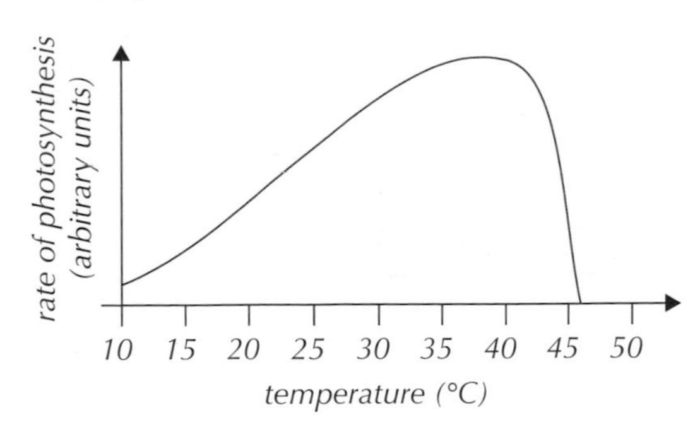

You could be asked to evaluate the benefits of growing plants in an artificial environment...
By looking at the graph you can see that the plants could photosynthesise faster if they were grown at temperatures higher than the average temperatures outside. If the farmer had a greenhouse he could control the temperature the plants were kept at, therefore increasing their rate of photosynthesis and growth. However, he would have to weigh up the cost of buying and running a greenhouse against this increased rate of growth.

You could be asked to make suggestions based on data from the graph...
For instance, you could be asked to suggest the best temperature for the farmer to keep his plants at and to explain why. Based on the graph, suggesting 37 °C would be a good idea, as this is the minimum temperature at which the rate of photosynthesis is highest. Keeping them at temperatures higher than this would just be a waste of money.

Exam Tip **H**
If you get a question like this in the exam, remember that the aim of commercial farming is to make money. Farmers try to create conditions to increase photosynthesis (and therefore growth) but at a minimum cost.

Practice Questions — Fact Recall

Q1 Explain why some farmers grow plants in an artificial environment.

Q2 Farmers can control the light and temperature in a greenhouse. Give one other condition they can control in a greenhouse.

Tip: H In the exam you might see experiments using rate of plant growth or increase in plant mass as an indirect way of measuring the rate of photosynthesis. This is because some of the glucose produced by photosynthesis is used for growth in plants (see page 160).

Practice Question — Application

Q1 A farmer grew the same type of plant in two greenhouses, A and B. He used a paraffin heater in Greenhouse A and an electric heater in Greenhouse B. Electric heaters don't release any carbon dioxide, whereas carbon dioxide is released when paraffin is burnt.

For 8 weeks the farmer recorded the average height of the plants in each greenhouse. He also monitored the temperature in both greenhouses. Plants in both greenhouses were exposed to the same amount of light and given the same amount of water and nutrients. His results are shown below:

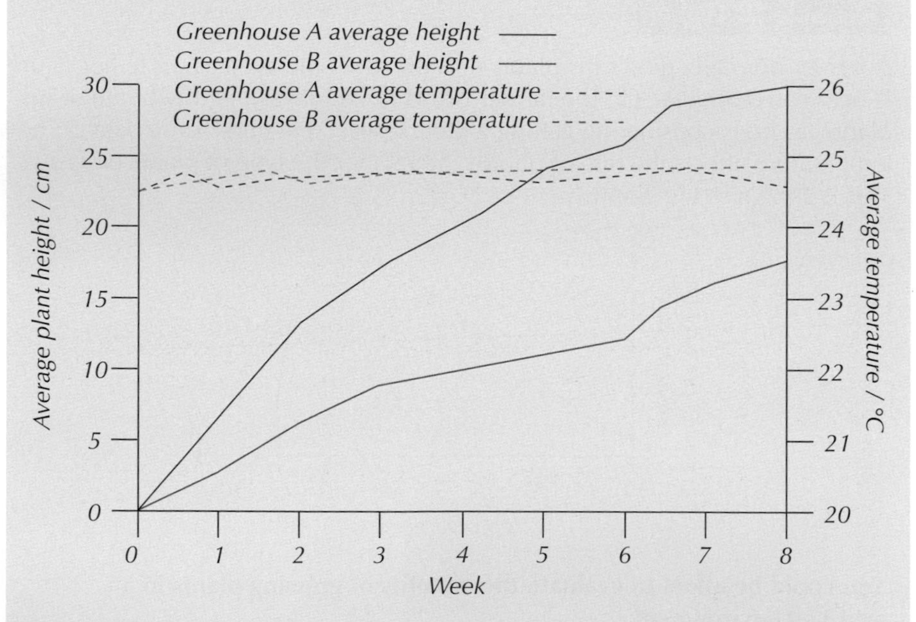

a) i) What was the average growth rate per week for the plants in Greenhouse A?

ii) What was the average growth rate per week for the plants in Greenhouse B?

b) What factor do you think is responsible for the difference between the average growth rates of the plants in Greenhouse A compared to Greenhouse B. Explain your answer.

c) The farmer is trying to decide which type of heater to use on a permanent basis. Other than the effect on growth rate, suggest what else the farmer will need to consider when choosing a heater.

Tip: H If you're stuck on part b, have a look back at the pages on limiting factors of photosynthesis (pages 162-165).

7. Aerobic Respiration

We, and other organisms, need energy to do... well... everything really. This energy comes from the reactions of respiration.

Respiration and enzymes

The chemical reactions that occur inside cells are controlled by enzymes. All of the reactions involved in respiration are catalysed by enzymes. These are really important reactions, as respiration transfers the energy that the cell needs to do just about everything — this energy is used for all living processes.

Respiration is not breathing in and breathing out, as you might think. Respiration is the process of transferring energy from the breakdown of glucose — and it goes on in every cell in your body continuously.

Respiration happens in plants too. All living things respire. It's how they release energy from their food.

Respiration is the process of transferring energy from glucose, which goes on in every cell.

Respiration is **exothermic** — it transfers energy to the environment.

What is aerobic respiration?

Aerobic respiration is respiration using oxygen. It's the most efficient way to transfer energy from glucose. Aerobic respiration goes on all the time in plants and animals.

Most of the reactions in aerobic respiration happen inside cell structures called **mitochondria** (see page 23).

Word equation

You need to learn the overall word equation for aerobic respiration:

glucose + oxygen → carbon dioxide + water

Symbol equation

There's a symbol equation for aerobic respiration too:

$$C_6H_{12}O_6 + 6O_2 \rightarrow 6CO_2 + 6H_2O$$

Learning Objectives:

- Know that all the energy needed for cellular processes is supplied by respiration.
- Know that respiration is the process of transferring energy from glucose, which occurs continuously in living cells.
- Know that respiration is an exothermic reaction.
- Know that respiration can take place aerobically (with oxygen) in cells to transfer energy from glucose.
- Know the word equation for aerobic respiration.
- Know the chemical symbols for glucose, oxygen, carbon dioxide and water.
- Know how organisms use the energy transferred from respiration.

Specification Reference 4.4.2.1

Tip: You can also have anaerobic respiration, which happens without oxygen, but that doesn't transfer nearly as much energy as aerobic respiration — see page 175.

Tip: You don't need to know this symbol equation, but you do need to recognise the symbols — lucky for you, they're the same as the ones you need to know for photosynthesis (see page 159).

Energy from respiration

Organisms use the energy transferred by respiration to fuel all sorts of processes. You need to know about the following ways in which an organism may use energy from respiration:

Figure 1: Animals, including humans, need energy from respiration to contract their muscles and move.

1. Organisms use energy to build up larger molecules from smaller ones (like proteins from amino acids).
2. Animals use energy to allow their muscles to contract (which in turn allows them to move about).
3. Mammals and birds use energy to keep their body temperature steady (unlike other animals, mammals and birds keep their bodies constantly warm).

Practice Questions — Fact Recall

Q1 What is aerobic respiration?

Q2 Aerobic respiration is an exothermic reaction. What does this mean?

Q3 Write the chemical symbols for the products of aerobic respiration.

Q4 Give three ways in which mammals use energy from respiration.

Exam Tip
Make sure you know the word equation for aerobic respiration off by heart, as it can get you easy marks in the exam.

Practice Question — Application

Q1 This diagram shows the equation for aerobic respiration.

Glucose + [A] → carbon dioxide + [B]

a) What is A in the equation?

b) What is B in the equation?

8. Anaerobic Respiration

When you're exercising hard, aerobic respiration isn't always enough to keep you going. Don't worry though, your body has another trick up its sleeve...

What is anaerobic respiration?

When you do vigorous exercise and your body can't supply enough oxygen to your muscles, they start doing anaerobic respiration as well as aerobic respiration.

"Anaerobic" just means "without oxygen". It's the incomplete breakdown of glucose, making **lactic acid**. Here's the word equation for anaerobic respiration in muscle cells:

glucose → lactic acid

Anaerobic respiration does not transfer nearly as much energy as aerobic respiration. This is because glucose isn't fully oxidised (because it doesn't combine with oxygen). So, anaerobic respiration is only useful in emergencies, e.g. during exercise when it allows you to keep on using your muscles for a while longer.

Oxygen debt

After resorting to anaerobic respiration, when you stop exercising you'll have an "oxygen debt". An oxygen debt is the amount of extra oxygen your body needs after exercise. In other words, you have to "repay" the oxygen that you didn't get to your muscles in time, because your lungs, heart and blood couldn't keep up with the demand earlier on.

This means you have to keep breathing hard for a while after you stop, to get more oxygen into your blood, which is transported to the muscle cells.

Removing lactic acid Higher

The extra oxygen that 'repays' the oxygen debt is needed to react with the lactic acid in the muscles to form harmless carbon dioxide and water. You can define oxygen debt like this:

Oxygen debt is the amount of extra oxygen the body needs after exercise to react with the build up of lactic acid and remove it from the cells.

The pulse and breathing rate stay high whilst there are high levels of lactic acid and carbon dioxide to deliver more oxygen to the cells and take away the carbon dioxide.

Your body also has another way of coping with the high level of lactic acid — the blood that enters your muscles transports the lactic acid to the liver. In the liver, the lactic acid is converted back to glucose.

Learning Objectives:

- Know that respiration can take place anaerobically (without oxygen) in cells to transfer energy from glucose.
- Know that anaerobic respiration in muscle cells produces lactic acid, and be able to write the word equation for this.
- Know that in anaerobic respiration, glucose doesn't get fully oxidised, which means that not as much energy is transferred as in aerobic respiration.
- Know that anaerobic respiration in muscle cells produces an oxygen debt which must be repaid after exercise.
- H Be able to define oxygen debt.
- H Know how lactic acid is removed from the muscles.
- Know the word equation for anaerobic respiration in plant and yeast cells.
- Know that anaerobic respiration in yeast cells is called fermentation, and that the process has economic importance in bread and alcoholic drink production.
- Be able to compare aerobic and anaerobic respiration in terms of the need for oxygen, the products and how much energy is transferred.

Specification References
4.4.2.1, 4.4.2.2

Anaerobic respiration in plants and yeast

Tip: Yeast are single-celled organisms.

Anaerobic respiration in plants and yeast is slightly different. Plants and yeast cells can respire without oxygen too, but they produce ethanol (alcohol) and carbon dioxide instead of lactic acid.

Word equation

You need to learn the overall word equation for anaerobic respiration in plants and yeast cells:

glucose → ethanol + carbon dioxide

Tip: Bread and alcoholic drinks make a lot of money, so yeast plays an important part in the economy.

Fermentation

Anaerobic respiration in yeast cells is called **fermentation**. In the food and drinks industry, fermentation by yeast is of great value because it's used to make bread and alcoholic drinks, e.g. beer and wine. In bread-making, it's the carbon dioxide from fermentation that makes bread rise. In beer and wine-making, it's the fermentation process that produces alcohol.

Figure 1: *Bread dough being left to rise in an industrial vat.*

Comparing aerobic and anaerobic respiration

You need to be able to compare aerobic and anaerobic respiration.

	Aerobic respiration	Anaerobic respiration
Is oxygen needed?	yes	no
What products are made?	CO_2 and water	lactic acid (muscles) / CO_2 and ethanol (plants & yeast)
How much energy is transferred?	A large amount.	A small amount.

Practice Questions — Fact Recall

Q1 Explain when the body uses anaerobic respiration.

Q2 Explain what an oxygen debt is.

Q3 How does the liver help to get rid of lactic acid?

Q4 Give the word equation for anaerobic respiration in yeast.

Q5 a) What is anaerobic respiration in yeast cells also known as?

b) Give two industrial uses of this reaction.

9. Exercise

When we exercise we need to get more glucose and oxygen to our muscles for respiration. The body has some clever ways of doing this...

Learning Objectives:

- Know that the body needs to react to provide more energy during exercise.
- Be able to describe how the heart rate, breathing rate and breath volume change during exercise to get more oxygen to the muscles.
- Know that long periods of exercise can cause muscle fatigue, where the muscles get tired and stop contracting efficiently.

Specification Reference 4.4.2.2

Energy for exercise

Muscle cells use oxygen to transfer energy from glucose (aerobic respiration — see page 173), which is used to contract the muscles.

When you exercise, some of your muscles contract more frequently than normal so you need more energy. This energy comes from increased respiration. The increase in respiration in your cells means you need to get more oxygen into them. For this to happen, the blood has to flow at a faster rate. Therefore physical activity:

- increases your breathing rate and breath volume to meet the demand for extra oxygen.
- increases your heart rate (the speed at which the heart pumps) to make your blood flow more quickly, delivering more oxygen and glucose to cells for respiration, and taking more carbon dioxide away.

When you do really vigorous exercise (like sprinting) your body can't supply oxygen to your muscles quickly enough, so they start respiring anaerobically (see page 175). This is not the best way to transfer energy from glucose because lactic acid builds up in the muscles, which gets painful. Long periods of exercise also cause **muscle fatigue** — the muscles get tired and then stop contracting efficiently.

Tip: Remember, lactic acid is formed from the incomplete oxidation of glucose.

Recovery period

After anaerobic exercise stops, there will be an "oxygen debt". Heart rate and breathing rate stay higher than normal for a while after exercise whilst the oxygen debt is being paid back. This is known as the recovery period — see Figure 1.

Exam Tip
In the exam you might be given some data on heart rate or breathing rate during exercise and asked to say what is happening and why. So make sure you learn this page really well.

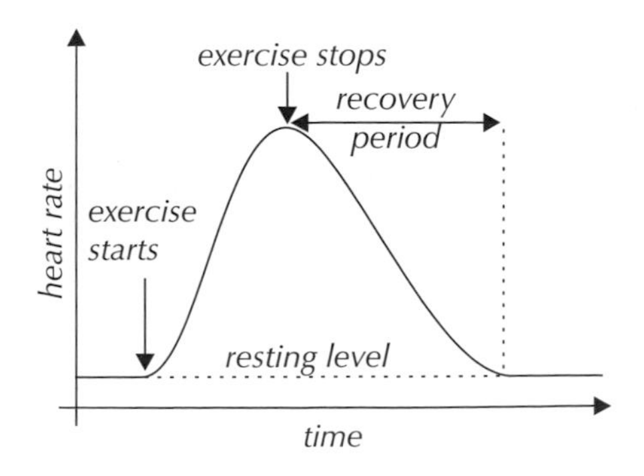

Figure 1: *Graph showing how a person's heart rate changes during exercise.*

Tip: The heart rate at resting level is just your heart rate when you're not exercising.

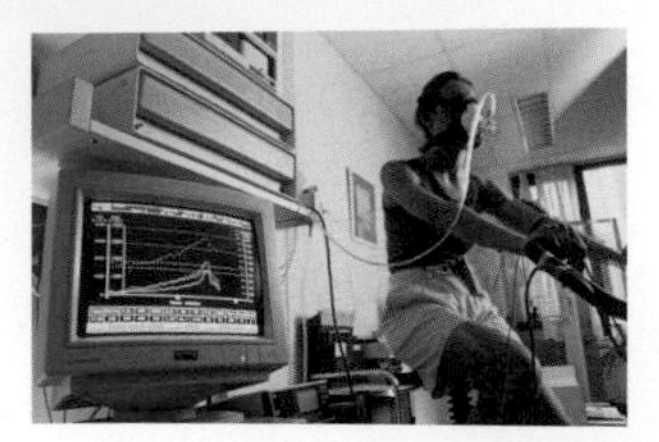

Figure 2: A person having their breathing rate and heart rate recorded during exercise.

Investigating the effect of exercise

You can measure breathing rate by counting breaths, and heart rate by taking the pulse.

Example

To take a pulse, place two fingers on the inside of the wrist (see Figure 3) and count the number of beats in 1 minute. You can also count the number of beats in 30 seconds and then double it.

You could take your pulse after:

- sitting down for 5 minutes,
- then after 5 minutes of gentle walking,
- then again after 5 minutes of slow jogging,
- then again after running for 5 minutes,

and plot your results in a bar chart.

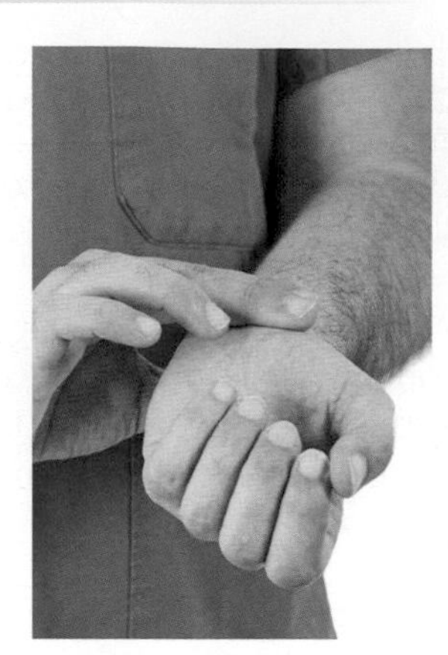

Figure 3: *Taking a pulse.*

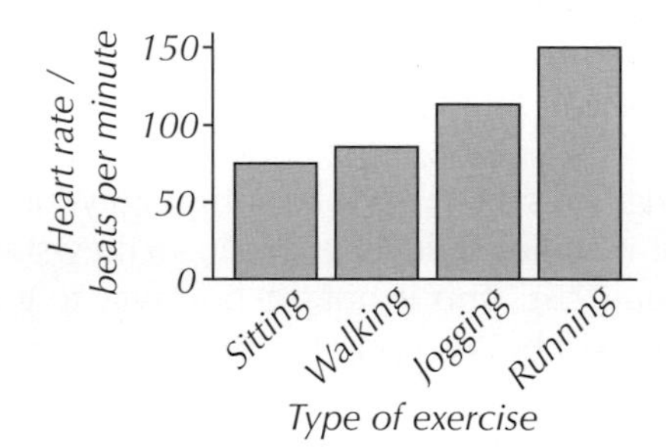

Figure 4: *Results from reading the heart rate after different activities.*

Your heart rate will increase the more intense the exercise is, as your body needs to get more oxygen to the muscles and take more carbon dioxide away from the muscles.

To reduce the effect of any random errors on your results, do it as a group and plot the average heart rate for each exercise.

Tip: You can also take a pulse from the side of the neck.

Tip: Beats per minute (bpm) is a unit used for measuring heart rate.

Tip: There's more about random error on page 13.

Practice Question — Application

Q1 Charlotte is working out on the cross trainer machine at the gym. As she's running, her heart rate increases from 80 to 146 bpm. Describe two other responses her body will have to the physical demands of running.

10. Metabolism

Metabolism is going on all of the time, so it's worthwhile knowing about it. That, and it could come up in your exam too...

What is metabolism?

In a cell there are lots of chemical reactions happening all the time, which are controlled by **enzymes**. Many of these reactions are linked together to form bigger reactions:

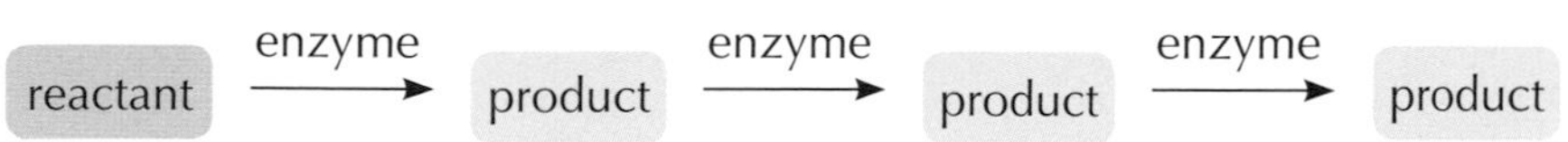

In some of these reactions, larger molecules are made from smaller ones. Overall these reactions are **endothermic** (they take in energy from the environment).

Examples

- Lots of small glucose molecules are joined together in reactions to form starch (a storage molecule in plant cells), glycogen (a storage molecule in animal cells) and cellulose (a component of plant cell walls).
- Lipid molecules are each made from one molecule of glycerol and three fatty acids.
- Glucose is combined with nitrate ions to make amino acids, which are then made into proteins.

In other reactions, larger molecules are broken down into smaller ones. Overall these reactions are **exothermic** (they transfer energy to the environment). For example:

Examples

- Glucose is broken down in respiration. Respiration transfers energy to power all the reactions in the body that make molecules.
- Excess protein is broken down in a reaction to produce urea. Urea is then excreted in urine.

The sum (total) of all of the reactions that happen in a cell or the body is called its metabolism.

Practice Questions — Fact Recall

Q1 What does a lipid molecule consist of?

Q2 What is formed when glucose combines with nitrate ions?

Q3 What component of urine is produced from the breakdown of excess protein?

Q4 What is meant by metabolism?

Learning Objectives:

- Know that the processes in metabolism use energy transferred by respiration and are controlled by enzymes.
- Be able to give examples of reactions involved in metabolism, including ones where larger molecules are made from smaller ones, and where larger molecules are broken down into smaller ones.
- Be able to explain how smaller molecules (glucose, amino acids, fatty acids and glycerol) are important in the formation and breakdown of larger ones (carbohydrates, proteins and lipids).
- Know that the sum of all of the reactions that happen in a cell, or in the body, is called its metabolism.

Specification Reference 4.4.2.3

Tip: Enzymes are biological catalysts — see page 115.

Topic 4 Checklist — Make sure you know...

The Basics of Photosynthesis

- ☐ That energy is transferred to chloroplasts when they absorb light for photosynthesis.
- ☐ That photosynthesis is an endothermic reaction because energy is transferred from the environment.
- ☐ The word equation for photosynthesis: carbon dioxide + water $\xrightarrow{\text{light}}$ glucose + oxygen
- ☐ The chemical symbols for carbon dioxide (CO_2), water (H_2O), oxygen (O_2) and glucose ($C_6H_{12}O_6$).

How Plants Use Glucose

- ☐ That plants use glucose produced during photosynthesis in many ways, including for respiration, to make cellulose for cell walls, to make amino acids for protein synthesis.
- ☐ That proteins in plants are made by combining glucose with nitrate ions from the soil to make amino acids, which are then joined together to form proteins.
- ☐ That glucose from photosynthesis can be converted to fats, oils or insoluble starch for storage.

The Rate of Photosynthesis

- ☐ How the rate of photosynthesis can be affected by temperature, the level of carbon dioxide, light intensity, or by the amount of chlorophyll in the plant and know that any of these factors can be the limiting factor (that stops photosynthesis from happening).
- ☐ How to draw, read from and interpret graphs that show the rate of photosynthesis linked to one limiting factor.
- ☐ H That the factors that affect the rate of photosynthesis can interact, and be able to work out which is the limiting factor from a graph that shows more than one factor.

Investigating Photosynthesis Rate

- ☐ How to investigate how light intensity affects the rate of photosynthesis of an aquatic plant (Required Practical 6), and how to calculate the rate of photosynthesis (e.g. by dividing the length of the bubble produced by the time taken to produce it).

The Inverse Square Law

- ☐ H That inverse proportion is when one variable increases as the other decreases at the same rate.
- ☐ H That the inverse square law is an example of inverse proportion that links together light intensity and distance between the light source and a plant, and be able to use it.

Artificially Controlling Plant Growth

- ☐ H That in greenhouses, the factors that affect the rate of photosynthesis can be controlled to increase the rate of photosynthesis, which in turn increases the profit made from the plants.
- ☐ H How to interpret data that relates to controlling the conditions in a greenhouse.

cont...

Aerobic Respiration

- ☐ That respiration is an exothermic reaction because it transfers energy to the environment, and that it transfers all the energy needed for living processes.
- ☐ That aerobic respiration is the process of transferring energy from glucose using oxygen, and that it occurs continuously in living cells.
- ☐ The word equation for aerobic respiration: glucose + oxygen → carbon dioxide + water
- ☐ The chemical symbols for glucose ($C_6H_{12}O_6$), oxygen (O_2), carbon dioxide (CO_2) and water (H_2O).
- ☐ That the energy transferred from respiration is used by organisms to make larger molecules from smaller ones, for muscle contraction and to keep warm.

Anaerobic Respiration

- ☐ That muscles will start to respire anaerobically (without oxygen) if they can't get enough oxygen to respire aerobically, and that this involves the incomplete oxidation of glucose, which forms lactic acid: glucose → lactic acid. Anaerobic respiration transfers less energy than aerobic respiration.
- ☐ That an oxygen debt occurs after anaerobic respiration in muscle cells — you have to 'repay' the oxygen you didn't manage to get to your muscles in time.
- ☐ H That lactic acid is removed from the body by being carried by the blood to the liver (where it's turned back into glucose), or by reacting with oxygen. An oxygen debt is the amount of extra oxygen needed to react with and remove lactic acid from muscle cells after exercise.
- ☐ The word equation for anaerobic respiration in plants and yeast: glucose → ethanol + carbon dioxide
- ☐ That anaerobic respiration in yeast cells is called fermentation, and that this process has economic importance in the bread and alcohol industries.
- ☐ The differences between aerobic and anaerobic respiration, including whether they need oxygen, what products are made, and how much energy is transferred.

Exercise

- ☐ That during exercise, there is a increased demand for energy.
- ☐ That breathing rate and breath volume increase during exercise to get more oxygen into the body, and a person's heart rate increases during exercise so that blood flow to their muscles increases.
- ☐ That long periods of exercise can cause muscle fatigue, where the muscles don't contract efficiently.

Metabolism

- ☐ That enzymes control the reactions of metabolism, which use energy transferred by respiration.
- ☐ That metabolism includes reactions that build up larger molecules from smaller ones, and reactions that break down larger molecules into smaller ones.
- ☐ How glucose, amino acids, fatty acids and glycerol are important in the formation and breakdown of carbohydrates, proteins and lipids.
- ☐ That metabolism is the sum of every reaction in a cell or the body.

Exam-style Questions

1 Tim read the following information in a science magazine:

> You can measure the rate of photosynthesis in plants by doing the following:
> Cut discs from a plant leaf and put them in a 0.2% sodium hydrogencarbonate solution (which serves as a source of carbon dioxide). Put the solution and leaf discs in a syringe and then remove all the air from the syringe — this removes any gases out of the air spaces in the leaf, which will cause the leaf discs to sink.
> Then put the syringe containing the leaf discs under a light source. As the leaf discs photosynthesise, their cells produce oxygen. The oxygen will fill the air spaces in the leaf — with enough oxygen in the air spaces they'll begin to float. You can use the time this takes as a measure of the rate of photosynthesis.

1.1 Where in a plant cell does photosynthesis take place?

(1 mark)

1.2 Complete the word equation for photosynthesis:

carbon dioxide + ... ⟶ glucose + oxygen

(1 mark)

Tim decided to use the method to investigate the effect of light intensity on the rate of photosynthesis. He conducted the experiment at three different light intensities. For each light intensity he cut 10 equally sized discs from a leaf, and set up the experiment as described in the magazine. He then timed how long it took for all of the leaf discs to float to the surface in the syringe. His results are shown in this table:

Light intensity	Time it took for all discs to float (minutes)
A	18
B	9
C	11

1.3 Which light intensity do you think was the highest, A, B or C? Explain your answer.

(2 marks)

1.4 Tim thinks that if he keeps increasing the light intensity, he will keep increasing the rate of photosynthesis. Is he right? Explain your answer.

(3 marks)

1.5 Tim conducts the experiment again using light intensity A. This time he uses a solution containing 0.8% sodium hydrogen carbonate. Do you think it will take more or less than 18 minutes for all of the discs to float? Explain your answer.

(2 marks)

1.6* The process of photosynthesis produces glucose.
Describe as fully as you can how this glucose may be used by a plant.

(6 marks)

2 A student wants to find out how his heart rate changes during exercise. The exercise he chooses is to run on a running machine. He records his heart rate before the run, then again after five minutes of running. He repeats the exercise three times.

His results are shown in this table:

Time of record	Heart rate (beats per minute)		
	1st go	2nd go	3rd go
At rest, before run	72	71	72
After 5 minutes of running	151	163	154

2.1 Calculate the student's mean heart rate at the end of his runs.

(2 marks)

Respiration gives the student energy to contract his muscles and run.

2.2 As the student was running, his heart rate increased from its resting level. Explain why.

(2 marks)

2.3 Give **two** other ways the student's body will use energy from respiration.

(2 marks)

3 A scientist is investigating the reactions of aerobic and anaerobic respiration.

3.1 What is the word equation for aerobic respiration?

(4 marks)

3.2 In a cell, energy is stored in a molecule called ATP.
The scientist conducts an experiment to find out how much ATP each type of respiration forms per glucose molecule.

His results are shown in this table:

Type of respiration	Number of ATP molecules released per glucose molecule
Aerobic	32
Anaerobic	2

Using the table, describe and explain the difference in the amount of energy transferred by aerobic and anaerobic respiration.

(2 marks)

1. Homeostasis

The internal conditions in the body must be kept constant — this stops cell damage that could happen if we didn't keep on top of things. Luckily there are some clever systems in place to keep everything plodding along steadily.

Learning Objectives:
- Understand that the internal conditions in the body must be carefully controlled so that cells and enzymes can function properly.
- Know what homeostasis means.
- Know that control systems can include nervous or chemical responses.
- Know that there are control systems in the human body that regulate body temperature, blood glucose concentration and water content.
- Know that a control system includes receptors, a coordination centre and effectors.
- Know that changes in the environment (stimuli) are detected by receptors.
- Know that information travels from receptors to coordination centres, which coordinate a response.
- Know that effectors respond to bring a system back to its normal level.

Specification Reference
4.5.1

What is homeostasis?

The conditions inside your body need to be kept steady, even when the external environment changes. This is really important because your cells need the right conditions in order to function properly, including the right conditions for enzyme action (see p. 116). Homeostasis is the way in which everything is kept at the right level.

> Homeostasis is the regulation of the conditions inside your body (and cells) to maintain a stable internal environment, in response to changes in both internal and external conditions.

You have loads of automatic **control systems** in your body that regulate your internal environment — these include both nervous and hormonal (chemical) communication systems. The conditions in your internal environment that need regulating include:

Examples
- Body temperature (see pages 200-202).
- Blood glucose content (see pages 210-211).
- Water content of the body (see pages 215-217).

All your automatic control systems are made up of three main components which work together to maintain a steady condition — cells called **receptors**, **coordination centres** (including the brain, spinal cord and pancreas) and **effectors** (muscles or glands — see page 186).

Stimuli and negative feedback

A change in your environment that you might need to respond to is called a **stimulus** (the plural is 'stimuli'). A stimulus can be light, sound, touch, pressure, pain, chemical, or a change in position or temperature.

The receptors of an automatic control system detect a stimulus when the level of something (e.g. water or temperature) is too high or too low.
They then send this information to the coordination centre, which processes the information and organises a response from the effectors. The effectors respond to counteract the change — bringing the level back to its optimum. The mechanism that restores the **optimum level** is called a **negative feedback mechanism** — see Figure 1 on the next page.

Tip: The optimum level is just the level that enables the body to work at its best — e.g. the body has an optimum level of water and temperature.

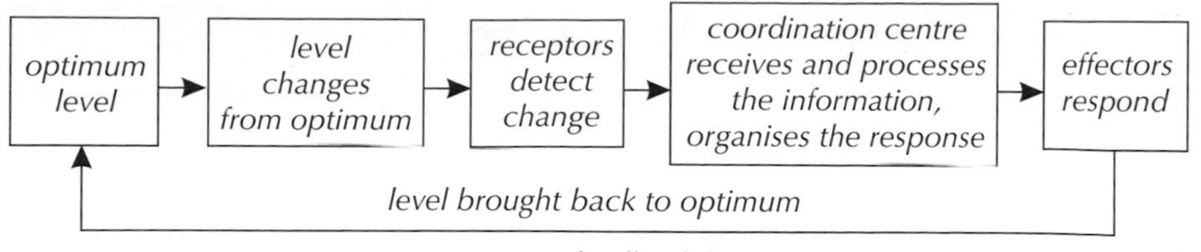

Figure 1: *A negative feedback mechanism.*

The effectors will just carry on producing the responses for as long as they're stimulated by the coordination centre. This might cause the opposite problem — making the level change too much (away from the ideal). Luckily the receptors detect if the level becomes too different and negative feedback starts again, so the level is kept at its optimum. This process happens without you thinking about it — it's all automatic.

Tip: A negative feedback system responds when a level changes from its optimum point, in order to bring the level back to optimum. It's a continuous, looping process.

Example

Body temperature is usually kept within 0.5 °C above or below 37 °C.

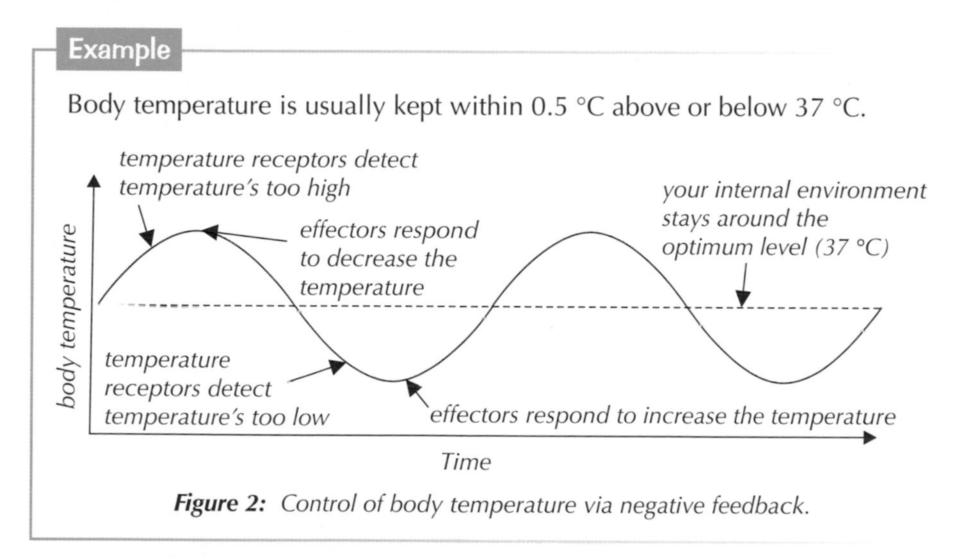

Figure 2: *Control of body temperature via negative feedback.*

Tip: Body temperature is kept level by several different effectors that work together to give a precise response to any changes — see pages 200-202.

Practice Questions — Fact Recall

Q1 Define homeostasis.

Q2 Give two examples of conditions in your internal environment that need regulating and maintaining.

Q3 What is a stimulus?

Q4 Name the part of an automatic control system that receives information about a stimulus and organises a response.

Practice Question — Application

Q1 The water content of the body is monitored by a control system. If it gets too low, a negative feedback mechanism is triggered. Briefly outline the stages in this negative feedback system.

2. The Nervous System

Next up, an overview of the nervous system...

Learning Objectives:
- Know that the nervous system allows humans to respond to their environment and coordinate their behaviour.
- Know that information from receptors travels via neurones as electrical impulses to the central nervous system (CNS), where the response of the effectors (muscles contracting or glands secreting hormones) is coordinated.
- Know that the CNS consists of the brain and spinal cord.
- Be able to explain how the structure of the nervous system is adapted to its functions.
- Know the path that a nervous impulse takes in response to a stimulus: stimulus – receptor – coordinator – effector – response
- Be able to read from and interpret tables, charts and graphs containing data about nervous system function.

Specification Reference 4.5.2.1

What is the nervous system?

Organisms need to respond to **stimuli** (changes in the environment) in order to survive. A single-celled organism can just respond to its environment, but the cells of multicellular organisms need to communicate with each other first. So as multicellular organisms evolved, they developed nervous and hormonal communication systems.

Your nervous system is what allows you to react to your surroundings. It also allows you to coordinate your behaviour. It's made up of all the **neurones** (nerve cells) in your body — there's more on these on the next page.

Receptors

Receptors are the cells that detect stimuli. There are many different types of receptors, such as:

Examples
- Taste receptors on the tongue.
- Sound receptors in the ears.
- Smell receptors in the nose.
- Light receptors in the eyes.

Receptors can form part of larger, complex organs, e.g. the retina of the eye is covered in light receptor cells.

The central nervous system (CNS)

The central nervous system (CNS) is where all the information from the receptors is sent, and where reflexes (see p. 189) and actions are coordinated.

In vertebrates (animals with backbones) this consists of the brain and spinal cord only (see Figure 1). In mammals, the CNS is connected to the body by sensory neurones and motor neurones. Neurones transmit information as electrical impulses to and from the CNS. This happens very quickly.

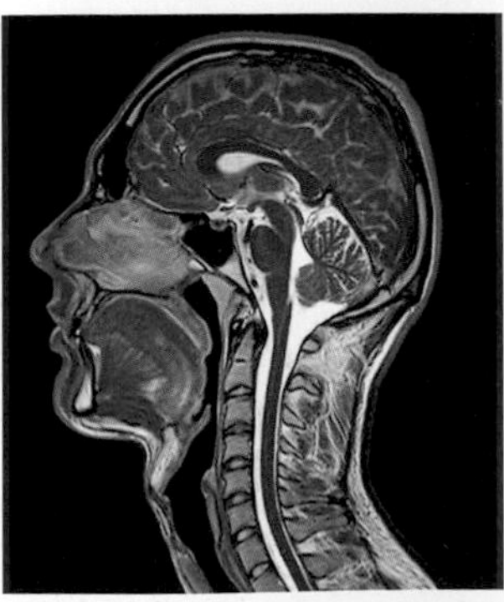

Figure 1: *Scan of the head and neck. The brain and upper spinal cord can be seen in orange/red.*

Effectors

'Instructions' from the CNS are sent along neurones to **effectors**. Effectors are **muscles** or **glands** which respond to nervous impulses and bring about a response to a stimulus. Muscles and glands respond to nervous impulses in different ways:

- Muscles contract.
- Glands secrete chemical substances called hormones (see page 207).

Different types of neurone

Different types of neurone are involved in the transfer of information to and from the CNS — see the next page.

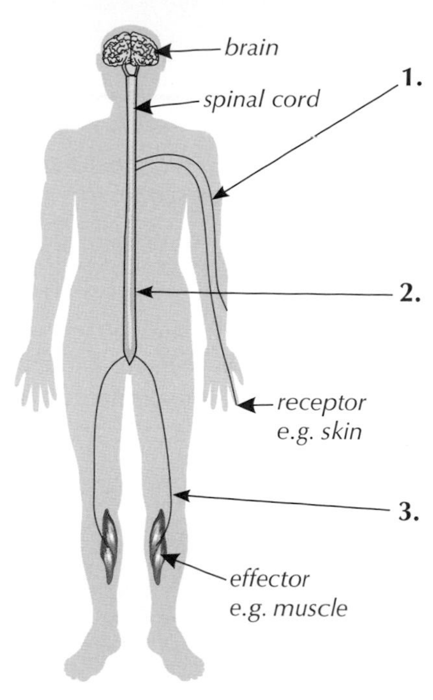

1. **Sensory neurones**
The neurones that carry information as electrical impulses from the receptors in the sense organs to the central nervous system.

2. **Relay neurones**
The neurones that carry electrical impulses from sensory neurones to motor neurones. They are found in the central nervous system.

3. **Motor neurones**
The neurones that carry electrical impulses from the central nervous system to the effectors.

Exam Tip
The exam isn't just a test of what you know — it's also a test of how well you can apply what you know. For instance, you might have to take what you know about a human and apply it to a horse (e.g. sound receptors in its ears send information to the brain via sensory neurones). The key is not to panic — just think carefully about the information that you are given.

The transmission of information to and from the CNS is summarised in Figure 2:

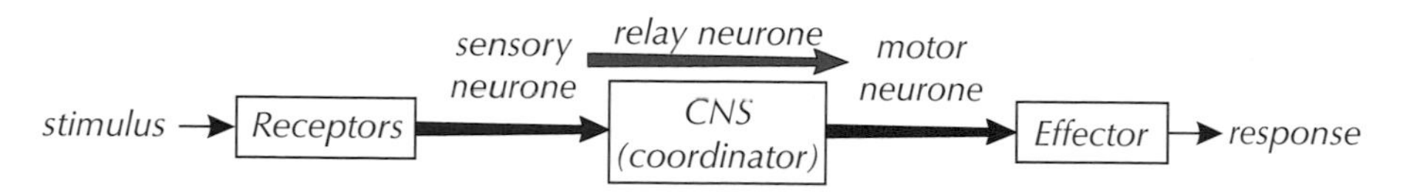

Figure 2: *Flow diagram showing the transmission of information to and from the CNS.*

Exam Tip
In your exam, you might need to read off values related to nervous system function from a table, graph or chart, or explain what the data is showing. Remember to always check any units or graph scales carefully to avoid throwing away marks.

Practice Questions — Fact Recall

Q1 Which type of effector secretes hormones?

Q2 A sensory neurone is a type of neurone.
Name two other types of neurone.

Practice Question — Application

Q1 A dog hears a cat moving in the garden, so runs towards it.

a) i) What is the stimulus in this situation?

ii) What detects the stimulus in this situation?

iii) What name is given to the type of neurone that transmits information about the stimulus to the central nervous system?

The dog's brain sends an impulse to the dog's muscles which act as an effector.

b) i) What type of neurone transmits information from the central nervous system to an effector?

ii) How do the dog's muscles respond to the nerve impulse?

Exam Tip
Make sure you know which type of neurone transfers information to which place in the nervous system — it's easy to get the different types of neurones confused and they often come up in exams.

3. Synapses and Reflexes

Learning Objectives:

- Know the structure and function of a synapse.
- Know that reflexes are fast, automatic responses to stimuli that don't involve the conscious brain.
- Understand the importance of reflex reactions.
- Know the order of events in a simple reflex arc and be able to explain how the structures of a reflex arc are related to their function.

Specification Reference 4.5.2.1

Reflexes are rapid responses to stimuli that happen without you having to think about them — they're automatic. The neurones involved in a reflex aren't all joined together though — they have gaps between them called synapses.

Synapses

The connection between two neurones is called a synapse. The nerve signal is transferred by chemicals which diffuse (move) across the gap. These chemicals then set off a new electrical signal in the next neurone. This is shown in Figure 1.

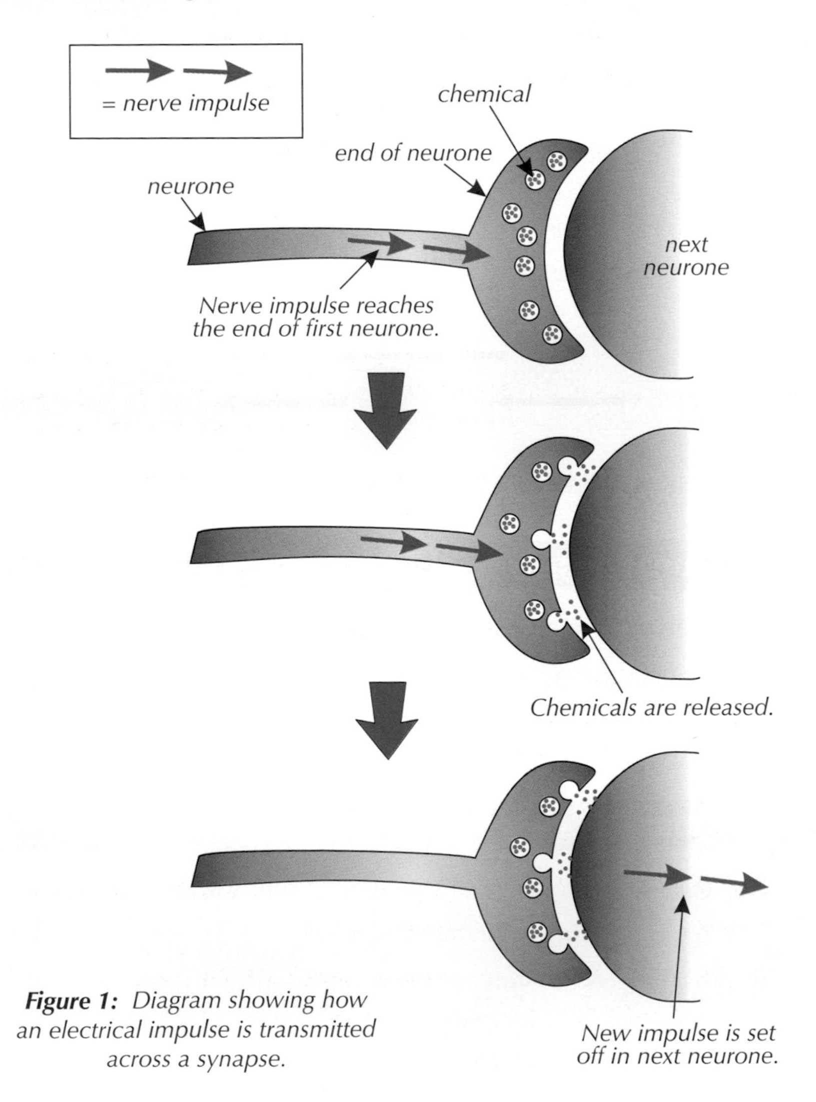

Figure 1: *Diagram showing how an electrical impulse is transmitted across a synapse.*

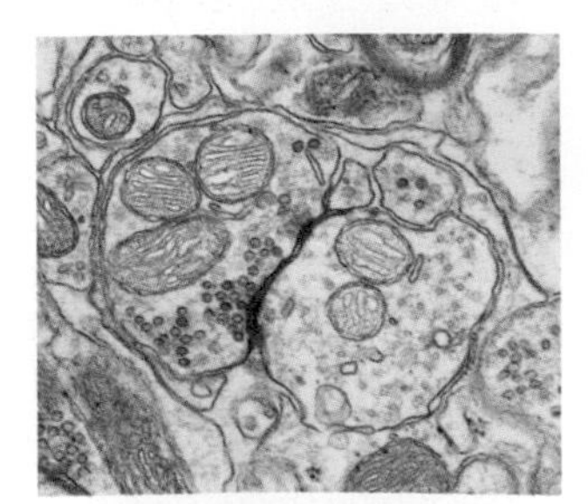

Figure 2: *A synapse viewed under a microscope. The neurones are shown in green. The red dots contain the chemicals that diffuse between the neurones.*

Neurones deliver information really quickly because the signal is transmitted by electrical impulses. Synapses slow down the transmission of a nervous impulse because the diffusion of chemicals across the gap takes time (it's still pretty fast though).

Reflexes

Reflexes are fast, automatic responses to certain stimuli. They bypass your conscious brain completely when a quick response is essential — your body just gets on with things. Reflexes can reduce your chance of being injured, although they have other roles as well.

Tip: 'Automatic' means done without thinking.

Examples

- If someone shines a bright light in your eyes, your pupils automatically get smaller. This means that less light gets into your eyes, which stops them getting damaged.
- Adrenaline is a hormone (see p. 235), which gets your body ready for action. If you get a shock, your body releases adrenaline automatically — it doesn't wait for you to decide that you're shocked.
- The knee-jerk reflex helps maintain posture and balance. Doctors test this reflex by tapping just below the knee with a small hammer. This stimulates pressure receptors, making a muscle in the upper leg contract, which causes the lower leg to rise up.

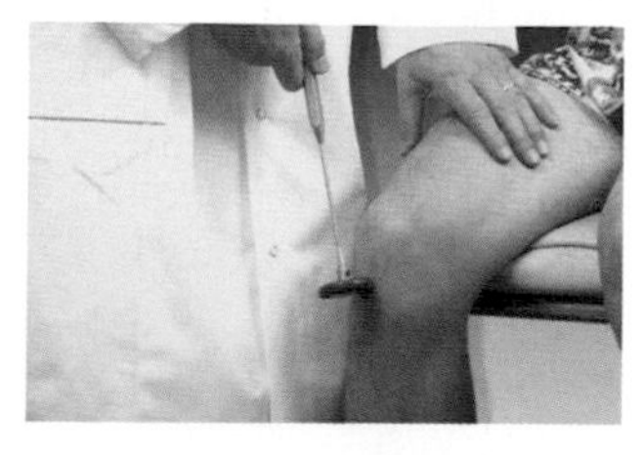

Figure 3: *The knee-jerk reflex. Doctors often use this test to see if a patient's reflexes are working properly.*

Reflex arcs

The passage of information in a reflex (from receptor to effector) is called a reflex arc. The neurones in reflex arcs go through the spinal cord or through an unconscious part of the brain. Here are the main stages in a reflex arc:

1. When a stimulus is detected by receptors, impulses are sent along a sensory neurone to a relay neurone in the CNS.
2. When the impulses reach a synapse between the sensory neurone and the relay neurone, they trigger chemicals to be released (see previous page). These chemicals cause impulses to be sent along the relay neurone.
3. When the impulses reach a synapse between the relay neurone and a motor neurone, the same thing happens. Chemicals are released and cause impulses to be sent along the motor neurone.
4. The impulses then travel along the motor neurone to the effector (which is usually a muscle).
5. If the effector is a muscle, it will respond to the impulse by contracting. If it's a gland, it will secrete a hormone.

Tip: Flick back to page 187 for more on sensory, relay and motor neurones.

Tip: Remember, electrical impulses pass between the different neurones via diffusion of chemicals at the synapse. They don't just jump between the neurones.

Because you don't have to think about the response (which takes time) a reflex is quicker than normal responses. Figure 4 summarises a reflex arc:

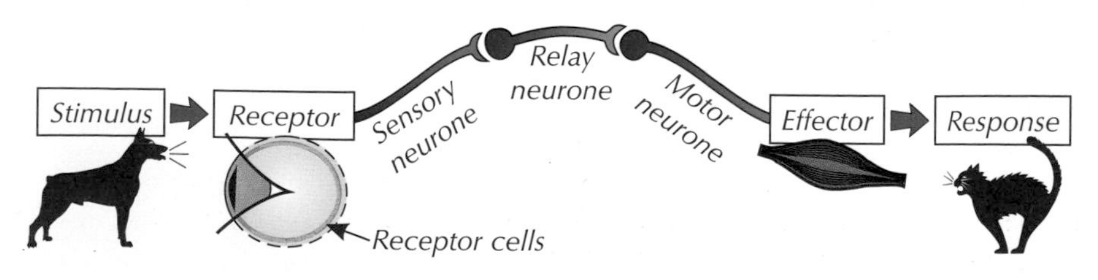

Figure 4: *Block diagram of a reflex arc.*

Tip: The important things to remember about reflexes are that they're automatic (you don't have to think about them) and rapid (so they're often able to prevent or reduce the chance of injury).

Example

If a bee stings a person's finger, the reflex response is that the hand moves away from the source of pain. Here's the pathway taken by this reflex arc:

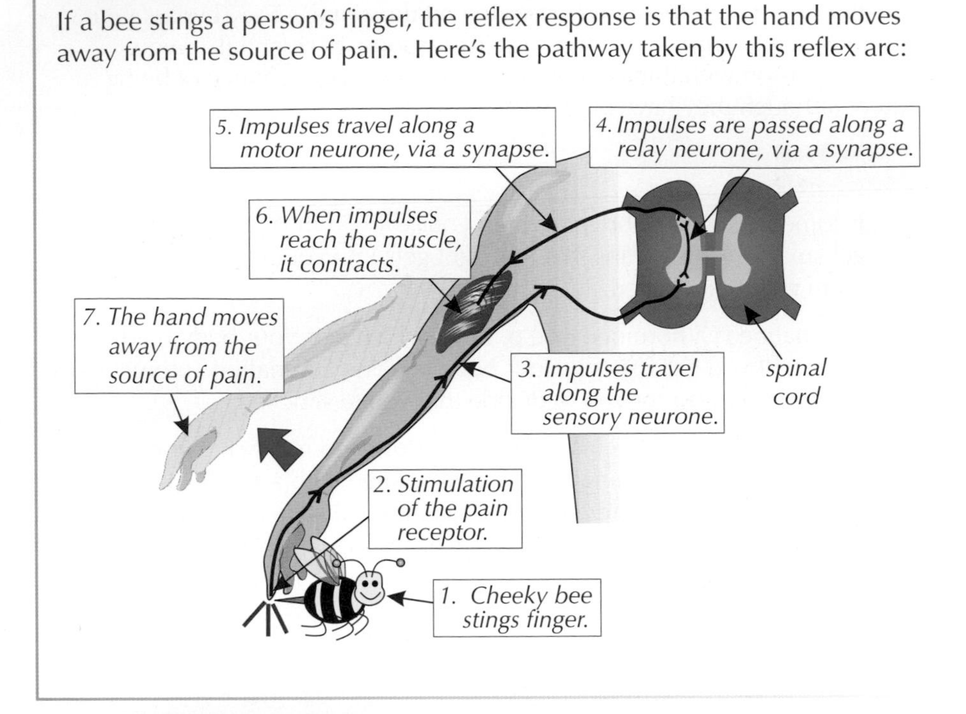

Practice Questions — Fact Recall

Q1 How do nerve impulses travel between two neurones?

Q2 What is a reflex?

Q3 Do reflex arcs travel through conscious parts of the brain?

Q4 What neurone comes after the sensory neurone in a reflex arc?

Q5 If the effector in a reflex arc is a muscle, what will the response of that reflex arc be?

Practice Question — Application

Q1 David steps on a drawing pin and immediately pulls his foot up off the pin.

a) What is the stimulus in this response?

b) What is the effector in this response? How does it respond?

c) Complete the pathway taken by this reflex arc.

Stimulus → Receptor → Sensory neurone...

Exam Tip
Questions on reflex arcs are quite common in the exam. Make sure you know the general pathway they take really well, and you'll be able to apply it to any reflex response you're given.

4. Investigating Reaction Time

The time it takes to react to a stimulus depends on how quickly the message travels from the receptor to the effector, via the central nervous system. This is called the reaction time, and it varies from person to person.

What is reaction time?

Reaction time is the time it takes to respond to a stimulus — it's often less than a second. It can be affected by factors such as age, gender or drugs.

Measuring reaction time

You need to know how to investigate the effect of a factor on a person's reaction time.

Example

Caffeine is a drug that can speed up a person's reaction time. The effect of caffeine on reaction time can be measured. You will need another person for this experiment — this is the person whose reaction time will be tested. You will provide the stimulus. Here's what to do:

1. The person being tested should sit with their arm resting on the edge of a table (this should stop them moving their arm up or down during the test).
2. Hold a ruler vertically between their thumb and forefinger. Make sure that the zero end of the ruler is level with their thumb and finger. Then let go without giving any warning.
3. The person being tested should try to catch the ruler as quickly as they can — as soon as they see it fall.
4. Reaction time is measured by the number on the ruler where it's caught, at the top of the thumb (see Figure 1) — the further down it's caught (i.e. the higher the number), the slower their reaction time.

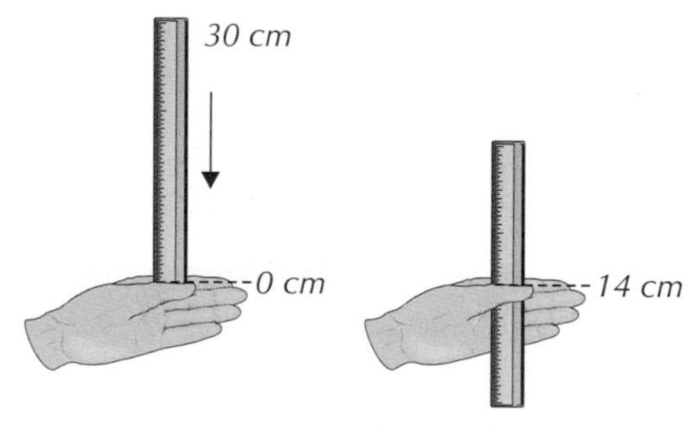

Figure 1: *A person catching a ruler in a drop test.*

5. Repeat the test several times then calculate the mean distance that the ruler fell.

The person being tested should then have a caffeinated drink (e.g. 300 ml of cola). After ten minutes, repeat steps 1–5.

Learning Objective:

- Be able to investigate how a factor affects reaction time (Required Practical 7).
- Be able to convert between reaction time data shown in a graph and in numerical form.

Specification Reference 4.5.2.1

Tip: The type of stimulus (e.g. whether you see, hear or touch it) can also affect reaction time.

Tip: Remember to do a risk assessment for this experiment. There are some examples of things you need to think about on the next page.

Tip: It's possible to work out the reaction time in seconds by popping the mean distance (in cm) into this formula:

$$\text{time} = \sqrt{\frac{\text{mean distance}}{490}}$$

But you don't need to learn this formula.

Tip: Reaction times can be plotted on a graph to show possible relationships, e.g.:

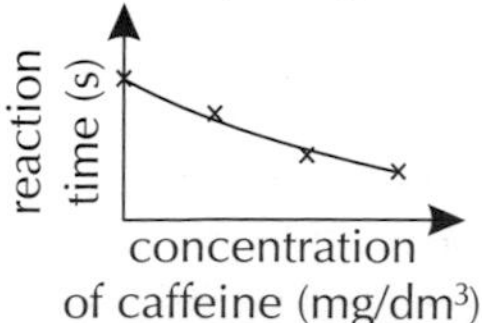

Make sure you know how to read off data from a graph like this, as well as how to draw one using reaction time data (see page 18).

Controlling variables

If you're carrying out the investigation on the previous page, you need to control any variables to make sure that it's a fair test.

Examples

- You should use the same person to catch the ruler each time.
- That person should always use the same hand to catch the ruler.
- The ruler should always be dropped from the same height.
- You should make sure that the person being tested has not had any caffeine (or anything else that may affect their reaction time) before the start of the experiment.

Tip: For more about variables, see page 10.

Safety issues

Testing reaction times seems simple enough, but there are still some important safety points you need to think about for this experiment. For example, any drinks need to be consumed outside of the lab, away from any chemicals or other hazards. Also, too much caffeine can cause unpleasant side-effects, so the person being tested should avoid drinking any more caffeine for the rest of the day after the experiment is completed.

Measuring reaction time using a computer

Simple computer tests can also be used to measure reaction time.

Example

The person being tested has to click the mouse (or press a key) as soon as they see a stimulus on the screen, e.g. a box change colour.

Figure 2: *The person clicks the mouse as quickly as they can when they see the stimulus on the computer screen.*

Computers can give a more precise reaction time because they remove the possibility of human error from the measurement. As the computer can record the reaction time in milliseconds, it can also give a more accurate measurement.

Using a computer can also remove the possibility that the person can predict when to respond — using the ruler test, the catcher may learn to anticipate the drop by reading the tester's body language.

Practice Questions — Application

Q1 A particular drug stops some of the chemical that transfers across a synapse from reaching the next neurone. Suggest what effect this drug would have on a person's reaction time.

Q2 a) Describe how you could use a ruler to investigate a person's reaction time.

b) Suggest two factors that could be tested to investigate their effect on reaction time.

5. The Brain

The brain contains a complex network of neurones. This is what makes the brain able to control complicated behaviours, and why it's so difficult to understand exactly what the brain does.

Learning Objectives:

- Know that the brain controls behaviour.
- Know that the brain contains billions of interlinked neurones and is split into regions that have different functions.
- Know the functions of the cerebral cortex, medulla and cerebellum, and be able to locate them on a diagram of the brain.
- H Know the methods used by neuroscientists to map regions of the brain.
- H Be able to explain why treating and investigating brain damage and disease is difficult.

Specification Reference 4.5.2.2

Brain function

Along with the spinal cord, the brain is part of the central nervous system. It's made up of billions of interconnected neurones (neurones that are connected together). The brain is in charge of all of our complex behaviours. It controls and coordinates everything you do — running, breathing, sleeping, remembering your gym kit...

We know that different regions of the brain carry out different functions — see below.

Regions of the brain

You need to know the function and location (see Figure 1) of these parts of the brain:

Cerebral cortex

The cerebral cortex is the outer wrinkly layer of the brain. It's responsible for things like consciousness, intelligence, memory and language.

Medulla

The medulla is at the base of the brain, at the top of the spinal cord. It controls unconscious activities (things you don't have to think about doing) like breathing and your heartbeat.

Cerebellum

The cerebellum is found at the back of the brain. It's responsible for muscle coordination.

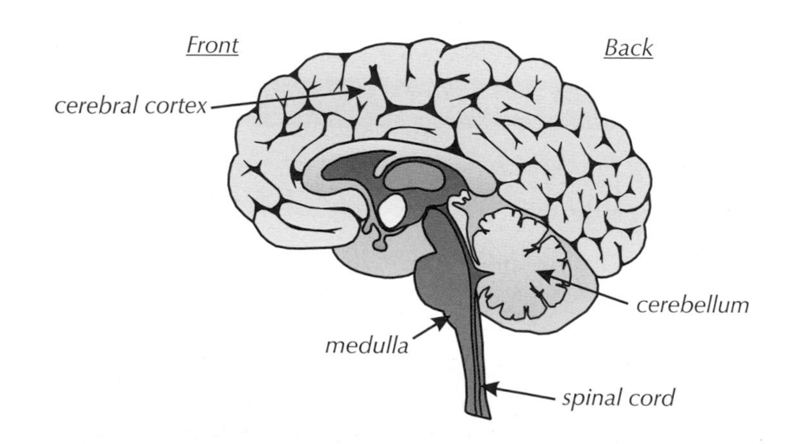

Figure 1: *Regions of the brain.*

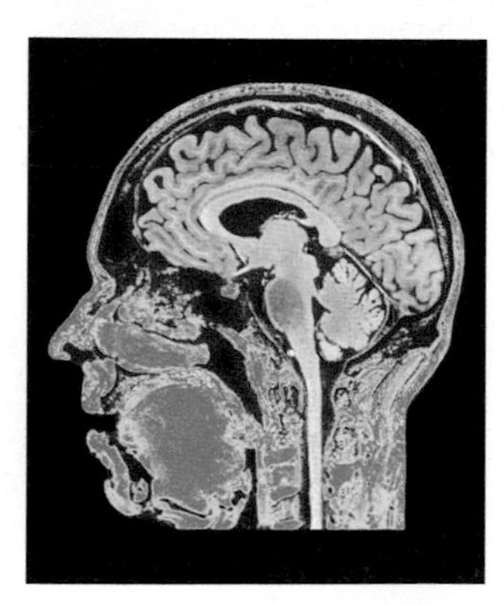

Figure 2: *A scan showing a section through the head. The cerebral cortex is the folded structure at the top (pink). The cerebellum (yellow) lies at the back of the head.*

Studying the brain Higher

Scientists that study the brain are called neuroscientists. Neuroscientists use a few different methods to study the brain and map out which bits do what:

Studying patients with brain damage

If a small part of the brain has been damaged, the effect this has on the patient can tell you a lot about what the damaged part of the brain does. E.g. if an area at the back of the brain was damaged by a stroke and the patient went blind, you know that that area has something to do with vision.

Tip: H A stroke is damage to or the death of brain cells due to lack of oxygen, caused by a disruption in the blood supply to the brain (e.g. by a blockage).

Electrically stimulating the brain

The brain can be stimulated electrically by pushing a tiny electrode into the tissue and giving it a small zap of electricity. By observing what stimulating different parts of the brain does, it's possible to get an idea of what those parts do. E.g. when a certain part of the brain (known as the motor area) is stimulated, it causes muscle contraction and movement.

MRI scans

A magnetic resonance imaging (MRI) scanner is a big fancy tube-like machine that can produce a very detailed picture of the brain's structures. Scientists use it to find out what areas of the brain are active when people are doing things like listening to music or trying to recall a memory.

Tip: H The image in Figure 1 on the previous page is an MRI scan of the brain.

Benefits and risks Higher

Knowledge of how the brain works has led to the development of treatments for disorders of the nervous system.

Examples Higher

- Electrical stimulation of the brain can help reduce muscle tremors caused by nervous system disorders such as Parkinson's disease.
- Surgery to remove a part of the brain that's causing seizures can be an option to treat epilepsy if it can't be controlled by medication.

However, the brain is incredibly complex and delicate — the investigation of brain function and any treatment of brain damage or disease is difficult. It also carries risks, such as physical damage to the brain or increased problems with brain function (e.g. difficulties with speech).

Exam Tip H
If you're asked to evaluate the benefits and risks of a procedure carried out on the brain, don't forget to give a balanced answer.

Practice Questions — Fact Recall

Q1 Name the region of the brain that is important for consciousness.

Q2 Where in the brain is the cerebellum located?

Q3 Give three ways that neuroscientists have been able to associate different parts of the brain with specific functions.

Practice Question — Application

Q1 Ataxia is a brain condition which can cause difficulty with walking. Suggest the part of the brain that may be affected in ataxia.

6. The Eye

The eye is a sense organ — it responds to changes in light (stimuli) by sending messages to the central nervous system. You need to know about several structures of the eye and how they work.

Learning Objectives:
- Know that the eye is a sense organ that detects light intensity and colour.
- Be able to identify the sclera, cornea, iris, retina, ciliary muscles, suspensory ligaments and optic nerve on a diagram of the eye, and know the functions of these structures.
- Know the role of the iris in adjusting the eye to changing amounts of light.
- Know that accommodation involves focussing on near or distant objects by changing the shape of the lens, and know how this is done.

Specification Reference 4.5.2.3

The structure of the eye

The eye has the following parts:

- **Sclera** — the tough, supporting wall of the eye.
- **Cornea** — the transparent outer layer found at the front of the eye. It refracts (bends) light into the eye.
- **Pupil** — the hole in the centre of the eye, through which light enters.
- **Iris** — contains muscles that allow it to control the diameter of the pupil and therefore how much light enters the eye.
- **Retina** — the layer at the back of the eye that contains two types of light receptor cells. One type is sensitive to light intensity and the other is sensitive to colour.
- **Lens** — focuses the light onto the retina.
- **Ciliary muscles** and **suspensory ligaments** — control the shape of the lens — see next page.
- **Optic nerve** — carries impulses from the receptors on the retina to the brain.

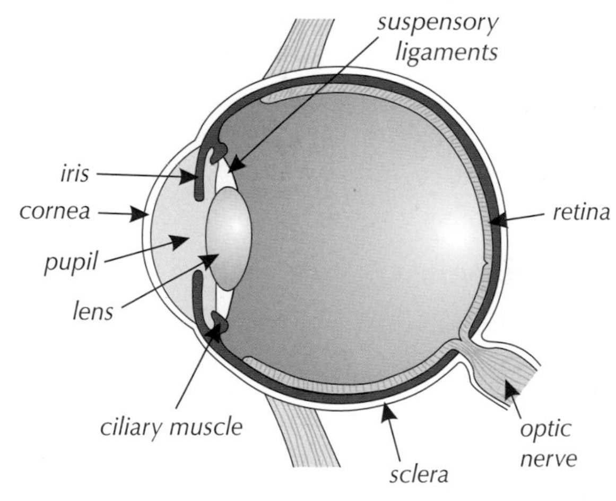

Figure 1: *Structures of the eye.*

The iris reflex

Very bright light can damage the retina — so you have a reflex to protect it.

When light receptors in the eye detect very bright light, a reflex is triggered that makes the pupil smaller. The circular muscles in the iris contract and the radial muscles relax. This reduces the amount of light that can enter the eye.

The opposite process happens in dim light. This time, the radial muscles contract and the circular muscles relax, which makes the pupil wider. The reflex reaction is shown in Figure 2 on the next page.

Tip: In the terms of a reflex reaction (p. 189), the light receptors in the retina detect the bright light and send a message along a sensory neurone to the CNS. The message then travels along a relay neurone in the CNS to a motor neurone, which tells circular muscles in the iris to contract, making the pupil smaller.

Tip: The pupil doesn't actually do anything during the iris reflex — it's just a hole, so although it looks like the pupil is changing size, it's really the iris that's actively doing the changing.

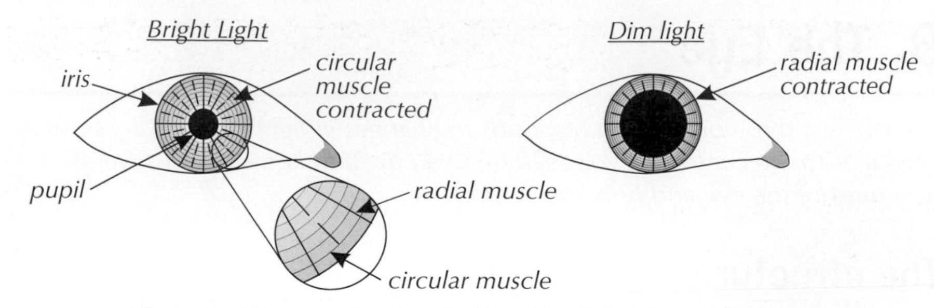

Figure 2: *Diagram showing the iris reflex in bright and dim light.*

Accommodation

The ability to look at near and distant objects is an example of a reflex — it's an automatic reaction, not controlled by the conscious brain (see page 189). The eye focuses light on the retina by changing the shape of the lens. This is known as accommodation.

Looking at near objects

To focus on near objects:

1. The ciliary muscles contract, which slackens the suspensory ligaments.
2. The lens becomes fat (more curved).
3. This increases the amount by which it refracts light.

Tip: The thicker the lens is, the more strongly it can bend rays of light.

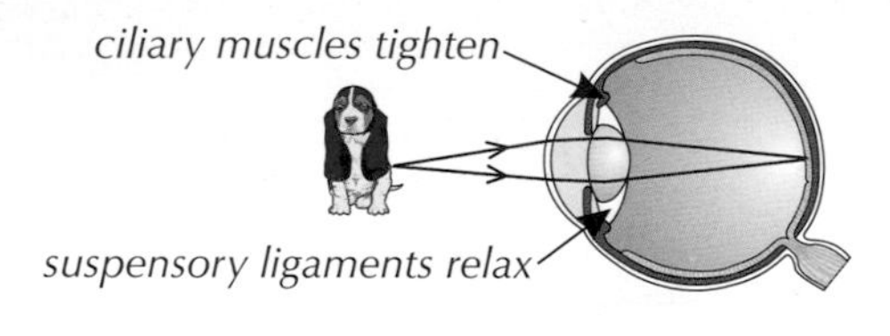

Figure 3: *Diagram showing the changes in the eye to look at a near object.*

Looking at distant objects

To focus on distant objects:

1. The ciliary muscles relax, which allows the suspensory ligaments to pull tight.
2. This makes the lens go thin (less curved).
3. So it refracts light by a smaller amount.

Tip: As you get older, your eye's lens loses flexibility, so it can't easily spring back to a round shape. This means light can't be focused well for near viewing, so older people often have to use reading glasses.

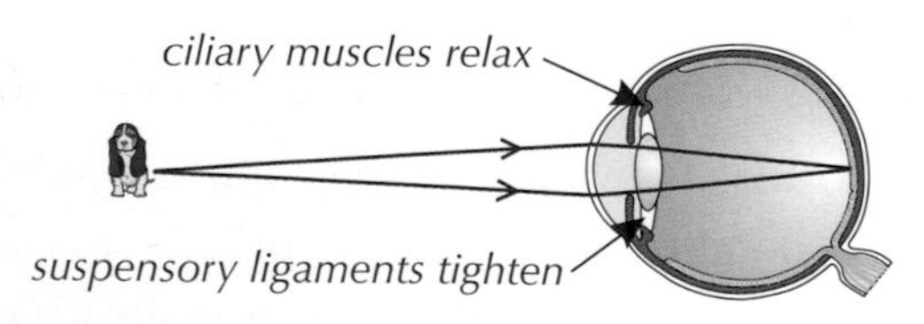

Figure 4: *Diagram showing the changes in the eye to look at a distant object.*

If the lens cannot refract the light by the right amount (so that it focuses on the retina), the person will be short- or long-sighted — see page 198 for more.

Practice Questions — Fact Recall

Q1 Name the tough, supporting wall of the eye.

Q2 What part of the eye refracts light into the eye?

Q3 What is the function of:

a) the retina?

b) the optic nerve?

Q4 a) Describe how the iris responds to bright light.

b) Explain how the iris reflex protects the eye.

Q5 The diagrams show the eye of someone who is looking at a distant object and a near object. Which diagram, A or B, shows the person looking at a distant object?

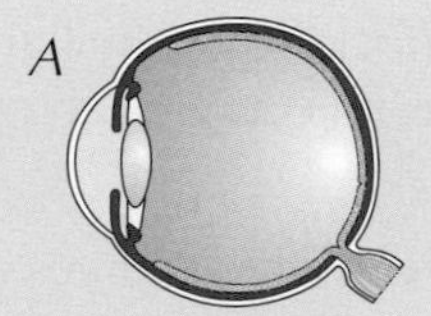

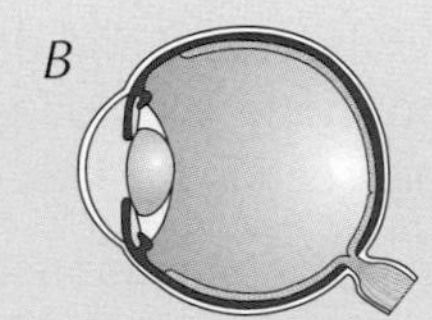

Practice Questions — Application

The diagram below shows a cross section of a human eye.

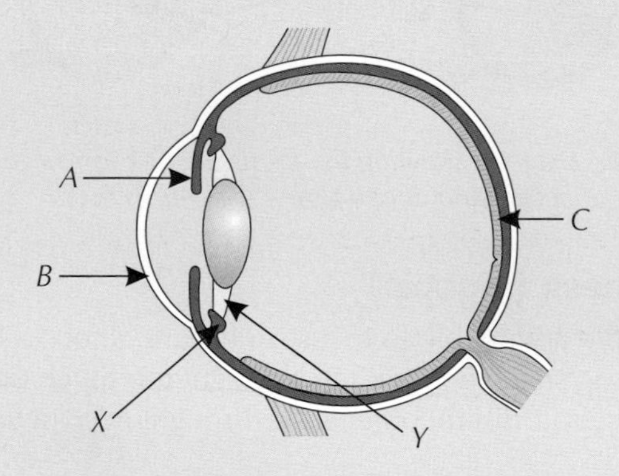

Q1 Name parts A to C on the diagram.

Q2 a) Name parts X and Y.

b) Describe how parts X and Y would change in order to focus on a nearby object.

c) Explain how these changes would allow the eye to focus on the nearby object?

Tip: If you're struggling with Q2b, think about the shape that the lens has to become.

Learning Objectives:

- Know that in the common vision defects long-sightedness (hyperopia) and short-sightedness (myopia), light rays focus either in front of or behind the retina.
- Be able to describe how these vision defects are treated with glasses in order to focus light on the retina.
- Be able to understand diagrams showing the treatment of long- and short-sightedness with glasses.
- Know that new treatments for long- and short-sightedness include soft and hard contact lenses, laser eye surgery and replacement lens surgery.

Specification Reference 4.5.2.3

7. Correcting Vision Defects

As you can see from the last few pages, the way the eye works is quite complex. Sometimes it doesn't work so well, but there are ways of correcting it.

Long- and short-sightedness

To produce a clear image, the light entering the eye needs to be focussed on the retina. The eye does this by changing the shape of the lens — see page 196. If the lens cannot focus the light in the correct place, the person will be long- or short-sighted.

Long-sightedness (hyperopia)

Long-sighted people are unable to focus on near objects. This occurs when the lens is the wrong shape and doesn't refract (bend) the light enough or the eyeball is too short. The images of near objects are brought into focus behind the retina (see Figure 1).

You can use glasses with a convex lens (a lens which curves outwards) to correct long-sightedness. The lens refracts the light rays so they focus on the retina. The medical term for long-sightedness is hyperopia.

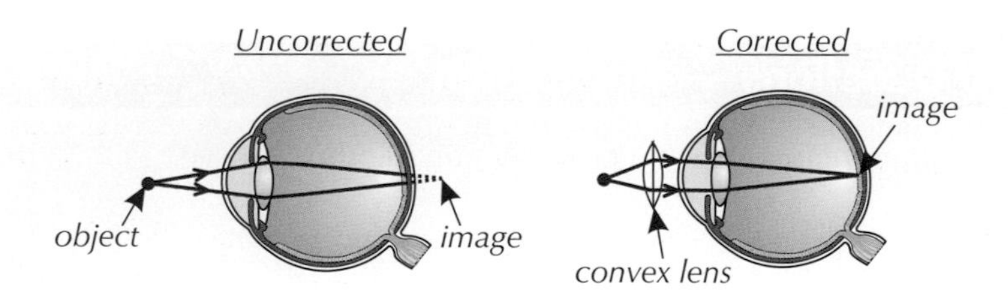

Figure 1: *Diagram showing the focussing of a long-sighted eye before and after a corrective lens is used.*

Figure 2: *Blurred vision caused by long-sightedness. The statue (close up) is out of focus.*

Short-sightedness (myopia)

Short-sighted people are unable to focus on distant objects. This occurs when the lens is the wrong shape and refracts the light too much or the eyeball is too long. The images of distant objects are brought into focus in front of the retina (see Figure 3).

You can use glasses with a concave lens (a lens which curves inwards) to correct short-sightedness, so that the light rays focus on the retina. The medical term for short-sightedness is myopia.

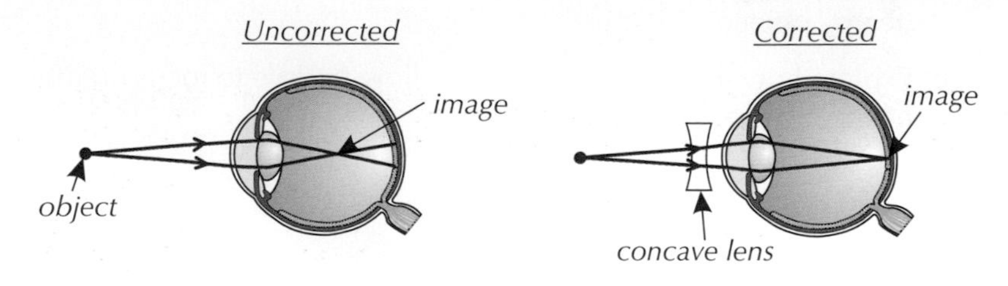

Figure 3: *Diagram showing the focussing of a short-sighted eye before and after a corrective lens is used.*

Exam Tip
You won't be asked to draw these diagrams in the exam, but you may be given a diagram and have to explain what's going on in it.

Alternative treatments for myopia and hyperopia

Wearing glasses isn't for everyone. You need to know about these alternatives:

Contact lenses

Contact lenses are thin lenses that sit on the surface of the eye and are shaped to compensate for the fault in focusing. They're popular because they are lightweight and almost invisible. They're also more convenient than glasses for activities like sports. The two main types of contact lenses are hard lenses and soft lenses. Soft lenses are generally more comfortable, but carry a higher risk of eye infections than hard lenses.

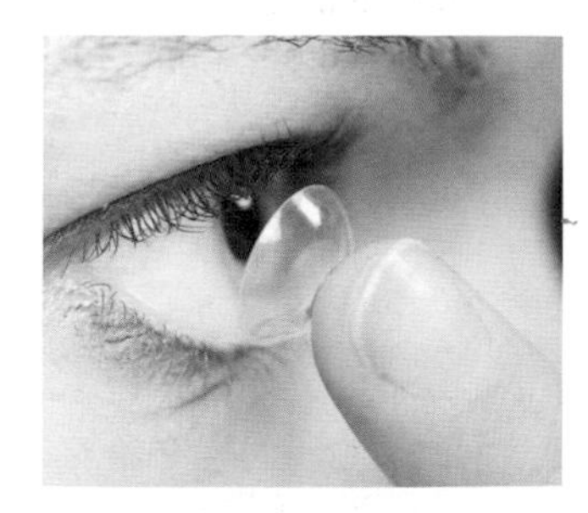

Figure 4: *Person inserting a contact lens into their eye.*

Laser eye surgery

Bad eyesight can sometimes be corrected with laser eye surgery. A laser can be used to vaporise tissue, changing the shape of the cornea (and so changing how strongly it refracts light into the eye). Slimming it down makes it less powerful and can improve short sight. Changing the shape so that it's more powerful will improve long sight. The surgeon can precisely control how much tissue the laser takes off, completely correcting the vision.

However, like all surgical procedures, there is a risk of complications, such as infection or the eye reacting in a way that makes your vision worse.

Tip: Surgery to fix long- or short-sightedness can offer a permanent solution, but it can be a lot more expensive than just wearing glasses.

Replacement lens surgery

Sometimes long-sightedness may be more effectively treated by replacing the lens of the eye (rather than altering the shape of the cornea with laser eye surgery). In replacement lens surgery, the natural lens of the eye is removed and an artificial lens, made of clear plastic, is inserted in its place.

As it involves work inside the eye, replacing a lens carries higher risks than laser eye surgery, including possible damage to the retina (which could lead to loss of sight).

Figure 5: *An artificial lens (held by forceps). The two arms are used to secure the lens in place inside the eye.*

Practice Questions — Fact Recall

Q1 Where is the image of a near object brought into focus in a long-sighted eye?

Q2 What type of lens is used to correct short-sightedness?

Q3 Give the medical term for: a) long-sightedness b) short-sightedness

Q4 What is a contact lens?

Q5 Name one surgical technique that can permanently correct long- or short-sightedness.

Practice Question — Application

Q1 Use the diagram to explain how glasses can help to improve short-sightedness.

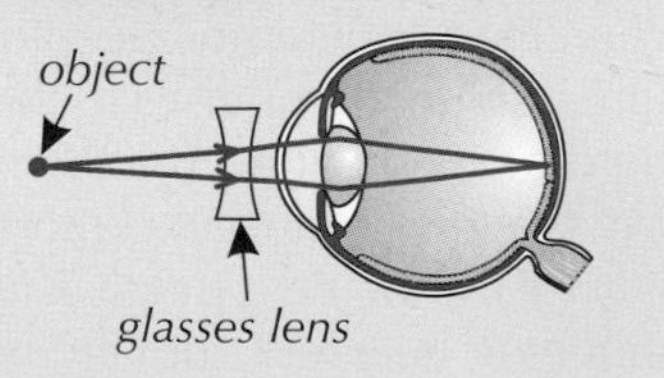

8. Controlling Body Temperature

Ever wondered why you get all sweaty when you're too hot, and shivery when you're too cold? It's just your body acting to control your temperature...

Learning Objectives:

- Know that the thermoregulatory centre in the brain monitors and controls body temperature.
- Know that the thermoregulatory centre has receptors which detect the temperature of the blood in the brain, and know how it receives information about skin temperature from receptors in the skin.
- Understand the response of the sweat glands and the blood vessels supplying the skin capillaries when core body temperature is too high.
- Understand the response of the sweat glands, muscles and the blood vessels supplying the skin capillaries when core body temperature is too low.
- H Be able to explain how the body's responses raise or lower its temperature in a given context.

Specification Reference
4.5.2.4

Monitoring body temperature

Body temperature needs to be kept at around 37 °C (because this is the temperature at which enzymes in the human body work best). In order to do this, body temperature must be carefully monitored and controlled — the body has to constantly balance the amount of energy gained (e.g. through respiration) and lost to keep the core body temperature constant.

This is where a part of the brain called the **thermoregulatory centre** comes in. It acts as your own personal thermostat. To do this, it receives information about body temperature from receptors — groups of cells which are sensitive to stimuli (see page 186). These include:

1. Receptors in the thermoregulatory centre that are sensitive to the temperature of the blood flowing through the brain.
2. Receptors in the skin that send information about skin temperature via nervous impulses.

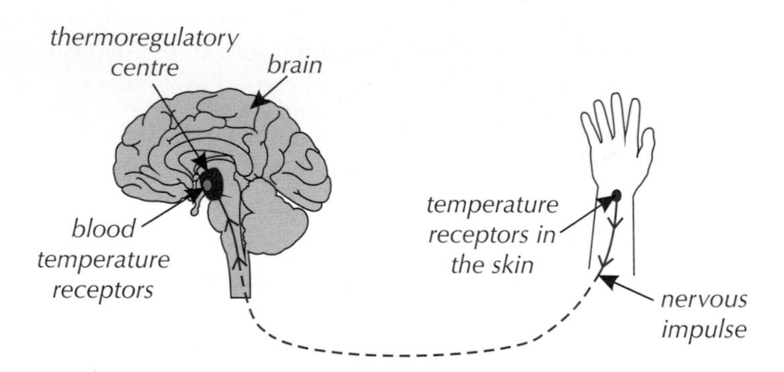

Figure 1: *Diagram showing how the thermoregulatory centre monitors temperature.*

How is body temperature controlled?

The temperature of the body is kept around a constant level due to a **negative feedback system** (see pages 184-185) — if it gets too high or too low, the body responds to bring the temperature back to optimum.

Responding to a rise in body temperature

When the temperature receptors detect that core body temperature is too high, they send impulses to the thermoregulatory centre, which acts as a coordination centre. The thermoregulatory centre processes the information from the temperature receptors and triggers the effectors automatically — see Figure 2 on the next page. The effectors produce a response that increases the amount of heat lost from the body and the body cools down.

Tip: Core body temperature refers to the temperature deep inside the body, where your internal organs are, rather than the temperature at the skin or in the limbs.

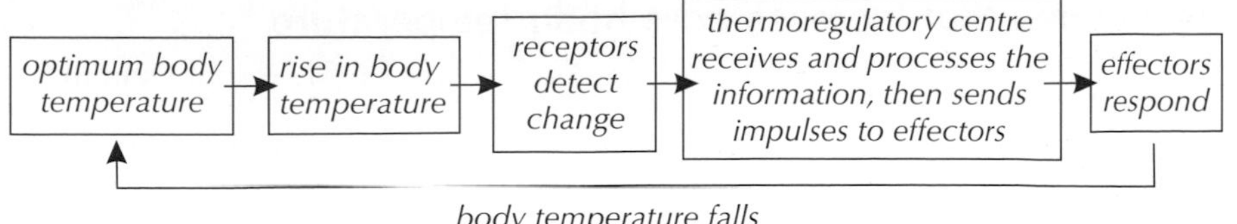

Figure 2: *Negative feedback mechanism activated by a rise in body temperature.*

Tip: Remember, negative feedback counteracts change (pages 184-185).

Responses that reduce core body temperature

When your core body temperature gets too high, your body responds in the following ways:

1. Hairs on the skin lie flat. This means less air is trapped near the surface of the skin, so there isn't a layer of insulating air surrounding the skin. This allows heat to be transferred to the environment more easily.
2. Sweat is produced by sweat glands. When sweat evaporates from the skin, it transfers energy to the environment, helping to reduce body temperature.
3. The blood vessels supplying the skin capillaries dilate (get wider) so more blood flows close to the surface of the skin. This is called **vasodilation**. It helps transfer energy from the skin to the environment.

Figure 3: *Exercise can increase body temperature. This can make the skin appear red, as blood flow to the skin increases in an attempt to cool the body.*

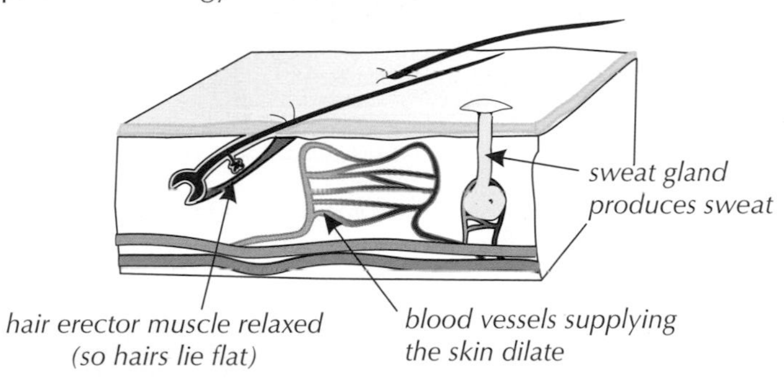

Figure 4: *Diagram showing some of the body's responses to a high core body temperature.*

Responding to a fall in body temperature

When the temperature receptors detect that core body temperature is too low, they send impulses to the thermoregulatory centre, which processes the information and triggers the effectors to reduce the amount of heat lost from the body (see next page).

The effectors, e.g. muscles, produce a response that counteracts the change and the body warms up — see Figure 5.

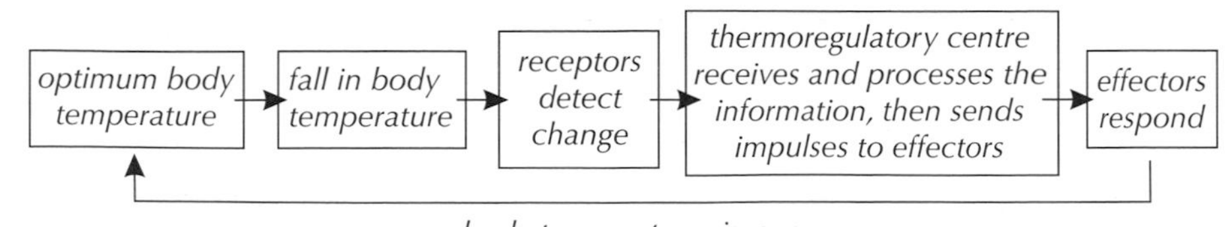

Figure 5: *Negative feedback mechanism activated by a fall in body temperature.*

Some effectors work antagonistically, e.g. one effector heats and another cools — they'll work at the same time to achieve a very precise temperature. This mechanism allows a more sensitive response.

Responses that increase core body temperature

When your core body temperature drops too low, your body responds in the following ways:

1. Hairs on the skin stand up. This traps an insulating layer of air next to the skin, reducing the amount of energy transferred to the environment.
2. No sweat is produced.
3. Blood vessels supplying skin capillaries constrict (get narrower) to reduce the skin's blood supply. This is called **vasoconstriction**. It reduces the amount of blood that flows close to the surface of the skin and so less energy is transferred from the skin to the environment.
4. When you're cold you **shiver** (your muscles contract automatically). This needs respiration, which transfers some energy to warm the body.

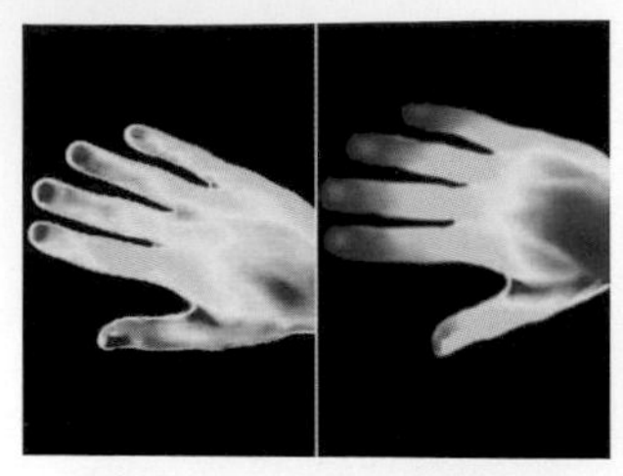

Figure 6: *A thermogram showing a hand before (left) and after (right) constriction of the blood vessels supplying the skin capillaries in the fingers. The colours show the variation in temperature in different parts of the hand — red is warmest, blue is coldest. After constriction of the blood vessels, the fingers are much cooler.*

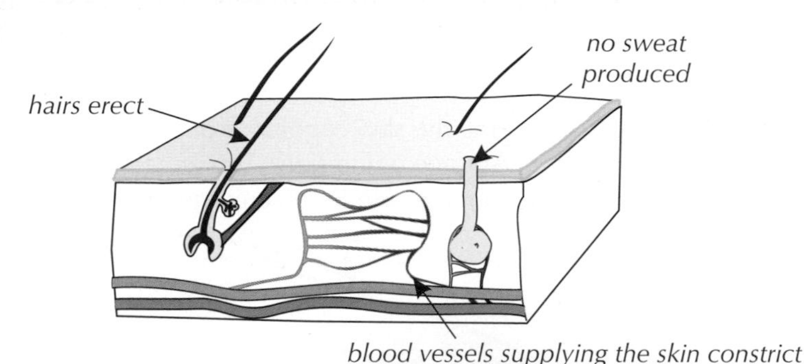

Figure 7: *Diagram showing some of the body's responses to a low core body temperature.*

Tip: Respiration transfers energy from glucose — see p. 173.

Practice Questions — Fact Recall

Q1 The thermoregulatory centre regulates body temperature. Where does the thermoregulatory centre receive inputs about body temperature from?

Q2 a) What happens to the blood vessels supplying the skin capillaries when you get too hot?

b) Explain how the response you gave in part a) helps to reduce body temperature.

Q3 Describe the pathway of the response from receptors to effectors when core body temperature falls.

Practice Question — Application

Tip: H In order to work out how the body will respond to a change in temperature, you first need to work out how the situation is affecting the body temperature — has the body temperature dropped or risen?

Q1 Gena is out walking and has forgotten her coat. The weather has changed and the temperature has begun to fall.

a) Describe the way Gena's body will respond in order to help her try to maintain her body temperature.

b) Explain how the responses you gave in part a) help Gena to maintain her body temperature.

Topic 5a Checklist — Make sure you know...

Homeostasis

- [ ] That internal conditions within the body need to be carefully controlled to maintain the optimum environment for enzymes and cell function.
- [ ] That homeostasis is the regulation of the conditions inside your body (and cells) to maintain a stable internal environment, in response to changes in both internal and external conditions. These internal conditions, including body temperature, blood glucose level and water content, are regulated by nervous and hormonal control systems.
- [ ] That a control system includes: cells called receptors that detect stimuli (changes in the environment), a coordination centre (e.g. the brain, spinal cord or pancreas) that receives and processes information from receptors and organises a response, and effectors (muscles or glands) that produce a response to restore the system back to its normal level.

The Nervous System

- [ ] That the nervous system allows humans to respond to changes in the environment (stimuli) and to coordinate their behaviour.
- [ ] That information from receptors travels as electrical impulses via neurones to the CNS (which consists of the brain and spinal cord), where a response is coordinated. Muscles respond to impulses by contracting, and glands respond by secreting hormones.
- [ ] How the nervous system is adapted to its functions.
- [ ] How to read from tables, charts and graphs containing data about nervous system function, and how to interpret this data.

Synapses and Reflexes

- [ ] That the connections between neurones are called synapses and that a nerve signal is transferred across a synapse by chemicals that diffuse across the gap.
- [ ] That reflexes are fast, automatic responses involving sensory neurones, synapses, relay neurones and motor neurones, that they don't involve the conscious brain, and that reflex reactions reduce the chance of being injured by providing a quick response to stimuli.
- [ ] That in a simple reflex, stimuli are detected by receptors and transmitted to the CNS as nervous impulses via sensory neurones. The impulses are then transferred via a relay neurone in the CNS to a motor neurone, which sends impulses to an effector. The effector then produces a response.

Investigating Reaction Time

- [ ] How to investigate how a factor, e.g. caffeine, affects reaction time (Required Practical 7), and how to convert between reaction time data shown in a graph and its numerical form.

The Brain

- [ ] That the brain controls all complex behaviour, and that it contains billions of connected neurones.

cont...

- [ ] That the brain is split into regions that have different functions, including the cerebral cortex (which controls consciousness, memory, intelligence and language), the medulla (which controls unconscious activities) and the cerebellum (which controls muscle coordination), and know where in the brain these regions are located.
- [ ] **H** That neuroscientists have mapped regions of the brain to their functions by studying patients with brain damage, electrically stimulating parts of the brain and by using MRI scanning, but that the brain is hard to study and treat due to its complexity and delicacy.

The Eye

- [ ] That the eye is a sense organ containing these structures: sclera, cornea, iris, retina, ciliary muscles, suspensory ligaments and optic nerve, and know the location and function of these structures.
- [ ] How the iris controls the size of the pupil in order to adjust the amounts of light that enters the eye.
- [ ] That changing the shape of the lens to focus on an object is called accommodation.
- [ ] That to focus on near objects, the ciliary muscles contract and the suspensory ligaments relax, which thickens the lens so that light rays are refracted more strongly (so they focus on the retina), and that to focus on distant objects, the ciliary muscles relax and the suspensory ligaments tighten, which pulls the lens thin so that light rays are refracted less strongly.

Correcting Vision Defects

- [ ] That long-sightedness (hyperopia), in which light rays focus behind the retina, is corrected using glasses with convex (0) lenses, and be able to understand diagrams showing this.
- [ ] That short-sightedness (myopia), in which light rays focus in front of the retina, is corrected using glasses with concave (I)lenses, and be able to understand diagrams showing this.
- [ ] Know that treatments for long- and short-sightedness include soft and hard contact lenses, laser eye surgery (which changes the shape of the cornea) and replacement lens surgery (which involves inserting an artificial lens).

Controlling Body Temperature

- [ ] That an area of the brain called the thermoregulatory centre monitors and controls body temperature.
- [ ] That receptors in the thermoregulatory centre detect the temperature of the blood in the brain, and that the thermoregulatory centre receives information about skin temperature from receptors in the skin, sent via nervous impulses.
- [ ] That when core body temperature is too high, dilation of the blood vessels supplying the skin capillaries (vasodilation) causes increased blood flow to the skin, increasing heat transfer to the environment. The sweat glands also produce more sweat, which transfers energy from the body to the environment as it evaporates.
- [ ] That when core body temperature is too low, the sweat glands stop producing sweat, and the blood vessels supplying the skin capillaries constrict (vasoconstriction), causing decreased blood flow to the skin, which decreases the transfer of energy to the environment. The muscles cause the body to shiver (they contract automatically) — this requires respiration, so some energy is transferred (from glucose) to the body, helping to raise body temperature.
- [ ] **H** How to explain the body's response to a change in temperature in a given context.

Exam-style Questions

1 A scientist conducted an experiment on reaction time.

Subjects had an electrode placed on their upper arm to detect when their muscle contracted. The subjects were asked to place their finger on a metal disc, which gave out a small electric shock at random. This shock caused the muscle in the upper arm to contract.

The scientist measured reaction time as the time it took between the initiation of the shock and the contraction of the muscle in the upper arm.

The diagram shows the pathway taken in this reflex response.

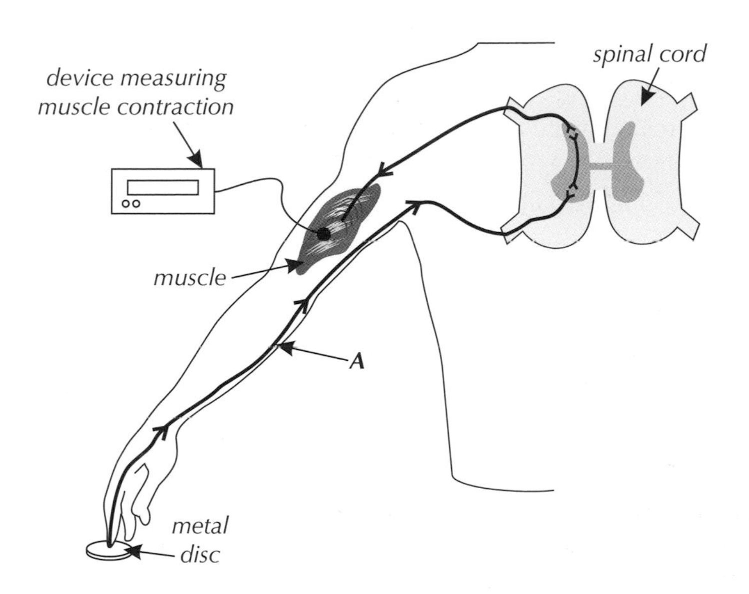

1.1 What is the effector in this response?

(1 mark)

1.2 Name the neurone labelled **A** in the diagram.

(1 mark)

1.3 The average reaction time measured during the experiment was 0.024 s.

After he had recorded their reaction times, the scientist gave the subjects a drug which is known to increase the amount of chemical released at the synapses. He then repeated the experiment.

What do you think would happen to the average reaction time after administration of the drug? Explain your answer.

(2 marks)

1.4 Reflexes often help to protect us from injury. Suggest and explain **two** features of reflexes that help them do this.

(4 marks)

2 A student was investigating the effect of exercise on reaction time.
He measured the reaction times of twenty people using a computer test, in which the person being tested had to click the mouse when a star appeared on the screen. The participants were then asked to complete five minutes of intense exercise, after which their reaction times were measured again.

The nervous system pathway for the reaction being tested is shown in the diagram below.

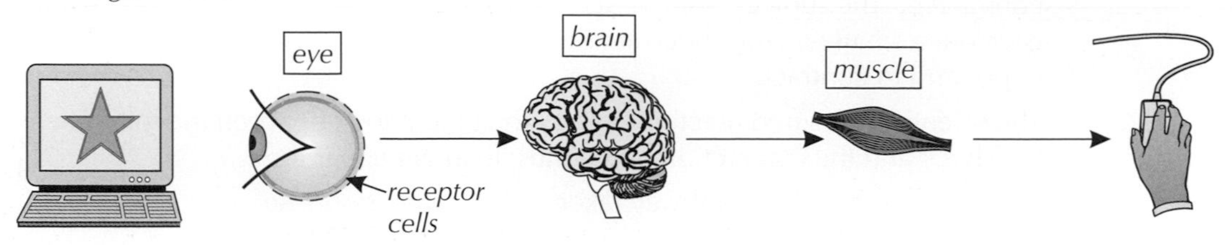

2.1 Name the structure of the eye that contains receptor cells.

(1 mark)

2.2 Name the region of the brain that coordinates the response in this pathway.

(1 mark)

The student summarised the results of the investigation in the histograms below.

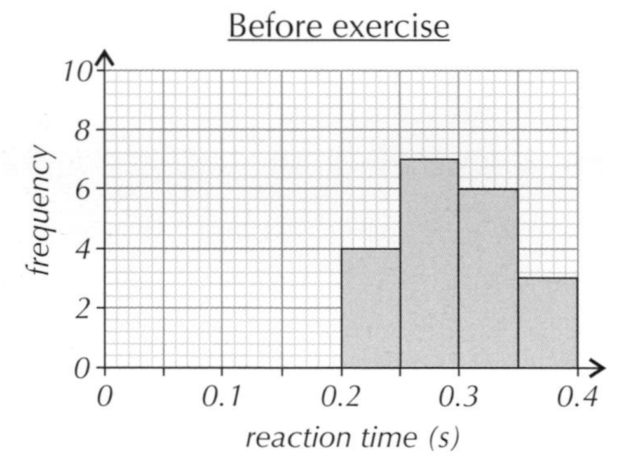

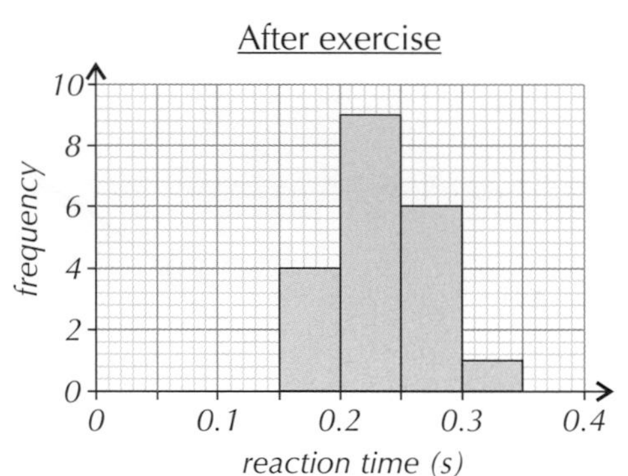

2.3 Compare the number of people who had a reaction time less than 0.3 s after exercise to those before exercise.

(3 marks)

2.4 Give a conclusion about the effect of exercise on reaction time based on these graphs.

(1 mark)

3 Richard has been for a run on a hot day. His body temperature has increased.

3.1 Describe how Richard's brain detects this increase in body temperature.

(3 marks)

3.2 Why is it important for Richard's body temperature to be kept around its optimum?

(1 mark)

3.3 Describe and explain how Richard's body will respond to the increase in body temperature.

(4 marks)

1. Hormones

Along with the nervous system, hormones allow us to react to changes in the environment and in our bodies in order to keep everything ticking over nicely. Hormones are secreted by the glands of the endocrine system.

Learning Objectives:
- Know that hormones are chemical molecules which are secreted into the bloodstream.
- Know that hormones are carried in the blood to their target organs, where they have an effect.
- Know that glands secrete hormones.
- Know that the endocrine system is made up of glands.
- Understand why the pituitary gland is sometimes called the 'master gland'.
- Know where certain endocrine glands are found in the body.
- Know that hormonal responses are slower but last longer than nervous impulses.

Specification Reference 4.5.3.1

What are hormones?

Hormones are chemical molecules released directly into the blood to regulate bodily processes. They are carried in the blood plasma (the liquid part of the blood) to other parts of the body, but only affect particular cells in particular organs (called **target organs**). Hormones control things in organs and cells that need constant adjustment.

Here's a definition of hormones:

> Hormones are chemical messengers which travel in the blood to activate target cells.

The endocrine system

Hormones are produced in (and secreted by) various glands, called endocrine **glands**. These glands make up the endocrine system.

The endocrine system uses hormones to react to changes in the environment or changes inside the body.

Endocrine glands

There are many glands in the endocrine system, including:

- The pituitary gland — this produces many hormones that regulate body conditions. It is sometimes called the 'master gland' because these hormones act on other glands, directing them to release hormones that bring about change.
- The pancreas — this produces insulin, which is used to regulate the blood glucose level (see page 210).
- The thyroid — this produces thyroxine, which is involved in regulating things like the rate of metabolism, heart rate and temperature (see p. 234).
- The adrenal glands — these produce adrenaline, which is used to prepare the body for a 'fight or flight' response (see page 235).

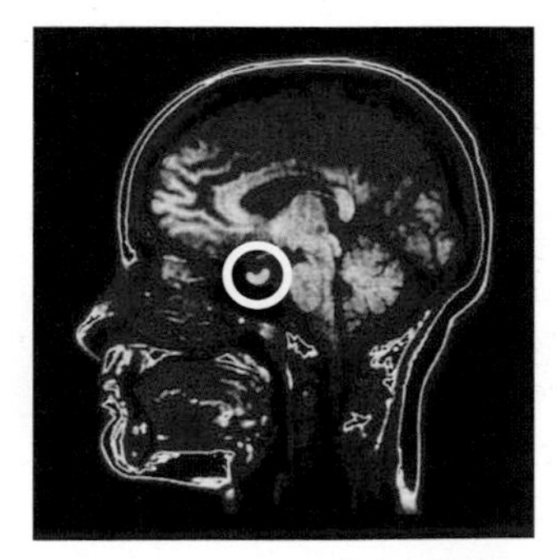

***Figure 1:** Scan of the brain. The pituitary gland (green structure) is circled in white.*

- The ovaries (females only) — these produce oestrogen, which is involved in the menstrual cycle (see pages 225-227).
- The testes (males only) — these produce testosterone, which controls puberty and sperm production in males (see page 225).

Exam Tip
In the exam, you might be asked to point out where a certain gland is found in the body, or be asked to name a gland from its location.

Exam Tip
These are the glands you need to learn for the exam. There are loads more glands in the body though, each doing its own thing.

Tip: Testis is the singular of testes.

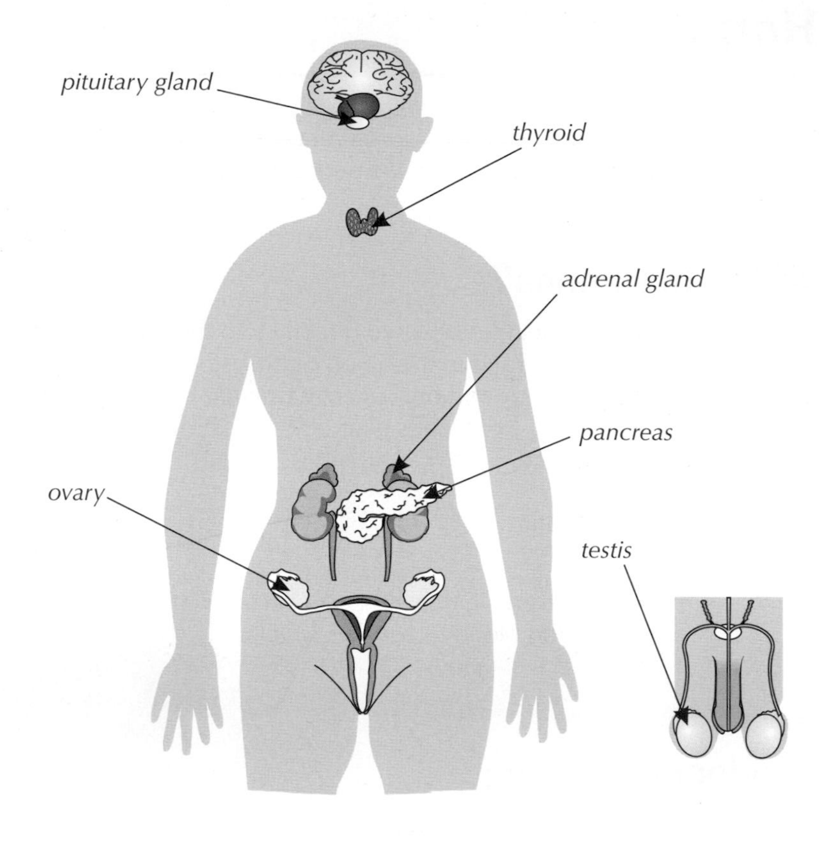

Figure 2: *Diagram showing the location of some of the endocrine glands.*

Comparing nerves and hormones

Tip: There's more about the nervous system on pages 186-187.

As hormones are carried in the blood, they tend to travel around the body relatively slowly (compared to nervous impulses anyway). They also tend to have relatively long-lasting effects. Hormones and nerves do similar jobs in the body, but with a few differences. These are summarised in Figure 3.

Nerves	Hormones
Fast action	Slower action
Act for a short time	Act for a long time
Acts on a very precise area	Acts in a more general way

Figure 3: *Table summarising the differences between nerves and hormones.*

If you're not sure whether a response is nervous or hormonal, have a think...

- If the response is really quick, it's probably nervous.
  Some information needs to be passed to effectors really quickly (e.g. pain signals, or information from your eyes telling you about a car heading your way), so it's no good using hormones to carry the message — they're too slow.

- If a response lasts for a long time, it's probably hormonal.
  For example, when you get a shock, a hormone called adrenaline is released into the body (causing the fight-or-flight response, where your body is hyped up ready for action). You can tell it's a hormonal response (even though it kicks in pretty quickly) because you feel a bit wobbly for a while afterwards.

Practice Questions — Fact Recall

Q1 How are hormones carried around the body?

Q2 True or false? Hormones affect all organs.

Q3 What are hormones secreted from?

Q4 Name a hormone secreted from the ovaries.

Q5 Which produces a faster response — nerves or hormones?

Exam Tip
Try to learn to spell difficult words like 'pituitary'. It'll help to make sure the examiner knows what you're talking about.

2. Controlling Blood Glucose

Your blood glucose level needs to carefully regulated. This is done with the help of hormones produced by the pancreas. However, if things don't work quite as they should, it can result in a condition called diabetes...

Learning Objectives:

- Know that the pancreas monitors and controls blood glucose level.
- Know that the pancreas produces insulin when the blood glucose level gets too high and understand how insulin reduces the blood glucose level.
- Know that excess glucose is stored as glycogen in liver and muscle cells.
- H Understand how glucagon from the pancreas increases the blood glucose level when it's too low, as part of a negative feedback cycle with insulin.
- Know what Type 1 diabetes is and understand why it can result in the blood glucose level being too high.
- Know how Type 1 diabetes can be controlled.
- Know what Type 2 diabetes is and why it can result in the blood glucose level being too high.
- Know that being obese increases a person's risk of getting Type 2 diabetes.
- Know how Type 2 diabetes can be controlled.
- Be able to compare Type 1 and Type 2 diabetes.
- Be able to read from and interpret graphs showing the effect of insulin on blood glucose concentration.

Specification Reference 4.5.3.2

Glucose concentration of the blood

Glucose is a type of sugar. Throughout the day the blood glucose level varies:

Examples

- Eating foods containing carbohydrate puts glucose into the blood from the digestive system.
- The normal metabolism of cells removes glucose from the blood.
- Vigorous exercise removes much more glucose from the blood.

Hormonal control of blood glucose

The level of glucose in the blood must be kept steady. Changes in the blood glucose level are monitored by the **pancreas**. The pancreas produces hormones which help to control the blood glucose level.

Insulin

Insulin is a hormone produced by the pancreas. It decreases the blood glucose level when it gets too high. Here's what happens:

1. After a meal containing carbohydrate, a person's blood glucose level rises. This rise is detected by the pancreas.
2. The pancreas responds by producing insulin, which is secreted into the blood.
3. Insulin causes body cells to take up more glucose from the blood. Cells in the liver and muscles can take up glucose and convert it into a storage molecule called **glycogen**.
4. This causes the blood glucose level to fall.

This process is shown in Figure 1.

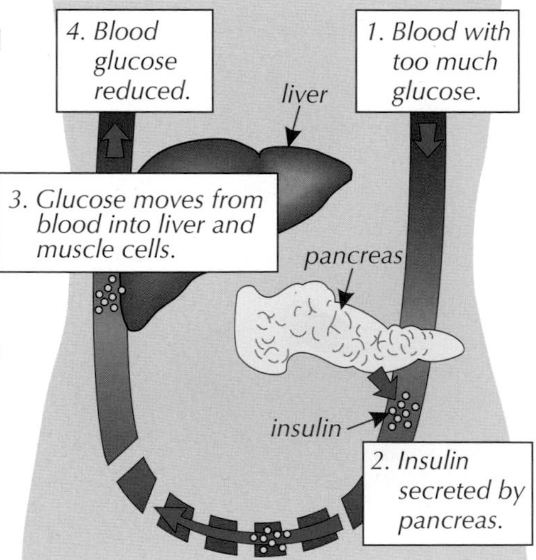

Figure 1: *Diagram showing the role of insulin in the control of blood glucose.*

Glucagon Higher

Glucagon is another hormone produced by the pancreas. It increases the blood glucose level when it gets too low. Here's what happens:

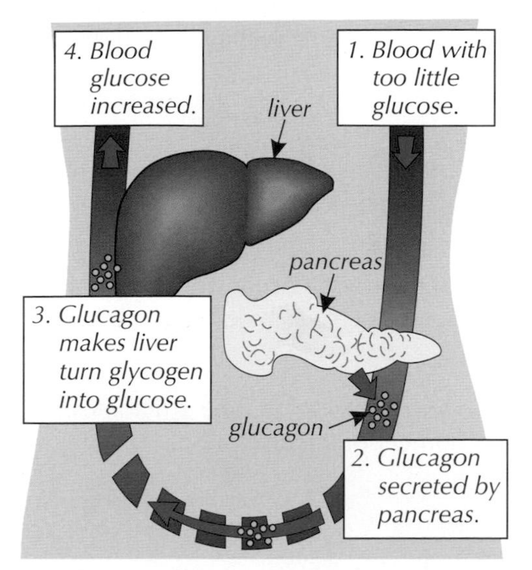

***Figure 2:** Diagram showing the role of glucagon in the control of blood glucose.*

1. If a person's blood glucose level decreases, the fall is detected by the pancreas.
2. The pancreas responds by producing glucagon, which is secreted into the blood.
3. Glucose can be stored in the muscles and liver as glycogen (see previous page). Glucagon causes the glycogen to be converted back into glucose, which enters the blood.
4. This causes the blood glucose level to rise.

This process is shown in Figure 2.

Exam Tip H
The words 'glucagon' and 'glycogen' look and sound very similar. You need to make sure you get the spelling of these words spot on in the exam. E.g. if you write 'glycogon' the examiner won't know whether you mean glucagon or glycogen so you won't get the marks.

Tip: H Glucagon works with insulin to control blood glucose level in a negative feedback cycle — together they keep blood glucose around its optimum. (See pages 184-185 for more on negative feedback.)

Type 1 diabetes

Type 1 diabetes is a condition where the pancreas produces little or no insulin. The result is that a person's blood glucose can rise to a level that can kill them.

Tip: Remember, insulin <u>reduces</u> blood glucose level.

Controlling Type 1 diabetes

Type 1 diabetes needs to be controlled in the following ways:

1. **Insulin therapy** — this usually involves injecting insulin into the blood. People with Type 1 diabetes usually have several injections of insulin throughout the day, which are likely to be at mealtimes. Insulin injections make sure glucose is removed from the blood quickly once the food has been digested. This stops the level of glucose in the blood from getting too high and is a very effective treatment.

   The amount of insulin that needs to be injected depends on the person's diet and how active they are, since these things will affect their blood glucose level.

2. **Limit the intake of foods rich in simple carbohydrates** — i.e. sugars (which cause the blood glucose level to rise rapidly). People with Type 1 diabetes are also advised to spread their intake of starchy carbohydrates (e.g. pasta, rice, bread, etc.) throughout the day and to pick varieties of these foods that are absorbed more slowly (so they don't cause such a sharp rise in the blood glucose level).

3. **Regular exercise** — this helps to lower the blood glucose level as the increased metabolism of cells during exercise removes more glucose from the blood.

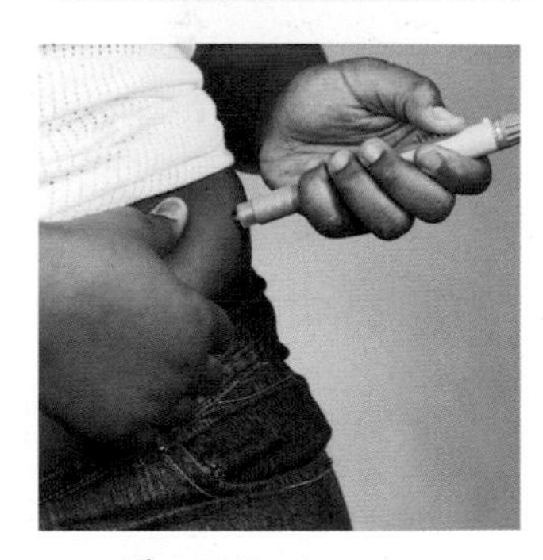

***Figure 3:** A person injecting insulin.*

Tip: Insulin can't be taken in a pill or tablet — the enzymes in the stomach completely destroy it before it reaches the bloodstream. That's why it's normally injected.

Tip: Prolonged periods of high blood glucose levels can cause damage to organs such as the eyes, heart and kidneys, resulting in long-term health problems.

Insulin used to be extracted from the pancreases of pigs or cows, but now human insulin is made by genetic engineering. This human insulin doesn't cause adverse reactions in patients, like animal insulin did.

Insulin injections help to control a person's blood glucose level, but it can't be controlled as accurately as having a normal working pancreas, so they may still have long-term health problems.

Type 2 diabetes

Type 2 diabetes is a condition where a person becomes resistant to their own insulin. They still produce insulin, but their body's cells don't respond properly to the hormone. This can also cause a person's blood glucose level to rise to a dangerous level.

Being overweight can increase your chance of developing Type 2 diabetes, as obesity is a major risk factor in the development of the disease.

Exam Tip
In the exam you might be asked to compare Type 1 and Type 2 diabetes. It might help to draw a table comparing the two. You could include what they are and how they can be controlled.

Controlling Type 2 diabetes

Type 2 diabetes can be controlled in the following ways:

1. **Eating a carbohydrate-controlled diet** — this includes controlling a person's intake of simple carbohydrates to avoid sudden rises in glucose.
2. **Regular exercise** — this helps to lower the blood glucose level.

Modern treatment options

There are some new treatment options available, and some currently being researched, for people with Type 1 diabetes:

1. Diabetics can have a pancreas transplant. A successful operation means they won't have to inject themselves with insulin again. But as with any organ transplant, your body can reject the tissue (see page 220). If this happens you have to take costly immunosuppressive drugs (drugs that suppress the immune system), which often have serious side-effects.
2. Modern research into artificial pancreases and stem cell research may mean the elimination of organ rejection, but there's a way to go yet.

Tip: You covered stem cells on page 44. They have the ability to turn into a range of different cell types. It's possible that stem cells could be made into pancreatic cells and used to replace the faulty insulin-producing cells in Type 1 diabetics.

Testing for diabetes

At a high blood glucose level, glucose begins to move into the urine. This means diabetes can be tested for by looking for glucose in a urine sample. A blood test for glucose can then confirm the diagnosis. The blood test is more reliable as it gives the current level of glucose in the blood (urine samples can be hours old), and can detect a high level of glucose when the level is not yet high enough for glucose in the urine.

Interpreting blood glucose level graphs

The changes in a person's blood glucose level over time can be shown by a graph. The shape of the graph can vary, depending on whether the person has diabetes — see Figure 4. You need to be able to explain what graphs like this are showing.

Example

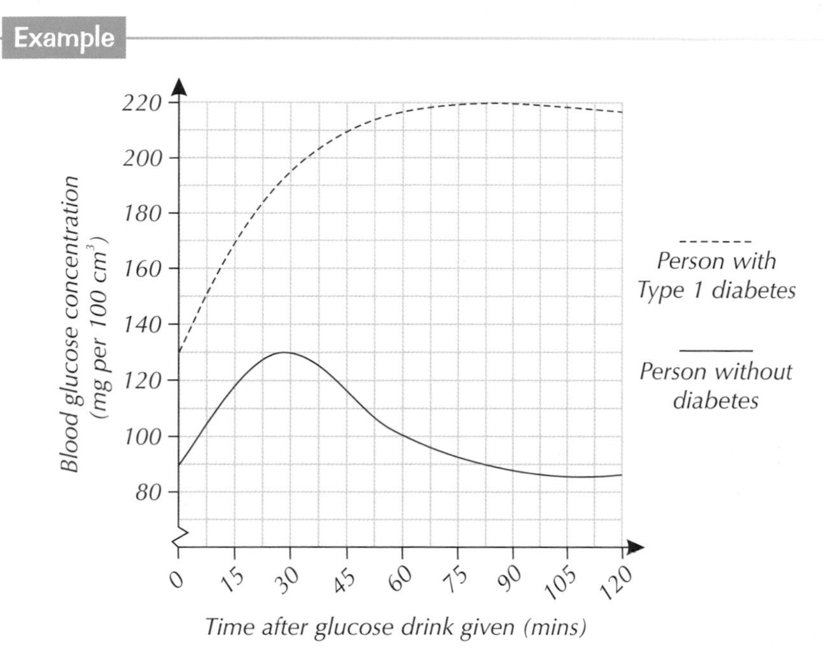

Figure 4: *Graph showing the change in blood glucose concentration over time after consumining a glucose drink for a person with Type 1 diabetes and a person without diabetes.*

Here's what Figure 4 is showing:

Person without diabetes

In the initial 30 minutes after consuming the glucose drink, their blood glucose concentration rises from 90 mg/100 cm³ to 130 mg/100 cm³. After 30 minutes, their blood glucose concentration begins to fall. This is because as the glucose level rises above normal, the pancreas responds by releasing insulin into the blood. The insulin causes body cells to take up the glucose.

After about 80 minutes, the blood glucose concentration is back to its normal level of 90 mg/100 cm³, so the pancreas stops releasing insulin. However, the blood glucose concentration then falls slightly below normal — this is due to some insulin still remaining in the blood.

Person with Type 1 diabetes

After the person with Type 1 diabetes consumes the glucose drink, their blood glucose concentration rises higher than that of the person without diabetes, from 130 mg/100 cm³ to 220 mg/100 cm³. Their glucose level then remains high.

This is because their pancreas produces little or no insulin in response to the rising amount of glucose in the blood, so they are not able to control their blood glucose level.

Tip: If the blood glucose concentration rises high then decreases slowly, the person may be resistant to insulin — this is Type 2 diabetes. E.g.:

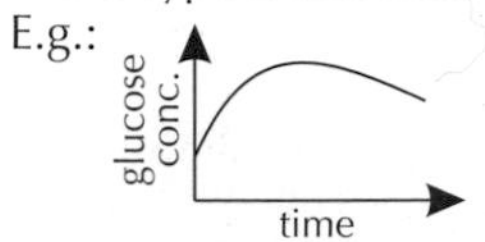

Their pancreas is producing insulin but it's not having the proper effect on the body cells.

Tip: There is a time delay between the release of insulin and the reduction of blood glucose concentration. This is because the insulin first needs to travel in the blood to its target cells.

Tip: H As the glucose level falls below normal, the pancreas responds by releasing glucagon into the blood. Glucagon works to bring the blood glucose back up to its normal level.

Practice Questions — Fact Recall

Q1 Where in the body is the hormone insulin produced?

Q2 What effect does insulin have on the body's cells?

Q3 Where in the body is the hormone glucagon produced?

Q4 How does glucagon increase blood glucose levels?

Q5 What is Type 1 diabetes?

Q6 Give two ways Type 2 diabetes can be controlled.

Practice Question — Application

Q1 In a study, a hormone was injected into a subject while their blood glucose level was monitored. The results are shown in the graph.

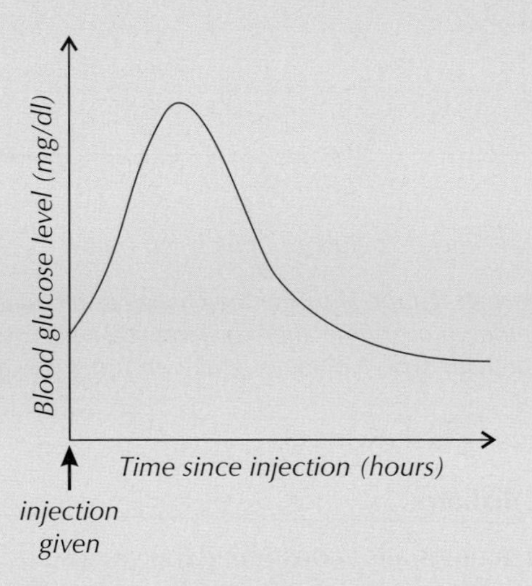

a) What hormone do you think was injected? Explain your answer.

b) i) Name the other main hormone that affects the blood glucose level.

ii) What do you think would happen to the blood glucose level if this hormone had been injected instead? Explain your answer.

Exam Tip
If you are given a graph on blood glucose level in the exam, it's a good idea to look at it carefully and try to work out what it is showing before you start answering the questions. It sounds obvious, but if you just jump straight in you might miss something important.

3. Controlling Water Content

The water level in cells is very important — organisms need to maintain the concentration of their cell contents at the correct level for cell activity. The kidneys are really important in regulating water content.

The role of the kidneys

The **kidneys** make **urine** by taking waste products (and other unwanted substances) out of your blood. Substances are filtered out of the blood as it passes through the kidneys. This process is called **filtration**. Useful substances, including all the glucose, some ions and the right amount of water, are then absorbed back into the blood. This process is called **selective reabsorption**. The substances that are removed from the body in urine include urea (a waste product from the breakdown of proteins) and excess ions and water.

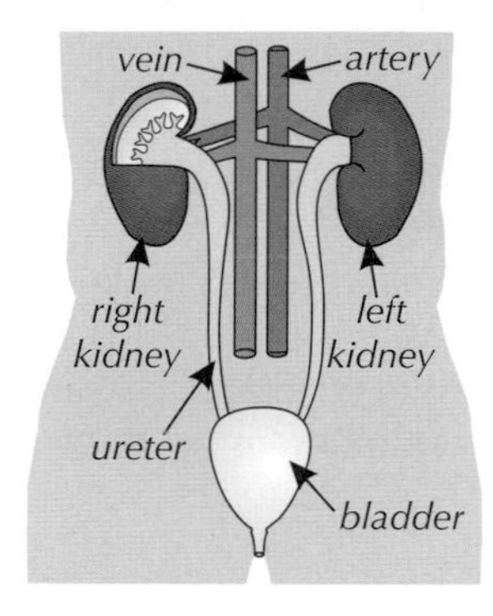

Figure 1: *Diagram showing the location of the kidneys and associated structures.*

Adjustment of water content

The body has to constantly balance the water coming in against the water going out.

We lose water from the skin in sweat and from the lungs when we exhale (breath out). We can't control how much we lose in these ways, so the amount of water is balanced by the amount we consume and the amount removed by the kidneys in urine.

Examples

On a cold day, if you don't sweat, you'll produce more urine which will be pale and dilute.

On a hot day, you sweat a lot, and you'll produce less urine which will be dark-coloured and concentrated. The water lost when it is hot has to be replaced with water from food and drink to restore the balance.

Learning Objectives:

- Be able to describe how the kidneys maintain the water content of the body.
- Know that urine is produced by filtering the blood.
- Know that selective reabsorption is where useful substances are absorbed back into the blood.
- Know that excess water and ions, and urea, are removed from the body in urine.
- Know that water is lost from the body in sweat and during exhalation.
- Know that there's no control over water loss from the lungs.
- Be able to explain how losing or gaining too much water can damage cells.
- Know that ions can be lost in sweat.
- Know that there's no control over water, ion and urea loss from the skin.
- Know that urea can be lost in sweat.
- H Know how excess amino acids are made and excreted.
- H Be able to describe how ADH, which is released by the pituitary gland, controls the body's water content.
- Be able to convert between bar charts and tables of data showing the levels of glucose, ions and urea before and after filtration.

Specification Reference 4.5.3.3

Adjustment of ion content

Ions such as sodium are taken into the body in food, and then absorbed into the blood.

If the ion (or water) content of the body is wrong, this could upset the balance between ions and water, meaning too much or too little water is drawn into cells by osmosis (see page 54). Having the wrong amount of water can damage cells or mean they don't work as well as normal.

Some ions are lost in sweat (which tastes salty, you may have noticed). However, this amount is not regulated, so the right balance of ions in the body must be maintained by the kidneys. The right amount of ions is reabsorbed into the blood after filtration and the rest is removed from the body in urine.

Example

A salty meal will contain far too much sodium and so the kidneys will remove the excess sodium ions from the blood.

Tip: If the ion content of the body is too low, or the water content is too high, water will move into the cells by osmosis. This will cause the cells to swell, and possibly burst. If the ion content of the body is too high, or the water content is too low, water will move out of the cells by osmosis and the cells will shrivel up.

Tip: Sports drinks contain water, sugar and ions. They are used by athletes to restore the balances of these substances after training — the water and ions replace those lost in sweat, while the sugar can replace the sugar that's used up by muscles during exercise.

Removal of urea Higher

Proteins (and the amino acids that they are broken down into) can't be stored by the body — so any excess amino acids are converted into fats and carbohydrates, which can be stored. This occurs in the liver and involves a process called deamination. Ammonia is produced as a waste product from this process.

Ammonia is toxic so it's converted to urea in the liver. Urea is then transported to the kidneys, where it's filtered out of the blood and excreted from the body in urine.

Tip: A small amount of urea is also lost from the skin in sweat. Like with ions and water, the amount of urea that's lost this way is not controlled by the body.

Anti-diuretic hormone (ADH) Higher

The structures in the kidney where filtration and selective reabsorption take place are called the kidney tubules. The kidney can let more or less water move out of the tubules to change the water content of the blood — the more water that leaves the tubules and is reabsorbed into the blood, the more concentrated the urine will be.

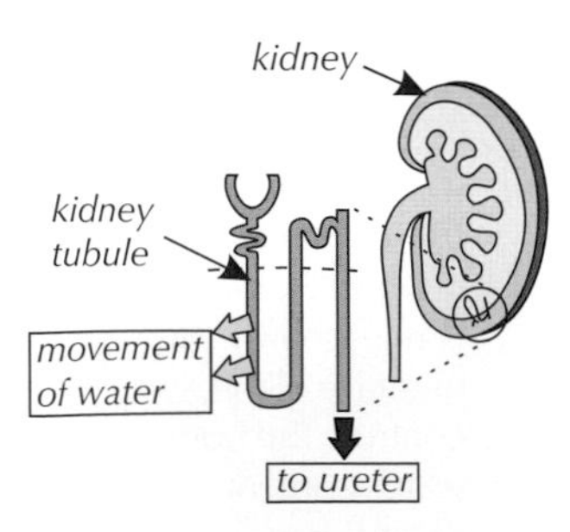

Figure 2: The location of one of the tubules in a kidney.

This process is controlled by a hormone called anti-diuretic hormone (ADH). It does this by acting on the cells in the kidney tubules, causing them to be more or less permeable to water and so changing the volume of water that can be reabsorbed.

ADH is released into the bloodstream by the pituitary gland. The brain monitors the water content of the blood and instructs the pituitary gland to release ADH into the blood according to how much is needed.

The whole process of water content regulation is controlled by negative feedback. This means that if the water content gets too high or too low a mechanism will be triggered that brings it back to normal (see Figure 3).

Tip: For more information on negative feedback go to pages 184-185.

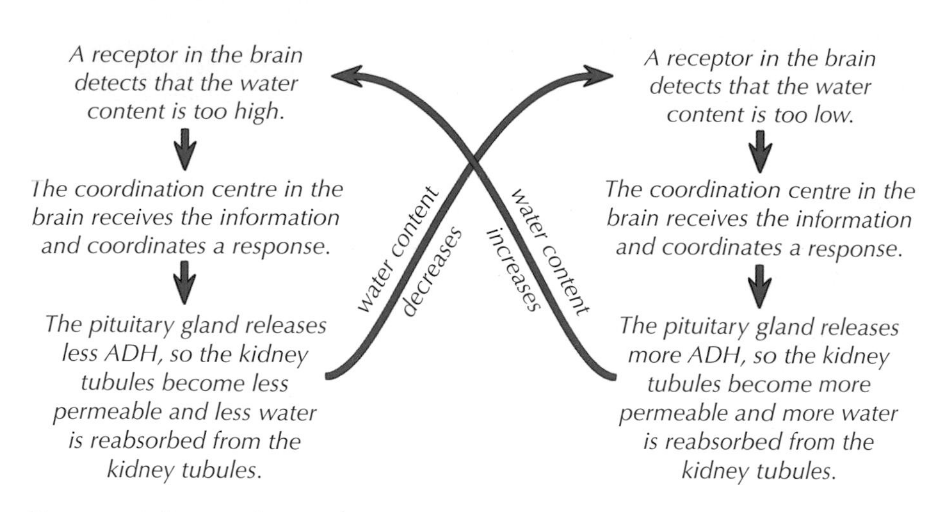

Figure 3: *A diagram showing how ADH secretion is controlled by negative feedback.*

Tip: H The more permeable the kidney tubules are, the more easily water can move out of them by osmosis (see page 54).

Practice Questions — Fact Recall

Q1 What is the name of the process in the kidneys where useful substances are absorbed back into the blood?

Q2 Name two substances that are absorbed back into the blood in the kidney.

Q3 Name three substances that should be present in the urine of a healthy person.

Q4 Give two ways that water can be lost from the body.

Q5 True or False? The hotter your body gets, the less water you need to take in.

Q6 Which process can cause cell damage if the ion or water content of the cell is unbalanced?

Practice Questions — Application

Q1 On a hot summer's day, Katherine has been running on the beach and hasn't drunk anything. Cate has drunk one small glass of orange juice whilst reading a book in the shade. Julia has been sat in her air conditioned office and has drunk four cups of tea.

Put one tick in each column to complete the table.

Person	Largest volume of urine	Smallest volume of urine	Most concentrated urine	Most dilute urine
Katherine				
Cate				
Julia				

Exam Tip
In the exam you could be given some figures to do with the kidneys and asked to do some calculations using them. Just make sure you read the information carefully, so that you're using the correct figures. And the good news is that you can take a calculator into the exam with you. Hooray.

Exam Tip
You may be given a bar chart in the exam that shows the levels of substances such as glucose, ions or urea in the blood before and after filtration. If you need to read off a value, just look at the where the top of the bar lines up with the y-axis. E.g.

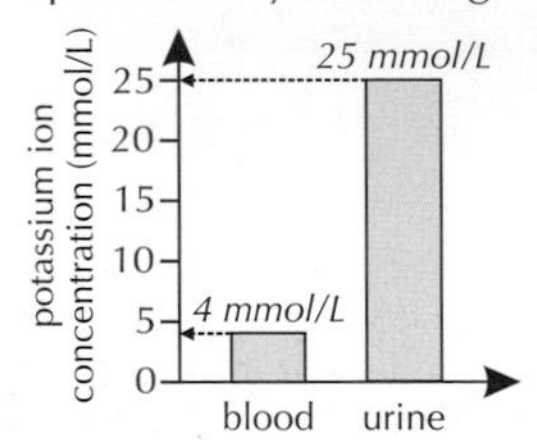

You could also be asked to draw a bar chart using data given to you in a table. You can read about how to draw bar charts on page 16.

Q2 This table shows the average rate of filtration and reabsorption of various substances found in the blood, in a person with healthy kidneys.

Substance	Filtration rate	Reabsorption rate
Water	180 litres per day	178.2 litres per day
Glucose	800 millimoles per day	?
Potassium ions	720 millimoles per day	500 millimoles per day

a) i) What percentage of the filtered water is reabsorbed?

ii) What do you think would happen to this percentage if the person went for a run on a hot day and didn't consume any extra liquid? Explain your answer.

iii) What effect do you think this would have on the urine produced by the person?

b) How many millimoles of glucose will be reabsorbed per day? Explain your answer.

c) Calculate the average amount of potassium ions found in the urine of this person per day.

4. Kidney Failure

As you've seen, the kidneys are a really important pair of organs. If they stop working properly it can cause big problems in the body...

Learning Objectives:
- Know that kidney dialysis is a treatment for kidney failure.
- Know how a dialysis machine works.
- Know that having a kidney transplant is a treatment for kidney failure.

Specification Reference 4.5.3.3

The effects of kidney failure

The kidneys play a number of important roles in the body (see page 215). If the kidneys don't work properly, waste substances build up in the blood and you lose your ability to control the levels of ions and water in your body. This can cause problems in the heart, bones, nervous system, stomach, mouth, etc. If left untreated, kidney failure will eventually result in death.

Fortunately there are treatment options available. People with kidney failure can be kept alive by:

1. Having **dialysis** treatment — where machines do the job of the kidneys.
2. Having a **kidney transplant** — where the diseased kidney is replaced by a healthy one.

Kidney dialysis

Dialysis machines take over the role of failing kidneys and filter the blood. Dialysis has to be done regularly to keep the concentrations of dissolved substances (e.g. glucose, ions, etc.) in the blood at normal levels, and to remove waste substances.

How does a dialysis machine work?

In a dialysis machine the person's blood flows alongside a partially permeable membrane, surrounded by **dialysis fluid** (see Figure 2).

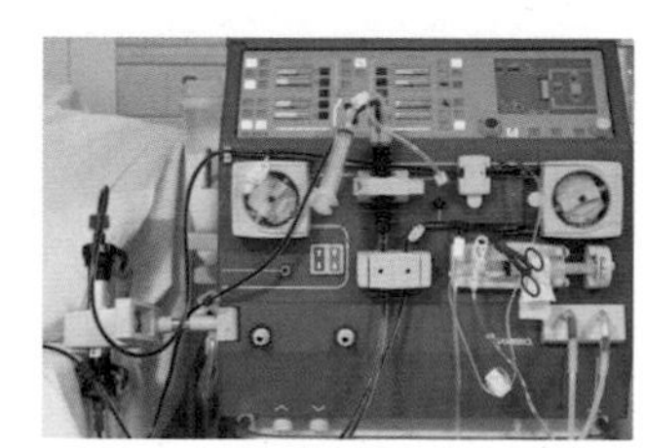

Figure 1: *A dialysis machine.*

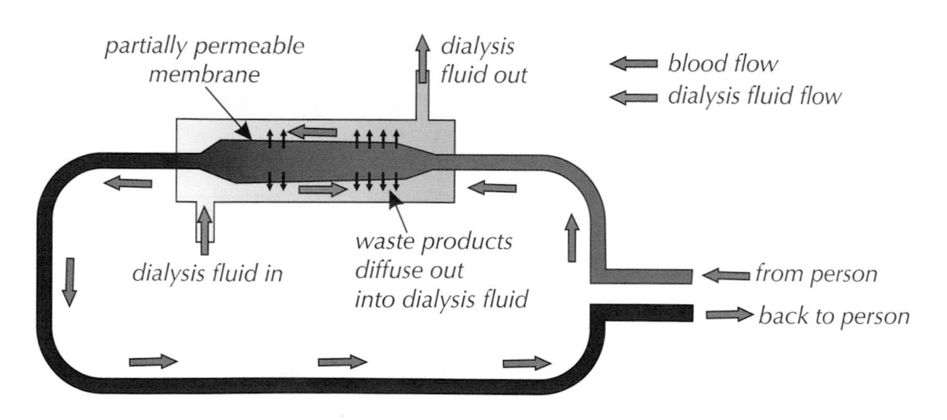

Figure 2: *Diagram showing how a dialysis machine works.*

The partially permeable membrane allows things like ions and waste substances through, but not big molecules like proteins (just like the membranes in the kidney). The dialysis fluid has the same concentration of dissolved ions and glucose as healthy blood. This means that useful dissolved ions and glucose won't be lost from the blood during dialysis. Only waste substances (such as urea) and excess ions and water diffuse across the membrane.

Tip: Substances diffuse from an area of higher concentration to an area of lower concentration. This means that if there's a higher concentration of a substance in the blood, it will diffuse across the partially permeable membrane into the dialysis fluid.

Problems with dialysis

Although kidney dialysis can keep a person alive until they can get a kidney transplant (see below), it has a number of disadvantages:

- Dialysis is not a pleasant experience and many patients with kidney failure have to have a dialysis session three times a week. Each session takes 3-4 hours — not much fun.
- Dialysis can lead to infections and can cause blood clots.
- Dialysis patients have to be careful about what they eat to avoid too much of a particular ion building up between dialysis sessions.
- Patients have to limit the amount of fluid they take in, as the kidneys play an important role in maintaining the water content of the body. When the kidneys aren't functioning properly, fluid can build up in the body, which can be dangerous — for example, it can cause the volume of blood to increase, leading to high blood pressure.
- Kidney dialysis machines are expensive things for the NHS to run.

Tip: Dialysis isn't a cure for kidney failure. It's a treatment which can keep people alive while they are on the waiting list for a donor kidney. It is also used for people who may be unable to have a kidney transplant — people can stay alive on dialysis for many years.

Kidney transplants

At the moment, the only cure for kidney disease is to have a kidney transplant. Healthy kidneys are usually transplanted from people who have died suddenly, say in a car accident, and who are on the organ donor register or carry a donor card (provided their relatives agree too). Kidneys can also be transplanted from people who are still alive (as we all have two kidneys, but can survive with just one) but there is a small risk to the person donating the kidney.

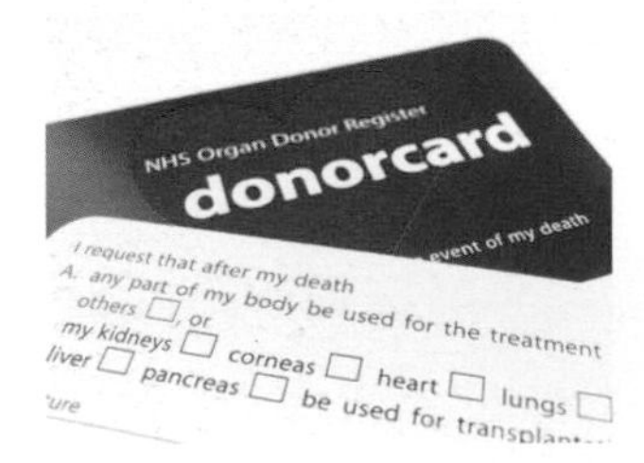

Figure 3: An NHS donor card. People carry this card to let medical staff know that they wish to donate their organs after they have died.

Rejection

A problem of kidney transplants is that the donor kidney can be rejected by the recipient's immune system — this happens when **antigens** on the donor kidney aren't recognised as being part of the body by the recipient's white blood cells. The white blood cells produce **antibodies** to attack the donor cells as a result (see Figure 4).

Tip: The recipient is the person who receives the kidney.

Tip: Antigens are unique molecules found on a cell's surface. Antibodies are produced by white blood cells in response to the presence of a foreign antigen — see page 136.

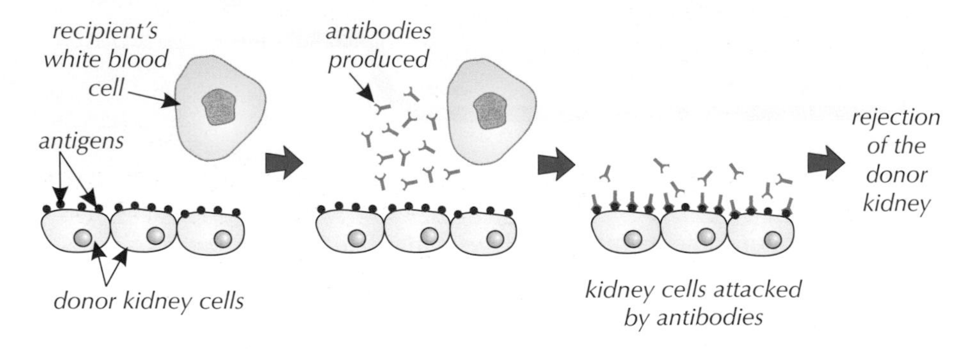

Figure 4: Diagram showing the possible reaction of a recipient's immune system in response to a donor kidney.

Preventing rejection

To help prevent rejection of the donor organ, precautions are taken:

- A donor with a tissue-type that closely matches the patient's is chosen. Tissue-type is based on a cell's antigens. The more similar the tissue-types of the donor and the patient, the more similar the antigens. This reduces the chance of the patient's white blood cells identifying the donor antigens as 'foreign' and producing antibodies to attack the donor organ.
- The recipient is treated with drugs that suppress the immune system. These drugs reduce the production and release of antibodies by the white blood cells, so that the immune system won't attack the transplanted kidney.

Tip: Donor kidneys are ideally matched by blood type (and a few other things) to the recipient, which make them less likely to be rejected. However, it means a potentially long waiting time for a suitable kidney. WORKING SCIENTIFICALLY

WORKING SCIENTIFICALLY

Other problems with kidney transplants

There are problems other than rejection associated with kidney transplants:

- There are long waiting lists for kidneys.
- Even if a kidney with a matching tissue-type is found, there's still the possibility that it'll be rejected.
- Taking drugs that suppress the immune system means the person is vulnerable to other illnesses and infections.
- A kidney transplant is a major operation, so it can be risky.

Despite these problems, kidney transplants can put an end to the hours spent on dialysis, allowing recipients to lead a relatively normal life. Transplants are also cheaper than long-term dialysis treatment.

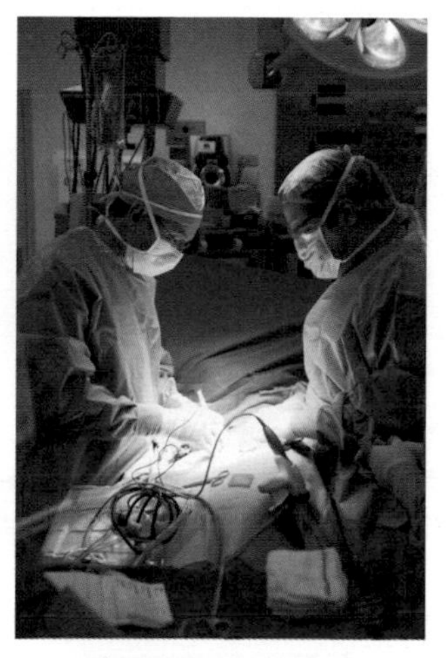

Figure 5: *Surgeons performing a kidney transplant.*

Practice Questions — Fact Recall

Q1 True or False? People with kidney failure only need dialysis once.

Q2 Give an example of a substance that can't pass through the partially permeable membrane used to filter the blood in a dialysis machine.

Q3 True or False? Everyone needing a kidney transplant gets a new kidney straight away.

Exam Tip
In the exam you might be asked to compare the treatment options available for someone with kidney failure. Make sure you know both the positive and negative aspects of being on dialysis or having a kidney transplant.

Practice Questions — Application

Q1 The level of dissolved sodium found in dialysis fluid before a dialysis session is similar to the typical level of dissolved sodium found in the blood plasma.

a) Why is the level of dissolved sodium in the dialysis fluid similar to the level in the blood plasma?

Tip: Plasma is the liquid part of the blood (see page 87). It carries lots of different substances.

Tip: Dissolved sodium is an ion.

Tip: A dialysis machine is like an artificial kidney. If you're struggling with Q1, think about the role of a healthy kidney in the body — it regulates the water and ion content of the blood and removes waste products.

Blood plasma contains urea. Dialysis fluid does not.

b) i) Why does the blood plasma contain urea?

ii) Explain why it's important that the dialysis fluid doesn't contain urea.

Blood plasma also contains glucose.

c) Suggest how the amount of glucose in the dialysis fluid would compare to the amount of glucose in the blood plasma. Explain your answer.

Q2 Nomia has kidney failure. She is currently having dialysis and she is hoping to have a kidney transplant in the future. Her doctor is explaining what a kidney transplant will involve.

a) Suggest three problems associated with having a kidney transplant that the doctor might tell Nomia about.

b) Suggest two reasons why having a kidney transplant may be a better option for Nomia than being on dialysis.

Topic 5b Checklist — Make sure you know...

Hormones

- ❑ That hormones are large chemical molecules which travel in the blood to activate target organs.
- ❑ That hormones are secreted by glands (e.g. the pituitary gland), and that glands make up the endocrine system.
- ❑ That the pituitary gland is called the 'master gland' because it releases hormones that act on other glands to stimulate the release of other hormones.
- ❑ Where the pituitary gland, pancreas, thyroid, adrenal glands, ovaries and testes are found in the body.
- ❑ That hormones have a slower action than nerve impulses but the effects last longer.

Controlling Blood Glucose

- ❑ That the pancreas monitors and controls the glucose level of the blood.
- ❑ That the pancreas produces insulin if blood glucose gets too high — insulin decreases the blood glucose level by causing body cells to take up more glucose from the blood.
- ❑ That any extra glucose that is not immediately needed by the body is stored as glycogen in liver and muscle cells.
- ❑ H That the pancreas produces glucagon if blood glucose gets too low — glucagon increases the blood glucose level by causing body cells to convert glycogen into glucose which then enters the blood.

cont...

- ☐ That Type 1 diabetes is a condition where the pancreas produces little to no insulin and that it can be controlled by insulin therapy, eating a carbohydrate-controlled diet and taking regular exercise.
- ☐ That Type 2 diabetes is a condition where the body's cells don't respond to insulin, and that obesity is a risk factor.
- ☐ That Type 2 diabetes can be controlled eating a carbohydrate-controlled diet and taking regular exercise.
- ☐ The differences between the causes and treatments of Type 1 and Type 2 diabetes.
- ☐ How to read from and interpret graphs that show the effect of insulin on the blood glucose concentration of a person with or without diabetes.

Controlling Water Content

- ☐ How the kidneys maintain the water level of the body — they filter the blood to remove urea and excess water and ions in urine, and use selective reabsorption to take useful substances (e.g. glucose, ions and water) back into the blood.
- ☐ That water is lost during exhalation and that the lungs have no control over this loss.
- ☐ That an imbalance of water and ions in the body can cause cells to lose or gain too much water by osmosis, which can cause damage to the cells.
- ☐ That water, ions and urea leave the body in sweat and there's no control for this loss.
- ☐ H Why there are excess amino acids in the body, and that they are broken down in the liver to produce toxic ammonia, which is converted to urea for excretion in urine.
- ☐ H That anti-diuretic hormone (ADH) is released from the pituitary gland to control water content in the body by controlling how permeable the kidney tubules are to water and so how much water is reabsorbed by the kidneys.
- ☐ H That the release of ADH is controlled by negative feedback.
- ☐ How to convert between tables of data and bar charts that show the levels of glucose, ions and urea before and after filtration by the kidneys.

Kidney Failure

- ☐ That people with kidney failure can be kept alive by kidney dialysis.
- ☐ How a dialysis machine filters a person's blood using a partially permeable membrane.
- ☐ That a kidney transplant is the only cure for kidney failure.

Exam-style Questions

1 Kaye has recently been diagnosed with Type 1 diabetes.

1.1 Which of the following statements about Type 1 diabetes is true?

A People with Type 1 diabetes are advised to eat a high sugar diet.

B Type 1 diabetes allows the blood glucose concentration to get too high.

C Type 1 diabetes is caused by the liver and stomach failing to produce vital hormones.

D Type 1 diabetes is directly caused by the overproduction of testosterone.

(1 mark)

Kaye's doctor prescribed her hormone injections to control her condition.

1.2 Name the hormone that the injections will contain.

(1 mark)

1.3 Explain why injecting this hormone can be an effective treatment for Type 1 diabetes.

(3 marks)

1.4 In cases where a patient has particularly poor control over their Type 1 diabetes, a pancreas transplant is sometimes recommended by doctors. Suggest why a pancreas transplant may be a suitable treatment for Type 1 diabetes but not for Type 2 diabetes.

(3 marks)

1.5 Outline and explain the treatment for Type 2 diabetes.

(4 marks)

2 The kidneys control the water and ion content of the blood.

2.1 Which hormone controls the regulation of water content?

A ADH

B luteinising hormone

C thyroxine

D adrenaline

(1 mark)

Urea is a waste product produced in the liver, which the body needs to get rid of.

2.2 How is urea produced by the body?

(3 marks)

This table shows average values for the way a healthy kidney handles sodium ions.

Ion	Amount filtered (g per day)	% reabsorbed
Sodium	575 g	99.5

2.3 Calculate the amount of sodium ions found in the urine per day.

(2 marks)

1. Puberty and the Menstrual Cycle

Hormones secreted by the ovaries and the pituitary gland are responsible for controlling the changes that occur during a woman's menstrual cycle.

Puberty

At puberty, your body starts releasing sex hormones that trigger off secondary sexual characteristics (such as the development of facial hair in men and breasts in women) and cause eggs to mature in women.

- In men, the main reproductive hormone is **testosterone**. It's produced by the testes and stimulates sperm production.
- In women, the main reproductive hormone is **oestrogen**. It's produced by the ovaries. As well as bringing about physical changes, oestrogen is also involved in the menstrual cycle.

The menstrual cycle

The menstrual cycle is the monthly sequence of events in which the female body releases an egg and prepares the uterus (womb) in case it receives a fertilised egg. The menstrual cycle has four stages:

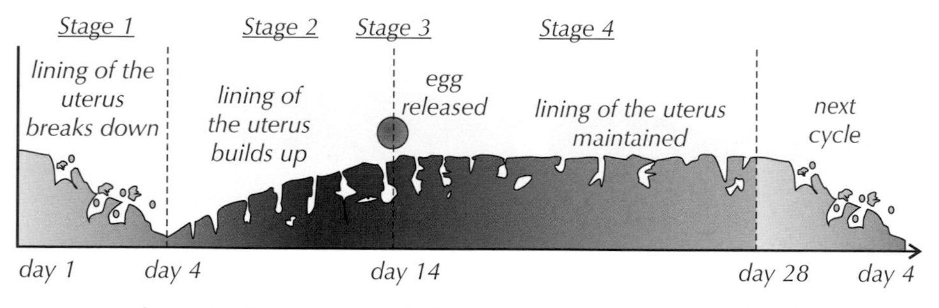

Figure 1: *Diagram showing the four stages of the menstrual cycle.*

- **Stage 1**
  Day 1 is when menstruation (bleeding) starts.
  The uterus lining breaks down for about four days.
- **Stage 2**
  The lining of the uterus builds up again, from day 4 to day 14, into a thick spongy layer full of blood vessels, ready to receive a fertilised egg.
- **Stage 3**
  An egg is released from the ovary at day 14. This is called **ovulation**.
- **Stage 4**
  The wall is then maintained for about 14 days, until day 28. If no fertilised egg has landed on the uterus wall by day 28, the spongy lining starts to break down again and the whole cycle starts again.

Learning Objectives:

- Know that reproductive hormones promote secondary sexual characteristics at puberty.
- Know that testosterone, produced in the testes, is the main male reproductive hormone and that it stimulates sperm production.
- Know that oestrogen, produced in the ovaries, is the main female reproductive hormone and that it is involved in the menstrual cycle.
- Know that an egg is released from the ovary roughly every 28 days and that this is known as ovulation.
- Know that hormones control the menstrual cycle.
- Know how follicle stimulating hormone (FSH), luteinising hormone (LH), oestrogen and progesterone are involved in the menstrual cycle.
- H Be able to explain how FSH, LH, oestrogen and progesterone interact to control the menstrual cycle.
- H Be able to read from and explain graphs showing changing hormone levels during the menstrual cycle.

Specification Reference 4.5.3.4

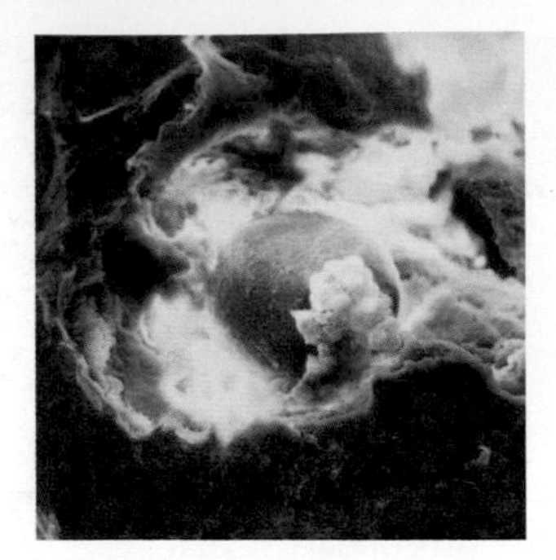

Figure 2: A microscope image showing an egg (pink oval) being released from an ovary (brown).

Hormonal control of the menstrual cycle

You need to know about four hormones that are involved in the menstrual cycle:

- **Follicle stimulating hormone (FSH)**
  FSH causes an egg to mature in one of the ovaries, in a structure called a follicle.
- **Luteinising hormone (LH)**
  LH stimulates the release of an egg at day 14 (ovulation).
- **Oestrogen** and **progesterone**
  These hormones are involved in the growth and maintenance of the uterus lining.

Exam Tip
Make sure you don't talk about egg 'production' or 'development' if you're asked about the menstrual cycle in the exam. Eggs already exist in a woman's ovaries at birth. FSH causes these eggs to mature and LH (see below) causes them to be released.

Hormonal interaction in the menstrual cycle Higher

The levels of the four hormones above fluctuate throughout the cycle (see Figure 3). They also interact with each other to promote or inhibit the release of other hormones.

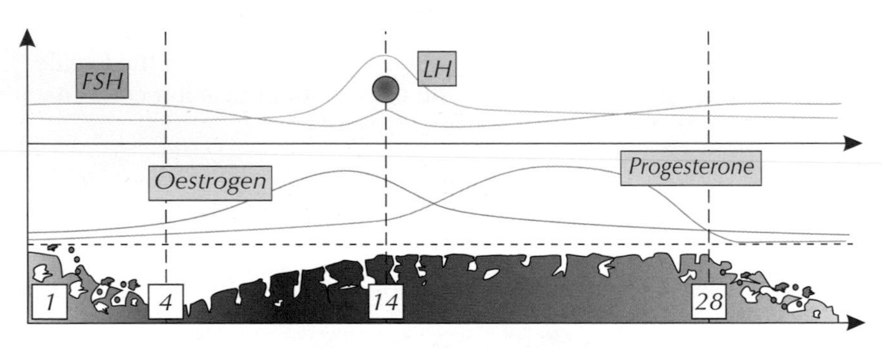

Figure 3: Diagram showing hormonal changes during the menstrual cycle

Exam Tip H
In the exam you might be asked to interpret a graph like the one in Figure 3. It might look confusing at first, but if you get your head around what each of the four hormones does, it should start to make more sense. A graph is a really useful way to visualise how the hormones interact when controlling the cycle.

1. **FSH** is produced in the pituitary gland and causes an egg to mature in an ovary. It also stimulates the ovaries to produce oestrogen.

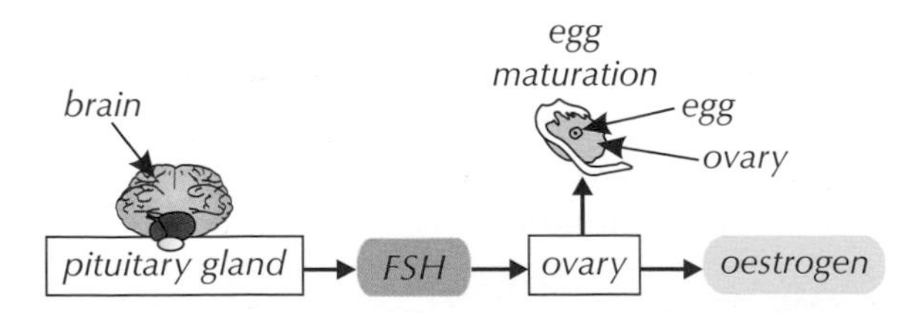

Figure 4: Diagram showing the role of FSH in the menstrual cycle.

2. **Oestrogen** is produced in the ovaries and causes the lining of the uterus to grow. It also stimulates the release of LH (which causes the release of an egg) and inhibits the release of FSH.

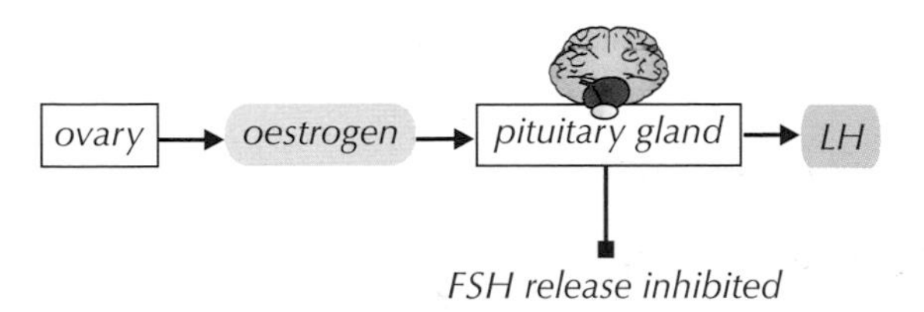

Figure 5: *Diagram showing the role of oestrogen in the menstrual cycle.*

Tip: H The inhibition of FSH by oestrogen makes sure that no more eggs mature during that month's cycle.

3. **LH** is produced by the pituitary gland and stimulates the release of an egg at day 14 (ovulation).

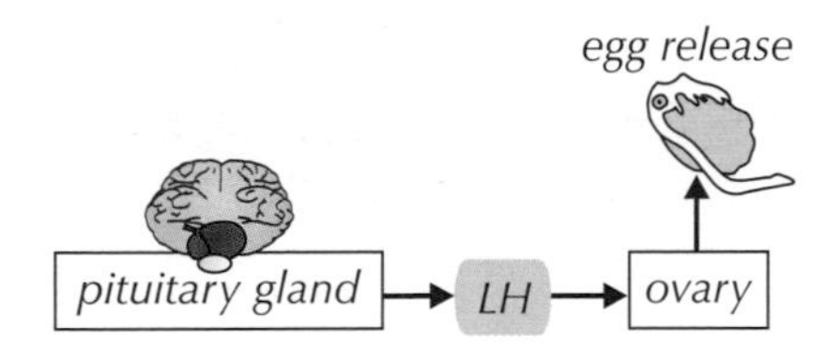

Figure 6: *Diagram showing the role of LH in the menstrual cycle.*

4. **Progesterone** is produced in the ovaries by the remains of the follicle after ovulation. It maintains the lining of the uterus during the second half of the cycle. When the level of progesterone falls, the lining breaks down. Progesterone also inhibits the release of LH and FSH.

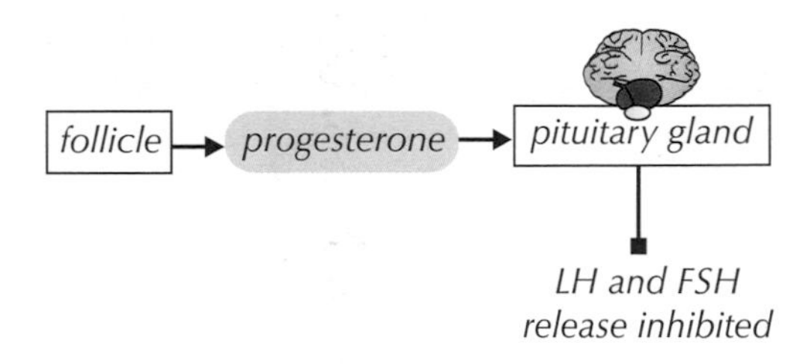

Figure 7: *Diagram showing the role of progesterone in the menstrual cycle.*

Levels of FSH, LH, oestrogen and progesterone change throughout the menstrual cycle. So the change in the level of one hormone can be used to predict the change in the level of another hormone:

Example **Higher**

FSH stimulates the ovaries to produce oestrogen, so if the FSH level rises, you'd expect the oestrogen level to rise too. The increasing oestrogen level will then inhibit FSH release, causing the FSH level to drop.

Exam Tip H
In the exam, you might be tested on how the hormones in the menstrual cycle interact by being shown a graph — don't worry, just apply what you know about the effects of each hormone.

Practice Questions — Fact Recall

Q1 Name the main female reproductive hormone.

Q2 Give one effect that testosterone has on the male body.

Q3 What effect does FSH have on an egg in the menstrual cycle?

Q4 What is the function of LH?

Q5 Which hormones inhibit the release of FSH?

Q6 Which two glands secrete the hormones that control the menstrual cycle?

Practice Question — Application

Q1 The graph below shows the level of a hormone measured in the bloodstream of one woman during her 28 day menstrual cycle.

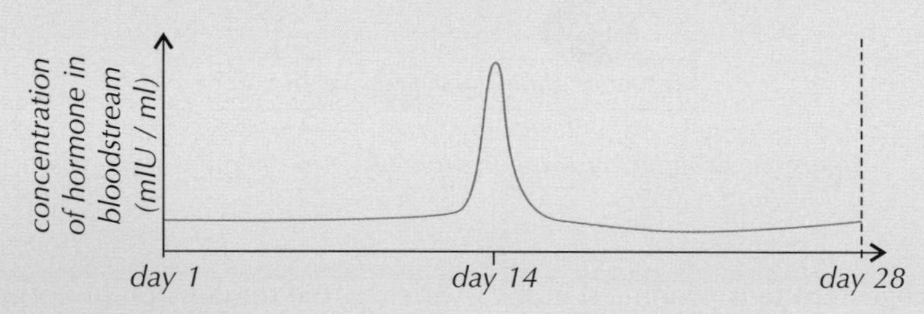

Tip: H Think about what happens around the middle of the menstrual cycle.

a) Which hormone do you think is shown on the graph? Give a reason for your answer.

b) Where is the hormone you gave in part a) produced?

c) The graph below shows the level of the same hormone measured in another woman during her 28 day menstrual cycle. This woman is struggling to have children. Suggest why this might be.

Tip: H Both graphs are drawn to the same scale.

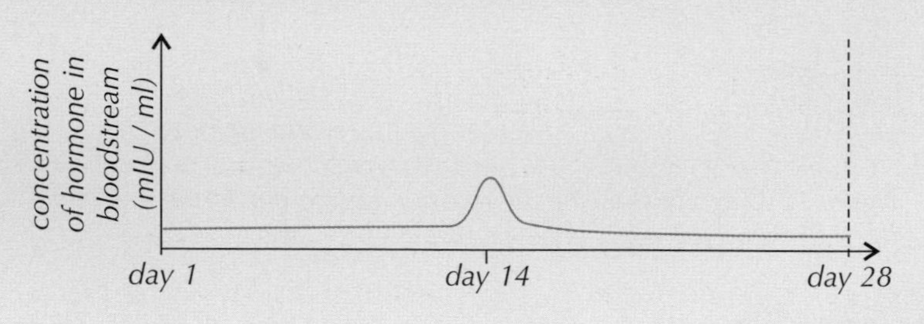

2. Contraceptives

Some of the hormones involved in the menstrual cycle can be used to decrease a woman's fertility and are used in a range of different contraceptives.

Learning Objectives:
- Understand that hormonal and non-hormonal methods can be used to control fertility.
- Understand how hormonal methods of contraception work, including oral contraceptives, skin patches, implants, injections and intrauterine devices.
- Understand how non-hormonal methods of contraception work, including barrier methods, spermicidal agents, intrauterine devices, surgery and abstinence.
- Be able to evaluate methods of contraception.

Specification Reference 4.5.3.5

Reducing fertility

Contraceptives are used to prevent pregnancy. The hormones oestrogen and progesterone can be taken by women to reduce their **fertility** (their ability to get pregnant) and so are often used as contraceptives.

Oestrogen

Oestrogen can be used to prevent egg release. This may seem kind of strange (since naturally oestrogen helps stimulate the release of eggs — see page 227). But if oestrogen is taken every day to keep the level of it permanently high, it inhibits FSH production, and after a while egg maturation and therefore egg release stop and stay stopped.

Progesterone

Progesterone also reduces fertility, e.g. by stimulating the production of thick cervical mucus which prevents any sperm getting through and reaching an egg. It can inhibit egg maturation and therefore the release of an egg too.

Oral contraceptives

The pill is an oral contraceptive (it can be taken by mouth to decrease fertility). The first version (known as the combined oral contraceptive pill) was made in the 1950s and contained high levels of oestrogen and progesterone.

However, there were concerns about a link between oestrogen in the pill and side effects like blood clots. The pill now contains lower doses of oestrogen so has fewer side effects.

There's also a progesterone-only pill — it has fewer side effects than the combined pill and is just as effective.

There are both positives and negatives associated with using the pill:

Benefits of the combined oral contraceptive pill

- The pill's over 99% effective at preventing pregnancy.
- It's also been shown to reduce the risk of getting some types of cancer.

Problems with the combined oral contraceptive pill

- It isn't 100% effective — there's still a very slight chance of getting pregnant.
- It can cause side effects like headaches, nausea, irregular menstrual bleeding, and fluid retention.
- It doesn't protect against STDs (sexually transmitted diseases).

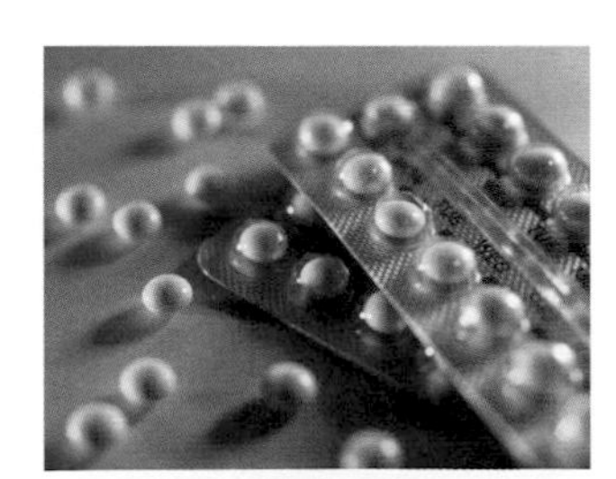

Figure 1: *The contraceptive pill is used to reduce fertility, decreasing the risk of pregnancy.*

Tip: Oral contraceptives are sometimes called birth-control pills.

Figure 2: *The contraceptive patch gets attached to the skin.*

Skin patches, implants and injections

- The contraceptive patch contains oestrogen and progesterone (the same as the combined pill). It's a small (5 cm × 5 cm) patch that's stuck to the skin. Each patch lasts one week.
- The contraceptive implant is inserted under the skin of the arm. It releases a continuous amount of progesterone, which stops the ovaries releasing eggs, makes it hard for sperm to swim to the egg and stops any fertilised egg implanting in the uterus. An implant can last for three years.
- The contraceptive injection also contains progesterone. Each dose lasts 2 to 3 months.

Barrier methods

Non-hormonal barrier forms of contraception are designed to stop the sperm from getting to the egg.

- Condoms are worn over the penis during intercourse to prevent the sperm entering the vagina. There are also female condoms that are worn inside the vagina. Condoms are the only form of contraception that will protect against sexually transmitted diseases.
- A diaphragm is a shallow plastic cup that fits over the cervix (the entrance to the uterus) to form a barrier. It has to be used with spermicide (a substance that disables or kills sperm). Spermicide can be used alone as a contraceptive, but it's not as effective as when used with a diaphragm.

Tip: When a diaphragm is used correctly with spermicide, it's 92-96% effective at preventing pregnancy (so between 4 and 8 women out of 100 that use this method will still become pregnant each year). Spermicide used by itself is only about 70-80% effective. Male condoms are the most effective barrier method at 98%.

Intrauterine devices

An intrauterine device (IUD) is a T-shaped device that is inserted into the uterus to kill sperm and prevent implantation of a fertilised egg. There are two main types — plastic IUDs that release progesterone and copper IUDs that prevent the sperm surviving in the uterus.

Surgical methods

Sterilisation involves cutting or tying the fallopian tubes (which connect the ovaries to the uterus) in a female, or the sperm duct (the tube between the testes and penis) in a male. This is a permanent procedure. However, there is a very small chance that the tubes can rejoin.

Tip: As with oral contraceptives, there are benefits and problems with all of the methods on this page. The contraceptive that is most suitable will differ based on the individual.

Natural methods

Pregnancy may be avoided by finding out when in the menstrual cycle the woman is most fertile and avoiding sexual intercourse on those days. It's popular with people who think that hormonal and barrier methods are unnatural, but it's not very effective. The only way to be completely sure that sperm and egg don't meet is to not have intercourse — this is abstinence.

Practice Questions — Fact Recall

Q1 Explain the function of oestrogen in the combined contraceptive pill.

Q2 Apart from oestrogen, what hormone is found in the combined pill?

Q3 Name one of the barrier methods of controlling fertility.

3. Increasing Fertility Higher

Scientific advances in understanding how hormones affect fertility have led to many infertile women being able to have babies. Read on for more...

Increasing fertility

Hormones can be taken by women to increase their fertility. For example, some women have an FSH level that is too low to cause their eggs to mature. This means that no eggs are released and the women can't get pregnant.

The hormones FSH and LH can be injected by these women to stimulate egg maturation and release in their ovaries. These 'fertility drugs' can help a lot of women to get pregnant when previously they couldn't. They are often used during **IVF** too — see below.

Problems with fertility drugs

Using fertility drugs like FSH and LH has its problems:

- They don't always work — some women may have to use the treatment many times, which can be expensive.
- Too many eggs could be stimulated, resulting in unexpected multiple pregnancies (twins, triplets, etc.).

In vitro fertilisation (IVF)

IVF is a process that can be used to help couples who are having difficulty having children.

IVF involves the following steps:

1. FSH and LH are given to the woman to stimulate the maturation of multiple eggs.
2. Eggs are then collected from the woman's ovaries.
3. The eggs are fertilised in a lab using the man's sperm.
4. The fertilised eggs then grow into embryos (small balls of cells) in a laboratory incubator.
5. Once the embryos have formed, one or two of them are transferred to the woman's uterus. Transferring more than one improves the chance of pregnancy.

This is summarised in Figure 1.

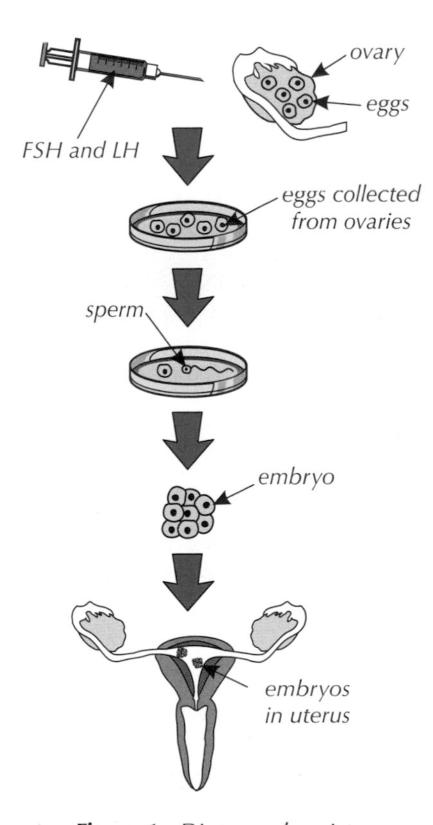

Figure 1: *Diagram showing the steps involved in IVF.*

Learning Objectives:

- H Understand how hormones can be used to treat infertility, including the use of FSH and LH in 'fertility drugs' for women.
- H Understand the steps involved in *in vitro* fertilisation (IVF).
- H Know about the disadvantages of using IVF to control fertility.

Specification Reference 4.5.3.6

Tip: H Flick back to page 226 for a reminder of what the hormones FSH and LH do.

Tip: H As well as being hard work, having multiple births puts a bigger stress on the mother and embryos' health during pregnancy.

Tip: H IVF treatment can also involve a technique called Intra-Cytoplasmic Sperm Injection (ICSI), where the sperm is injected directly into an egg. It's useful if the man has a very low sperm count.

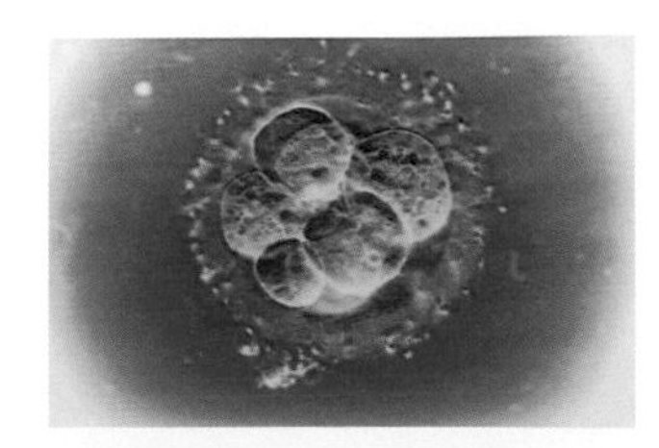

Figure 2: *An embryo ready for transfer into a uterus.*

Tip: H These pros and cons must be taken into consideration by the patients being treated for infertility but also by the doctor who is treating them.

The pros and cons of IVF

The main benefit of IVF is that it can give an infertile couple a child — a pretty obvious benefit. However there are negative sides to the treatment:

- Multiple births can happen if more than one embryo grows into a baby — these are risky for the mother and babies (there's a higher risk of miscarriage, stillbirth...).
- The success rate of IVF is low — the average success rate in the UK is about 26%. This makes the process incredibly stressful and often upsetting, especially if it ends in multiple failures.
- As well as being emotionally stressful, the process is also physically stressful for the woman. Some women have a strong reaction to the hormones — e.g. abdominal pain, vomiting, dehydration.

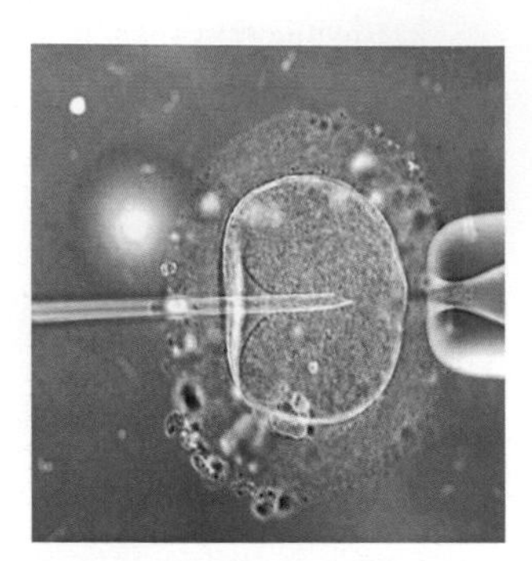

Figure 3: *A light microscope image of the fertilisation of a human egg using a micro-needle.*

Developments in microscopy

Advances in microscope techniques have helped to improve the techniques (and therefore the success rate) of IVF.

- Specialised micro-tools (see Figure 3) have been developed to use on the eggs and sperm under the microscope. They're also used to remove single cells from the embryo for genetic testing (to check that it is healthy — see page 273).
- More recently, the development of time-lapse imaging (using a microscope and camera built into the incubator) means that the growth of the embryos can be continuously monitored to help identify those that are more likely to result in a successful pregnancy.

Tip: H You find can out more about microscopy on page 25.

Social and ethical issues

The process of IVF often results in unused embryos that are eventually destroyed. Because of this, some people think it is unethical because each embryo is a potential human life.

The genetic testing of embryos before implantation also raises ethical issues as some people think it could lead to the selection of preferred characteristics, such as gender or eye colour.

Practice Questions — Fact Recall

Q1 Why is a mother given FSH and LH as the first stage of IVF treatment?

Q2 What is the next step after egg collection in IVF?

Q3 Give two downsides of using IVF treatment.

Q4 Give one reason why a person may believe that IVF is unethical.

Practice Question — Application

Q1 The table shows some data about women aged 18-34 undergoing IVF in the UK in 2013 and 2014 using their own fresh eggs.

Year	Number of embryo transfers during IVF	Number of IVF pregnancies
2013	17 892	
2014	17 765	7768

a) Calculate the percentage of embryo transfers that resulted in pregnancy for 18-34 year olds undergoing IVF in 2014.

b) The percentage of embryo transfers that resulted in pregnancy for 18-34 year olds undergoing IVF in 2013 was 41.74%. Calculate the number of these embryo transfers that resulted in pregnancy.

c) The average multiple pregnancy rate of women undergoing IVF in 2014 in the UK with their own fresh eggs was 15.7%. The multiple pregnancy rate of normal pregnancies was lower than this. Give a reason why this might be.

Tip: H IVF can be done using a woman's own fresh eggs (eggs freshly collected from the woman), frozen eggs (eggs which have been extracted at an earlier date and frozen until needed), or using eggs from another woman.

Exam Tip H
Read any data given to you carefully. In this question, make sure you're using the correct numbers from the table.

4. Thyroxine and Adrenaline Higher

Adrenaline and thyroxine are both pretty useful hormones for us humans.

Learning Objectives:
- H Know that thyroxine is a hormone produced in the thyroid gland.
- H Understand the role thyroxine plays in metabolism, growth and development.
- H Understand how thyroxine is controlled by negative feedback to keep it at the right level in the blood.
- H Know that adrenaline is a hormone produced at times of fear and stress in the adrenal glands.
- H Understand the role adrenaline plays in the 'flight or fight' response.

Specification Reference 4.5.3.7

Thyroxine

Thyroxine is a hormone released by the **thyroid gland**, which is found in the neck.

It plays an important role in regulating the basal metabolic rate — the speed at which chemical reactions in the body occur while the body is at rest. It is also important for loads of processes in the body, such as stimulating protein synthesis for growth and development.

Thyroxine is released in response to thyroid stimulating hormone (TSH), which is released from the pituitary gland.

Negative feedback

As you know from page 184, when the levels of certain substances in the body go above or below a normal level, the body triggers responses that help to bring these levels back into a normal range. This is called negative feedback.

A negative feedback system keeps the amount of thyroxine in the blood at the right level (see Figure 1).

- When the level of thyroxine in the blood is higher than normal, the secretion of TSH from the pituitary gland is inhibited (stopped). This reduces the amount of thyroxine released from the thyroid gland, so the level in the blood falls back towards normal.
- When the level of thyroxine in the blood is lower than normal, the secretion of TSH from the pituitary gland is stimulated (started up) again. This increases the amount of thyroxine released from the thyroid gland, so the level in the blood rises back towards normal.

Tip: H You can think about negative feedback working like a thermostat — if the temperature gets too low, the thermostat will turn the heating on, then if the temperature gets too high, it'll turn the heating off again.

Exam Tip H
In the exam you could be asked to interpret graphs like Figure 1. Even if they look different, just apply your understanding of negative feedback to help you work out what the diagram is showing you.

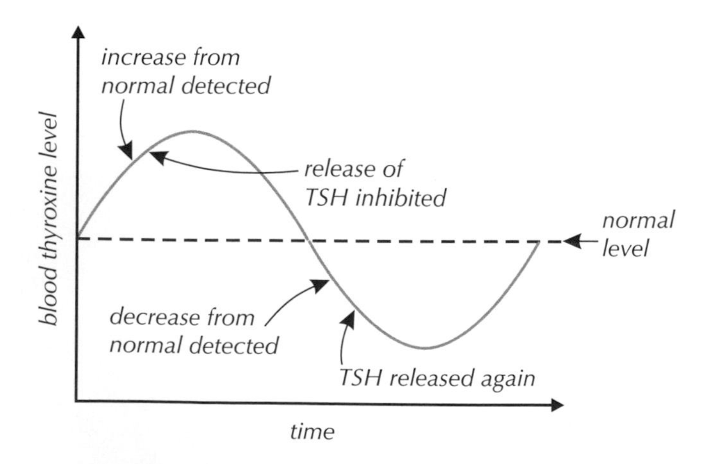

***Figure 1:** Graph showing how the level of thyroxine in the blood is controlled by negative feedback.*

Adrenaline

Adrenaline is a hormone released by the **adrenal glands** (found just above the kidneys — see the diagram on page 208).

It is released in response to stressful or scary situations. Your brain detects fear or stress and sends nervous impulses to the adrenal glands, which respond by secreting adrenaline. Adrenaline gets the body ready for 'fight or flight' by triggering mechanisms that increase the supply of oxygen and glucose to cells in the brain and muscles. For example, it increases heart rate.

Practice Questions — Fact Recall

Q1 Give two roles of thyroxine in the body.

Q2 Name the glands that release adrenaline.

Practice Question — Application

Q1 Pierre is surprised when a dog jumps up at him.
His heart starts to beat faster.

a) Namc thc hormonc that has caused this response.

b) Explain the purpose of this response.

5. Plant Hormones

Learning Objectives:
- Understand that the hormone auxin controls phototropism and gravitropism (geotropism) in plants.
- Know that an uneven distribution of auxin causes uneven growth rates in the shoots and roots of plants.
- Be able to investigate plant growth responses to light and gravity (Required Practical 8).

Specification Reference 4.5.4.1

Like animals, plants have to respond to stimuli. Plant hormones control and coordinate the response of plants to light and gravity.

A plant's needs

Plants need to be able to detect and respond to stimuli (changes in the environment) in order to survive:

Examples

- Plants need light to make their own food. Plants can sense light, and grow towards it in order to maximise the amount of light they receive.
- Plants can sense and respond to gravity. This makes sure that their roots and shoots grow in the right direction.

Auxin

Auxin is a plant hormone that controls growth near the tips of shoots and roots. It controls the growth of a plant in response to different stimuli.

Figure 1: *A plant displaying phototropism — its shoot is growing towards the light.*

Auxin controls:

- **Phototropism** — plant growth in response to light.
- **Gravitropism** (also known as **geotropism**) — plant growth in response to gravity.

Tip: Cell elongation just means that the cells of the plant get bigger (longer).

Auxin is produced in the tips of roots and shoots and moves backwards to stimulate the cell elongation process which occurs in the cells just behind the tips (see Figure 2). If the tip of a shoot is removed, no auxin is available and the shoot may stop growing.

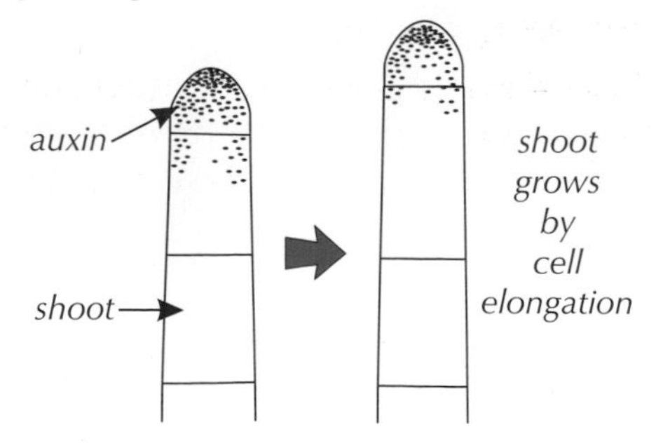

Figure 2: *Auxin release from the tips of shoots results in cell elongation and shoot growth.*

Shoot and root growth

Extra auxin promotes growth in the shoot but inhibits growth in the root. This, coupled with the unequal distribution of auxin in the shoot or root tip, produces the following results:

1. Shoots grow towards light

When a shoot tip is exposed to light, more auxin accumulates on the side that's in the shade than the side that's in the light. This makes the cells grow (elongate) faster on the shaded side, so the shoot bends towards the light (see Figure 3).

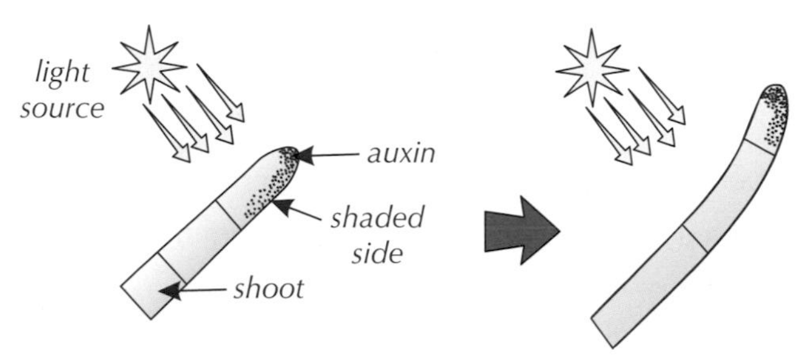

Figure 3: *Diagram to show how auxin causes shoot growth towards the light.*

Tip: In shoots, more auxin on one side means that the cells on that side will grow faster. This will cause the plant to bend away from that side, e.g.

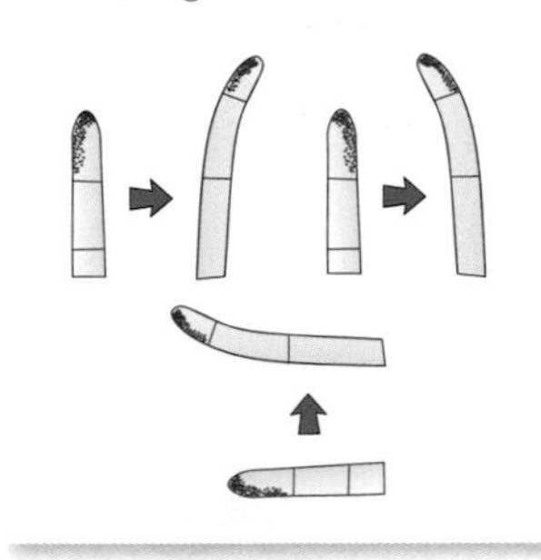

2. Shoots grow away from gravity

When a shoot is growing sideways, gravity produces an unequal distribution of auxin in the tip, with more auxin on the lower side. This causes the lower side to grow faster, bending the shoot upwards, as shown in Figure 5.

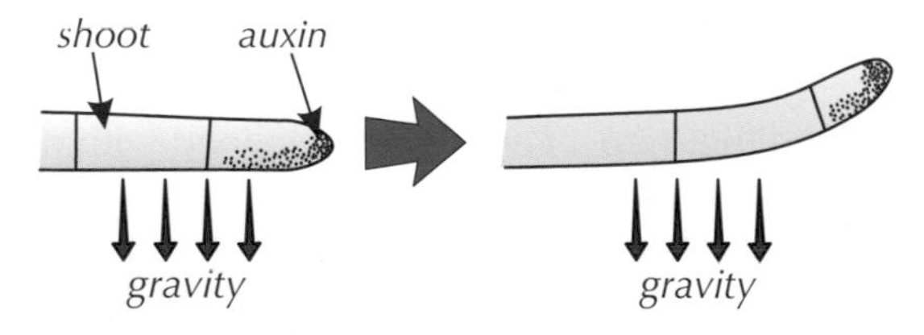

Figure 5: *Diagram to show how auxin causes shoot growth away from gravity.*

Figure 4: *A plant displaying gravitropism — its shoots are growing away from gravity.*

The distribution of auxin in response to gravity means that the shoot should always grow in the right direction (i.e. upwards), even in the absence of light.

3. Roots grow towards gravity

When a root is growing sideways, more auxin will accumulate on its lower side. In a root the extra auxin inhibits growth. This means the cells on top elongate faster, and the root bends downwards (see Figure 6).

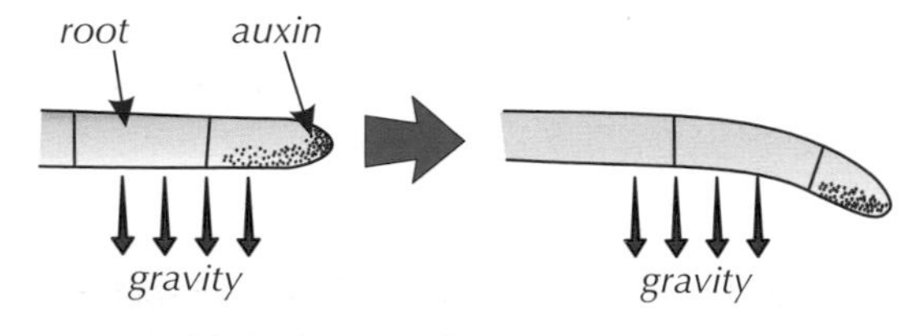

Figure 6: *Diagram to show how auxin causes root growth towards gravity.*

Tip: In roots, more auxin on one side means that the cells on that side will grow slower. This will cause the plant to bend towards that side, e.g.

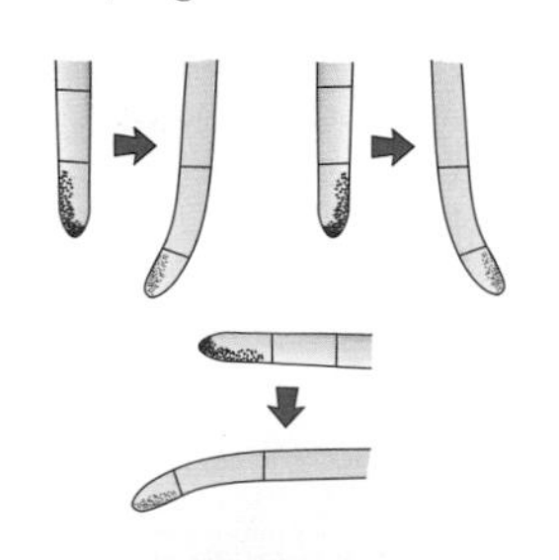

Investigating plant growth responses

Investigating the effect of light

You can investigate the effect of light on the growth of cress seeds like this:

1. Put 10 cress seeds into three different Petri dishes, each lined with moist filter paper. (Remember to label your dishes, e.g. A, B, C.)
2. Shine a light onto one of the dishes from above and two of the dishes from different directions (see Figure 7).
3. Leave your cress seeds alone for one week until you can observe their responses — you'll find the seedlings grow towards the light.
4. You know that the growth response of the cress seedlings is due to light only, if you control all other variables. Here are some examples:
   - Number of seeds — use the same number of seeds in each dish.
   - Type of seed — use seeds that all come from the same packet.
   - Temperature — keep your Petri dishes in a place where the temperature is stable (i.e. away from heat sources and draughts).
   - Water — use a measuring cylinder to add the same amount of water to each dish.
   - Light intensity — keep the distance between the bulb and dish the same for each dish.

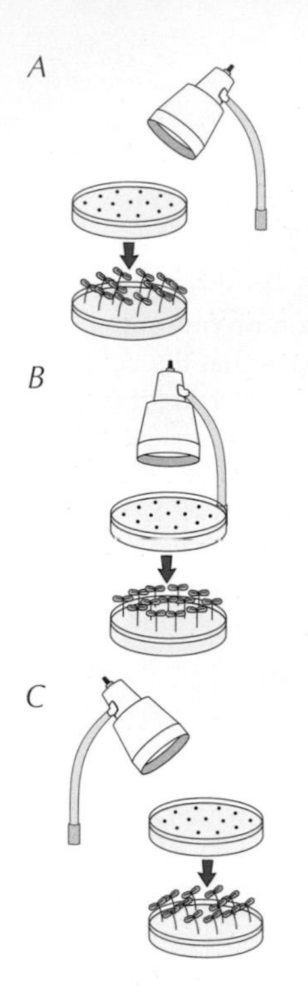

Figure 7: *Diagram showing plant growth in response to light.*

Investigating the effect of gravity

You can do another experiment to investigate the effect of gravity on the growth of cress seeds, like this:

1. Put four cress seedlings into a Petri dish that's lined with damp cotton wool. The roots of each cress seedling should be pointing in a different direction (see Figure 8).
2. Store the Petri dish vertically for a few days in the dark.
3. You should find that the roots of each seedling grow downwards (towards gravity) and the shoots grow upwards (away from gravity).

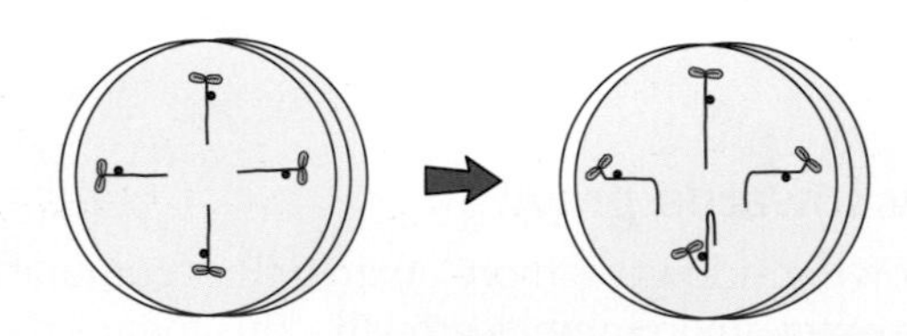

Figure 8: *Diagram showing plant growth in response to gravity.*

Tip: By storing the Petri dish in the dark, you'll know that the result of the experiment won't have been affected by light.

Tip: You need to be able to draw labelled diagrams to show the results of your experiment — make sure you draw them neatly.

Practice Questions — Fact Recall

Q1 What is gravitropism?

Q2 Explain how plant roots grow in response to gravity.

6. Uses of Plant Hormones **Higher**

Plant hormones can be extracted and used by people in agriculture (farming) or horticulture (gardening). Artificial versions can also be made. You need to know about the uses of auxins, gibberellins and ethene.

Learning Objectives:

- H Know that gibberellins stimulate seed germination.
- H Be able to describe how auxins and gibberellins are used in agriculture and horticulture.
- H Know that ethene causes cell division and stimulates the ripening of fruit, and is used in the food industry.

Specification References
4.5.4.1, 4.5.4.2

Auxins

Auxins are useful for controlling plant growth. They are used for:

1. Killing weeds

Most weeds growing in fields of crops or in a lawn are broad-leaved, in contrast to grasses and cereals which have very narrow leaves. Selective weedkillers have been developed using auxins, which only affect the broad-leaved plants. They totally disrupt their normal growth patterns, which soon kills them, whilst leaving the grass and crops untouched.

Example **Higher**

A farmer was studying the effect of selective weedkillers on barley yield. He grew the same type of barley in three fields of the same size and soil type. He used one type of selective weedkiller in Field A, another type in Field B, and left Field C untreated. He measured the crop yield of each field after one year. The results are shown in the graph.

You could be asked to draw conclusions from the data...
From this graph, you can conclude that the crop yield for Field B was higher than the yields for Fields A and C. This could be because the weedkiller used to treat Field B was the most effective at removing weeds, which meant the crops had less competition, so the yield was greater.

Crop yield (tonnes per hectare): 0, 2, 4, 6, 8, 10
Field A *Field B* *Field C*

You could be asked why Field C was left untreated...
It was a control (see page 10) — the conditions were the same in Field C as the other fields (e.g. same sized field, same soil type, etc.) but no weedkiller was applied. This shows that the increased crop yield displayed in the other fields was likely to be due to the presence of weedkiller and nothing else.

Tip: H Crop yield is a way of measuring the amount of crop produced by an area of land.

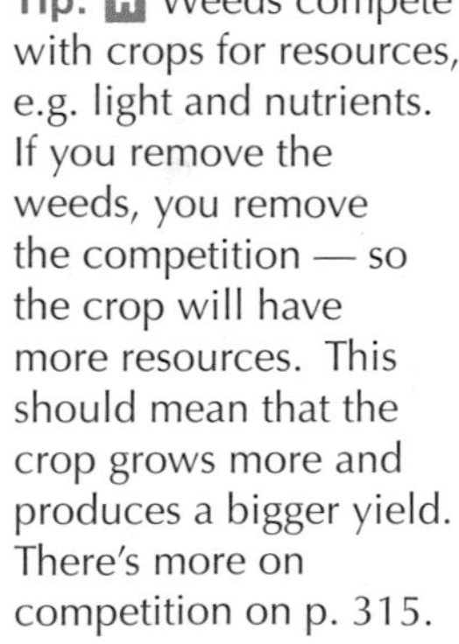

Tip: H Weeds compete with crops for resources, e.g. light and nutrients. If you remove the weeds, you remove the competition — so the crop will have more resources. This should mean that the crop grows more and produces a bigger yield. There's more on competition on p. 315.

2. Growing from cuttings with rooting powder

A cutting is part of a plant that has been cut off it, like the end of a branch with a few leaves on it. Normally, if you stick cuttings in the soil they won't grow, but if you add **rooting powder,** which contains auxins, they will produce roots rapidly and start growing as new plants. This enables growers to produce lots of clones (exact copies) of a really good plant very quickly.

Figure 1: *A gardener using rooting powder when planting cuttings.*

3. Growing cells in tissue culture

Tissue culture (see page 297) can be used to grow clones of a plant from a few of its cells. To do this, hormones such as auxins need to be added to the growth medium (along with nutrients) to stimulate the cells to divide to form both roots and shoots.

Gibberellins

Gibberellins are another type of plant growth hormone. They stimulate seed germination, stem growth and flowering. Uses include:

1. Controlling dormancy

Tip: H Seed germination is when a seed starts to grow into a plant.

Lots of seeds won't germinate until they've been through certain conditions (e.g. a period of cold or of dryness). This is called dormancy. Seeds can be treated with gibberellins to alter dormancy and make them germinate at times of year that they wouldn't normally. It also helps to make sure all the seeds in a batch germinate at the same time.

2. Inducing flowering

Some plants require certain conditions to flower, such as longer days or low temperatures. If these plants are treated with gibberellins, they will flower without any change in their environment. Gibberellins can also be used to grow bigger flowers.

3. Growing larger fruit

Seedless varieties of fruit (e.g. seedless grapes) often do not grow as large as seeded fruit. However, if gibberellins are added to these fruit, they will grow larger to match the normal types.

Ethene

Ethene is a gas produced by aging parts of a plant. It influences the growth of the plant by controlling cell division. It also stimulates enzymes that cause fruit to ripen.

Tip: H Some fruit will produce more ethene as it ripens.

Commercially, it can be used to speed up the ripening of fruits — either while they are still on the plant, or during transport to the shops. This means that fruit can be picked while it's still unripe (and therefore firmer and less easily damaged). The gas is then added to the fruit on the way to the supermarket so that it will be perfect just as it reaches the shelves.

Ripening can also be delayed while the fruit is in storage by adding chemicals that block ethene's effect on the fruit or reduce the amount of ethene that the fruit can produce. Alternatively, some chemicals can be used that react with ethene to remove it from the air.

Practice Questions — Fact Recall

Q1 Name the type of plant hormone that is found in rooting powders.

Q2 Name the type of plant hormone that controls seed dormancy.

Q3 What effect does ethene have on fruit?

Practice Question — Application

Q1 Dan grows plants for a garden centre. He is trying to work out the best rooting powder to use for his cuttings. From the same type of plant, he grows 50 cuttings in rooting powder A, 50 in rooting powder B and 50 without rooting powder. He grows all the cuttings under the same conditions. His results are shown in the table.

		Powder A	Powder B	No powder
Average increase in root length (mm)	After 1 week	8	10	6
	After 2 weeks	17	19	14
	After 3 weeks	24	28	20

a) From these results, what can you conclude about which rooting powder is the best to use? Explain your answer.

b) Explain why Dan grew some cuttings without rooting powder.

Topic 5c Checklist — Make sure you know...

Puberty and the Menstrual Cycle

- ☐ That sex hormones trigger the development of secondary sexual characteristics during puberty.
- ☐ That testosterone (produced in the testes) is the main male reproductive hormone and it stimulates sperm production.
- ☐ That oestrogen (produced in the ovaries) is the main female reproductive hormone and it is involved in the menstrual cycle.
- ☐ That an egg is released from the ovary approximately every 28 days and that this is called ovulation.
- ☐ The roles that four hormones (follicle stimulating hormone (FSH), luteinising hormone (LH), oestrogen and progesterone) have in the menstrual cycle.
- ☐ H How FSH, LH, oestrogen and progesterone interact to control the menstrual cycle, including the stimulation and inhibition of the release of other hormones.
- ☐ H How to read from and interpret graphs that show the changing levels of FSH, LH, oestrogen and progesterone during the menstrual cycle.

cont...

Contraceptives

- [ ] That hormonal and non-hormonal methods can be used to control fertility.
- [ ] How hormonal methods of contraception work, including how oral contraceptives, skin patches, implants, injections and intrauterine devices use oestrogen and/or progesterone to reduce fertility.
- [ ] How non-hormonal methods of contraception work, including barrier methods, spermicidal agents, intrauterine devices, surgery and abstinence.
- [ ] How to evaluate different methods of contraception, using their advantages and disadvantages.

Increasing Fertility

- [ ] H That hormones can be used to treat infertility, e.g. the use of fertility drugs involving FSH and LH.
- [ ] H The steps involved in IVF and how the process increases the chance of pregnancy.
- [ ] H That although IVF can give a couple a child, it has disadvantages — it can cause multiple births (which are risky), the success rate is low, and it is emotionally and physically stressful.

Thyroxine and Adrenaline

- [ ] H That thyroxine is produced in the thyroid gland and helps regulate basal metabolic rate and stimulates protein synthesis for growth and development.
- [ ] H How negative feedback is used to control the levels of thyroxine in the blood.
- [ ] H That adrenaline is produced in the adrenal glands and prepares the body for 'fight or flight'.

Plant Hormones

- [ ] That auxin controls plant responses to light (phototropism) and gravity (gravitropism/geotropism).
- [ ] That plant shoots grow towards light and away from gravity, and that roots grow towards gravity, due to the unequal distribution of auxin causing unequal growth rates in shoots and roots.
- [ ] How to investigate plant growth responses to light and gravity (Required Practical 8).

Uses of Plant Hormones

- [ ] H The effects of auxins, gibberellins and ethene on plants.
- [ ] H The ways that plant hormones (auxins, gibberellins and ethene) can be used in agriculture and horticulture.

Exam-style Questions

1 The diagram shows a plant that has been kept in a cupboard for one week. While it was in the cupboard, it had three possible light sources (labelled **A**, **B** and **C** in the diagram).

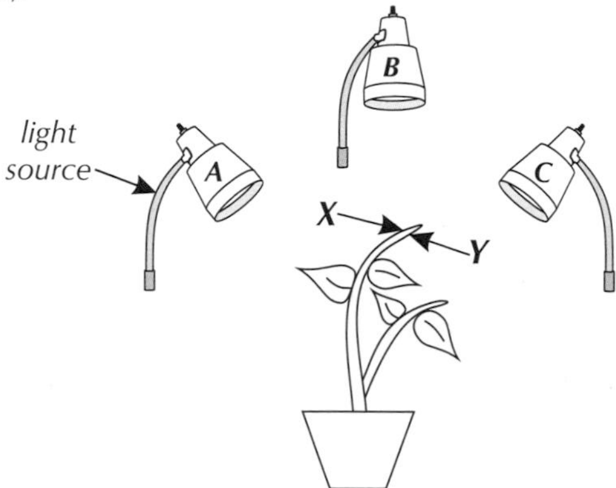

1.1 The plant only had one light source on during the week. Which light source do you think it was, **A**, **B** or **C**? Explain your answer.

(1 mark)

1.2 Samples of the plant shoot are taken from point **X** and point **Y** and analysed for the presence of auxin. Which sample do you think will contain most auxin, sample **X** or sample **Y**? Explain your answer.

(3 marks)

1.3 This plant was produced by taking a cutting from another plant and applying rooting powder to it before planting it in a pot. Explain why rooting powder was used.

(2 marks)

2 Thyroxine is an important hormone of the endocrine system.

2.1 In which of the following glands of the endocrine system is thyroxine produced?

A pituitary **B** ovary **C** thyroid **D** pancreas

(1 mark)

2.2* Describe how the level of thyroxine in the blood is kept at its optimum.

(4 marks)

One role of thyroxine is the regulation of the basal metabolic rate. Basal metabolic rate can be given as the number of calories burned at rest in a certain time.

2.3 A man's basal metabolic rate is 1860 kcal/day. How many calories does he burn at rest during six hours?

(2 marks)

1. DNA

DNA is a pretty important molecule because it's what makes us unique. Therefore it's really important that you learn all about it...

Learning Objectives:
- Know that genetic material is made up from DNA.
- Know that DNA is found in the nucleus of cells, and that it is found in chromosomes.
- Be able to explain the term 'chromosome'.
- Be able to describe the structure of DNA — it is a polymer and has two strands which are coiled together in a double helix shape.
- Be able to explain that a gene is a small section of DNA, and that each one codes for a specific chain of amino acids.
- Know that specific chains of amino acids make specific proteins.
- Know what is meant by the term 'genome'.
- Know why understanding the human genome is a really important tool.

Specification References
4.6.1.4, 4.6.1.6

Chromosomes and DNA

DNA stands for deoxyribonucleic acid. It's the chemical that all of the genetic material in a cell is made up from. It contains coded information — basically all the instructions to put an organism together and make it work. So it's what's in your DNA that determines what inherited characteristics you have.

DNA is found in the nucleus of animal and plant cells, in really long structures called **chromosomes**. The DNA is coiled up to form the 'arms' of the chromosome. Chromosomes normally come in pairs.

DNA is a polymer — a large molecule built from a chain of smaller molecules. As you can see from Figure 1, DNA is made up of two strands coiled together in the shape of a **double helix**.

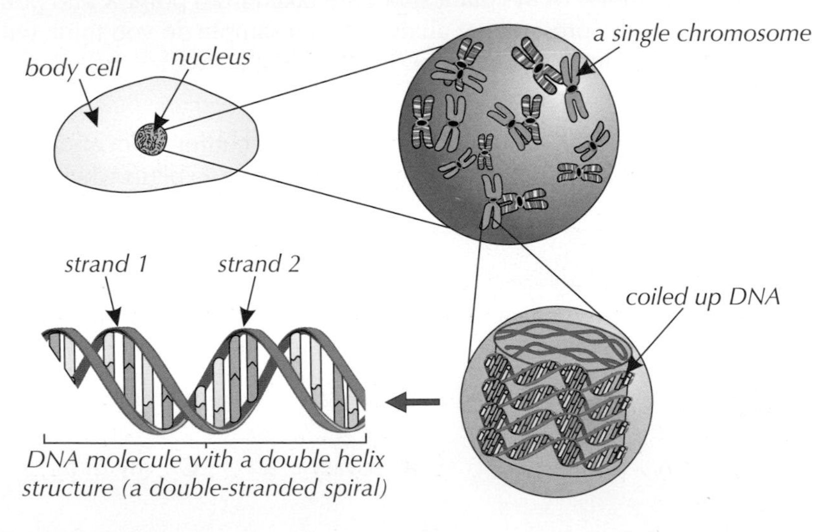

Figure 1: *Diagram showing the structure of DNA and where it is found in a cell.*

Tip: You might remember learning about genes and chromosomes on pages 33-34 — however, there's a bit more you need to know now.

Genes and proteins

A **gene** is a small section of DNA found on a chromosome. Each gene codes for (tells the cells to make) a particular sequence of amino acids which are put together to make a specific **protein**. Only 20 amino acids are used, but they make up thousands of different proteins. Genes simply tell cells in what order to put the amino acids together (more on this on page 246).

Tip: Amino acids are the building blocks that make up proteins.

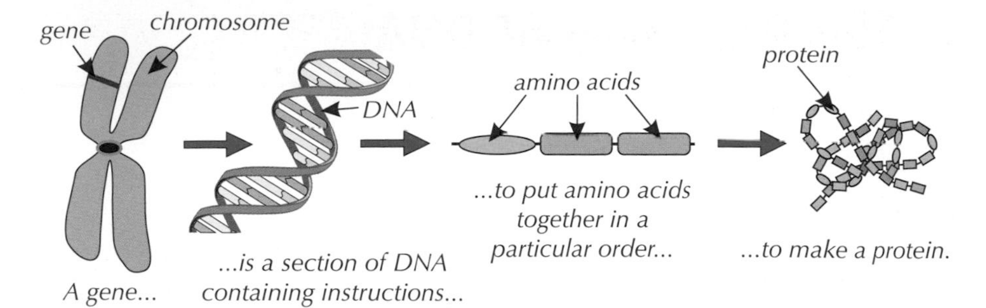

Figure 2: *Diagram showing how a gene codes for a protein.*

Tip: Another way of saying a gene contains the instructions for a protein is to say a gene 'codes for' a protein.

DNA also determines what proteins the cell produces, e.g. haemoglobin, keratin. That in turn determines what type of cell it is, e.g. red blood cell, skin cell.

Genomes

You need to know the definition of **genome**:

A genome is the entire set of genetic material in an organism.

Scientists have worked out the complete human genome. Understanding the human genome is a really important tool for science and medicine for many reasons.

Tip: A huge research project called the Human Genome Project mapped out the genes that make up human chromosomes.

Examples

- It allows scientists to identify genes in the genome that are linked to different types of disease.
- Knowing which genes are linked to inherited diseases could help us to understand them better and could help us to develop effective treatments for them.
- Scientists can look at genomes to trace the migration of certain populations of people around the world. All modern humans are descended from a common ancestor who lived in Africa, but humans can now be found all over the planet. The human genome is mostly identical in all individuals, but as different populations of people migrated away from Africa, they gradually developed tiny differences in their genomes. By investigating these differences, scientists can work out when new populations split off in a different direction and what route they took.

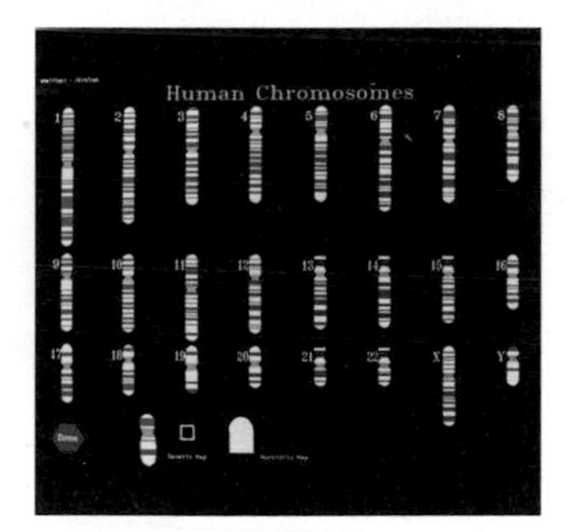

Figure 3: *Analysis of human chromosomes during the Human Genome Project.*

Practice Questions — Fact Recall

Q1 Where in a cell is DNA found?

Q2 What are chromosomes made up of?

Q3 Explain why DNA molecules are described as a double helix.

Q4 How does a gene help a cell to make a specific protein?

Q5 What is meant by the term 'genome'?

2. The Structure of DNA

Learning Objectives:
- Know that each strand of DNA is a polymer made up of a chain of repeating nucleotide units of which there are four different types.
- Know that a nucleotide is made up of a sugar and phosphate group, with one of four different bases joined to the sugar.
- Know that in each DNA strand, sugar and phosphates alternate.
- Know that the four bases in DNA are A, T, C and G.
- Know that each amino acid is coded for by a sequence of three bases, and that a sequence of amino acids is joined together to make a specific protein.
- H Know that in complementary base pairing, A always pairs with T, and C always pairs with G on opposite strands.
- H Know that DNA has non-coding parts that can switch genes on and off.

Specification Reference 4.6.1.5

Now that you've met DNA, it's time for a close-up on what makes up each strand within the double helix structure.

Nucleotides

DNA strands are polymers made up of lots of repeating units called **nucleotides**. There are four different nucleotides — each nucleotide consists of one sugar molecule, one phosphate molecule and one of four different 'bases'. The sugar and phosphate molecules in the nucleotides form a 'backbone' to the DNA strands. The sugar and phosphate molecules alternate. The four different bases are A, T, C and G — one joins to each sugar.

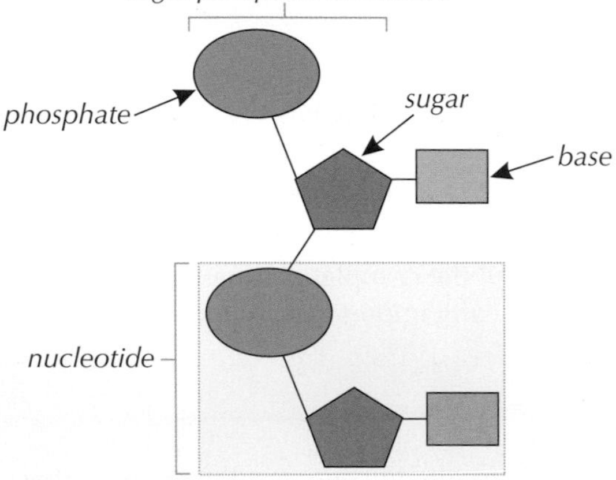

sugar-phosphate backbone

phosphate

sugar

base

nucleotide

Figure 1: *Diagram showing part of a DNA strand.*

Bases

It's the order of bases in a gene that decides the order of amino acids in a protein. Each amino acid is coded for by a sequence of three bases in the gene. The amino acids are joined together to make various proteins, depending on the order of the gene's bases.

Complementary base pairing Higher

Each base in a DNA strand links to a base on the opposite strand in the helix.

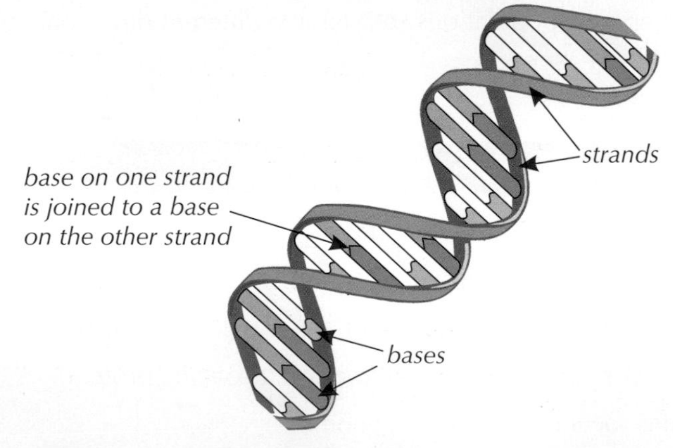

Figure 2: *Diagram showing part of a DNA molecule.*

Exam Tip
As well as remembering what a nucleotide is made up of, make sure you know how the components join together — the base is always attached to a sugar, and the sugar is attached to a phosphate.

A always pairs up with T, and C always pairs up with G (see Figure 3). This is called complementary base pairing.

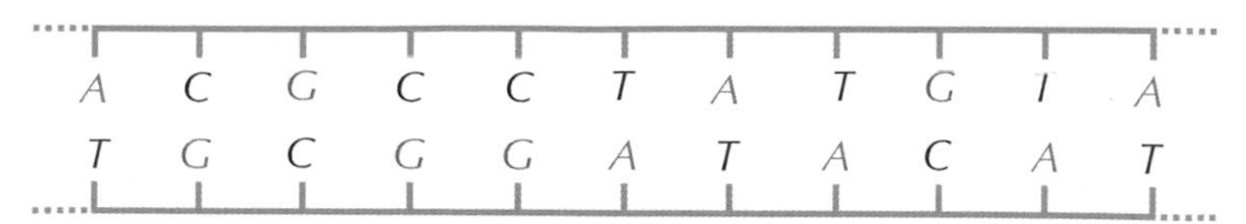

Figure 3: *Diagram showing complementary pairing of bases.*

Exam Tip H
Make sure you memorise which bases link with which — it could be easy marks in the exam.

Non-coding parts of DNA Higher

There are parts of DNA that don't code for proteins. Some of these non-coding parts switch genes on and off, so they control whether or not a gene is expressed (used to make a protein).

Tip: H If a gene isn't expressed, the protein that it codes for won't be made.

Practice Questions — Fact Recall

Q1 a) What is a nucleotide made up of?

b) Which parts of a nucleotide alternate to form a DNA strand?

Q2 How many bases code for an amino acid?

Q3 How does the sequence of bases in a gene control which particular protein is made?

Q4 Give one function of non-coding parts of DNA.

Practice Questions — Application

Q1 Use the table to determine the amino acids that are coded for by the base sequence below.

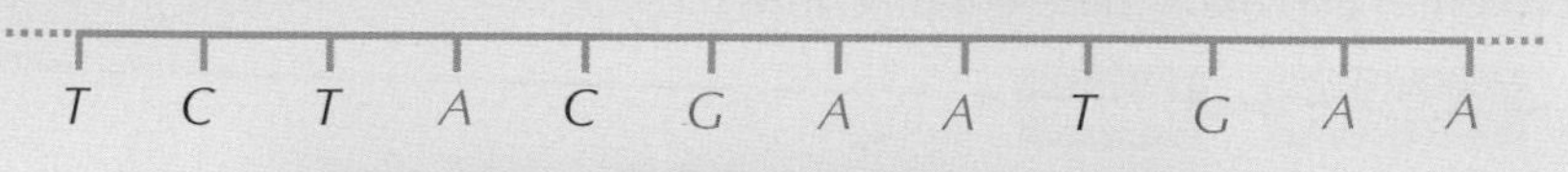

Base Sequence	Amino Acid
AAT	Asparagine
TCT	Serine
GTC	Valine
TGG	Tryptophan
TTT	Phenylalanine
GAA	Glutamic acid
ACG	Threonine
CCA	Proline

Q2 Write the base sequence for the complementary strand to the base sequence above.

3. Protein Synthesis Higher

Learning Objectives:

- H Know how to describe protein synthesis, including:
  - Know that proteins are made on ribosomes using a template made from the DNA in the nucleus.
  - Know that carrier molecules bring amino acids to the ribosomes, so that they can join together in the correct order to make proteins.
  - Know that when a protein has been made, it folds up into a particular shape so that it can carry out its specific role, e.g. as an enzyme, a hormone or a structural protein (such as collagen).
- H Be able to explain how the structure of DNA affects what protein is made.

Specification Reference 4.6.1.5

Proteins are chains of amino acids, and they're assembled with a bit of help.

Making proteins

Proteins are made in the cell cytoplasm on tiny structures called ribosomes. To make proteins, ribosomes use the code in the DNA. DNA is found in the cell nucleus and can't move out of it because it's really big. So the cell needs to get the code from the DNA to the ribosome. This is done using a molecule called mRNA — which is made by copying the code from DNA. The mRNA acts as a messenger between the DNA and the ribosome — it carries the code between the two and acts as a template.

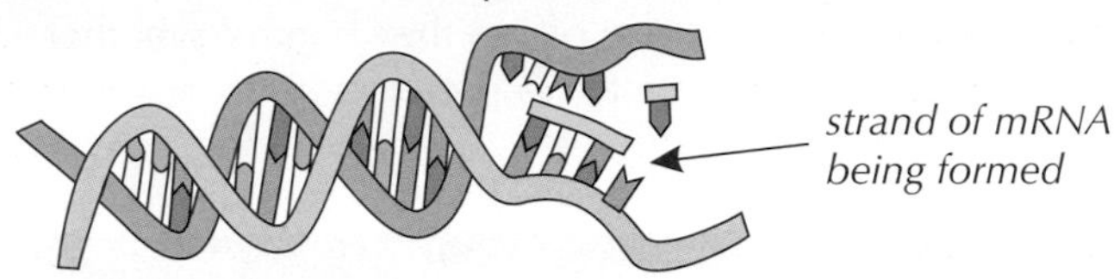

Figure 1: *Diagram showing mRNA being formed.*

The correct amino acids are brought to the ribosomes in the correct order by carrier molecules.

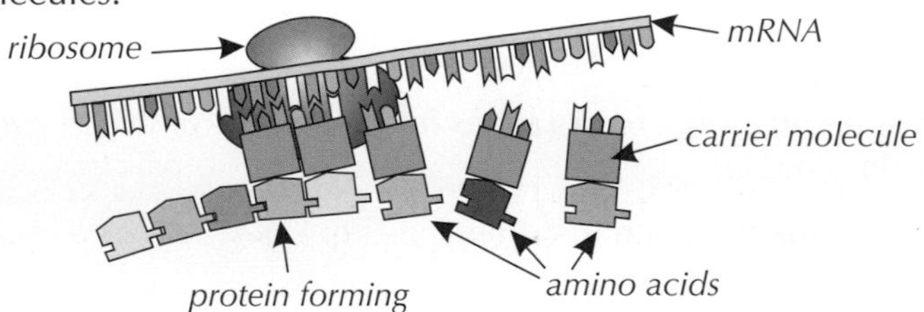

Figure 2: *Diagram showing mRNA being used to assemble amino acids.*

Protein functions

When a chain of amino acids has been assembled, it folds into a unique shape which allows the protein to perform the task it's meant to do. Proteins have many different functions:

Examples **Higher**

- Enzymes — act as biological catalysts to speed up chemical reactions in the body (see page 115).
- Hormones — used to carry messages around the body. E.g. insulin is a hormone released into the blood by the pancreas to regulate the blood sugar level.
- Structural proteins — are physically strong. E.g. collagen is a structural protein that strengthens connective tissues (like ligaments and cartilage).

Practice Questions — Fact Recall

Q1 Where are proteins assembled?

Q2 What do carrier molecules do?

Q3 What happens when a protein chain is complete?

4. Mutations **Higher**

You've seen how proteins are made — but it's not always problem free.

What are mutations?

Occasionally a gene may mutate. A mutation is a random change in an organism's DNA. They can sometimes be inherited.

Mutations occur continuously. They can occur spontaneously, e.g. when a chromosome isn't quite replicated properly. However, the chance of mutation is increased by exposure to certain substances or some types of radiation.

Mutations change the sequence of the DNA bases in a gene, which produces a genetic variant (a different form of the gene). As the sequence of DNA bases codes for the sequence of amino acids that make up a protein (see page 246), mutations to a gene sometimes lead to changes in the protein that it codes for. Most mutations have very little or no effect on the protein. Some will change it to such a small extent that its function or appearance is unaffected. However, some mutations can seriously affect a protein. Sometimes, the mutation will code for an altered protein with a change in its shape. This could affect its ability to perform its function.

Examples **Higher**

- If the shape of an enzyme's active site is changed, its substrate may no longer be able to bind to it.
- Structural proteins like collagen could lose their strength if their shape is changed, making them pretty useless at providing structure and support.

If there's a mutation in the non-coding DNA, it can alter how genes are expressed.

Learning Objectives:

- H Know that mutations happen continuously.
- H Know that most mutations have very little or no effect at all on the protein the gene codes for.
- H Know that a small number of mutations mean that the protein that's coded for is changed.
- H Understand why a change to the DNA base sequence can result in a change to the protein it codes for, including why an enzyme may no longer fit with its substrate, and why a structural protein might lose its strength.
- H Know that a mutation in non-coding DNA can alter how a gene is expressed.

Specification References
4.6.1.5

Tip: H Enzymes are proteins — see p. 115.

Tip: H There's more on non-coding DNA on page 247.

Types of mutation

There are different ways that mutations can change the DNA base sequence.

Insertions

Insertions are where a new base is inserted into the DNA base sequence where it shouldn't be. You should remember from page 246 that every three bases in a DNA base sequence codes for a particular amino acid. An insertion changes the way the groups of three bases are 'read', which can change the amino acids that they code for. Insertions can change more than one amino acid as they have a knock-on effect on the bases further on in the sequence.

Example **Higher**

Original gene: T A T | A G T | C T T
Amino acid coded for: Tyrosine, Serine, Leucine

insertion here

Mutated gene: T A T | A G G | T C T | T
Amino acid coded for: Tyrosine, Arginine, Serine

Figure 1: Diagram showing an insertion of an extra base.

Deletions

Deletions are when a random base is deleted from the DNA base sequence. Like insertions, they change the way that the base sequence is 'read' and have knock-on effects further down the sequence.

Example **Higher**

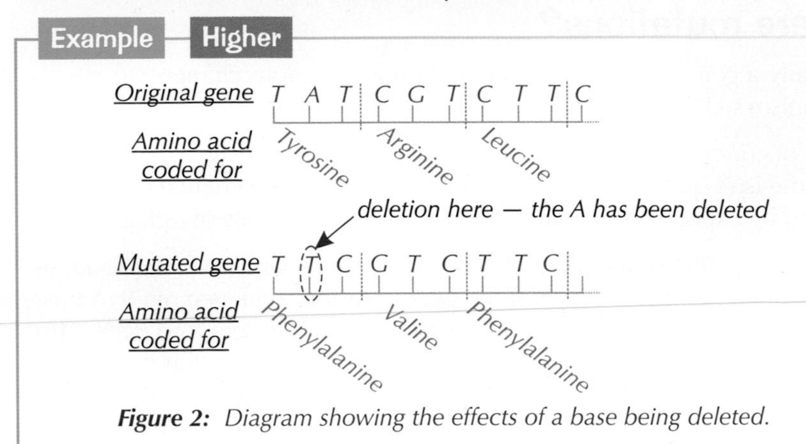

Figure 2: *Diagram showing the effects of a base being deleted.*

Tip: H You don't need to remember these specific examples, but make sure you understand what's happening.

Substitutions

Substitution mutations are when a random base in the DNA base sequence is changed to a different base.

Example **Higher**

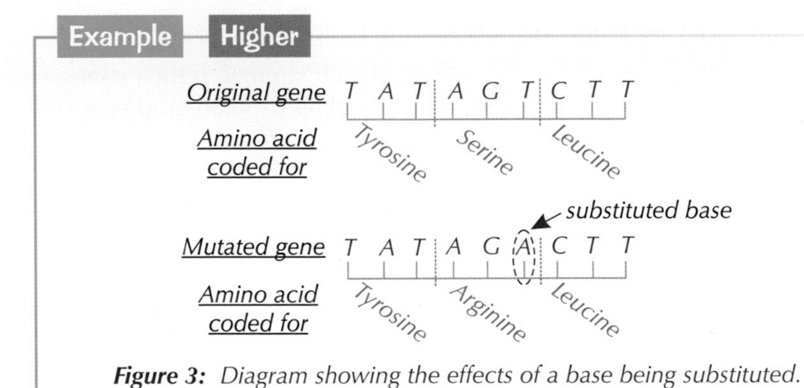

Figure 3: *Diagram showing the effects of a base being substituted.*

Practice Questions — Fact Recall

Q1 True or False? Mutations occur continuously.

Q2 True or False? All mutations change the protein that the gene codes for.

Practice Question — Application

Q1 A gene that codes for an enzyme has a mutation. Suggest why the enzyme may no longer be able to bind with its substrate.

5. Reproduction

Organisms make more of themselves through reproduction.
There are two types: sexual and asexual. You need to know about both.

Learning Objectives:

- Be able to explain the term 'gamete'.
- Know that gametes are produced by meiosis.
- Know that in animals, the gametes are sperm and egg cells.
- Know that in sexual reproduction, male and female gametes are fused.
- Know that in sexual reproduction, genetic information from the mother and father is mixed, which leads to variation in the offspring.
- Know that in flowering plants, the gametes are pollen and egg cells.
- Know that in asexual reproduction there is only one parent, and the offspring are genetical identical to the parent (clones).
- Know that asexual reproduction happens by mitosis, and that identical cells are made.
- Know that in asexual reproduction, gametes do not fuse and genetic information isn't mixed.

Specification References
4.6.1.1, 4.6.1.6

Sexual reproduction

Sexual reproduction is where genetic information from two organisms (a father and a mother) is combined to produce offspring which are genetically different to either parent.

Gametes

In sexual reproduction the mother and father produce gametes by meiosis (see page 253) — e.g. egg and sperm cells in animals (see Figure 1). In humans, each gamete contains 23 chromosomes — half the number of chromosomes in a normal cell. (Instead of having two of each chromosome, a gamete has just one of each.)

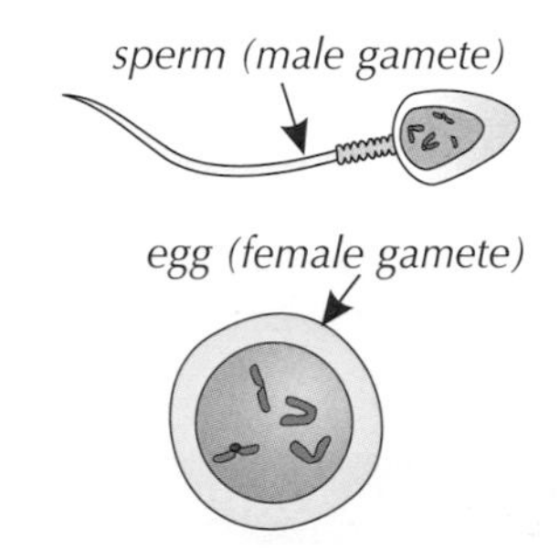

Figure 1: *The sperm and egg cells.*

Fertilisation

The egg (from the mother) and the sperm cell (from the father) fuse together to form a cell with the full number of chromosomes (half from the father, half from the mother). The fusion of gametes is known as fertilisation (see Figure 2).

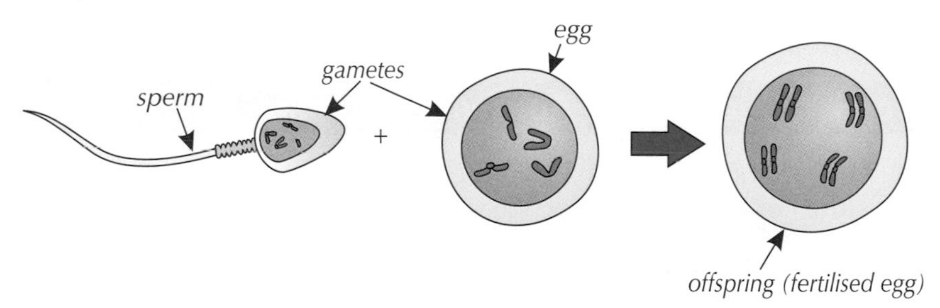

Figure 2: *Diagram showing fertilisation.*

This is why the offspring inherits features from both parents — it's received a mixture of chromosomes from its mum and its dad (and it's the chromosomes that decide how you turn out). This mixture of genetic information produces variation in the offspring. Pretty cool, eh.

Here's the main thing to remember about sexual reproduction:

> Sexual reproduction involves the fusion of male and female gametes. Because there are two parents, the offspring contain a mixture of their parents' genes and are genetically different to their parents.

Flowering plants can reproduce in this way too. They also have egg cells, but their version of sperm is known as pollen.

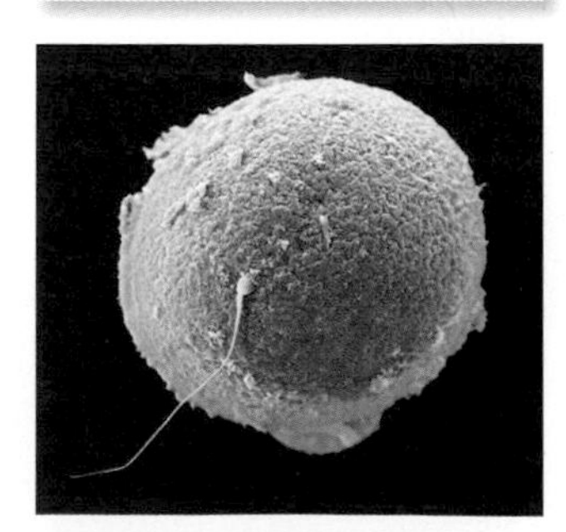

Figure 3: *A sperm (shown in blue) fertilising an egg (shown in yellow) as seen under a microscope.*

Asexual reproduction

In asexual reproduction there's only one parent so the offspring are genetically identical to that parent. Asexual reproduction happens by mitosis — an ordinary cell makes a new cell by dividing in two (see pages 35-36). The new cell has exactly the same genetic information (i.e. genes) as the parent cell — it's called a clone.

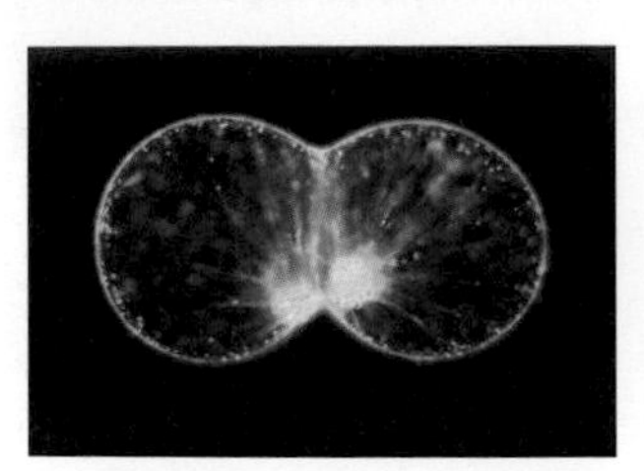

Figure 4: *A single-celled organism called a 'sea sparkle' undergoing asexual reproduction.*

Here's the main thing to remember about asexual reproduction:

In asexual reproduction there's only one parent. There's no fusion of gametes, no mixing of chromosomes and no genetic variation between parent and offspring. The offspring are genetically identical to the parent — they're **clones**.

Bacteria, some plants and some animals reproduce asexually.

Practice Questions — Fact Recall

Q1 Are gametes produced by meiosis or by mitosis?

Q2 Name the male and female gametes in humans.

Q3 How many parents are there in asexual reproduction?

Practice Question — Application

Q1 For each of the following examples, write down whether it's a case of sexual or asexual reproduction, and explain your answer.

a) A single-celled amoeba splits in two to form two genetically identical offspring.

b) Gametes from a male pea plant are crossed with gametes from a female pea plant to produce genetically varied offspring.

c) A lion and a tiger mate to produce an animal known as a 'liger'. The liger shares features with both its parents.

d) A single Brahminy blind snake lays a batch of unfertilised eggs. The offspring that hatch are clones of the mother snake.

Tip: To answer these questions, you need to think about the main differences between sexual and asexual reproduction — look back at the green summary boxes if you need clues.

6. Meiosis

You met mitosis back on pages 35-36. Meiosis is another type of cell division. It only happens in the cells of the reproductive organs — it produces gametes.

Gamete production

As you know from page 251, gametes only have one copy of each chromosome, so that when gamete fusion takes place, you get the right amount of chromosomes again (two copies of each).

To make gametes which only have half the original number of chromosomes, cells divide by meiosis. This process involves two cell divisions. In humans, it only happens in the reproductive organs (the ovaries in females and testes in males).

Meiosis

Meiosis produces cells which have half the normal number of chromosomes. Here are the steps involved in meiosis:

1. Before the cell starts to divide, it duplicates its genetic information, forming two armed chromosomes — one arm of each chromosome is an exact copy of the other arm. After replication, the chromosomes arrange themselves into pairs.
2. In the first division in meiosis the chromosome pairs line up in the centre of the cell.
3. The pairs are then pulled apart so each new cell only has one copy of each chromosome. Some of the father's chromosomes (shown in blue) and some of the mother's chromosomes (shown in red) go into each new cell.
4. In the second division, the chromosomes line up again in the centre of the cell. The arms of the chromosomes are pulled apart.
5. You get four gametes, each with only a single set of chromosomes in it. Each of the gametes (sperm or egg cells) is genetically different from the others because the chromosomes all get shuffled up during meiosis and each gamete only gets half of them, at random.

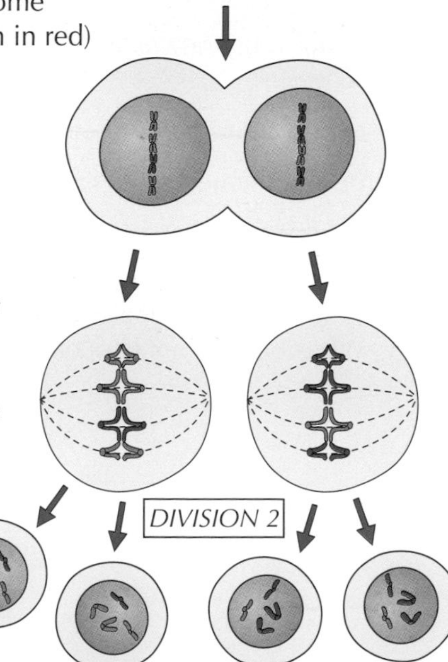

Figure 1: *Diagram showing the stages in meiosis.*

Learning Objectives:

- Know that meiosis produces gametes with only half the normal number of chromosomes, and that when gametes fuse at fertilisation, the full number of chromosomes is restored.
- Know that meiosis only happens in the reproductive organs.
- Be able to recall what happens when a cell divides in meiosis to form genetically different gametes (sperm or egg cells).
- Know that when gametes fuse during fertilisation, the cell that's created divides by mitosis to produce more cells, which start to differentiate as the embryo develops.

Specification Reference 4.6.1.2

Tip: The genetic information is stored in DNA — see page 244.

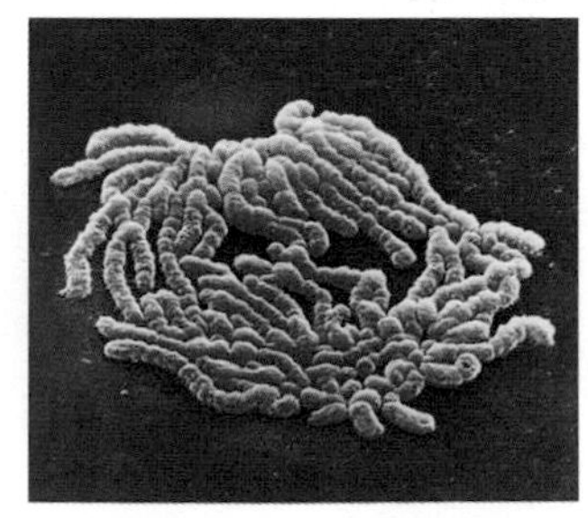

Figure 2: *Microscope image showing pairs of chromosomes being pulled apart during the first division in meiosis.*

Tip: There's loads on mitosis on pages 35-36.

Embryo development

After two gametes have fused during fertilisation, the resulting new cell divides by mitosis to make a copy of itself. Mitosis repeats many times to produce lots of new cells in an embryo. As the embryo develops, these cells then start to differentiate (see page 30) into the different types of specialised cell that make up a whole organism.

Practice Questions — Fact Recall

Q1 How many sets of chromosomes do gametes contain?

Q2 Where are gametes produced?

Q3 How many cell divisions occur in meiosis?

Q4 How many new cells are produced when a cell divides by meiosis?

Q5 True or False? Gametes are all genetically identical.

Q6 Following fertilisation, how does the resulting cell grow?

Practice Questions — Application

Q1 Different animals have different numbers of chromosomes in their body cells and in their gametes, as shown in the table:

Type of animal cell	Number of chromosomes
Dog body cell	78
Cat egg cell	19
Horse sperm cell	32

Using your knowledge and the information provided in the table, suggest:

a) how many chromosomes there are in a dog sperm cell.

b) how many chromosomes there are in the body cell of a cat.

c) how many chromosomes there are in a horse egg cell.

Q2 Mary thinks that after the first division in meiosis, cells will have half a set of chromosomes. Why does this not happen when a cell divides by meiosis?

Exam Tip
The words 'mitosis' and 'meiosis' look and sound very similar. Make sure that you know how to spell each one. You won't gain marks in the exam if the examiner can't tell which type of cell division you are talking about.

7. More on Reproduction

Now you know all about the different methods of reproduction, it's time to look at them side by side and make some comparisons.

Learning Objectives:

- Be able to give the advantages of sexual reproduction over asexual reproduction.
- Be able to give the advantages of asexual reproduction over sexual reproduction.
- Be able to explain, using given data for a particular organism, the advantages and disadvantages of asexual and sexual reproduction.
- Know that some organisms can reproduce by both asexual and sexual reproduction, including malarial parasites, many fungi, and many plant species (such as strawberry plants and daffodils).

Specification Reference
4.6.1.3

Advantages of sexual reproduction

Sexual reproduction has advantages over asexual reproduction.

Offspring from sexual reproduction have a mixture of two sets of chromosomes. The organism inherits genes (and therefore features) from both parents, which produces variation in the offspring (see p. 282).

Variation increases the chance of a species surviving a change in the environment. While a change in the environment could kill some individuals, it's likely that variation will have led to some of the offspring being able to survive in the new environment. They have a survival advantage. Because individuals with characteristics that make them better adapted to the environment have a better chance of survival, they are more likely to breed successfully and pass the genes for the characteristics on. This is known as natural selection (see p. 287).

We can use selective breeding to speed up natural selection. This allows us to produce animals with desirable characteristics. Selective breeding is where individuals with a desirable characteristic are bred to produce offspring that have the desirable characteristic too (see p. 292). This means that we can increase food production.

Examples

Food production can be increased by:

- Breeding animals that produce a lot of meat.
- Breeding cows that have a high milk yield.

Figure 1: *The Belgian Blue breed of cow has been produced by selective breeding to produce lots of meat.*

Advantages of asexual reproduction

Asexual reproduction has advantages over sexual reproduction.

There only needs to be one parent. This means that asexual reproduction uses less energy than sexual reproduction, because organisms don't have to find a mate. This also means that asexual reproduction is faster than sexual reproduction.

Another advantage is that many identical offspring can be produced in favourable conditions.

Example

Dandelion plants are able to reproduce asexually. When conditions in the environment are ideal for growth, being able to produce many identical offspring quickly means that the dandelions can spread fast to take over an area. Because the offspring all have the same genes as the parent, they're all likely to survive in the environmental conditions.

Exam Tip
In the exam, you might be asked about the advantages and disadvantages of asexual or sexual reproduction for a particular organism. Don't panic — just use what you've learnt here and apply it to the information you've been given.

Reproducing using both methods

Some organisms can reproduce sexually or asexually depending on their circumstances.

You need to know these examples:

Examples

- Malaria is caused by a parasite that's spread by mosquitoes. When a mosquito carrying the parasite bites a human, the parasite can be transferred to the human. The parasite reproduces sexually when it's in the mosquito and asexually when it's in the human host.
- Many species of fungus can reproduce both sexually and asexually. These species release spores, which can become new fungi when they land in a suitable place. Spores can be produced sexually and asexually. Asexually-produced spores form fungi that are genetically identical to the parent fungus. Sexually-produced spores introduce variation and are often produced in response to an unfavourable change in the environment, increasing the chance that the population will survive the change.
- Loads of species of plant produce seeds sexually, but can also reproduce asexually. Asexual reproduction can take place in different ways. For example, strawberry plants produce 'runners'. These are stems that grow horizontally on the surface of the soil away from a plant. At various points along the runner, a new strawberry plant forms that is identical to the original plant — see Figure 2. Another example is in plants that grow from bulbs (e.g. daffodils). New bulbs can form from the main bulb and divide off — see Figure 3. Each new bulb can grow into a new identical plant.

Tip: There's more about malaria on page 132.

Figure 2: *Runners from a strawberry plant. New plants are starting to grow from the runners.*

Figure 3: *A daffodil bulb dividing.*

Practice Questions — Fact Recall

Q1 Sexual reproduction introduces variation. Why is this an advantage?

Q2 Why does only requiring one parent give asexual reproduction an advantage over sexual reproduction?

Q3 When does a mosquito reproduce asexually?

Q4 Describe how a strawberry plant reproduces asexually.

Topic 6a Checklist — Make sure you know...

DNA

- ☐ That DNA makes up all of the genetic material in cells, and that it's found in the nucleus.
- ☐ That DNA is found in really long structures called chromosomes.
- ☐ The structure of DNA — that DNA is a polymer, and that it's made up of two strands coiled together to form a double helix.
- ☐ That a gene is a small section of DNA, and that each gene in DNA codes for a particular sequence of amino acids which are put together to make a particular protein.
- ☐ That a genome is the entire set of genetic material in an organism.
- ☐ That understanding the human genome is important because it means that scientists can link genes to particular diseases, understand and treat inherited disorders effectively, and also trace the migration patterns of human populations.

The Structure of DNA

- ☐ That chains of repeating nucleotide units make up a DNA polymer, and that each nucleotide is made up of a sugar and phosphate group, with one of four bases attached to the sugar.
- ☐ That in a strand of DNA, the sugar and phosphate molecules alternate to form a 'backbone'.
- ☐ That DNA has four bases — A, T, C and G.
- ☐ That sequences of three bases in a certain order code for particular amino acids.
- ☐ That the amino acids coded for by the order of the DNA bases join together to form particular proteins.
- ☐ H That the bases in DNA match up to particular bases on the opposite strand — A always pairs with T, and C always pairs with G.
- ☐ H That non-coding parts of DNA switch genes on and off and control which genes are expressed.

Protein Synthesis

- ☐ H That proteins are made on ribosomes found in the cell cytoplasm, and that a template made from the DNA in the nucleus is used to get the genetic code from the nucleus to the ribosomes.
- ☐ H That amino acids are brought to the ribosomes by carrier molecules so that they can be joined in the correct order to make a protein.
- ☐ H That when a chain of amino acids is complete, the protein it forms folds up into the unique shape it needs to do its job, e.g. as an enzyme, a hormone or a structural protein (such as collagen).

Mutations

- ☐ H That gene mutations are random changes in DNA that occur continuously.
- ☐ H That most mutations have very little or no effect on the protein that the gene codes for.

cont...

- [ ] H That a small number of mutations can result in a different protein being coded for, with a different shape, and that this can affect its function (e.g. an enzyme may no longer fit with its substrate, or a structural protein may lose its strength).
- [ ] H That a mutation in non-coding DNA can alter how a gene is expressed.

Reproduction

- [ ] That gametes (sex cells) are made by meiosis, and that in animals they are sperm and egg cells.
- [ ] That in sexual reproduction, male and female gametes fuse together.
- [ ] That in sexual reproduction, genetic information from the male and female gametes is mixed, which results in variation in the offspring.
- [ ] That in flowering plants, the gametes are egg cells and pollen.
- [ ] That in asexual reproduction, there is only one parent, so offspring that are genetically identical to the parent are produced (called clones).
- [ ] That asexual reproduction happens by mitosis, and produces identical cells.
- [ ] That in mitosis, there's no fusing of gametes and no mixing of genetic material, which means that there is no variation in the offspring.

Meiosis

- [ ] That meiosis produces gametes that have half the normal chromosome number, and so when fertilisation happens, the cell that is produced has the full number of chromosomes again.
- [ ] That in humans, meiosis only happens in the reproductive organs (the ovaries and testes).
- [ ] That during meiosis, the cell first duplicates its genetic material before dividing twice to produce four gametes.
- [ ] That each of the gametes produced during meiosis is genetically different because the chromosomes all get shuffled up during meiosis and each gamete only gets half of them, at random — so meiosis produces genetically different cells being formed.
- [ ] That after fertilisation, the cell that's created divides by mitosis over and over again to produce lots of new cells which differentiate as the embryo develops.

More on Reproduction

- [ ] That sexual reproduction has advantages over asexual reproduction in that it produces variation in the offspring, which increases the chances of a species surviving an environmental change.
- [ ] That selective breeding can be used to speed up natural selection to increase food production.
- [ ] That asexual reproduction has advantages over sexual reproduction in that only one parent is needed, which means that it uses less time and energy than sexual reproduction because organisms don't need to find a mate. It also means that many identical offspring can be produced when conditions in the environment are favourable.
- [ ] How to explain the pros and cons of both types of reproduction when given data about an organism.
- [ ] That some organisms can reproduce sexually or asexually depending on the circumstances.

Exam-style Questions

1 A farmer selects a male pig and a female pig and breeds them together. **Figure 1** shows the process.

Figure 1

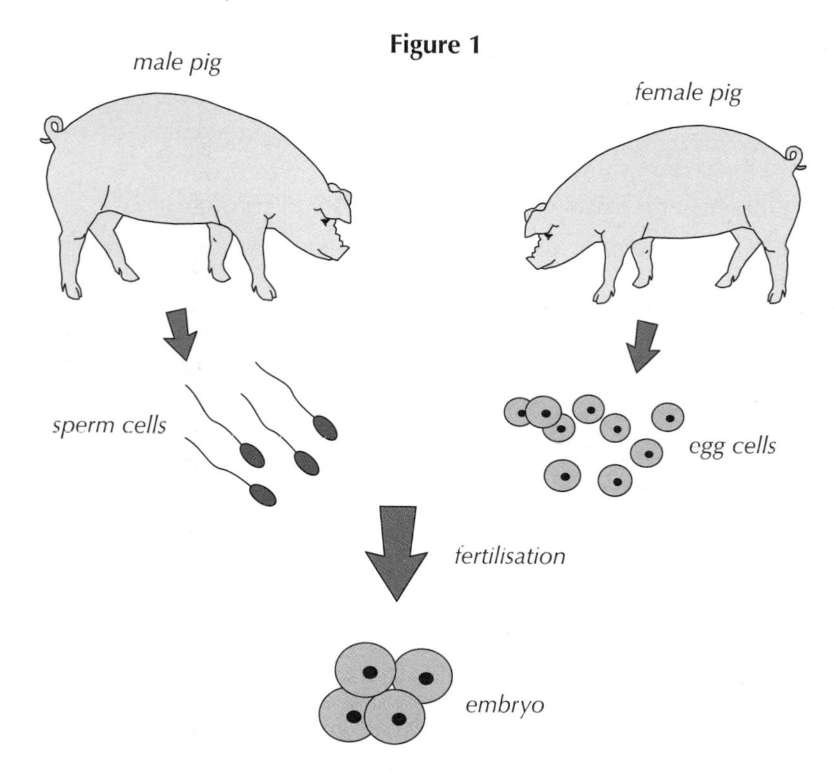

1.1 Is the reproduction shown in **Figure 1** an example of sexual or asexual reproduction? Give a reason for your answer.

(1 mark)

1.2 Outline **one** advantage of the type of reproduction shown in **Figure 1**.

(4 marks)

1.3 The pigs that are born as a result of this process will share characteristics with both the male pig and the female pig. Explain why.

(2 marks)

1.4 The sperm cells are gametes.
Explain why it is important that the sperm cells only have one set of chromosomes.

(1 mark)

1.5 What type of cell division occurs in the embryo in order for it to grow?

(1 mark)

2 The Human Genome Project was a major genetics project that revealed the genes in the human body.

2.1 Outline what is meant by the term 'human genome'.

(1 mark)

2.2 Give **two** reasons why research into the human genome is important.

(2 marks)

3 DNA in human cells exists in the form of genes and chromosomes.

3.1* Describe what is meant by these terms and outline how they are linked with each other.

(4 marks)

3.2 **Figure 2** shows the sequence of bases on part of a strand of DNA.

Figure 2

A T C A G C C T A G T T

Name the part of a cell where strands of human DNA are found.

(1 mark)

3.3 A strand of DNA is made up from repeating nucleotide units.
Name the component in a nucleotide unit that a base is joined to.

(1 mark)

3.4 How many amino acids does the base sequence in **Figure 2** code for?

(1 mark)

3.5 Write out the base sequence for the strand of DNA that would be complementary to the strand in **Figure 2**.

(2 marks)

3.6 A mutation occurs in part of the DNA strand shown in **Figure 2**.
Figure 3 shows the new strand.

Figure 3

A T C A G G C T A G T T

Explain how the mutation could cause a different protein to be coded for.

(2 marks)

1. X and Y Chromosomes

We all know that there are loads of differences between males and females. It's all to do with two little chromosomes — X and Y.

Sex chromosomes

There are 23 pairs of chromosomes in every human body cell. Of these, 22 are matched pairs of chromosomes — these just control your characteristics. The 23rd pair are labelled XY or XX. They're the two chromosomes that decide your sex — whether you turn out male or female.

Males have an **X** and a **Y** chromosome: **XY**
The Y chromosome causes male characteristics.

Females have **two X** chromosomes: **XX**
The XX combination allows female characteristics to develop.

When making sperm, the X and Y chromosomes from the original male cell are drawn apart in the first division of meiosis (see page 253) for more on meiosis). There's a 50% chance each sperm cell gets an X chromosome and a 50% chance it gets a Y chromosome (see Figure 1).

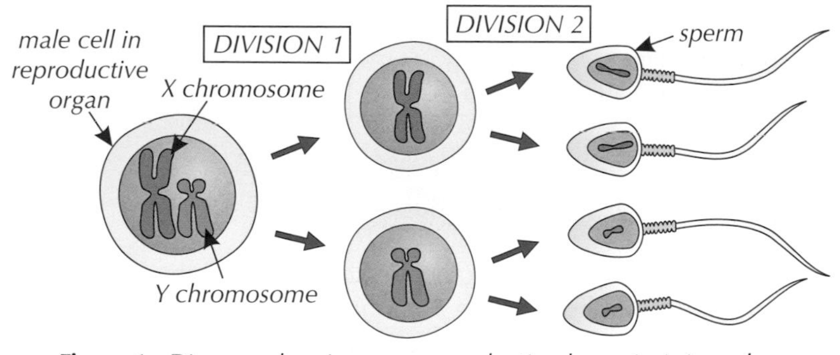

Figure 1: *Diagram showing sperm production by meiosis in males.*

A similar thing happens when making eggs. But the original cell has two X chromosomes (as it's from a female), so all the eggs end up with one X chromosome (see Figure 3).

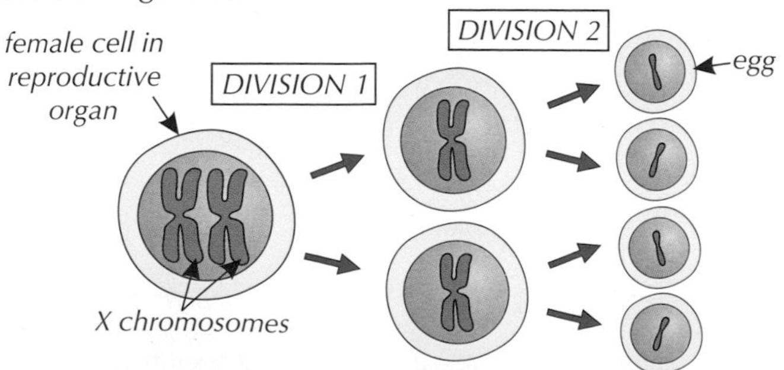

Figure 2: *Diagram showing egg production by meiosis in females.*

Learning Objectives:

- Know that there are 23 pairs of chromosomes found in ordinary human body cells and that 22 of these pairs just control characteristics.
- Know that the genes which determine sex are carried on the 23rd pair of chromosomes — the sex chromosomes.
- Know that females have two of the same sex chromosome (XX) whereas males have two different sex chromosomes (XY).
- Be able to complete and interpret genetic diagrams to show sex inheritance.
- Be able to understand probability in terms of the outcome of a genetic cross.
- Be able to express the outcome of a genetic cross using ratios or direct proportion.
- H Be able to construct genetic diagrams.

Specification References 4.6.1.6, 4.6.1.8

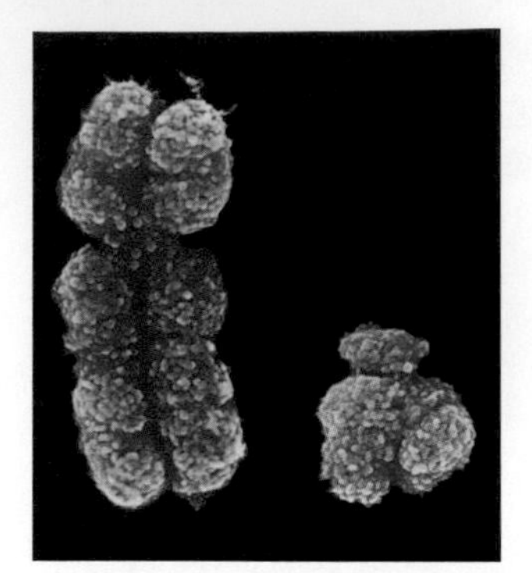

Figure 3: *Here's what the sex chromosomes actually look like. The Y chromosome (shown in blue) is smaller than the (pink) X chromosome.*

Sex inheritance and genetic diagrams

At fertilisation, the sperm fertilises the egg, and the chromosomes from the gametes combine, forming a new individual with the correct number of chromosomes (see page 251). Whether the individual is male or female depends on the combination of sex chromosomes it receives — this is sex inheritance.

Genetic diagrams can be used to show sex inheritance. They are just models that are used to show all the possible genetic outcomes when you cross together different genes or chromosomes.

Exam Tip
Most genetic diagrams you'll see in exams concentrate on a gene, instead of a chromosome. But the principle's the same. Don't worry — there are loads of other examples on pages 266-272.

Interpreting genetic diagrams

You need to be able to interpret genetic diagrams showing sex inheritance.

Examples

Figure 4 is a type of genetic diagram called a Punnett square, showing sex inheritance. The pairs of letters in the middle show the possible combinations of the gametes.

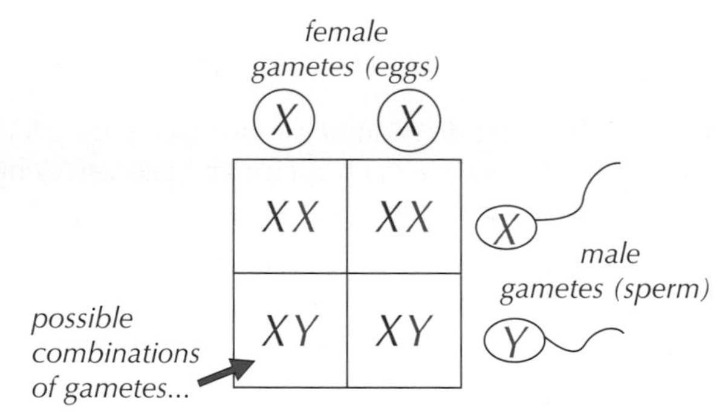

Figure 4: *Punnett square showing sex inheritance.*

Tip: Only one of these possible combinations would actually happen for any one offspring.

Figure 5 is another type of genetic diagram showing sex inheritance. It looks a bit more complicated than a Punnett square, but it shows exactly the same thing. The possible combinations of gametes are shown in the bottom circles.

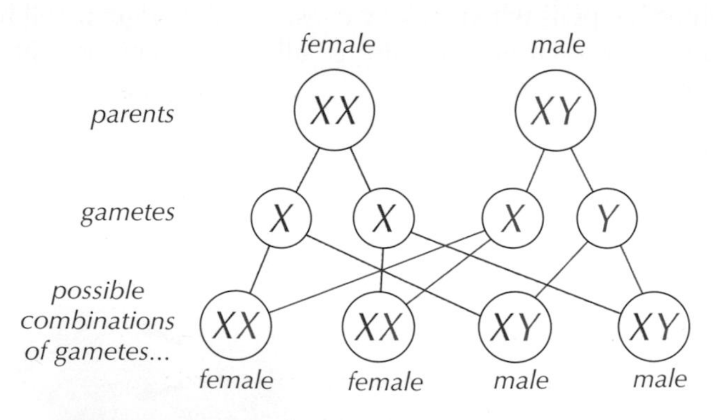

Figure 5: *Genetic diagram showing sex inheritance.*

You can use genetic diagrams to find the probability of any offspring inheriting either combination of sex chromosomes. (You can also use them to find the probability of offspring inheriting a certain characteristic — see pages 266-269.)

Probability is a measure of how likely something is to happen. In maths, the probability of a certain event happening is written as a number between 0 (impossible) and 1 (certain). It can also be written as a fraction or a percentage.

Example

Both Figure 4 and Figure 5 show two XX results and two XY results, so there's an equal probability of getting a boy or a girl. This can be written as a 50:50 ratio, which is the same as 1:1. Alternatively, you could say the probability of getting a boy is 1 in 2, 50%, 0.5 or ½.

Tip: Don't forget that this 50:50 ratio is only a probability at each pregnancy. If you had four kids they could all be boys.

Tip: There's more about ratios on page 377.

Constructing genetic diagrams **Higher**

You can construct genetic diagrams to show sex inheritance.

Example 1 **Higher**

Drawing a Punnett square to show sex inheritance in humans.

1. First, draw a grid with four squares.
2. Put the possible gametes (eggs or sperm) from one parent down the side, and those from the other parent along the top.
3. Then in each middle square you fill in the letters from the top and side that line up with that square — the pairs of letters in the middle show the possible combinations of the gametes.

Figure 7 shows these steps:

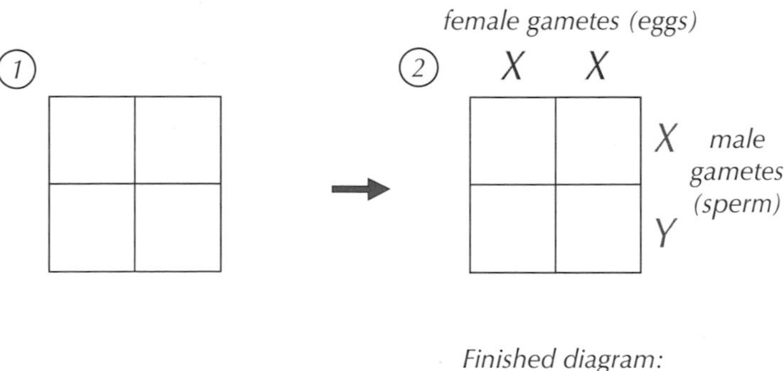

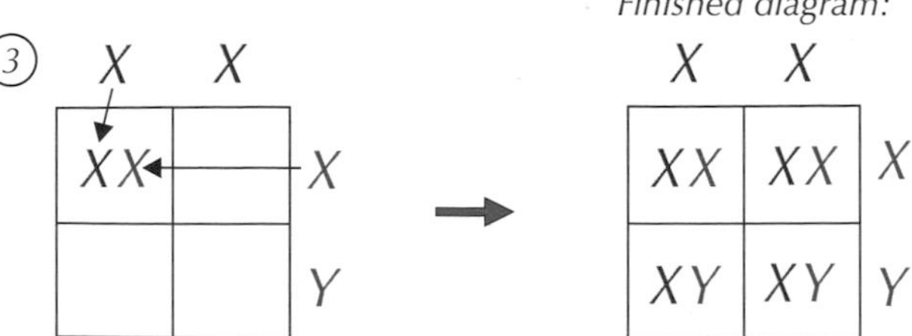

Figure 7: *How to draw a Punnett square to show sex inheritance in humans.*

Figure 6: *Having lots of sons doesn't increase the chances of having a daughter in the next pregnancy. There's still a 50:50 chance of having a boy or a girl at each pregnancy.*

Example 2 — Higher

Drawing a different type of genetic diagram to show sex inheritance in humans.

1. Draw two circles at the top of the diagram to represent the parents. Put the female sex chromosomes in one and the male sex chromosomes in the other.

female parent *male parent*

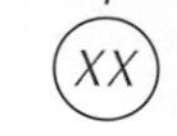

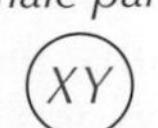

2. Draw two circles below each of the parent circles to represent the possible gametes. Put a single chromosome from each parent in each circle. Draw lines to show which chromosomes come from each parent.

female parent *male parent*

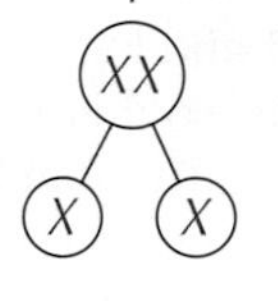

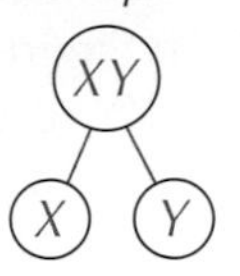

3. One gamete from the female combines with one gamete from the male during fertilisation, so draw criss-cross lines to show all the possible ways the X and Y chromosomes could combine.

female parent *male parent*

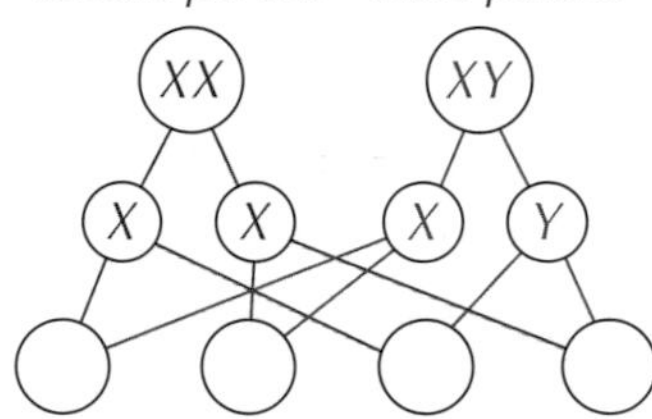

4. Then write the possible combinations of gametes in the bottom circles.

female parent *male parent*

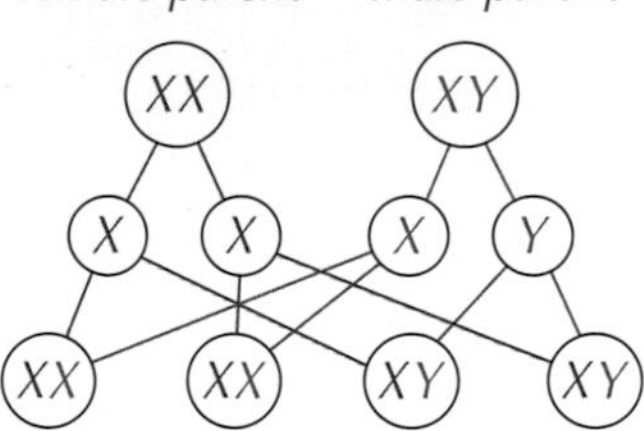

Tip: H Remember, one gamete from the female combines with one gamete from the male during fertilisation. It's quite easy to get confused, so check you aren't drawing lines that put both the male's gametes or both the female's gametes into the same circle.

Practice Questions — Fact Recall

Q1 How many of the 23 pairs of human chromosomes determine sex?

Q2 True or False? All sperm cells carry the Y chromosome.

Practice Questions — Application

Rachael and her husband Luca are expecting their first child.

Q1 Complete this genetic diagram to show the possible combinations of sex chromosomes that the baby could have.

Rachael *Luca*

XX XY

X X X Y

female *female* *male* *male*

Q2 What is the probability that their child will be a boy?

Q3 Draw a Punnett square to show the possible combinations of sex chromosomes that Rachael and Luca's baby could have.

2. Alleles and Genetic Diagrams

The next few pages are all about how our genes determine our characteristics...

Genes and characteristics

What genes (see p. 34) you inherit control what characteristics you develop. Different genes control different characteristics. Most characteristics are controlled by several genes interacting. However, some characteristics are controlled by a single gene.

Examples

- Mouse fur colour — see Figure 1.
- Red-green colour blindness in humans.

Figure 1: *Two mice with different coloured fur.*

What are alleles?

All genes exist in different versions called alleles. Gametes only have one allele, but all the other cells in an organism have two — one on each chromosome in a pair (see Figure 2). This is because we inherit half of our alleles from our mother and half from our father. In genetic diagrams, letters are usually used to represent alleles.

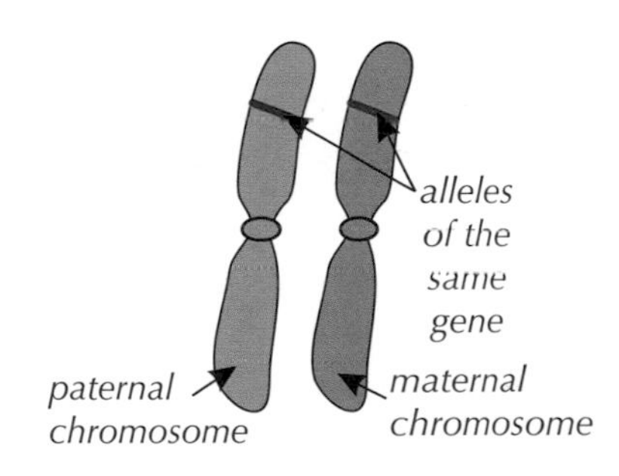

Figure 2: *Diagram showing two alleles for the same gene.*

For characteristics controlled by a single gene, if an organism has two alleles for that particular gene that are the same, then it's **homozygous** for that trait. If its two alleles for a particular gene are different, then it's **heterozygous**. If the two alleles are different, only one can determine what characteristic is present. The allele for the characteristic that's shown is called the **dominant** allele (use a capital letter for dominant alleles — e.g. 'C'). The other one is called **recessive** (and you show these with small letters — e.g. 'c').

For an organism to display a recessive characteristic, both its alleles must be recessive (e.g. cc). But to display a dominant characteristic the organism can be either CC or Cc, because the dominant allele overrules the recessive one if the plant/animal/other organism is heterozygous.

Your **genotype** is the combination of alleles you have, e.g. you could have the genotype Bb for hair colour. Your alleles work at a molecular level (by coding for slightly different proteins — see page 244) to determine what characteristics you have — your **phenotype**, e.g. brown eyes, blonde hair, etc.

Learning Objectives:

- Know that most characteristics are controlled by several genes but some characteristics are controlled by just one gene — e.g. fur colour in mice and red-green colour blindness in humans.
- Be able to explain the terms 'allele', 'homozygous', heterozygous', 'dominant' and 'recessive'.
- Understand that if two different alleles of the same gene are present, then the allele for the characteristic displayed is dominant, and the allele for the characteristic that isn't displayed is recessive.
- Know that two recessive alleles are needed for a recessive characteristic to be displayed.
- Be able to explain the terms 'genotype' and 'phenotype'.
- Know that your genotype determines your phenotype.
- Be able to complete and interpret genetic diagrams showing single gene inheritance.
- Be able to express the outcome of a single gene cross using ratios or direct proportion.
- **H** Be able to construct genetic diagrams and use them to predict what will happen in single gene crosses.

Specification References
4.6.1.6

Single gene inheritance

Characteristics that are determined by a single gene can be studied using **monohybrid crosses**. This is where you cross two parents to look at just one characteristic. In the exam they could ask you about the inheritance of any characteristic controlled by a single gene, as the principle's always the same.

Tip: 'Cross' just means 'breed together'.

Genetic diagrams of single gene inheritance

Genetic diagrams allow you to see how certain characteristics are inherited. The inheritance of round or wrinkly peas from pea plants is an example of single gene inheritance and can be shown using genetic diagrams.

Figure 3: *The peas produced by pea plants can either be wrinkly (left) or round (right).*

Example 1

The gene which causes wrinkly peas is recessive, so you can use a small 'r' to represent it. Round peas are due to a dominant gene, which you can represent with a capital 'R'. If you cross one pea plant which produces wrinkly peas (rr) and one that produces round peas (in this case RR), all of the offspring will produce round peas — but they'll have the alleles Rr.

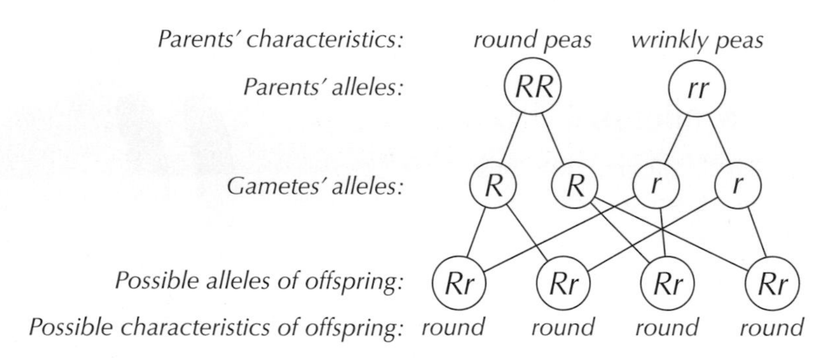

You can also show this genetic cross in a Punnett square:

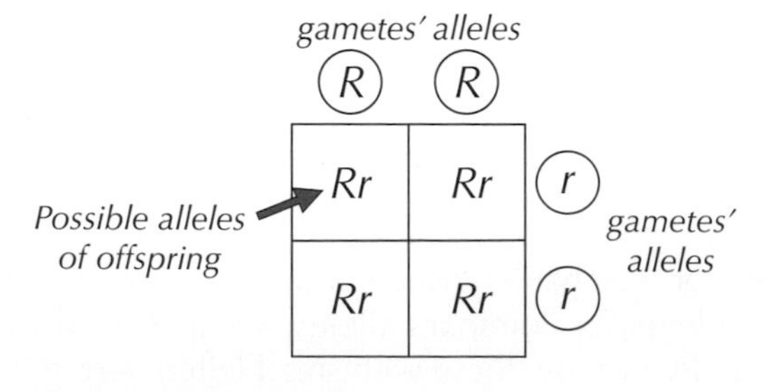

Tip: In this example, a plant producing wrinkly peas must be homozygous recessive (it must have 'rr' alleles). However, a plant producing round peas could be homozygous dominant or heterozygous — it can have two possible combinations of alleles — RR or Rr.

You need to be able to use genetic diagrams to predict and explain the outcomes of single gene crosses between individuals for lots of different combinations of alleles. The outcomes are given as ratios and can be used to work out the probability of having offspring with a certain characteristic.

A 3:1 ratio in the offspring

A cross could produce a 3:1 ratio of certain characteristics in the offspring. On the next page there's an example showing that this would happen if two of the heterozygous offspring from Example 1 were crossed.

Example 2

Parents' characteristics: *round peas* *round peas*

Parents' alleles: Rr Rr

Gametes' alleles: R r R r

Possible alleles of offspring: RR Rr Rr rr

Possible characteristics of offspring: *round* *round* *round* *wrinkly*

Again, this cross can be shown in a Punnett square:

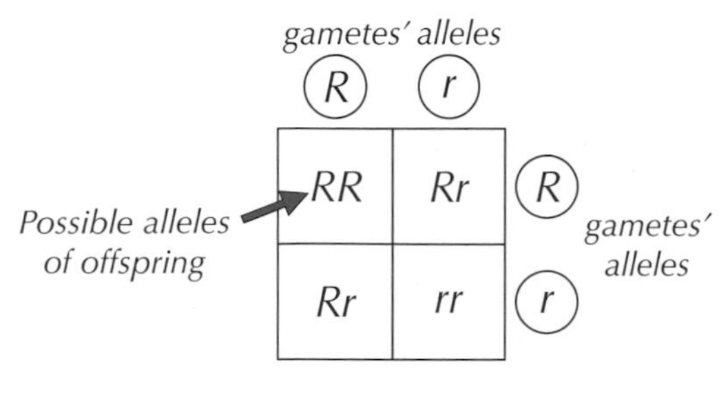

MATHS SKILLS

This cross gives a 3:1 ratio of plants producing round peas to plants producing wrinkly peas. This means there's a 1 in 4, 25%, 0.25 or ¼ chance of any new pea plant having wrinkly peas. Remember that "results" like this are only probabilities — they don't say definitely what'll happen.

Exam Tip
In the exam you might be given the results of a breeding experiment and asked to say whether a characteristic is dominant or recessive. To figure it out, look at the ratios of the characteristic in different generations. For example, in Example 2 here the 3:1 ratio of round to wrinkly peas shows that the round allele is dominant.

Exam Tip
A 3:1 ratio of plants producing round peas to plants producing wrinkly peas means there's a 1 in 4 chance of a new plant having wrinkly peas, not a 1 in 3 chance — it's an easy mistake to make, so be careful when you're talking about proportions and probabilities.

All the offspring are the same

More than one cross could result in all of the offspring showing the same characteristic. Here you have to do some detective work to find out what's gone on:

Example 3

If you cross a homozygous pea plant that produces round peas (so has two dominant alleles — RR), with a homozygous pea plant that produces wrinkly peas (rr), all the offspring will be heterozygous (Rr), so produce round peas:

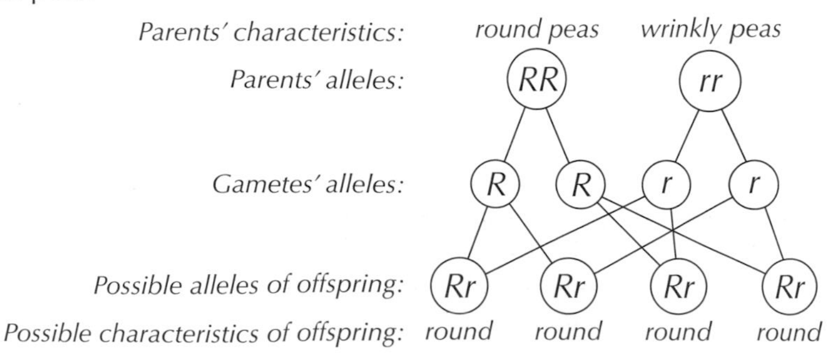

This means there's a 100% probability of any new pea plant having round peas.

But, if you crossed a homozygous pea plant that produces round peas (RR), with a heterozygous pea plant that produces round peas (so has a dominant

Tip: You can see a Punnett square showing this cross in Example 1 on the previous page.

and a recessive allele — Rr), you would also only get offspring that produce round peas:

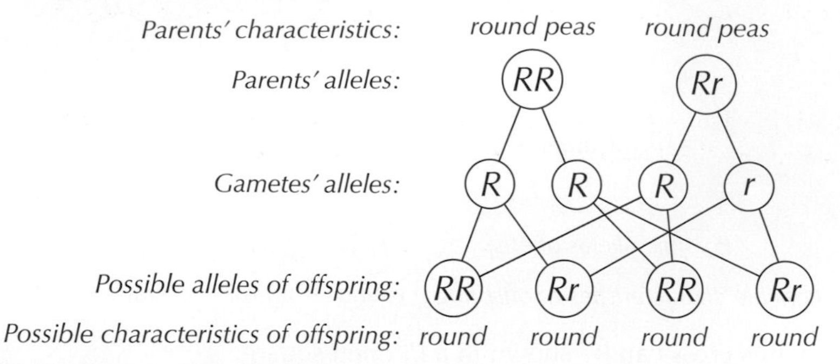

To find out which cross you'd done, you'd have to breed the offspring together and see what kind of ratio you got — then you'd have a good idea. If it was a 3:1 ratio of round to wrinkly in the offspring, it's likely that you originally had RR and rr plants (see Example 2 on the previous page).

Tip: You know the drill by now — here's the Punnett square showing this cross:

	R	R
R	RR	RR
r	Rr	Rr

A 1:1 ratio in the offspring

Next up is an example of when a cross produces a 1:1 ratio in the offspring — half the offspring are likely to show one characteristic and half are likely to show another characteristic. This time we're using cats with long and short hair. (Peas are a bit dull...)

Example

A cat's long hair is caused by a dominant allele 'H'. Short hair is caused by a recessive allele 'h'. A heterozygous cat with long hair (Hh) was bred with a homozygous cat with short hair (hh):

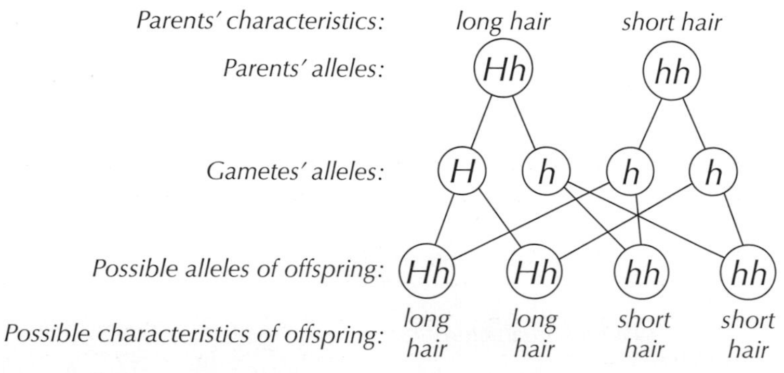

Here's the Punnett square for this cross:

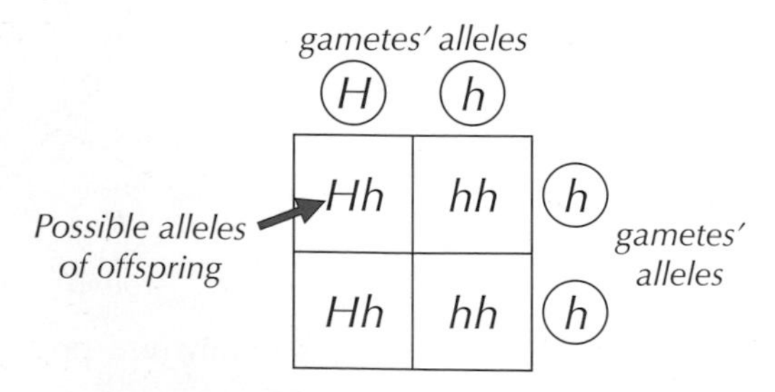

This is a 1:1 ratio, which gives a 50% probability of a new cat being born to these parents having long hair.

Exam Tip
Half the offspring of this cross are homozygous recessive — their genotype is 'hh', which produces a short-haired phenotype. The other half are heterozygous — their genotype is 'Hh', producing a long-haired phenotype. Get used to using terms such as 'homozygous', 'heterozygous', 'dominant', 'recessive' 'genotype' and 'phenotype' now — it'll really help you out in the exams.

Constructing genetic diagrams Higher

You need to be able to construct genetic diagrams for single gene crosses.

Example 1 Higher

Drawing a Punnett square to show a cross between a pea plant with round peas (Rr) and a pea plant with wrinkly peas (rr).

1. First, draw a grid with four squares.
2. Put the possible alleles from one parent down the side, and those from the other parent along the top.
3. Then in each middle square you fill in the letters from the top and side that line up with that square — the pairs of letters in the middle show the possible combinations of the alleles.

The steps are shown here:

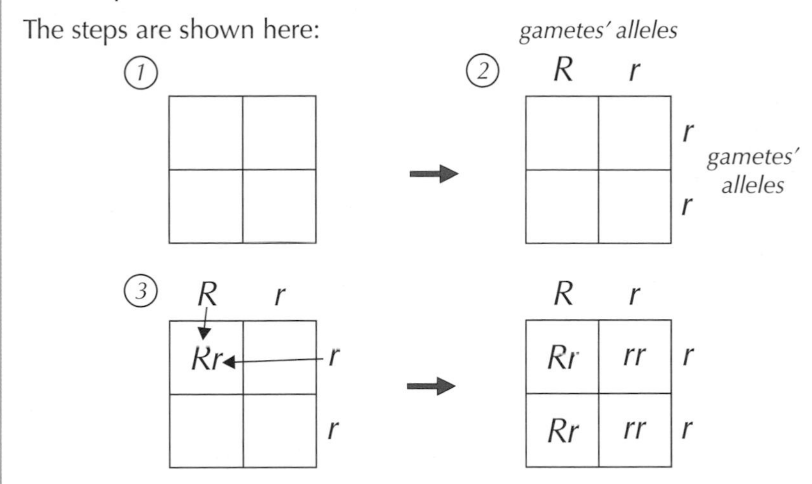

Example 2 Higher

Drawing a genetic diagram to show a cross between a pea plant with wrinkly peas (rr) and a pea plant with round peas (RR).

1. Draw two circles at the top of the diagram to represent the parents. Put the round genotype in one and the wrinkly genotype in the other.
2. Draw two circles below each of the parent circles to represent the possible gametes. Put a single allele from each parent in each circle.
3. One gamete from the female combines with one gamete from the male during fertilisation, so draw criss-cross lines to show all the possible ways the alleles could combine.
4. Then write the possible combinations of alleles in the offspring in the bottom circles.

These steps are illustrated here:

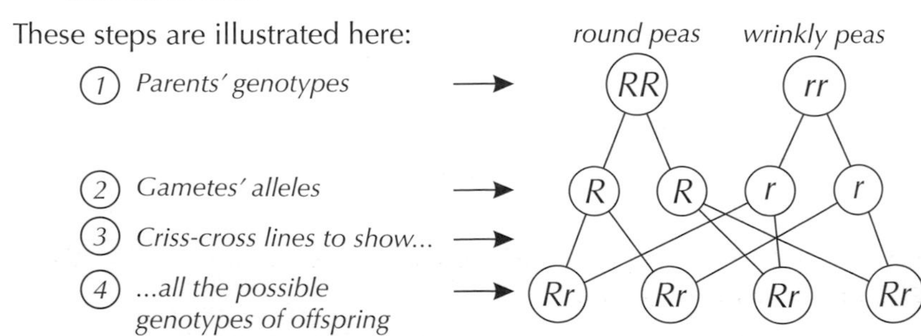

Tip: H Constructing the genetic diagrams shown on this page is just the same as constructing the sex inheritance diagrams on pages 263-264 — the only difference is that you are showing alleles here rather than chromosomes.

Practice Questions — Fact Recall

Q1 Give an example of a characteristic controlled by a single gene.

Q2 Explain what is meant by each of these terms:

a) alleles

b) homozygous

c) dominant allele

d) phenotype

Practice Questions — Application

Q1 A female cat has the genotype Ss. 'S' is the allele for spots and is dominant. 's' is the allele for no spots and is recessive.

a) Is the cat homozygous or heterozygous for the spot allele?

b) What genotype would a cat without spots have?

c) What other possible genotype could a spotty cat have?

Q2 Charlotte has some guinea pigs with rough coats and some guinea pigs with smooth coats. The allele for a rough coat is represented by 'R'. The allele for a smooth coat is represented by 'r'. The rough coat allele is dominant to the smooth coat allele. She crosses a rough coated guinea pig (RR) with a smooth coated guinea pig (rr). Here is a genetic diagram of the cross:

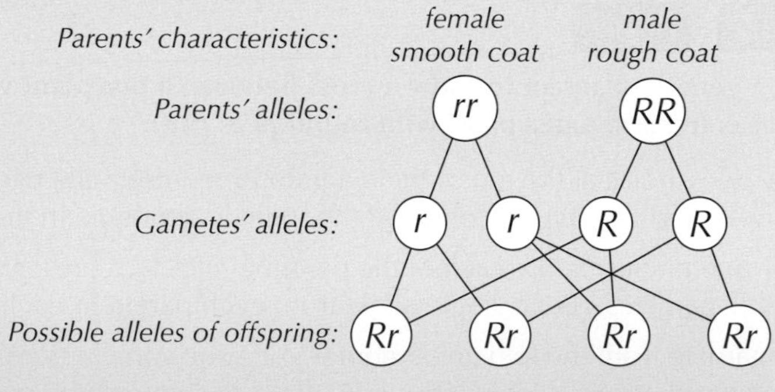

Parents' characteristics: *female smooth coat* *male rough coat*

Parents' alleles: *rr* *RR*

Gametes' alleles: *r* *r* *R* *R*

Possible alleles of offspring: *Rr* *Rr* *Rr* *Rr*

a) What type of coat will the offspring have?

b) i) Charlotte breeds two guinea pigs together that are both heterozygous for a rough coat. Construct a Punnett square to show the possible genotypes their offspring could have.

ii) From the Punnett square, what is the probability of the offspring having a smooth coat?

Exam Tip H
Make sure you read exam questions carefully to avoid throwing away easy marks. For example, in Q2 you're asked to construct a Punnett square — if you were asked this question in the exam and you produced a different type of genetic diagram you wouldn't get all the marks available to you.

3. Inherited Disorders

It's not just an organism's characteristics that can be passed on to its offspring — some disorders can be too. These are known as inherited disorders.

Learning Objectives:

- Know that some disorders are inherited, due to certain alleles being passed on to offspring.
- Know that cystic fibrosis is an inherited disorder affecting the cell membranes.
- Understand that people will only have cystic fibrosis if both their parents possess the faulty allele because it's a recessive disorder.
- Know that polydactyly is an inherited disorder that results in a person having extra fingers or toes.
- Understand that people will have polydactyly if only one parent has the allele because it is a dominant disorder.
- Be able to interpret family trees.
- Appreciate the economic, social and ethical issues surrounding the screening of embryos.

Specification References 4.6.1.6, 4.6.1.7

What are inherited disorders?

Inherited disorders are disorders that are caused by a faulty allele, which can be passed on to an individual's offspring. You need to know about two inherited disorders — **cystic fibrosis** and **polydactyly**.

Cystic fibrosis

Cystic fibrosis is an inherited disorder of the cell membranes. It results in the body producing a lot of thick sticky mucus in the air passages (which makes breathing difficult) and in the pancreas.

The allele which causes cystic fibrosis is a recessive allele, carried by about 1 person in 25. Because it's recessive, people with only one copy of the allele won't have the disorder — they're known as **carriers** (they carry the faulty allele, but don't have any symptoms).

For a child to have the disorder, both parents must be either carriers or have the disorder themselves. There's a 1 in 4 chance of a child having the disorder if both parents are carriers. This is shown in Figure 1, where 'f' is used for the recessive cystic fibrosis allele.

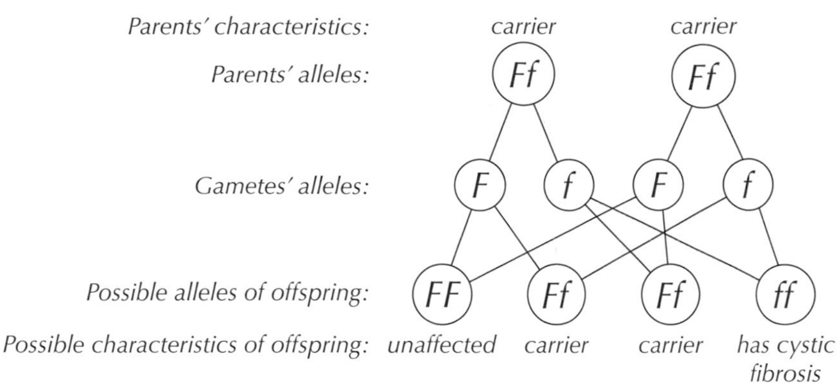

***Figure 1:** Genetic diagram to show the inheritance of cystic fibrosis from two carriers.*

Polydactyly

Polydactyly is an inherited disorder where a baby's born with extra fingers or toes (see Figure 2). It doesn't usually cause any other problems so isn't life-threatening.

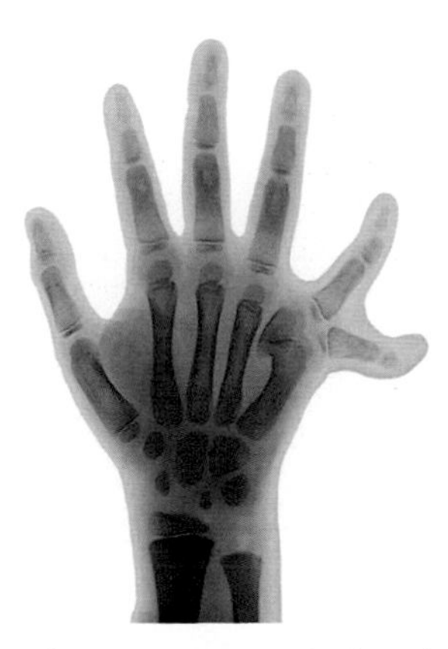

***Figure 2:** X ray of a hand with an extra finger. This person has polydactyly.*

The disorder is caused by a dominant allele and so can be inherited if just one parent carries the defective allele. The parent that has the defective allele will have the condition too since the allele is dominant.

There's a 50% chance of a child having the disorder if one parent has the polydactyly allele. This is shown in Figure 3, where 'D' is used for the dominant polydactyly allele.

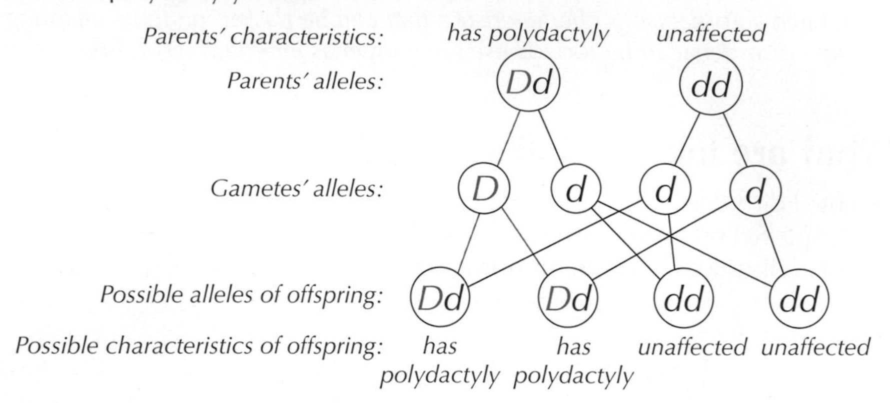

Figure 3: *Genetic diagram to show the inheritance of polydactyly from a person with the condition and an unaffected individual.*

Tip: Remember that there are no carriers for polydactyly. If you carry the faulty gene then you will be affected because the disorder is caused by a dominant allele, so you only need one copy to have the disorder.

Family trees

In genetics, a family tree is a diagram that shows how a characteristic (or disorder) is inherited in a group of related people. In the exam you might be asked to interpret a family tree.

Exam Tip
In the exam you might get a family tree showing the inheritance of a dominant allele — in this case there won't be any carriers shown.

Example

Here is a family tree for cystic fibrosis:

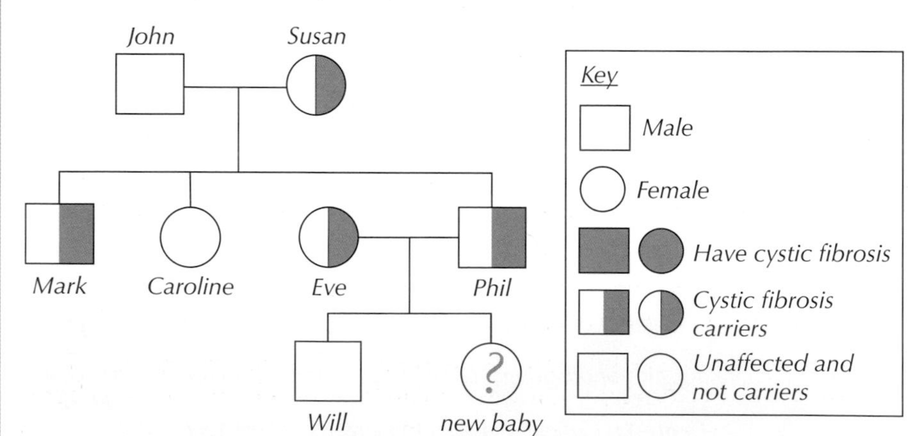

From the family tree, you can tell that:

1. The allele for cystic fibrosis isn't dominant because plenty of the family carry the allele but don't have the disorder.
2. There is a 25% chance that Eve and Phil's new baby will have the disorder and a 50% chance that it will be a carrier, as Eve and Phil are carriers but are unaffected. This is because the case of the new baby is just the same as in the genetic diagram on the previous page — so the baby could be unaffected (FF), a carrier (Ff) or have cystic fibrosis (ff).

MATHS SKILLS

Exam Tip
A good way to work out a family tree is to write the genotype of each person onto it.

Embryonic screening

Embryonic screening is a way of detecting inherited disorders, such as cystic fibrosis, in embryos. There are different methods used to do this:

1. **Pre-implantation genetic diagnosis (PGD).** During IVF, embryos are fertilised in a laboratory, and then implanted into the mother's womb. Before being implanted, it's possible to remove a cell from each embryo and analyse its genes (see Figure 4) — this is called pre-implantation genetic diagnosis (PGD). Embryos with 'healthy' alleles would be implanted into the mother — the ones with 'faulty' alleles destroyed.
2. **Chorionic villus sampling (CVS).** CVS is usually carried out between 10 and 13 weeks of pregnancy. It involves taking a sample of cells from part of the placenta and analysing their genes. The part of the placenta that's taken and the embryo develop from the same original cell — so they have the same genes. If the embryo is found to have an inherited disorder, the parents can decide whether or not to terminate (end) the pregnancy.

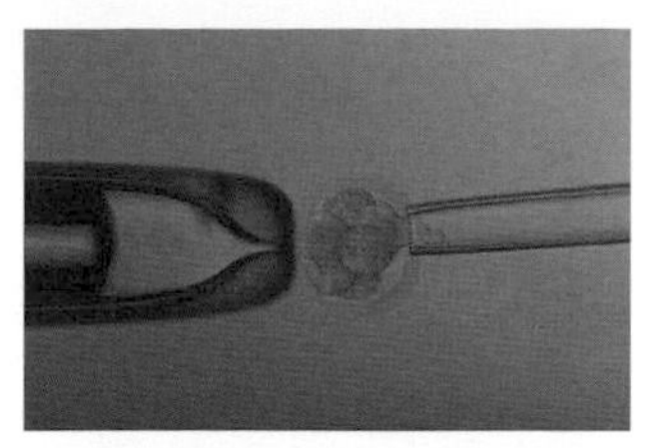

Figure 4: *A single cell being extracted from an embryo in order to be screened for inherited disorders.*

Tip: The placenta is an organ that attaches the embryo to the lining of the womb. It provides the developing embryo with nutrients and takes away waste.

Issues surrounding embryonic screening

There is a huge debate raging about embryonic screening. Here's why:

Arguments for screening

- It helps to stop people suffering from certain inherited disorders.
- Treating disorders costs the Government (and the taxpayers) a lot of money, so screening embryos could reduce healthcare costs.
- During IVF, most of the embryos are destroyed anyway — PGD just ensures that the selected one is healthy.
- If an inherited disorder is diagnosed through CVS, parents don't have to have a termination — but it does give them the choice.

Arguments against screening

- There may come a point where everyone wants to screen their embryos so they can pick the most 'desirable' one, e.g. they want a blue-eyed, blond-haired, intelligent boy.
- It implies that people with genetic problems are 'undesirable' — this could increase prejudice.
- After PGD, the rejected embryos are destroyed — they could have developed into humans, so some people think destroying them is unethical.
- There's a risk that CVS could cause a miscarriage. And if an inherited disorder is diagnosed through CVS, it could lead to a termination (abortion).
- Screening embryos is expensive.

Tip: There are laws to stop embryonic screening going too far. At the moment, parents can't even select their baby's sex (unless it's for health reasons).

Many people think that screening isn't justified for inherited disorders that don't affect a person's health:

Example

Polydactyly causes a physical disfigurement but it isn't life-threatening, so lots of people don't agree with screening for it. In comparison, conditions such as cystic fibrosis are potentially life-threatening and so more people agree to screening for them.

Exam Tip
You could be given some data about screening for inherited disorders in the exam, and then asked questions about it. You'll just need to read the information given to you carefully, and apply what you know.

Practice Questions — Fact Recall

Q1 What subcellular structure does cystic fibrosis affect?

Q2 Is cystic fibrosis caused by a dominant or recessive allele?

Q3 Why would an embryo be screened before it is inserted into the mother's uterus during IVF?

Practice Questions — Application

Q1 If 'f' represents the recessive allele for cystic fibrosis and 'F' represents the dominant allele:

a) give the alleles a carrier of cystic fibrosis would have.

b) give the alleles that a person with cystic fibrosis would have.

Q2 The family tree below shows the inheritance of the inherited disorder polydactyly in a family.

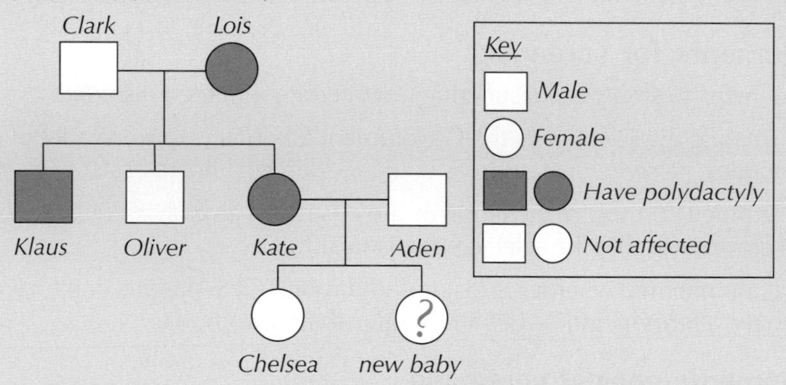

Using the family tree, answer the following questions:

a) From the diagram, how do you know that the allele for polydactyly is dominant?

b) If the new baby was a girl with polydactyly, what symbol would she have on the family tree?

For the following questions, use 'D' to represent the allele for polydactyly and 'd' to represent the allele for an unaffected person.

c) Give the genotype of the following individuals:

i) Clark ii) Kate

d) Kate and Aden are having a baby.

i) Draw a Punnett square to show the inheritance of polydactyly from Kate and Aden.

ii) What is the probability the new baby will have polydactyly?

Exam Tip
Sometimes you can't tell a person's genotype from looking straight at the family tree and the key — you might need to do a bit of detective work and figure out the genotypes of other people on the tree.

4. The Work of Mendel

We haven't always known as much about genetics and inheritance as we do now. The work of a monk called Gregor Mendel helped us on our way...

Learning Objectives:
- Understand how Mendel's work in the mid 1800s led him to develop the idea of inherited units.
- Know that these units are what we now know as genes.
- Understand why Mendel's work was not appreciated until after he had died.
- Know that scientists first observed how chromosomes act during cell division in the late 1800s.
- Know that the observation of similarities between the behaviour of chromosomes and Mendel's units led to the development of the idea in the early 1900s that the units are located on chromosomes.
- Know that determining DNA structure in the mid 1900s, and subsequent work by scientists, means that we now have a greater understanding of how genes work.

Specification Reference 4.6.3.3

Who was Gregor Mendel?

Gregor Mendel was an Austrian monk who trained in mathematics and natural history at the University of Vienna. On his garden plot at the monastery, Mendel noted how characteristics in plants were passed on from one generation to the next. The results of his research were published in 1866 and eventually became the foundation of modern genetics.

Figure 2: *Gregor Mendel.*

Mendel's work

Mendel did lots of experiments with pea plants. In one experiment he crossed two pea plants of different heights — a tall pea plant and a dwarf pea plant. The offspring produced were all tall pea plants (see Figure 1).

A tall pea plant and a dwarf pea plant are crossed...

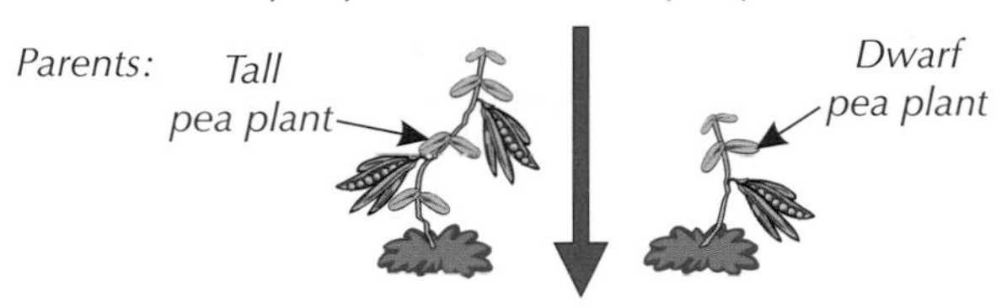

...resulting in all tall pea plants.

Figure 1: *Diagram of the first cross in Mendel's pea plant height experiment.*

He then bred two of these tall pea plants together. The resulting offspring consisted of three tall pea plants and one dwarf pea plant (see Figure 3).

Two pea plants from the 1st set of offspring are crossed...

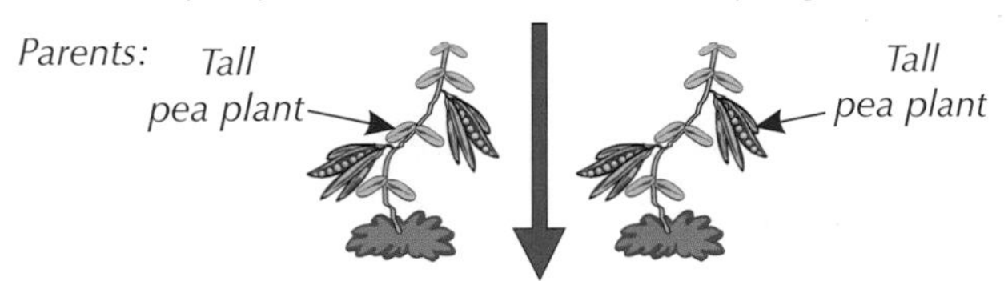

...resulting in three tall pea plants and one dwarf pea plant.

Figure 3: *Diagram of the second cross in Mendel's pea plant height experiment.*

Explaining Mendel's results

From his pea plant height experiment, Mendel had shown that the height characteristic in pea plants was determined by separately inherited "hereditary units" passed on from each parent.

This can be explained using genetic diagrams, where **T** represents the hereditary unit for tall plants and **t** represents the hereditary unit for dwarf plants.

First cross

As shown in Figure 4, in the first of Mendel's crosses, a tall and a dwarf pea plant are crossed together.

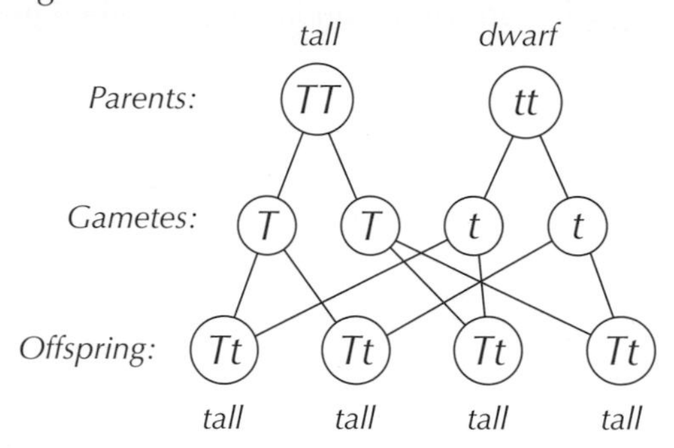

Figure 4: *Genetic diagram of the first cross in Mendel's pea plant height experiment.*

The resulting offspring are **Tt** — they are all tall plants, but they all carry the hereditary unit for dwarf plants.

Exam Tip

In Mendel's pea plant experiments he also investigated lots of other pea plant characteristics, such as flower colour:

So don't be put off if you have a question in the exam about one of Mendel's other experiments — just use what you've learnt here to answer the question.

Second cross

In the second cross, two tall pea plants from the first cross are crossed together (see Figure 5).

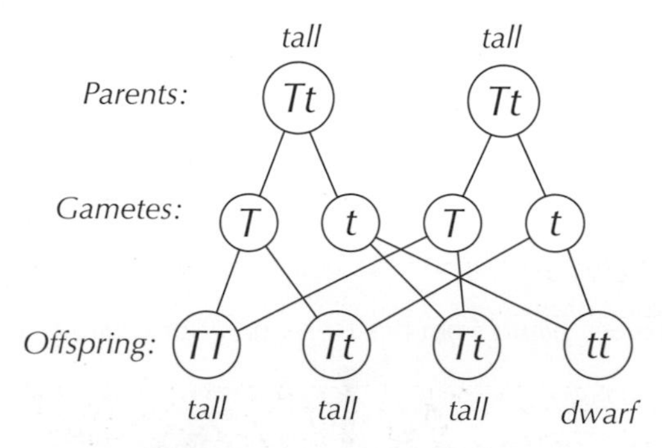

Figure 5: *Genetic diagram of the second cross in Mendel's pea plant height experiment.*

The resulting offspring are **TT**, **Tt**, **Tt** and **tt** — this gives three tall plants and one dwarf plant.

The ratios of tall and dwarf plants in the offspring show that the hereditary unit for tall plants, **T**, is dominant over the hereditary unit for dwarf plants, **t**, i.e. plants with hereditary units for both types of plant (**Tt**) will be tall plants.

Tip: Figure 5 shows that when two tall (Tt) pea plants are crossed, there's a 3:1 ratio of tall:dwarf pea plants in the offspring, or a 25% chance of getting a dwarf plant from this cross. There's more about ratios on p. 377 and more about probabilities on p. 263.

Mendel's conclusions

From all his experiments on pea plants, Mendel reached these three important conclusions about heredity in plants:

1. Characteristics in plants are determined by "hereditary units".
2. Hereditary units are passed on to offspring unchanged from both parents, one unit from each parent.
3. Hereditary units can be dominant or recessive — if an individual has both the dominant and the recessive unit for a characteristic, the dominant characteristic will be expressed.

The importance of Mendel's work

WORKING SCIENTIFICALLY

We now know that the "hereditary units" are of course genes. But Mendel's work was cutting edge and new to the scientists of the day. They didn't have the background knowledge to properly understand his findings — they had no idea about genes, DNA and chromosomes.

It wasn't until after his death that people realised how significant his work was. Using Mendel's work as a starting point, the observations of many different scientists have contributed to the understanding of genes that we have today.

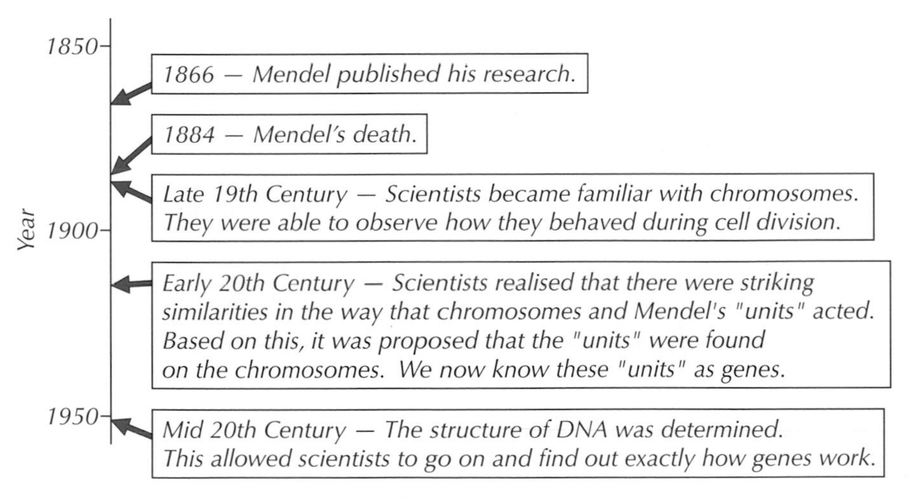

Figure 6: *Timeline showing how Mendel's work has contributed to our understanding of genetics.*

Tip: The idea that genes are the units of inheritance is called 'gene theory'.

Exam Tip
Mendel is one of the great scientists, who found out really important things about genetics. In the exam don't get him mixed up with other great scientists, like Darwin.

Tip: Have a look at page 246 for more on DNA structure and page 244 for more on how genes work.

Practice Questions — Fact Recall

Q1 True or false? One of the conclusions that Mendel reached from his experiments was that hereditary units are changed before being passed on from parents to offspring.

Q2 Why didn't Mendel get any credit for his work while he was alive?

Q3 Put the following events in chronological order:
A. The structure of DNA being determined.
B. Mendel's research on pea plants.
C. The observation of chromosomes during cell division.

Practice Question — Application

Q1 Mendel did experiments on pea plants with different seed colours. From these experiments he found that there is a hereditary unit in pea plants that gives them green seeds and one that gives them yellow seeds. The hereditary unit for yellow pea seeds is dominant.

Tick or cross the boxes in the table to show which hereditary units pea plants with the following seed colours could have. (You can tick more than one option for each seed colour.)

Seed colour of pea plant	Type of hereditary unit		
	Just green	Just yellow	Both green and yellow
Green			
Yellow			

Topic 6b Checklist — Make sure you know...

X and Y Chromosomes

- ☐ That 22 of the 23 pairs of chromosomes in a human body cell are just responsible for controlling your characteristics and the 23rd pair of chromosomes carry the genes that determine sex. The 23rd pair of chromosomes are called sex chromosomes — females have two X sex chromosomes and males have an X sex chromosome and a Y sex chromosome.
- ☐ How to complete and interpret genetic diagrams to show sex inheritance.
- ☐ How to use probability to predict the likelihood of a result of a genetic diagram, and how to express the results of a genetic diagram using ratios or direct proportion.
- ☐ H How to construct genetic diagrams to show sex inheritance.

Alleles and Genetic Diagrams

- ☐ That most characteristics are controlled by several genes, but some are controlled by just one gene — e.g. fur colour in mice and red-green colour blindness in humans.
- ☐ That all genes exist in different forms called alleles, and that if an organism has two alleles for a gene that are the same then it is homozygous. If the two alleles are different then it's heterozygous.
- ☐ That if an individual is heterozygous for a characteristic, the allele for the characteristic that is displayed is dominant, and the other allele is recessive. You need to be homozygous recessive for a recessive characteristic to be displayed.
- ☐ That your genotype is the combination of alleles you have (e.g. Gg, bb) and that these alleles work at the molecular level to determine your phenotype — the characteristics that you have (e.g. blue eyes, blonde hair).
- ☐ That single gene inheritance is where the alleles of a single gene control a characteristic.
- ☐ How to complete and interpret genetic diagrams showing the inheritance of a single gene for all combinations of dominant and recessive alleles.

cont...

- [ ] How to express the results of a single gene cross (a cross between two parents that looks at just one characteristic inherited by a single gene) as a ratio or a directly proportional relationship.
- [ ] **H** How to construct genetic diagrams to predict the results of single gene crosses.

Inherited Disorders

- [ ] That inherited disorders result from faulty alleles, which can be passed onto an individual's offspring.
- [ ] That cystic fibrosis is an inherited disorder of the cell membranes caused by a recessive allele, meaning that both parents of a person with cystic fibrosis must have the faulty allele for that person to inherit the condition. If a person inherits just one allele for cystic fibrosis, they will be a carrier.
- [ ] That polydactyly is an inherited disorder where people have extra fingers and toes, and it's caused by a dominant allele, meaning that anyone who inherits a single polydactyly allele from one of their parents will have the condition.
- [ ] How to interpret family trees (diagrams showing how a characteristic is inherited in a group of related people).
- [ ] What the economic, social and ethical issues surrounding embryonic screening are and be able to form opinions regarding these given relevant information.

The Work of Mendel

- [ ] How Mendel's experiments with pea plants in the mid 1800s led him to develop the idea of hereditary units and what that meant — characteristics are determined by separate units, that organisms receive one unit from each parent and that these units can be dominant or recessive.
- [ ] That what Mendel called "hereditary units" are what we now know as genes.
- [ ] That Mendel's work was not appreciated until after his death because people were not aware of things like genes, DNA and chromosomes at the time.
- [ ] That chromosome behaviour during cell division was observed in the late 1800s and that this led some scientists in the early 1900s to realise that Mendel's units are located on chromosomes.
- [ ] That the determination of the structure of DNA in the mid 1900s and further work by scientists has led to the understanding of genes that we have today.

Exam-style Questions

1 What are Mendel's inherited units now better known as?

A DNA
B chromosomes
C genes
D gametes

(1 mark)

2 Polydactyly is an example of an inherited disorder.

2.1 What effect does polydactyly have on a person?

(1 mark)

2.2 If 'D' represents the allele for polydactyly and 'd' represents the allele for an unaffected person, give the possible genotype(s) of a person with polydactyly.

(1 mark)

2.3 Greg is homozygous for the polydactyly allele.
Which of the following shows Greg's genotype?

A DD
B dd
C Dd
D dDd

(1 mark)

2.4 Greg's parents are both heterozygous for the polydactyly allele.
Draw a Punnett square and identify any offspring with polydactyly.
Write the ratio of affected to unaffected offspring.

(4 marks)

3 Kaye and Mark are expecting a baby.

3.1 Complete the sex inheritance Punnett square in **Figure 1** for Kaye and Mark's baby.

Kaye's gametes

	X	◯
X	*XX*	
◯		

Mark's gametes (rows)

Figure 1

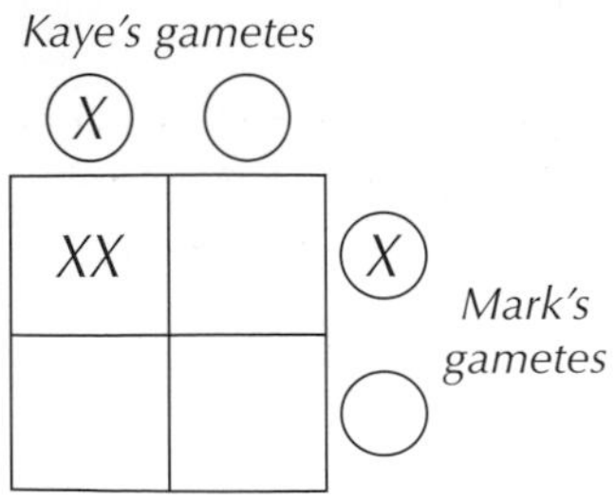

(2 marks)

3.2 Kaye and Mark have found out that they are expecting a baby boy.
What combination of sex chromosomes will the baby have?

(1 mark)

3.3 Kaye has dimples in her cheeks, but Mark does not.
The presence of dimples is thought to be caused by a dominant allele represented by the letter **D**. The recessive allele is represented by the letter **d**.
Kaye is heterozygous for the dimples gene.

What is the probability that the baby will have dimples?

Draw a genetic diagram to show the possible inheritance of the dimples gene by the baby. Identify the phenotype of each possible outcome.

(5 marks)

4 Family trees can show the inheritance of characteristics, such as cystic fibrosis.

Figure 2 is a family tree showing the inheritance of cystic fibrosis in a family.

Figure 2

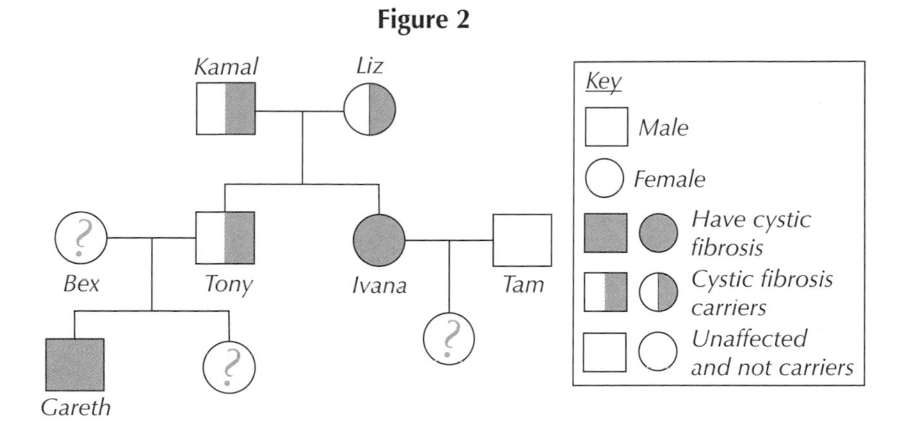

4.1 What is cystic fibrosis?

(2 marks)

4.2 Is Gareth heterozygous or homozygous for the cystic fibrosis allele? Explain your answer.

(1 mark)

4.3 Bex and Tony have a child with cystic fibrosis. However, Bex doesn't have cystic fibrosis. Therefore, what combination of alleles must Bex have for the cystic fibrosis gene? Explain your answer.

(3 marks)

4.4 Bex and Tony are considering having another child.
They want to use *in vitro* fertilisation (IVF) so that they can select an embryo without cystic fibrosis to be implanted into Bex's womb. The remaining embryos that aren't implanted will be destroyed.

Give **two** reasons why some people are against this use of embryonic screening.

(2 marks)

4.5 Looking at the family tree, could Ivana and Tam have a child with cystic fibrosis? Draw a Punnett square to explain your answer. Use **f** to represent the allele for cystic fibrosis and **F** to represent the healthy allele.

(4 marks)

1. Variation

These pages are all about the differences between you, me, and well... everyone else really. It's fascinating stuff...

Learning Objectives:
- Know that variation within a species is the differences in the characteristics of the individuals.
- Know that variation can be a result of genes, the environment, or a combination of both.
- Know that there is usually a lot of genetic variation within a population.
- Know that mutations occur continuously.
- Understand that mutations lead to new genetic variants.
- Know that most genetic variants have no effect on phenotype.
- Know that some genetic variants can influence phenotype, and a few can lead to a new phenotype.
- Know that if a new phenotype makes an individual more suited to the environment, it can become common throughout the species quickly.

Specification Reference 4.6.2.1

What is variation?

Different species look... well... different — my dog definitely doesn't look like a daisy. But even organisms of the same species will usually look at least slightly different — e.g. in a room full of people you'll see different colour hair, individually shaped noses, a variety of heights, etc.

These differences are called the variation within a species — and there are two types of variation: **genetic variation** and **environmental variation**.

Genetic variation

All plants and animals have characteristics that are in some ways similar to their parents' (e.g. I've got my dad's nose, apparently). This is because an organism's characteristics are determined by the **genes** inherited from their parents.

Genes are the codes inside your cells that control how you're made (there's more about genes on page 244). These genes are passed on in **sex cells** (**gametes**), which the offspring develop from (see page 251).

Most animals (and quite a lot of plants) get some genes from the mother and some from the father. This combining of genes from two parents causes genetic variation — no two of the same species are genetically identical (other than identical twins). This means that there is a lot of genetic variation within a population.

Some characteristics are determined only by genes.

Examples

- Violet flower colour.
- Eye colour.
- Blood group.
- Inherited disorders (e.g. haemophilia or cystic fibrosis).

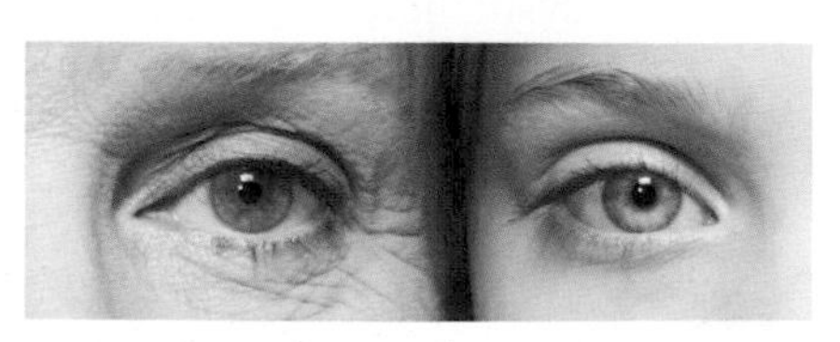

Figure 1: *Identical twins have exactly the same genes, which is why they look so alike.*

Environmental variation

The environment that organisms live and grow in also causes differences between members of the same species — this is called environmental variation.

Environmental variation covers a wide range of differences — from losing your toes in a piranha attack, to getting a suntan, to having yellow leaves and so on. Basically, any difference that has been caused by the conditions something lives in is an environmental variation.

Tip: Plants are strongly influenced by environmental factors, e.g. sunlight, moisture level, temperature and soil mineral content.

Example

A plant grown on a nice sunny windowsill would grow luscious and green.

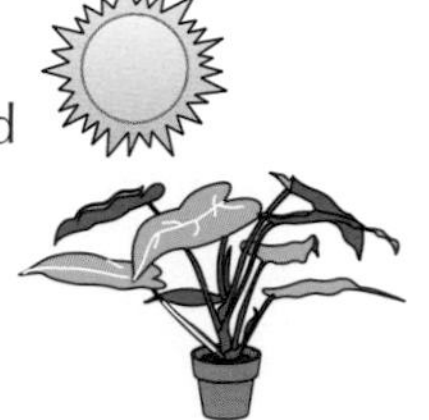

The same plant grown in darkness would grow tall and spindly and its leaves would turn yellow — these are environmental variations.

Genetic and environmental variation

Most characteristics are determined by a mixture of genetic and environmental factors.

Examples

- Height — the maximum height that an animal or plant could grow to is determined by its genes. But whether it actually grows that tall depends on its environment, e.g. how much food it gets.
- Intelligence — one theory is that although your maximum possible IQ might be determined by your genes, whether or not you get to it depends on your environment, e.g. your upbringing and school life.
- Health — some people are more likely to get certain diseases (such as cancer or heart disease) because of their genes. But lifestyle also affects the risk, e.g. whether you smoke or how much junk food you eat.

Tip: Environmental factors aren't just the physical things around you. They can include things like the way you were brought up too.

Mutations

As you know from page 249, occasionally a gene may mutate. A mutation is a random change in an organism's DNA. Mutations occur continuously.

Mutations mean that the gene is altered, which produces a **genetic variant** (a different form of the gene). As the gene codes for the sequence of amino acids that make up a protein, gene mutations sometimes lead to changes in the protein that it codes for.

Most genetic variants have very little or no effect on the protein the gene codes for. Some will change it to such a small extent that its function is unaffected. This means that most mutations have no effect on an organism's phenotype.

Tip: Mutations in DNA happen at random — it doesn't matter whether or not they'll result in a beneficial change.

Examples

- A mutation to a gene may produce a genetic variant that codes for the same sequence of amino acids, so the same protein is produced.
- The mutation may cause a change one amino acid in a protein, but the change has no effect on the protein's function.

Some variants have a small influence on the organism's phenotype — they alter the individual's characteristics but only slightly.

Example

Some characteristics, e.g. eye colour, are controlled by more than one gene. A mutation in one of the genes may change the eye colour a bit, but the difference might not be huge.

Very occasionally, variants can have such a dramatic effect that they determine phenotype.

Example

The genetic disorder, cystic fibrosis, is caused by a mutation that has a huge effect on phenotype. The gene codes for a protein that controls the movement of salt and water into and out of cells. However, the protein produced by the mutated gene doesn't work properly. This leads to excess mucus production in the lungs and digestive system, which can make it difficult to breathe and to digest food.

Tip: Mutations that result in a new phenotype are very rare.

If the environment changes, and the new phenotype makes an individual more suited to the new environment, it can become common throughout the species relatively quickly by natural selection — see page 287.

Example

A mutation may make a bacterium resistant to an antibiotic (see page 306). If the antibiotic appears in the bacterium's environment (e.g. a person with a bacterial infection starts taking antibiotics), the mutation may give the bacterium a better chance of surviving, reproducing and passing the beneficial mutation on to future generations. Over time, the mutation will accumulate in the bacterial population.

Practice Questions — Fact Recall

Q1 True or False? Only differences in genes can cause variation within a population.

Q2 a) What is a mutation?

b) Explain why most mutations have no effect on an organism's phenotype.

c) Outline how mutations can result in variation.

Practice Questions — Application

Q1 For the animal characteristics below, say whether they are determined by genes only, the environment only, or both.

a) eye colour

b) height

c) sickle cell anaemia, a genetic disorder

Q2 For the plant characteristics below, say whether they are determined by genes only, the environment only, or both.

a) stem height

b) spots caused by a fungus

Q3 Your sporting ability may be affected by your genes and your environment. Suggest one environmental factor that may affect your sporting ability.

Q4 Identical twins have exactly the same genes.
Non-identical twins don't. Studies have shown that:

- identical twins tend to have more similar IQs than non-identical twins (Study 1).
- identical twins who are brought up together tend to have more similar IQs than identical twins who are brought up separately (Study 2).

What do the results of Studies 1 and 2 suggest about the influence of genes and the environment on IQ? Explain your answer.

2. Evolution and Extinction

This is it. The 'Big One' — how life on Earth began, and, um, kept on going...

Learning Objectives:

- Know that, according to the theory of evolution, all of today's species have evolved from simple life forms that first started to develop over three billion years ago.
- Be able to give examples of the knowledge and observations that Darwin used to come up with his theory of evolution by natural selection.
- Understand Darwin's theory of evolution by natural selection.
- Understand and be able to explain how evolution by natural selection happens.
- Know that evolution is the changing of the inherited characteristics of a population over time.
- Know that the theory of evolution by natural selection has largely been accepted.
- Be able to describe evidence that supports Darwin's theory.
- Know that if the phenotypes of two populations of the same species change so much that they can no longer interbreed to produce fertile offspring, they have become two new species.
- Know that extinction has occurred when no individuals of a species remain and be able to describe possible causes.

Specification References
4.6.2.2, 4.6.3.1, 4.6.3.4, 4.6.3.6

The theory of evolution

The theory of evolution is this:

> All of today's species have evolved from simple life forms that first started to develop over three billion years ago.

It means that rather than just popping into existence, complex organisms like animals and plants 'developed' over billions of years from simpler single-celled organisms, such as bacteria.

Charles Darwin

Charles Darwin came up with a really important theory about evolution. He used the observations he made on a huge round-the-world trip, along with experiments, discussions and new knowledge of fossils and geology, to suggest the theory of evolution by natural selection. It works like this:

1. All of the individuals within a species show a wide range of variation for a particular characteristic. For example, they might have a wide range of fur lengths.
2. Because of this range of variation, some of these individuals will be more suited to the environment than others — e.g. in a warm environment, those with a short fur length may be more suited to the conditions.
3. The individuals who are most suited to the environment will be more likely to survive, and therefore more likely to breed successfully.
4. The useful characteristic will be passed on to the offspring.

New discoveries

Darwin's theory wasn't perfect. Because the relevant scientific knowledge wasn't available at the time, he couldn't give a good explanation for why new characteristics appeared or exactly how individual organisms passed on beneficial adaptations to their offspring.

We now know that phenotype is controlled by genes. New genetic variants (see page 283) arise because of mutations (changes in DNA), and these genetic variants can give rise to phenotypes that are suited to the environment. Beneficial genetic variants are passed on to future generations in the DNA that parents contribute to their offspring.

These new discoveries have meant that scientists have been able to develop Darwin's original ideas into the theory of evolution by natural selection that is accepted today.

Today's theory of evolution by natural selection

Scientists have developed Darwin's theory using what we now know about genetics. It works like this:

1. Individuals within a species show wide variation in their characteristics (phenotypic variation) because of the mix of genetic variants present in the population.

Example

Some rabbits have big ears and some have small ones.

2. Some genetic variants give rise to characteristics that are better suited to the environment. Organisms with these characteristics have a better chance of survival and so are more likely to breed successfully.

Example

Big-eared rabbits are more likely to hear a fox approaching them, and so are more likely to survive and have lots of offspring. Small-eared rabbits are more likely to get eaten.

3. So, the genetic variants that are responsible for the useful characteristics are more likely to be passed on to the next generation.

Example

All the baby rabbits are born with big ears.

Over time, the genetic variant for a useful characteristic will become more common (accumulate) in a population. This will lead to a change in the species. The changing of the inherited characteristics of a population over time is evolution.

An accepted hypothesis

WORKING SCIENTIFICALLY

There's so much evidence for Darwin's idea that it's now an accepted hypothesis (a theory). As well as the discovery of genetics, other evidence was also found by looking at fossils of different ages (the fossil record) — this allows you to see how changes in organisms developed slowly over time. The relatively recent discovery of how bacteria are able to evolve to become resistant to antibiotics also further supports evolution by natural selection.

Tip: Remember, genetic variants are different forms of genes. They arise when DNA mutates (see page 283).

Tip: Genetic differences between individuals of a species are also caused by sexual reproduction (see page 251).

Tip: Remember, it's the genetic variants that get passed on to the next generation (not the characteristics themselves).

Exam Tip
Don't get natural selection and evolution mixed up. Natural selection is when a useful characteristic (i.e. one that gives a better chance of survival) becomes more common in a population. Evolution is the gradual changing of a species over time (which happens as a result of natural selection).

Tip: There's more about fossils on p. 301-303.

Tip: You can read more about antibiotic resistance on p. 306.

Speciation

Tip: There's more on speciation on page 304.

Over a long period of time, the phenotype of organisms can change so much because of natural selection that a completely new species is formed. This is called speciation.

Speciation happens when populations of the same species change enough to become reproductively isolated — this means that they can't interbreed to produce fertile offspring.

Extinction

Tip: If an environment changes too rapidly, a species may not be able to evolve quickly enough to survive. If so, the species will eventually die out.

Tip: If lots of species die out at the same time (e.g. due to a catastrophic event), it's known as a mass extinction.

Extinction is when no living individuals of a species remain. The fossil record contains many species that don't exist any more — these species are said to be **extinct**. Dinosaurs and mammoths are extinct animals, with only fossils to tell us they existed at all.

Species become extinct for these reasons:

- The environment changes too quickly (e.g. destruction of habitat).
- A new predator kills them all (e.g. humans hunting them).
- A new disease kills them all.
- They can't compete with another (new) species for food.
- A catastrophic event happens that kills them all (e.g. a volcanic eruption or a collision with an asteroid).

Example

Dodos (a type of large, flightless bird) are now extinct. Humans not only hunted them, but introduced other animals which ate all their eggs, and also destroyed the forest where they lived — they really didn't stand a chance.

Figure 1: *An artist's impression of the dodo.*

Practice Questions — Fact Recall

Q1 According to the theory of evolution, how long ago did life on Earth begin?

Q2 According to the theory of evolution, how did complex organisms come to exist on Earth?

Q3 a) Who came up with the idea of natural selection?

b) What information did this scientist use to come up with this theory?

c) What is natural selection?

Q4 What does it mean if a species has become extinct?

Practice Questions — Application

Q1 Warfarin™ is a chemical that was commonly used to kill rats. It is now used less often as many rats have become resistant to it.

Use the theory of evolution by natural selection to explain how most rats have become warfarin-resistant.

Q2 In 1810, a herd of reindeer were taken from the Arctic to an area with a warmer climate. The herd were then left to live and reproduce in the area and were revisited in 1960. Some information about the herd is shown in the graph.

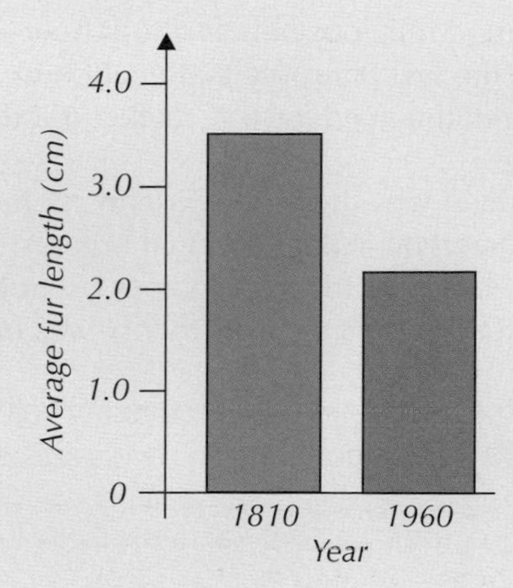

a) By roughly how much did the average fur length of the herd change between 1810 and 1960?

b) Explain this change in terms of natural selection.

Tip: To help you answer Q2 b), think about how the change in fur length might affect the reindeer in their new environment.

3. Ideas About Evolution

Learning Objectives:
- Know that in 1859 Darwin published "On the Origin of Species" to propose his theory.
- Understand why Darwin's ideas caused controversy when they were published, and why his ideas were only gradually accepted.
- Understand that there have been other hypotheses about evolution, including Lamarck's — which was based on the idea that 'acquired characteristics' could be inherited.
- Understand that we now know Lamarck's hypothesis was wrong.

Specification Reference 4.6.3.1

Lamarck had different ideas from Darwin, but they were eventually rejected — leaving Darwin's ideas to form the theory of evolution.

Controversy

Darwin's theory of evolution by natural selection is widely accepted today. But when he proposed his theory in his book "On the Origin of Species" in 1859, his idea was very controversial for various reasons...

- It went against common religious beliefs about how life on Earth developed — it was the first plausible explanation for the existence of life on earth without the need for a "Creator" (God).
- Darwin couldn't explain why these new, useful characteristics appeared or how they were passed on from individual organisms to their offspring. But then he didn't know anything about genes or mutations — they weren't discovered 'til 50 years after his theory was published.
- There wasn't enough evidence to convince many scientists, because not many other studies had been done into how organisms change over time.

For these reasons, Darwin's idea was only accepted gradually, as more and more evidence came to light.

Figure 1: *Darwin.*

Lamarck

Darwin wasn't the only person who tried to explain evolution. There were different scientific hypotheses about evolution around at the same time, such as Lamarck's:

Jean-Baptiste Lamarck (1744-1829) argued that changes that an organism acquires during its lifetime will be passed on to its offspring — e.g. he thought that if a characteristic was used a lot by an organism, then it would become more developed during its lifetime, and the organism's offspring would inherit the acquired characteristic.

Example

Using Lamarck's theory, if a rabbit used its legs to run a lot (to escape predators), then its legs would get longer. The offspring of that rabbit would then be born with longer legs.

Figure 2: *Lamarck.*

Accepting or rejecting hypotheses

WORKING SCIENTIFICALLY

Often scientists come up with different hypotheses to explain similar observations. Scientists might develop different hypotheses because they have different beliefs (e.g. religious) or they have been influenced by different people (e.g. other scientists and their way of thinking)... or they just think differently.

The only way to find out whose hypothesis is right is to find evidence to support or disprove each one.

Tip: There's more about hypotheses on page 2.

Example

Lamarck and Darwin both had different hypotheses to explain how evolution happens.

In the end Lamarck's hypothesis was rejected because experiments didn't support his hypothesis. You can see it for yourself, e.g. if you dye a hamster's fur bright pink (not recommended), its offspring will still be born with the normal fur colour because the new characteristic won't have been passed on.

The discovery of genetics supported Darwin's idea because it provided an explanation of how organisms born with beneficial characteristics can pass them on (i.e. via their genes).

There's now so much evidence for Darwin's idea that it's an accepted hypothesis (a theory).

Tip: Evidence is really important when it comes to accepting or rejecting scientific hypotheses — if there's no experimental evidence to support a hypothesis, it won't become a theory.

WORKING SCIENTIFICALLY

Practice Questions — Application

Q1 A farmer clips the flight feathers on the wings of his chickens. This makes the feathers shorter and stops the birds being able to fly. The offspring of these birds develop normal flight feathers and are able to fly. Explain how this scenario helps to disprove Lamarck's hypothesis about evolution.

Q2 Anteaters feed on insects such as ants. They have evolved extremely long tongues, which help them to reach inside ant nests and get at the ants.

a) Suggest how Lamarck may have explained the evolution of long tongues in anteaters.

b) How would scientists explain the evolution of long tongues in anteaters using the idea of natural selection?

Tip: To answer Q2b), you'll need to look back at the detail of the theory of evolution by natural selection on page 287.

4. Selective Breeding

Selective breeding is all about getting the characteristics you want.

Learning Objectives:

- Know what selective breeding is.
- Know some characteristics that might be selectively bred for.
- Understand the process of selective breeding.
- Know that humans have been selectively breeding plants and animals for thousands of years.
- Be able to explain how selective breeding has affected crops and domesticated animals.
- Understand that selective breeding can lead to inbreeding, which can mean that some breeds are more susceptible to certain defects or diseases.

Specification Reference 4.6.2.3

What is selective breeding?

Selective breeding is when humans artificially select the plants or animals that are going to breed so that the genes for particular characteristics remain in the population. Organisms are selectively bred to develop features that are useful or attractive:

Examples

- Animals that produce more meat or milk.
- Crops with disease resistance.
- Dogs with a good, gentle temperament.
- Decorative plants with big or unusual flowers.

The process of selective breeding

This is the basic process involved in selective breeding:

1. From your existing stock, select the ones which have the characteristics you're after.
2. Breed them with each other.
3. Select the best of the offspring, and breed them together.
4. Continue this process over several generations, and the desirable trait gets stronger and stronger. Eventually, all the offspring will have the characteristic.

Tip: Selective breeding is also known as 'artificial selection'.

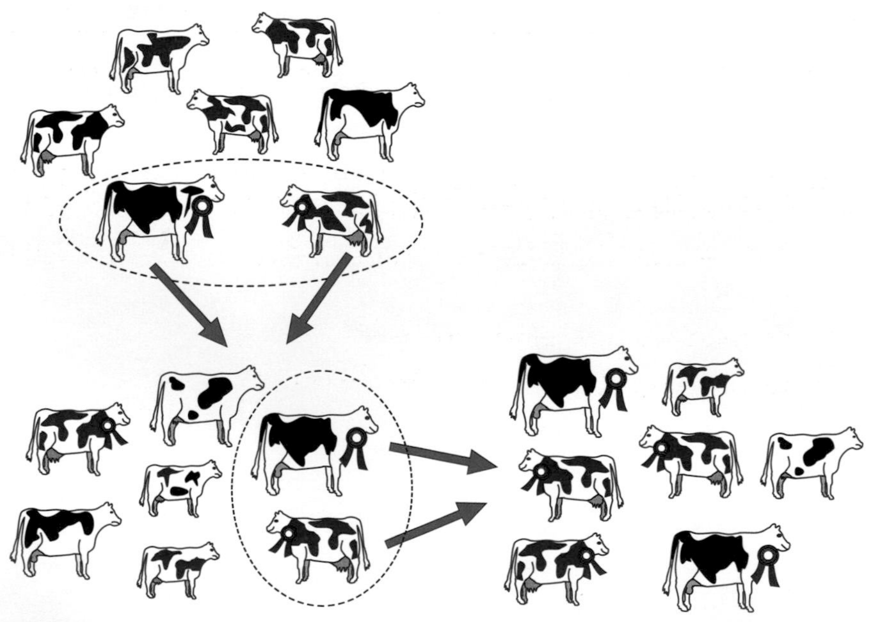

Figure 1: *Diagram showing the best offspring being selectively bred together over many generations — the desirable trait becomes more common.*

In agriculture (farming), selective breeding can be used to improve yields.

> **Example**
>
> To improve meat yields, a farmer could breed together the cows and bulls with the best characteristics for producing meat, e.g. large size. After doing this for several generations the farmer would get cows with a very high meat yield.

Exam Tip
You could be asked how an organism would be selectively bred for a different characteristic than this. Don't worry — the general process would still be exactly the same.

Selective breeding is nothing new — people have been doing it for thousands of years. It's how we ended up with edible crops from wild plants and how we got domesticated animals like cows and dogs.

Reducing the gene pool

The main problem with selective breeding is that it reduces the gene pool — the number of different alleles (forms of a gene) in a population. This is because the farmer keeps breeding from the "best" animals or plants — which are all closely related. This is known as **inbreeding**.

Inbreeding can cause health problems because there's more chance of the organisms inheriting harmful genetic defects when the gene pool is limited. Some dog breeds are particularly susceptible to certain defects because of inbreeding — e.g. pugs often have breathing problems.

There can also be serious problems if a new disease appears, because there's not much variation in the population. All the stock are closely related to each other, so if one of them is going to be killed by a new disease, the others are also likely to succumb to it.

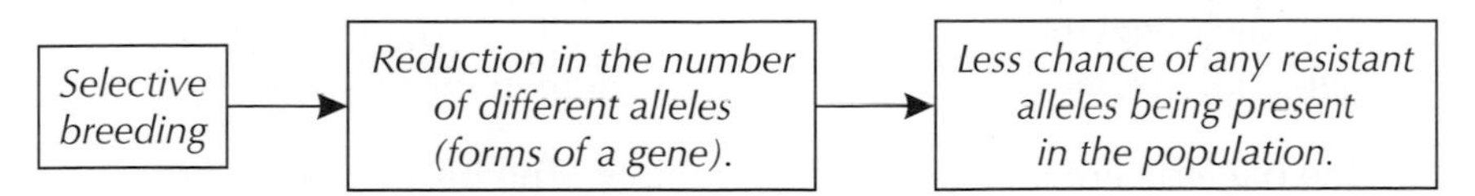

Figure 2: *Flow diagram showing a problem associated with reducing a gene pool.*

Practice Questions — Fact Recall

Q1 Give an example of a feature that an organism might be selectively bred for.

Q2 Outline the process of selective breeding.

Q3 Explain one problem associated with selective breeding.

5. Genetic Engineering

Humans are able to change an organism's genes through a process called genetic engineering. It's really quite clever...

What is genetic engineering?

The basic idea of genetic engineering is to transfer a gene responsible for a desirable characteristic from one organism's genome into another organism, so that it also has the desired characteristic. This is done by 'cutting out' the desired gene and transferring it into the cells of another organism.

How genetic engineering works Higher

1. A useful gene is isolated (cut) from one organism's genome using enzymes and is inserted into a **vector**.
2. The vector is usually a virus or a bacterial plasmid (a small ring of DNA), depending on the type of organism that the gene is being transferred to.
3. When the vector is introduced to the target organism, the useful gene is inserted into its cell(s).

Example **Higher**

The human insulin gene can be inserted into bacteria to produce human insulin:

1. The insulin gene is first cut out of human DNA using enzymes.
2. The same enzymes are then used to cut the bacterial DNA and different enzymes are used to insert the human insulin gene.
3. The bacteria are then allowed to multiply. The insulin they produce while they grow is purified and used by people with diabetes.

This is summarised in Figure 1.

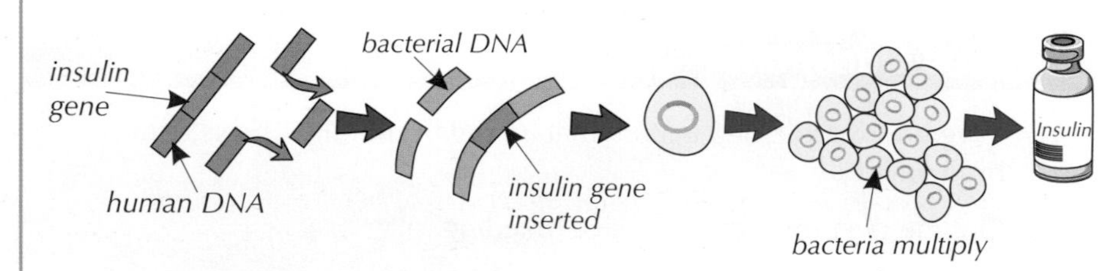

Figure 1: *Diagram showing how bacteria can be genetically engineered to produce insulin.*

In some cases, the transfer of the gene is carried out when the organism receiving the gene is at an early stage of development (e.g. egg or embryo). This means that the organism develops with the characteristic coded for by the gene.

Learning Objectives:

- Know that genetic engineering involves 'cutting out' and transferring a gene for a desired characteristic from one organism's genome to another organism's genome so that it also has the desired characteristic.
- H Understand that in genetic engineering, enzymes are used to cut out the desired gene, which is transferred to the target organism using a vector.
- H Know that 'new' genes are often inserted at an early stage of an organism's development.
- Know that genetic engineering can be used to produce human insulin, and also to produce crops with better fruit or resistance to disease, insects or herbicides.
- Know that genetically modified (GM) crops are ones that have had their genes modified.
- Understand why GM crops usually have an increased yield.
- Know that scientists are investigating genetic modification as a treatment for some inherited disorders.
- Be able to explain the benefits and concerns surrounding genetic engineering in medicine and GM crops.

Specification Reference 4.6.2.4

Uses of genetic engineering

Scientists use genetic engineering to do all sorts of things:

Examples

- Bacteria have been genetically modified to produce human insulin that can be used to treat diabetes.
- Genetically modified (GM) crops have had their genes modified, e.g. to improve the size and quality of their fruit, or make them resistant to disease, insects and herbicides (chemicals used to kill weeds).
- Sheep have been genetically engineered to produce substances, like drugs, in their milk that can be used to treat human diseases.
- Scientists are researching genetic modification treatments for inherited disorders caused by faulty genes, e.g. by inserting working genes into people with the disorder. This is called gene therapy.

Tip: Herbicides kill weeds, but they can also end up killing crops. 'Selective herbicides' are usually used to stop this from happening, but they don't always kill all the weeds. If a crop is herbicide-resistant, more effective herbicides can be used to get rid of weeds — without risk to the crop.

The issues surrounding genetic engineering

Genetic engineering is an exciting new area in science which has the potential for solving many of our problems (e.g. treating diseases, more efficient food production, etc.) but not everyone thinks it's a great idea.

There are worries about the long-term effects of genetic engineering — that changing a person's genes might accidentally create unplanned problems, which could then get passed on to future generations.

It's the same with GM crops...

The issues surrounding GM crops

Benefits

- On the plus side, the characteristics chosen for GM crops can increase the yield, making more food.
- People living in developing nations often lack nutrients in their diets. GM crops could be engineered to contain the nutrient that's missing. For example, 'golden rice' is a GM rice crop that contains beta-carotene — lack of this substance causes blindness.

Tip: 'Yield' just means the amount of product made. So the yield of a wheat field would be the amount of wheat produced. Insects eating the crop, disease, and competition with weeds all reduce crop yield.

Concerns

- Some people say that growing GM crops will affect the number of wild flowers (and so the population of insects) that live in and around the crops — reducing farmland biodiversity.
- Not everyone is convinced that GM crops are safe and some people are concerned that we might not fully understand the effects of eating them on human health. E.g. people are worried they may develop allergies to the food — although there's probably no more risk for this than for eating usual foods.
- A big concern is that transplanted genes may get out into the natural environment. For example, the herbicide resistance gene may be picked up by weeds, creating a new 'superweed' variety.

Tip: GM crops are already being grown elsewhere in the world (not the UK) often without any problems.

Practice Questions — Fact Recall

Q1 What does genetic engineering involve?

Q2 In genetic engineering, useful genes are sometimes transferred into animals and plants in the early stages of their development. Explain why.

Q3 Give an example of a characteristic that bacterial cells have been genetically engineered to have.

Q4 a) What does the 'GM' stand for in 'GM crop'?

b) Some GM crops are resistant to viruses. What else can they be made resistant to? Give two examples.

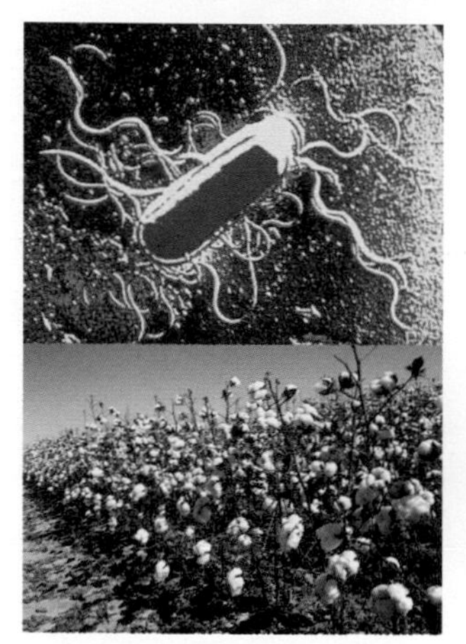

Figure 2: *A* B. thuringiensis *bacterium (top). GM cotton which produces the* Bt *crystal protein (bottom).*

Practice Questions — Application

B. thuringiensis is a species of bacteria. It produces a crystal protein, which is poisonous to insects when eaten. Some crop plants, including cotton and potatoes, have been genetically engineered to produce the *Bt* crystal protein.

Q1 Explain how enzymes would be used to make a cotton plant that can produce the *Bt* crystal protein.

Q2 Suggest a benefit of genetically engineering crop plants to produce the *Bt* crystal protein. Explain your answer.

Q3 Suggest a possible risk of growing crop plants that have been genetically engineered to produce the *Bt* crystal protein.

6. Cloning

This topic is all about how humans can control reproduction. Read on...

Learning Objectives:

- Know how plant clones can be made using tissue culture.
- Know that tissue culture allows rare plant species to be preserved, and also means that nurseries can quickly produce lots of stock.
- Know that plant clones can be made using cuttings, and that this is an older, simpler method than tissue culture.
- Know how cloning can be carried out by embryo transplants.
- Understand the process of adult cell cloning.

Specification Reference 4.6.2.5

What is cloning?

Cloning means making an exact (genetically identical) copy of an organism. It's basically just another term for asexual reproduction (see page 252). Plants which reproduce asexually are able to clone themselves naturally. Cloning can also be done artificially (by humans) — which is what the next few pages are all about.

Cloning plants

It's pretty easy to clone plants. Gardeners have been doing it for years. There are two ways of doing it:

1. Tissue culture

A few plant cells are put in a growth medium with hormones, and they grow into new plants — clones of the parent plant. These plants can be made very quickly, in very little space, and be grown all year. Tissue culture is used by scientists to preserve rare plants that are hard to reproduce naturally and by plant nurseries to produce lots of stock quickly.

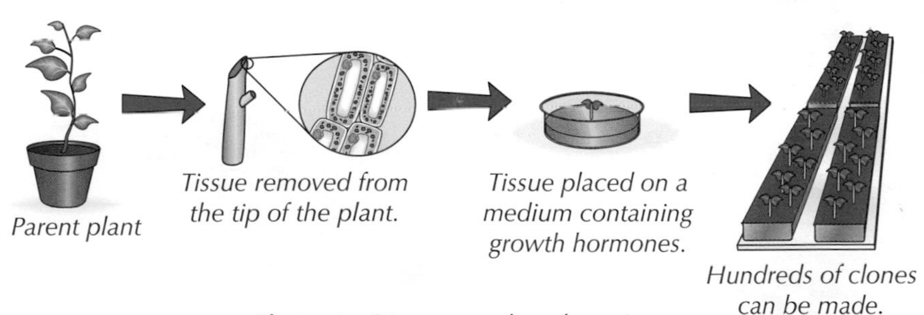

Figure 1: *Diagram to show how tissue culture can be used to clone plants.*

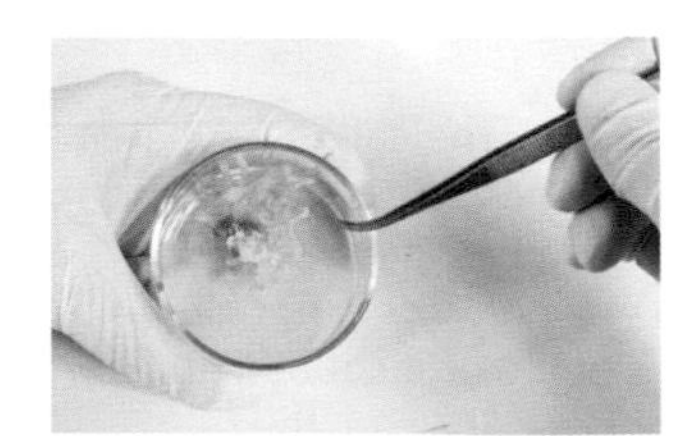

Figure 2: *A tobacco plant being grown by tissue culture.*

2. Cuttings

Gardeners can take cuttings from good parent plants, and then plant them to produce genetically identical copies (clones) of the parent plant. These plants can be produced quickly and cheaply. This is an older, simpler method than tissue culture.

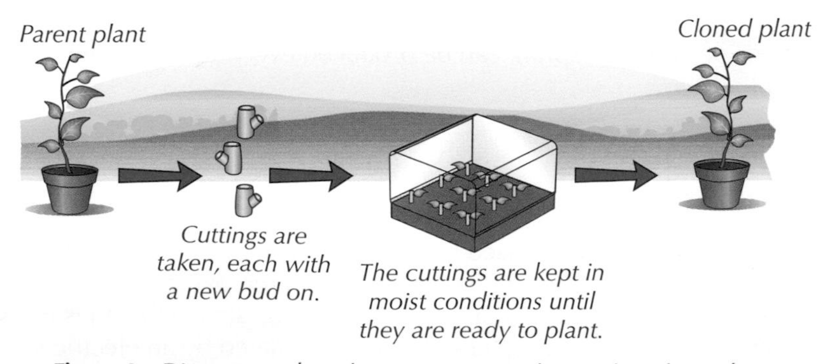

Figure 3: *Diagram to show how cuttings can be used to clone plants.*

> **Tip:** Once an egg cell has been fertilised and starts dividing, it becomes an embryo. The embryo grows and develops into a baby.

> **Tip:** A specialised cell is one that performs a specific function, e.g. a white blood cell defends against pathogens.

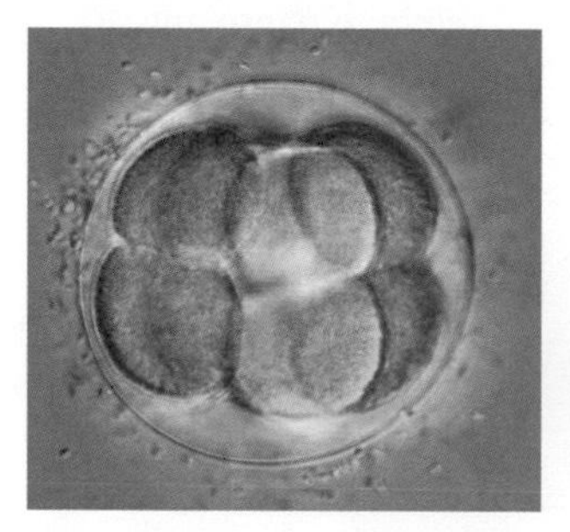

Figure 4: An early embryo under the microscope — each ball is an individual cell. It's at this stage that the cells are separated in an embryo transplant.

> **Tip:** The calves here are clones of each other and of the original embryo — but they're not clones of the original bull and cow. Instead, they contain a mixture of genes from both parents.

Embryo transplants

You can produce animal clones using embryo transplants. An embryo is created, then split many times in the early stages to form clones. The cloned embryos are then implanted (inserted) into host mothers to continue developing.

Example

Farmers can use embryo transplants to produce cloned offspring from their best bull and cow:

1. Sperm cells are taken from a prize bull and egg cells are taken from a prize cow. The sperm are then used to artificially fertilise an egg cell.
2. The embryo that develops is then split many times (to form clones) before any cells become specialised.
3. These cloned embryos can then be implanted into lots of other cows...
4. ...where they grow into calves (which will all be genetically identical to each other).

This is shown in Figure 5.

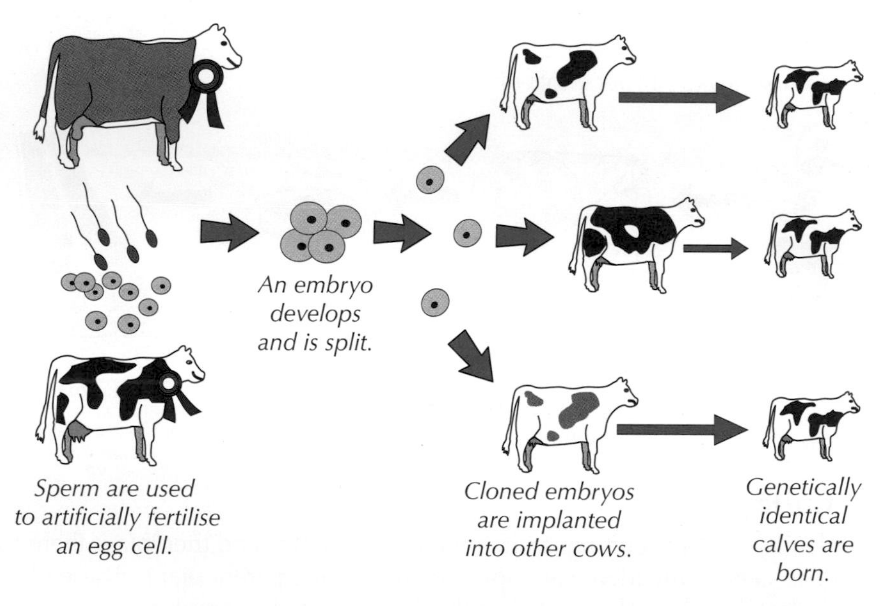

Figure 5: Diagram to show how genetically identical calves can be produced through embryo transplants.

Hundreds of "ideal" offspring can be produced every year from the best bull and cow.

Adult cell cloning

Adult cell cloning can also be used to make animal clones. Adult cell cloning involves taking an unfertilised egg cell and removing its nucleus. The nucleus is then removed from an adult body cell (e.g. skin cell) and is inserted into the 'empty' egg cell. The egg cell is then stimulated by an electric shock — this makes it divide, just like a normal embryo.

When the embryo is a ball of cells, it's implanted into the uterus (womb) of an adult female (the surrogate mother). Here the embryo grows into a clone of the original adult body cell (see Figure 7) as it has the same genetic information. This technique was used to create Dolly — the famous cloned sheep.

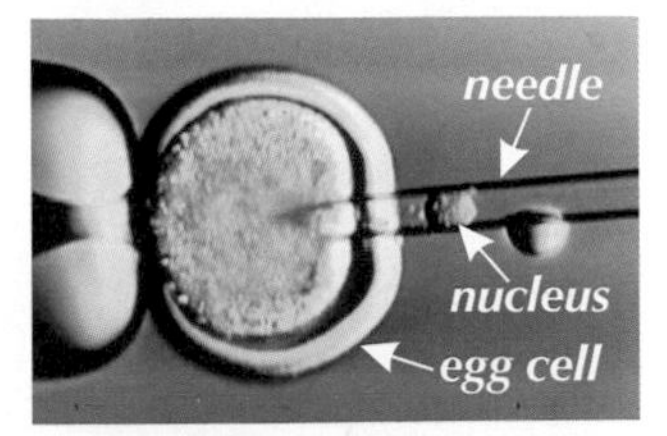

Figure 6: A nucleus being injected into an egg cell during adult cell cloning.

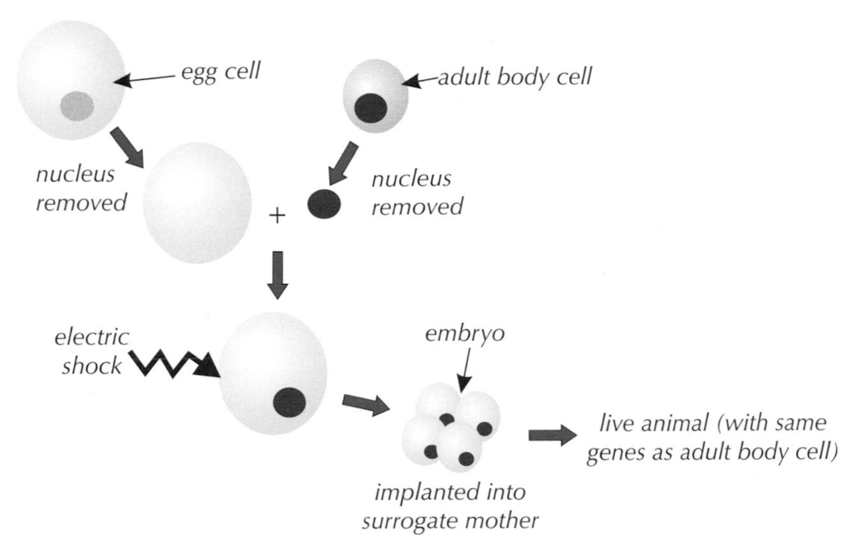

Figure 7: *Diagram to show how an animal can be cloned using adult cell cloning.*

Tip: Remember, it's only the nucleus from the adult body cell that gets inserted into the empty egg cell.

Issues surrounding cloning

There are many social, economic and ethical issues surrounding cloning.

Benefits

- Cloning quickly gets you lots of "ideal" offspring with known characteristics. This can benefit farmers, e.g. if a farmer has a cow that produces lots of milk, he could clone it and create a whole herd of cows that all produce lots of milk relatively quickly.
- The study of animal clones could lead to greater understanding of the development of the embryo, and of ageing and age-related disorders.
- Cloning could be used to help preserve endangered species.

Concerns

- Cloning gives you a "reduced gene pool" — this means there are fewer different alleles in a population. If a population are all closely related and a new disease appears, they could all be wiped out — there may be no allele in the population giving resistance to the disease.
- It's possible that cloned animals might not be as healthy as normal ones, e.g. Dolly the sheep had arthritis, which tends to occur in older sheep (but the jury's still out on if this was due to cloning).
- Some people worry that humans might be cloned in the future. If it was allowed, any success may follow many unsuccessful attempts, e.g. children born severely disabled. Also, you'd need to consider the human rights of the clone — the clone wouldn't have a say in whether it wanted to be a clone or not, so is it fair to produce one?

Tip: Alleles are different forms of a gene (see page 265).

Tip: It's currently illegal to clone a human in the UK.

Practice Questions — Fact Recall

Q1 Plants can be cloned by taking cuttings.

a) Give two benefits of taking cuttings to clone plants.

b) Give one other way of cloning plants.

Q2 Name two methods of animal cloning.

Practice Questions — Application

Figure 8: *The African black-footed cat.*

Read the following passage and then answer the questions that follow.

Scientists in the US have been experimenting with cloning the African black-footed cat — an animal classed as 'threatened' on endangered species lists. Their aim is to take a skin cell nucleus from an adult black-footed cat and implant it into an empty egg cell provided by a domestic cat. Once dividing, the embryo will be implanted in the domestic cat in order to develop.

Q1 Once the black-footed cat nucleus has been implanted into the egg cell of the domestic cat, how will it be stimulated to divide?

Q2 Would an embryo created by the scientists' method share any genetic material with the domestic cat? Explain your answer.

Although no black-footed kittens have yet been born as a result of this method, the US team behind the project are confident that it will help to save endangered species.

Q3 Suggest how each of the following could help to save the black-footed cat:

a) using common domestic cats as surrogate mothers rather than black-footed cats.

b) being able to take skin cells from black-footed cats that have already died as well as ones that are still alive.

Tip: Think about the number of domestic cats compared to how many black-footed cats there are likely to be.

Many wildlife conservationists are against using cloning to save endangered species. Some argue that the money would be better spent on conserving the places in which endangered species live.

Q4 Suggest two more concerns people may have of using cloning to save an endangered species.

7. Fossils

Fossils might sound a bit dull, but they're actually dead interesting. Without them, we wouldn't know that dinosaurs ever existed.

Learning Objectives:
- Know what fossils are.
- Know that fossils show how much organisms have evolved as life on Earth developed.
- Understand the main ways in which fossils form, including by gradual replacement by minerals, from casts and impressions, and from parts of organisms that haven't decayed.
- Understand why scientists have different views on how life on Earth began.

Specification Reference 4.6.3.5

What are fossils?

Fossils are the remains of organisms from many years ago, which are found in rocks. They provide the evidence that organisms lived ages ago. Fossils can tell us a lot about how much or how little organisms have changed (evolved) over time. By comparing fossils with species that are alive today, we can see that many of today's species have developed from much simpler organisms over millions of years.

How do fossils form?

Fossils form in rocks in one of three ways:

1. From gradual replacement by minerals

Things like teeth, shells, bones, etc., which don't **decay** easily, can last a long time when buried. They're eventually replaced by minerals as they decay, forming a rock-like substance shaped like the original hard part (see Figures 1 and 2).

The surrounding sediments also turn to rock, but the fossil stays distinct inside the rock and eventually someone digs it up. Most fossils are made this way.

Tip: When organisms die, they usually get broken down (digested) by microorganisms such as bacteria and fungi. This process is known as decay. Decay can take place at different rates. Very occasionally, it doesn't happen at all (see next page).

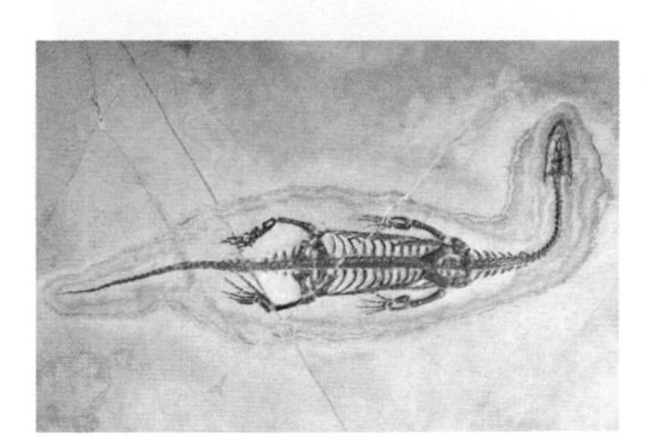

Figure 1: *Fossil of a reptile skeleton, dug out of the rock in China.*

Figure 2: *Fossils of ammonite shells. Ammonites were ancient marine organisms.*

Tip: The soft tissue of this reptile (Figure 1) and the soft bodies of the ammonites (Figure 2) decayed away quickly, so they haven't formed part of the fossils.

2. From casts and impressions

Sometimes, fossils are formed when an organism is buried in a soft material like clay. The clay later hardens around it and the organism decays, leaving a **cast** of itself. An animal's **burrow** (see Figure 3, next page) or a plant's **rootlet traces** can be preserved as casts. Things like **footprints** can also be pressed into soft materials, leaving an impression when the material hardens (see Figure 4, next page).

Tip: Rootlet traces are the impressions made by plant roots.

Figure 3: Fossilised burrows made in the sea bed by small organisms millions of years ago.

Figure 4: The footprint of an early human-like primate, preserved in hardened volcanic ash.

3. From preservation in places where no decay happens

The microbes involved in decay need the right conditions in order to break down material — that means plenty of oxygen (for aerobic respiration, see p. 173), enough moisture, and the right temperature and pH for their enzymes to work properly (see pages 116-117). If the conditions aren't right, decay won't take place and the dead organism's remains will be preserved.

Tip: As it happens, the conditions needed for decay are usually right — so it's very rare for dead organisms to be preserved in this way.

Examples

- In amber (a clear yellow 'stone' made from fossilised tree sap) and tar pits there's no oxygen or moisture so decay microbes can't survive (see Figure 5).
- In glaciers it's too cold for the decay microbes to work.
- Peat bogs are too acidic for decay microbes. Whole human bodies have been preserved as fossils in peat bogs (see Figure 6).

Figure 5: Fossil of a midge, preserved in amber.

Figure 6: Photo of the 'Tollund Man'. This well preserved fossil of a man was discovered in a bog.

Tip: The 'Tollund Man' lived over 2000 years ago. He was so well preserved in the bog that scientists were able to tell what he ate for his last meal from the contents of his stomach (barley gruel, yum).

The origins of life on Earth

There are various **hypotheses** suggesting how life first came into being, but no one really knows the answer.

Maybe the first life forms came into existence in a primordial swamp (or under the sea) here on Earth. Maybe simple organic molecules were brought to Earth on comets — these could have then become more complex organic molecules, and eventually very simple life forms.

Tip: 'Primordial' means 'original' or 'first'. So primordial swamps were the first swamps.

Tip: Organic molecules are molecules containing carbon, e.g. amino acids.

These hypotheses can't be supported or disproved because there's a lack of good, **valid evidence**. There's a lack of evidence because:

- Scientists believe many early organisms were soft-bodied, and soft tissue tends to decay away completely, without forming fossils.
- Fossils that did form millions of years ago may have been destroyed by **geological activity**, e.g. the movement of tectonic plates may have crushed fossils already formed in the rock.

This means that the **fossil record** is incomplete — in other words, we don't have fossils of every organism or even every type of organism that has ever lived.

Tip: Validity is explained on page 9.

Tip: Tectonic plates are the large 'pieces' of rock that make up the Earth's crust.

Practice Questions — Fact Recall

Q1 What is a fossil?

Q2 What do fossils provide evidence of?

Q3 Explain why there is a lack of valid evidence to support any hypothesis about how life on Earth began.

Practice Question — Application

Q1 Suggest how each of the following fossils was formed:

A

Fossilised animal burrows preserved in rock.

B

A fossilised snail shell.

C

A fossilised leaf imprint in rock.

D

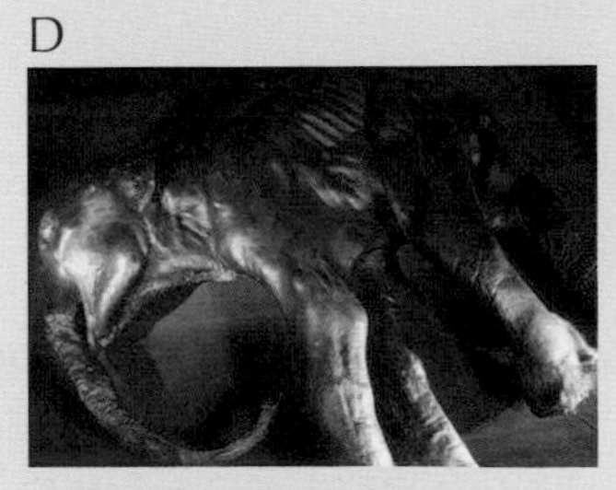

A fossilised baby mammoth, found in frozen ground in Siberia.

8. Speciation

Life on Earth is changing all the time. Species die out, leaving fossils if scientists are lucky. New species also emerge — this happens by speciation...

Learning Objectives:

- Understand the steps that lead to the formation of a new species.
- Know that Alfred Russel Wallace carried out work which contributed, along with newer evidence, to our understanding of speciation.
- Know that Alfred Russel Wallace independently came up with the idea of natural selection, and published work with Darwin in 1858.
- Know that Darwin's work with Wallace prompted Darwin to publish 'On the Origin of Species' in 1859.
- Know that Wallace worked all over the world to gain evidence for natural selection, such as his research into animal warning colours.

Specification Reference 4.6.3.2

Speciation

A **species** is a group of similar organisms that can reproduce to give fertile offspring (offspring that are able to breed themselves). **Speciation** is the development of a new species. Speciation occurs when populations of the same species become so different that they can no longer successfully interbreed to produce fertile offspring.

Isolation

Isolation is where populations of a species are separated. This can happen due to a physical barrier. E.g. floods and earthquakes can cause barriers that geographically isolate some individuals from the main population. Isolation can eventually lead to speciation.

Example

The chimpanzee and bonobo are two separate species of ape that evolved from a common ancestor. It's thought that two populations of the ancestor became isolated from each other when the Congo River formed. The population to the north of the river became chimpanzees and the population to the south became bonobos (see Figure 1).

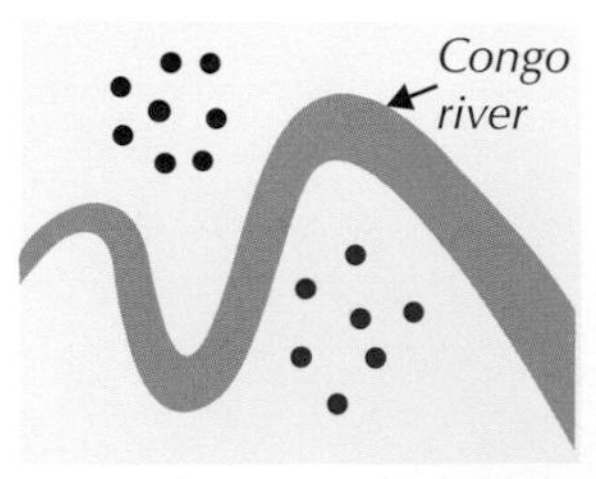

Figure 1: *Diagram to show the isolation, and so speciation, of chimp and bonobo populations.*

Natural selection

When two populations are isolated due to a physical barrier, conditions on either side of the barrier will be slightly different, e.g. the climate may be different. Because the environment is different on each side, different characteristics will become more common in each population due to natural selection. Here's how:

1. Each population shows genetic variation because they have a wide range of **alleles**.
2. In each population, individuals with characteristics that make them better adapted to their environment have a better chance of survival and so are more likely to breed successfully.
3. So the alleles that control the beneficial characteristics are more likely to be passed on to the next generation.

Eventually, individuals from the different populations will have changed so much that they won't be able to breed with one another to produce fertile offspring. This means the two groups have become separate species — in other words, speciation has occurred.

The whole process is shown in Figure 2.

Tip: Alleles are genetic variants — different forms of a gene. See page 265 for more.

Key:

- ● = *individual organism*
- ● = *individual organism of new species*

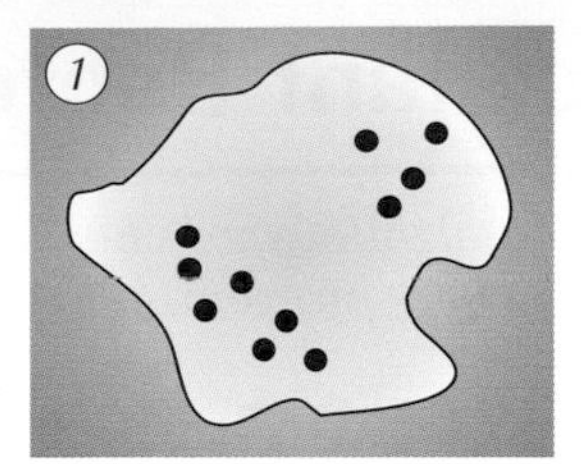

Two populations of the same species.

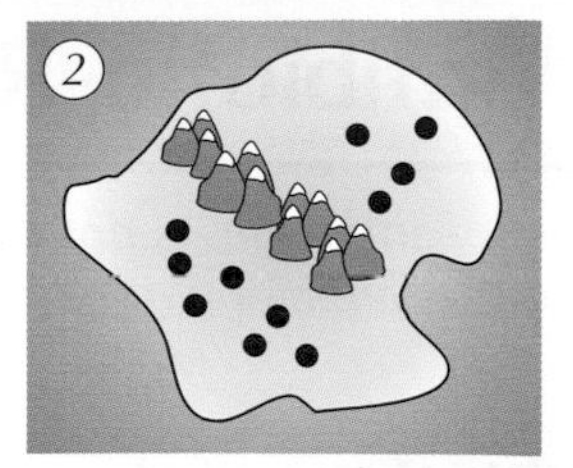

Physical barriers separate populations.

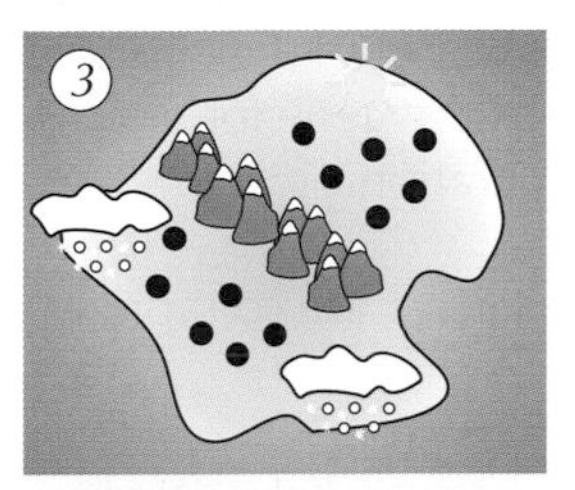

Populations adapt to new environments.

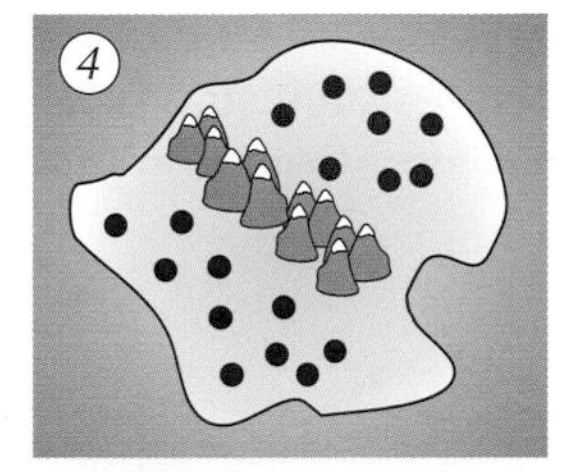

Development of a new species.

Figure 2: *Diagram to show how speciation occurs.*

Tip: Populations can sometimes split up into more than two groups — this could result in the formation of several new species.

Exam Tip
In the exam, you could be asked to apply your knowledge of how speciation occurs to explain the formation of any new species. So make sure you know the main stages involved — isolation, then genetic variation and natural selection lead to speciation.

Alfred Russel Wallace

Alfred Russel Wallace was a scientist working at the same time as Charles Darwin. He was one of the early scientists working on the idea of speciation. His observations greatly contributed to how we understand speciation today. Our current understanding developed as more evidence became available over time.

During his career, Wallace independently came up with the idea of natural selection and published work on the subject together with Darwin in 1858. This then prompted Darwin to publish 'On the Origin of Species' in 1859.

Observations made by Wallace as he travelled the world provided lots of evidence to support the theory of evolution by natural selection. For example, he realised that warning colours are used by some species (e.g. butterflies) to deter predators from eating them and that this was an example of a beneficial characteristic that had evolved by natural selection. It's this work on warning colours and his work on speciation that he's most famous for.

Tip: There's more on Darwin and natural selection on p. 286-287.

Practice Questions — Fact Recall

Q1 What is speciation?

Q2 Why do populations of the same species show variation?

Q3 How can you tell that speciation has taken place?

Q4 When did Wallace publish joint work with Darwin?

Q5 What is the name of the book that Darwin published in 1959 as a result of working with Wallace?

Q6 As well as his work on speciation, which research is Alfred Russel Wallace best known for?

9. Antibiotic-Resistant Bacteria

The use of antibiotics is not always straightforward. Because bacteria can mutate, antibiotic resistance can be a problem...

Learning Objectives:

- Know that bacteria are able to evolve quickly because they reproduce rapidly.
- Understand how antibiotic-resistant strains of bacteria develop.
- Know that antibiotic-resistant strains spread because people don't have immunity to it and there's no effective treatment.
- Know that MRSA is an antibiotic-resistant bacteria.
- Know ways to reduce the rate that antibiotic-resistant strains develop.
- Understand why the development of new antibiotics is not likely to keep up with the rate that new resistant strains emerge.

Specification Reference 4.6.3.7

Antibiotic resistance

Like all organisms, bacteria sometimes develop random mutations (see p. 249) in their DNA. These can lead to changes in the bacteria's characteristics, e.g. being less affected by a particular antibiotic. This can lead to antibiotic-resistant strains forming as the gene for antibiotic resistance becomes more common in the population.

To make matters worse, because bacteria are so rapid at reproducing, they can evolve quite quickly.

For the bacterium, the ability to resist antibiotics is a big advantage. Non-resistant bacteria will be killed by antibiotics, but a resistant bacterium is better able to survive, even in a host who's being treated to get rid of the infection. This means it lives for longer and reproduces many more times. This increases the population size of the antibiotic-resistant strain.

Antibiotic-resistant strains are a problem for people who become infected with these bacteria because they aren't immune to the new strain and there is no effective treatment. This means that the infection easily spreads between people. Sometimes drug companies can come up with a new antibiotic that's effective, but 'superbugs' that are resistant to most known antibiotics are becoming more common.

Examples

- MRSA (methicillin-resistant *Staphylococcus aureus*) is a relatively common 'superbug' that's really hard to get rid of. It often affects people in hospitals and can be fatal if it enters their bloodstream.
- Multi-drug-resistant TB (also known as MDR-TB) is a form of TB that has developed resistance to multiple antibiotics, making it hard to treat. Scientists in China have reported an MDR-TB epidemic there — about 63 000 Chinese people develop MDR-TB each year.

Tip: The gene for antibiotic resistance becomes more common in the population because of natural selection — there's more about how this happens on page 287.

Tip: TB is a very serious lung disease caused by a bacterial infection.

The spread of antibiotic resistance

For the last few decades, we've been able to deal with bacterial infections pretty easily using antibiotics. The death rate from infectious bacterial diseases (e.g. pneumonia) has fallen dramatically.

But the problem of antibiotic resistance is getting worse — partly because of the overuse and inappropriate use of antibiotics.

Examples

- Doctors prescribing antibiotics for non-serious conditions.
- Doctors prescribing antibiotics for infections caused by viruses.

Tip: Antibiotics don't kill viruses.

The more often antibiotics are used, the bigger the problem of antibiotic resistance becomes, so it's important that doctors only prescribe antibiotics when they really need to.

It's not that antibiotics actually cause resistance — they create a situation where naturally resistant bacteria have an advantage and so increase in numbers. If they're not doing you any good, it's pointless to take antibiotics — and it could be harmful for everyone else.

It's also important that you take all the antibiotics a doctor prescribes for you. Lots of people stop taking their antibiotics as soon as they feel better, but this can increase the risk of antibiotic-resistant bacteria emerging. Taking the full course makes sure that all the bacteria are destroyed, which means that there are none left to mutate and develop into antibiotic-resistant strains.

In farming, antibiotics can be given to animals to prevent them becoming ill and to make them grow faster. This can lead to the development of antibiotic-resistant bacteria in the animals which can then spread to humans, e.g. during meat preparation and consumption. Increasing concern about the overuse of antibiotics in agriculture has led to some countries restricting their use.

The increase in antibiotic resistance has encouraged drug companies to work on developing new antibiotics that are effective against the resistant strains. Unfortunately, the rate of development is slow, which means we're unlikely to be able to keep up with the demand for new drugs as more antibiotic-resistant strains develop and spread. It's also a very costly process.

Practice Questions — Fact Recall

Q1 Why can bacteria evolve rapidly?

Q2 Name one type of antibiotic-resistant bacteria.

Q3 Why is the development of new antibiotics unlikely to keep up with the appearance of new antibiotic-resistant strains of bacteria?

Practice Questions — Application

Q1 James has a mild bacterial infection. James' doctor tells him it will clear up on its own and does not prescribe him antibiotics. Suggest why James' doctor did this.

Q2 Many strains of the bacteria *Streptococcus pneumoniae* are now resistant to the antibiotic penicillin. Describe how populations of penicillin-resistant *Streptococcus pneumoniae* may have increased.

10. Classification

Classification means sorting things into groups. Like organisms, for example...

Learning Objectives:
- Know that the system of classification proposed by Carl Linnaeus grouped organisms depending on their characteristics and structure.
- Know that the Linnaean system divided organisms into kingdom, phylum, class, order, family, genus and species.
- Be able to use given information to demonstrate understanding of the Linnaean system.
- Know that as microscopes improved and scientists understood more about biochemical processes, new classification models were proposed.
- Know that Carl Woese proposed the three-domain system, where organisms are divided into Archaea, Bacteria and Eukaryota, and know what these domains contain.
- Know that in the binomial system, organisms are named using their genus and species.
- Understand what evolutionary trees are and how they are worked out.
- Be able to read from and interpret evolutionary trees.

Specification Reference
4.6.3.5, 4.6.4

The Linnaean system

Traditionally, organisms have been classified according to a system first proposed in the 1700s by Carl Linnaeus, which groups living things according to their characteristics and the structures that make them up.

In this system (known as the Linnaean system), living things are first divided into kingdoms (e.g. the plant kingdom). The kingdoms are then subdivided into smaller and smaller groups — phylum, class, order, family, genus, species.

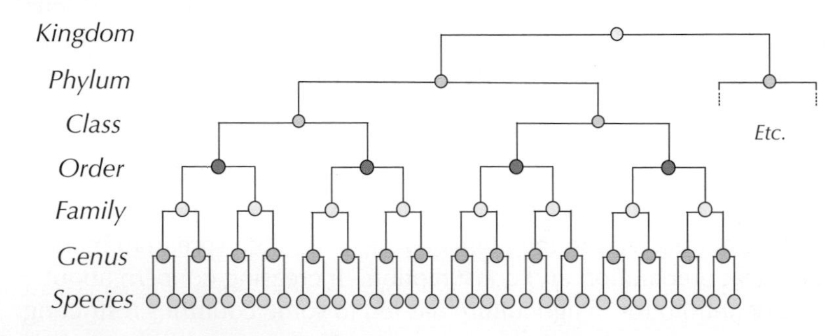

***Figure 1:** Diagram to show the Linnaean system of classification.*

Example

Using the Linnaean system, humans are classified as follows:

Kingdom = *Animalia*, Phylum = *Chordata*, Class = *Mammalia*, Order = *Primates*, Family = *Hominidae*, Genus = *Homo*, Species = *sapiens*.

The three-domain system

As knowledge of the biochemical processes taking place inside organisms developed and microscopes improved (which allowed us to find out more about the internal structures of organisms), scientists put forward new models of classification.

In 1990, Carl Woese proposed the three-domain system. Using evidence gathered from new chemical analysis techniques such as RNA sequence analysis, he found that in some cases, species thought to be closely related in traditional classification systems are in fact not as closely related as first thought.

Tip: Biochemical processes are all the chemical reactions that go on within the body.

In the three-domain system, organisms are first of all split into three large groups called domains:

1. Archaea

Organisms in this domain are primitive bacteria. They're often found in extreme places such as hot springs and salt lakes.

2. Bacteria

This domain contains true bacteria like *E. coli* and *Staphylococcus*. Although they often look similar to Archaea, there are lots of biochemical differences between them.

3. Eukaryota

This domain includes a broad range of organisms including fungi, plants, animals and protists (page 130).

These are then subdivided into smaller groups — kingdom, phylum, class, order, family, genus, species.

Exam Tip
Make sure you know what each of the domains includes in the three-domain system, as you may be asked to describe their features in the exam.

The binomial system

In the binomial system, every organism is given its own two-part Latin name.

The first part refers to the genus that the organism belongs to. This gives you information on the organism's ancestry. The second part refers to the species.

Tip: Genus and species are the final two groups in the Linnaean and three-domain systems of classification.

Example

Humans are known as *Homo sapiens*.
'*Homo*' is the genus and '*sapiens*' is the species.

The binomial system is used worldwide and means that scientists in different countries or who speak different languages all refer to a particular species by the same name — avoiding potential confusion.

Exam Tip
You might have to apply your knowledge of how the Linnaean system works to information given to you in the exam — don't worry, just make sure you understand what's going on on the previous page.

Evolutionary trees

Evolutionary trees show how scientists think different species are related to each other.

They show common ancestors and relationships between species. The more recent the common ancestor, the more closely related the two species — and the more characteristics they're likely to share.

Scientists analyse lots of different types of data to work out evolutionary relationships. For living organisms, they use the current classification data (e.g. DNA analysis and structural similarities). For extinct species, they use information from the fossil record (see page 288).

Tip: An evolutionary tree is a bit like a family tree — but instead of showing a mum, dad, grandparents, etc., it shows ancestors that existed many thousands or even millions of years ago.

Examples

- This evolutionary tree shows that whales and dolphins have a recent common ancestor so are closely related. They're both more distantly related to sharks:

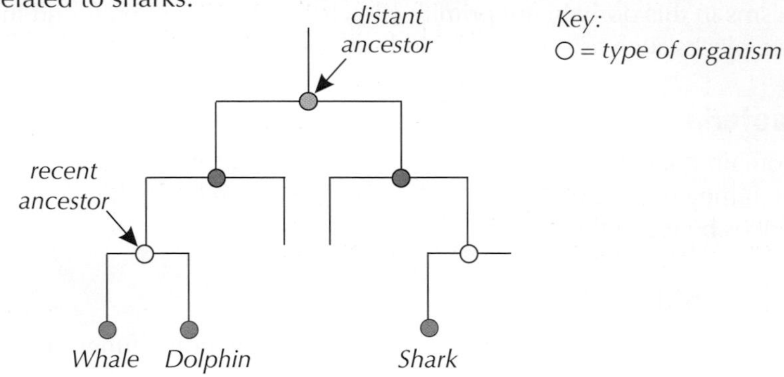

- This evolutionary tree shows the relationships between humans and some of the great apes. Each point at which the lines meet is a common ancestor.

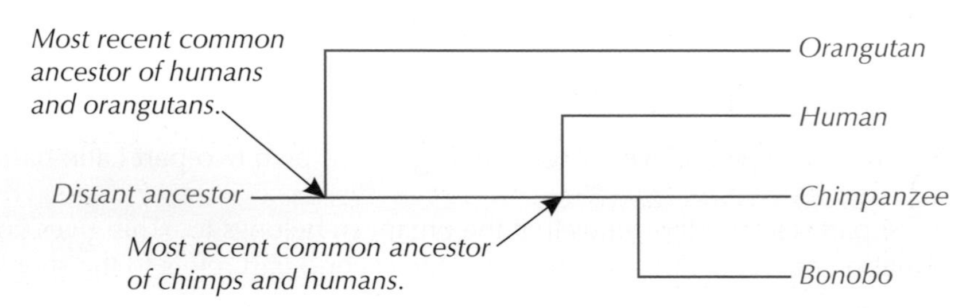

You can see that humans are more closely related to chimpanzees than to orangutans because they share a more recent common ancestor.

Exam Tip
You may need to interpret other graphs, tables or charts relating to evolutionary data in the exam.

Tip: The distant ancestor is the 'oldest' organism on this evolutionary tree (i.e. it evolved first).

Practice Question — Application

Q1 The tree below shows the evolutionary relationships between some of the big cats.

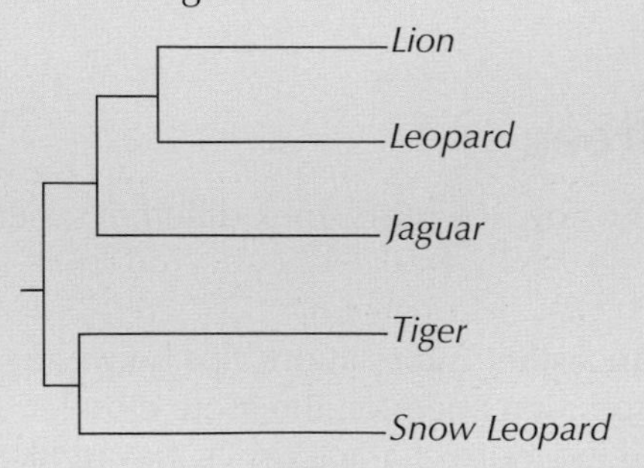

a) Which animal is most closely related to:

i) the leopard? ii) the tiger?

b) Which animal evolved first, the jaguar or the lion?

c) Do the lion and the snow leopard share a common ancestor?

Topic 6c Checklist — Make sure you know...

Variation

- [ ] That variation within a species is the differences in characteristics between individuals, and that these differences are caused by genetic variation, environmental variation, or a combination of both.
- [ ] That there is usually a lot of genetic variation within a population of a species.
- [ ] That mutations (changes to DNA) occur continuously, and lead to new genetic variants.
- [ ] That although most genetic variants have no effect on phenotype, some can influence phenotype and a few can lead to a new phenotype. This new phenotype can become more common if it makes the organism more suited to its environment.

Evolution

- [ ] The theory of evolution — all of today's species have evolved from simple life forms that first started to develop over three billion years ago.
- [ ] Darwin's theory of evolution by natural selection, and the information he used to come up with it.
- [ ] How evolution by natural selection happens.
- [ ] That evolution is the changing of a species' inherited characteristics over time.
- [ ] That Darwin's theory has largely been accepted, and the reasons why.
- [ ] That if two populations of one species change so much that they can no longer interbreed to produce fertile offspring, they have become two new species.
- [ ] What is meant by extinction, and the reasons why species become extinct.

Ideas About Evolution

- [ ] That Darwin's ideas of evolution were published in "On the Origin of Species" in 1859, and why his theory of natural selection wasn't accepted straight away.
- [ ] That there have been other hypotheses to explain how evolution occurs, including Lamarck's (which was based on the idea that 'acquired characteristics' could be inherited).

Selective Breeding

- [ ] What selective breeding is, including examples of its uses, and how it is done.
- [ ] That humans have been selectively breeding plants and animals for thousands of years, and how this has affected food plants and domesticated animals.
- [ ] The problems linked to selective breeding.

Genetic Engineering

- [ ] That genetic engineering involves 'cutting out' and transferring a gene for a desired characteristic from one organism's genome to another organism's genome, so that it also has that characteristic.
- [ ] **H** The processes used in genetic engineering.

cont...

- [ ] Some uses of genetic engineering, including the possibility of treating inherited disorders, and that GM is short for genetically modified (e.g. GM crops).
- [ ] The reasons for why GM crops usually have an increased yield.
- [ ] The benefits and concerns of using genetic engineering in medicine and GM crop production.

Cloning

- [ ] That plants can be cloned through tissue culture (a method of producing whole plants from plant cells grown on a growth medium), and why this is useful.
- [ ] That plants can be cloned using cuttings, which allows plants to be produced quickly and cheaply.
- [ ] How cloned animals can be produced using embryo transplants and by adult cell cloning.

Fossils

- [ ] That fossils are the remains of organisms from many years ago, which are found in rocks.
- [ ] That fossils provide evidence of how species evolved.
- [ ] That fossils can be made when hard body parts (that don't decay easily) are gradually replaced by minerals, from casts and impressions, and from preservation in places where no decay happens.
- [ ] That there are various hypotheses as to how life on Earth began, but there's a lack of valid evidence because there are so few fossils of early organisms — their soft bodies decayed easily and didn't form fossils, and geological activity has destroyed many fossils that did form.

Speciation

- [ ] How, following isolation, genetic variation and natural selection can lead to speciation.
- [ ] Who Alfred Russel Wallace was, that he published joint work with Darwin in 1858 (which prompted Darwin to publish "On the Origin of Species" in 1859), and know the areas that he researched.

Antibiotic-Resistant Bacteria

- [ ] That bacteria can evolve quickly because they reproduce rapidly.
- [ ] How antibiotic-resistant strains of bacteria (e.g. MRSA) develop, and why they spread.
- [ ] Ways to reduce the rate that antibiotic-resistant strains develop.
- [ ] Why antibiotic development is unlikely to keep up with the rate that new resistant strains emerge.

Classification

- [ ] About the classification system proposed by Carl Linnaeus and that as microscopes and scientific knowledge improved, new classification models were proposed — such as that of Carl Woese.
- [ ] That in the binomial system, organisms are named using their genus and their species.
- [ ] What evolutionary trees are, how they're worked out, and how to read and interpret data from them.

Exam-style Questions

1 The evolutionary tree shows how some groups of animals, including snakes, are related.

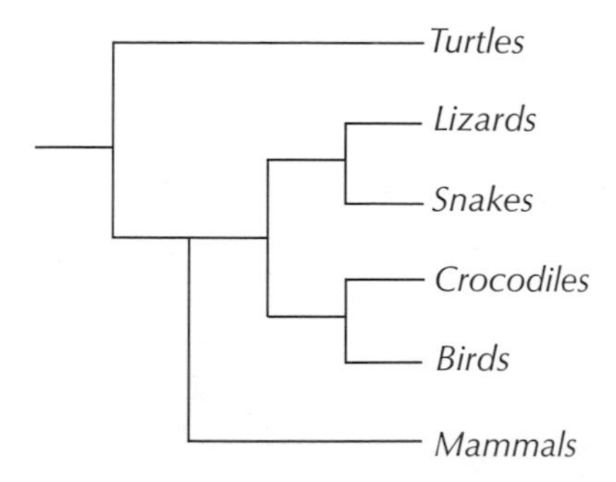

1.1 Which of the following groups of animals are most closely related?

A turtles and lizards

B snakes and crocodiles

C crocodiles and birds

D birds and mammals

(1 mark)

1.2 According to the tree, which group of animals evolved first?

(1 mark)

The picture shows a type of snake called a Sinaloan milk snake.

The milk snake is relatively harmless, but it has evolved to have a very similar pattern and colouring to the deadly poisonous coral snake, which lives in the same environment. The coral snake's colouring warns predators that it is dangerous to eat.

1.3 Using the idea of natural selection, explain how the milk snake may have evolved to have a similar colouring to the coral snake.

(3 marks)

1.4 The theory of evolution by natural selection was proposed by Charles Darwin. Darwin's ideas about natural selection were only gradually accepted.

Give **three** reasons why this was the case.

(3 marks)

2 The photograph below shows part of the Grand Canyon in Arizona, USA. The canyon is several kilometres wide and nearly two kilometres deep in places.

Two closely-related, but separate squirrel species live either side of the canyon. They are thought to be descended from a single original squirrel species, present in the area before the canyon formed.

2.1 Use your knowledge of speciation to explain how the two separate squirrel species formed.

(5 marks)

2.2 A bird species is found on both sides of the canyon.

Suggest why the formation of the canyon did not cause the bird species to form two separate species.

(2 marks)

Fossils of shelled organisms are often found in the Grand Canyon.

2.3 Suggest how these fossils may have formed.

(2 marks)

2.4 Why didn't the soft bodies of these shelled organisms form part of the fossils?

(1 mark)

2.5 Give **one** way in which some fossils may have been destroyed over time.

(1 mark)

1. Competition

There is only a limited amount of resources to go around, so organisms have to compete with each other to get what they need from their environment.

Learning Objectives:
- Know what an ecosystem is.
- Know the levels of organisation that make up an ecosystem.
- Know that organisms need resources from both their surroundings and other organisms in their environment to survive and reproduce.
- Know that organisms compete with each other for the resources they need, and know what animals and plants compete for.
- Be able to identify the resources being competed for in a habitat when given relevant information.
- Be able to read from graphs, charts and tables relating to how organisms in a community interact with each other and interpret the information presented.
- Know that species in a community are interdependent — they rely on each other for the resources they need.
- Know that the removal of one species can have wide-reaching effects on the whole community.
- Know what a stable community is.

Specification Reference 4.7.1.1

Important definitions

Topic 7 will make a lot more sense if you become familiar with these terms:

Habitat	The place where an organism lives.
Population	All the organisms of one species living in a habitat.
Community	The populations of different species living in a habitat.
Abiotic factors	Non-living factors of the environment, e.g. temperature.
Biotic factors	Living factors of the environment, e.g. food.
Ecosystem	The interaction of a community of living organisms (biotic) with the non-living (abiotic) parts of their environment.

Competition for resources

Organisms need things from their environment and from other organisms in order to survive and reproduce. Plants need light and space, as well as water and mineral ions (nutrients) from the soil. Animals need space (territory), food, water and mates. Organisms compete with other species (and members of their own species) for the same resources.

Examples

- Weeds compete with crop plants for light and nutrients. Farmers use weedkillers to kill weeds so there is less competition and the crops get all the light and nutrients, which enables them to grow better.
- Lions and hyenas share the same habitat in Africa. They compete with each other for food, water and territory, which results in frequent aggression between the two species.
- Male peacocks compete with each other for mates by displaying their eye-catching tail feathers to females during mating season.

Identifying competition

In the exam you might be given information about several different organisms and the habitat that they share and be asked to analyse the information and work out what resources they are in competition for. Take a look at the example on the next page.

Exam Tip
You need to be able to read from graphs, charts and tables relating to how organisms in a community interact with each other, and make sense of the data.

Exam Tip
Questions like these can be quite tricky because the answer isn't always obvious straight away. If you think logically and work through the information, you should be able to work out the right answer.

Exam Tip
The introduction to a question can sometimes contain vital information. Make sure you read everything you're given carefully or you might throw away easy marks.

Tip: Organisms that are better adapted to their environment are better at competing for resources and will be more likely to survive. See page 321 for more on adaptations.

Example

In 2007, an invasive species of bush was introduced to a riverbank and was in competition with a similar, native species. The river usually floods the riverbank each January, covering it in nutrient-rich silt. This didn't occur in 2012. The graph on the right shows how the populations of the two species changed in an area of the riverbank over an 8 year period. The table below gives some information about the bushes.

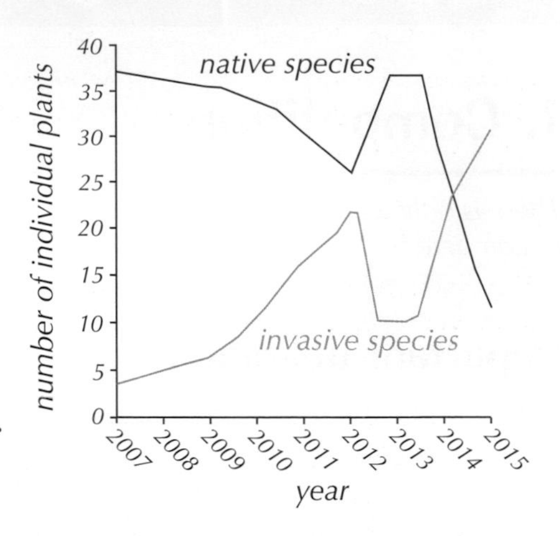

Feature	Invasive species	Native species
Average growth rate of main stem	0.8 cm/day	0.4 cm/day
Average surface area of leaves	11.5 cm^2	6.2 cm^2
Average volume of soil where roots have spread	729 cm^3	5832 cm^3

Between 2007 and 2012, the invasive species outcompeted the native species. Use the information above and your own knowledge of plants to suggest the resource that the invasive species was more successful at competing for during this time. Explain your answer.

You need to take a close look at the information and identify any differences in it. You'll notice that the invasive species was able to grow much faster on average than the native species and has leaves with a much bigger surface area. Leaves with a bigger surface area can absorb more light for photosynthesis and, coupled with the extra height, block the light from getting to shorter plants like the native bush. So the resource that the invasive species was much better at competing for is likely to have been light.

Between January 2012 and mid-2013, the pattern in the population sizes changed. This was due to competition between the species for a particular resource. Suggest which resource may have been responsible for the changes in population sizes during this period. Explain your answer.

During this time period, the population of the native species increased, while the population of the invasive species decreased. During the same time period, the nutrient content of the soil is likely to have been lower than normal because the river didn't flood in January 2012 (so the riverbank didn't receive any nutrient-rich silt). As the roots of the native species spread much further than those of the invasive species, the native species is likely to have been able to absorb more nutrients from the soil, enabling it to outcompete the invasive species while the soil nutrient content was low. So the short supply of nutrients was likely to have been responsible for the changes in population sizes.

Interdependence of species

In a community, each species depends on other species for things such as food, shelter, pollination and seed dispersal — this is called **interdependence**.

The interdependence of all the living things in an ecosystem means that any major change in the ecosystem (such as one species being removed) can have far-reaching effects.

Example

The diagram below shows part of a food web (a diagram of what eats what) from a stream.

Pike
Frog
Stickleback
Diving beetle
Waterboatman
Water spider
Stonefly larvae
Blackfly larvae
Mayfly larvae
Algae

Stonefly larvae are particularly sensitive to pollution. Suppose pollution killed them in this stream. The table below shows some of the effects this might have on some of the other organisms in the food web.

Organism	Effect of loss of stonefly larvae	Effect on population
Blackfly larvae	Less competition for algae	Increase
	More likely to be eaten by predators	Decrease
Water spider	Less food	Decrease
Stickleback	Less food (if water spider or mayfly larvae numbers decrease)	Decrease

Tip: Remember that food webs are very complex and that these effects are difficult to predict accurately.

In some communities, all the species and environmental factors are in balance so that the population sizes are roughly constant (they may go up and down in cycles — see pages 323-324). These are called **stable communities**. Stable communities include tropical rainforests and ancient oak woodlands.

Figure 1: *Tropical rainforests (top) and ancient oak woodlands (bottom) are both examples of stable communities.*

Practice Questions — Fact Recall

Q1 What is a community?

Q2 What is an ecosystem?

Q3 Give four resources that animals compete for.

Q4 What is meant by a stable community?

2. Abiotic and Biotic Factors

Habitats are really fluid environments — they change all the time. It's the abiotic (non-living) and biotic (living) factors that change.

Learning Objectives:
- Be able to recall the following examples of abiotic (non-living) factors:
  - moisture level,
  - light intensity,
  - temperature,
  - carbon dioxide level,
  - direction and intensity of the wind,
  - oxygen level,
  - pH and mineral content of the soil.
- Understand how a community can be affected by changes in the abiotic factors of an ecosystem.
- Be able to recall the following examples of biotic (living) factors:
  - the arrival of new predators,
  - competition from other species,
  - the arrival of new pathogens,
  - availability of food.
- Understand how a community can be affected by changes in the biotic (living) factors of an ecosystem.
- Be able to read and make sense of data about abiotic and biotic factors presented in graphs, charts and tables.

Specification References 4.7.1.2, 4.7.1.3

Abiotic factors

Abiotic factors are the non-living factors in an ecosystem.

Examples

- Moisture level
- Light intensity
- Temperature
- Carbon dioxide level (for plants)
- Wind intensity and direction
- Oxygen level (for aquatic animals)
- Soil pH and mineral content

Abiotic factors can vary in an ecosystem. A change in the environment could be an increase or decrease in an abiotic factor. These changes can affect the size of populations in a community. This means they can also affect the population sizes of other organisms that depend on them (see previous page).

Examples

- A decrease in light intensity, temperature or level of carbon dioxide could decrease the rate of photosynthesis in a plant species (see pages 162-164). This could affect plant growth and cause a decrease in the population size.
- A decrease in the mineral content of the soil (e.g. a lack of nitrates) could cause nutrient deficiencies (see p. 151). This could also affect plant growth and cause a decrease in the population size.

Tip: Animals depend on plants for food, so a decrease in a plant population could affect the animal species in a community.

Biotic factors

Biotic factors are the living factors in an ecosystem.

Examples

- New predators arriving
- Competition — one species may outcompete another so that numbers are too low to breed
- New pathogens
- Availability of food

Biotic factors can also vary in an ecosystem. A change in the environment could be the introduction of a new biotic factor, e.g. a new predator or pathogen. These changes can also affect the size of populations in a community, which can have knock-on effects because of interdependence (see page 317).

Examples

- A new predator could cause a decrease in the prey population. There's more about predator-prey populations on p. 323-324.
- Red squirrels are native to the UK. Grey squirrels were introduced in the Victorian times. Grey squirrels quickly took to their new environment and are now found in large numbers in most parts of the UK. Red and grey squirrels live in the same habitat and eat the same food. The greys are able to outcompete the reds for the same resources — there's not enough food for the reds, so the population is decreasing.

Figure 1: *The population size of red squirrels in Britain is decreasing due to increased competition from grey squirrels.*

Interpreting data on abiotic and biotic factors

In the exam, you might be given a graph, chart or table containing information relating to abiotic or biotic factors and be asked to make sense of it. Take a look at this example:

Example

In an investigation into the effects of fertiliser on the environment, fertiliser was added to a field which contained a pond. The plants and wildlife in and around the pond were then studied for 40 days. The table below shows the size of the fish population in the pond. The graph below shows the percentage cover of the pond by algae over time. A higher percentage cover means that less light is able to reach the plants below the surface of the water.

		Days after fertiliser applied				
		0	10	20	30	40
Number of fish	Species A	78	77	56	16	3
	Species B	19	19	16	4	1
	Species C	58	58	42	12	5
	Total	155	154	114	32	9

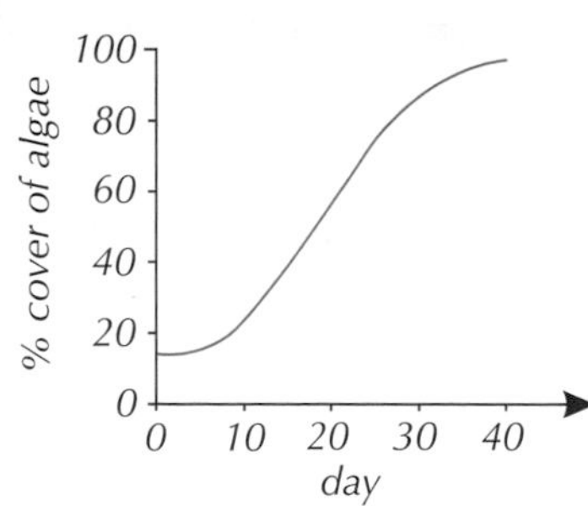

The amount of algae increased because the fertiliser increased the amount of nutrients in the water. Explain whether the increase in algae was caused by a biotic or abiotic factor.

An abiotic factor because the amount of nutrients in the water is a non-living factor.

Describe the relationship between the percentage cover of algae and the size of the fish population.

Between days 0 and 10, the percentage cover of algae increases, but the fish population barely changes. Between days 10 and 40, as percentage cover continues to increase, the size of the fish population decreases rapidly.

Tip: You need to identify the pattern in both the graph and the table to answer the second question.

Suggest an abiotic factor that could be responsible for the change in the size of the fish population. Explain your answer.

A reduced oxygen level. The algae cover reduces the amount of light reaching the plants below the surface of the water, so the plants can't carry out photosynthesis and die off. One of the products of photosynthesis is oxygen, so if there are fewer plants carrying out photosynthesis, less oxygen is released into the water. This lack of oxygen means that the fish can't survive, so the population decreases.

Tip: Head back to page 158 for a recap on photosynthesis.

The graph below shows how the size of the fish population and a population of fish-eating birds changed over the 40 days of observation.

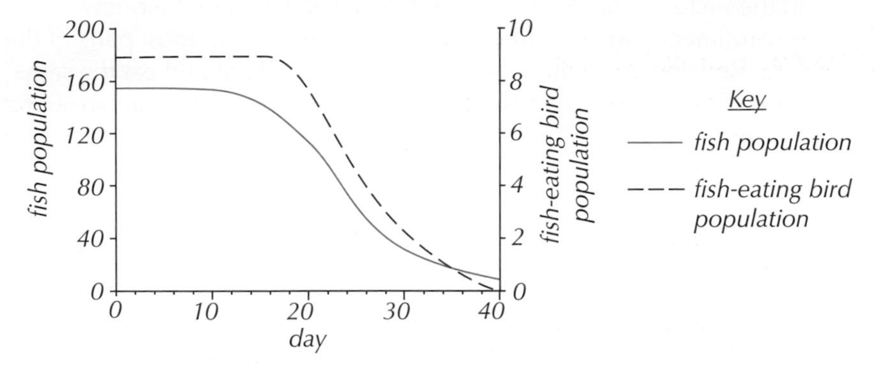

Suggest a biotic factor that could be responsible for the trend in the size of the fish-eating bird population between day 10 and day 40.

A shortage of food for the birds. The size of the bird population starts to decrease shortly after the size of the fish population starts to decrease. Because the size of the fish population decreases, there is less food available to the birds, so the birds either die or move elsewhere to find food.

Practice Questions — Fact Recall

Q1 Give three examples of abiotic factors.

Q2 Give three examples of biotic factors.

Practice Question — Application

Q1 An environmental study was done on a lake over a number of years. Over the course of the study, the concentration of oxygen in the lake gradually decreased due to the effect of pollution. A certain species of fish found in the lake cannot survive in water with a low oxygen concentration. Herons feed on these fish.

a) The population size of the fish changed during the study.

i) Suggest how the population size changed.

ii) Was this change due to a biotic or abiotic factor?

b) The population size of the herons also changed during the study. Suggest how and why the population size changed.

3. Adaptations

Different organisms can survive in different environments because they are adapted to them and can compete for resources.

What are adaptations?

Organisms, including microorganisms, are adapted to live in the conditions of their natural environment. The features or characteristics that allow them to do this are called **adaptations**. Adaptations can be structural, behavioural or functional.

Structural adaptations

These are features of an organism's body structure — such as shape or colour.

Examples

- Arctic animals like the arctic fox have white fur so they're camouflaged against the snow. This helps them avoid predators and sneak up on prey.
- Animals that live in cold places (like whales) have a thick layer of blubber (fat) and a low surface area to volume ratio (see page 60) to help them retain heat.
- Animals that live in hot places (like camels) have a thin layer of fat and a large surface area to volume ratio to help them lose heat.
- Plants that live in deserts have a small surface area compared to volume (e.g. cacti have spines instead of leaves). Plants lose water vapour from the surface of their leaves, so this reduces water loss.

Behavioural adaptations

These are ways that organisms behave.

Examples

Many species (e.g. swallows) migrate to warmer climates during the winter to avoid the problems of living in cold conditions.

Functional adaptations

These are things that go on inside an organism's body that can be related to processes like reproduction and metabolism (all the chemical reactions happening in the body).

Examples

- Desert animals conserve water by producing very little sweat and small amounts of concentrated urine.
- Brown bears hibernate over winter. They lower their metabolism which conserves energy, so they don't have to hunt when there's not much food about.

Learning Objectives:

- Know that organisms have adaptations that make them suited for survival in their natural living environment.
- Know that adaptations can be structural, behavioural or functional.
- Know that organisms that live in very extreme environments (e.g. at high temperatures or pressures, or with a high salt concentration) are known as extremophiles (e.g. species of bacteria that live in deep sea vents).
- Be able to identify adaptations in an organism and understand how they help the organism to survive in its environment.

Specification References
4.7.1.1, 4.7.1.4

Figure 1: *An arctic fox camouflaged in the snow.*

> **Tip:** Not all extremophiles are microorganisms. Some other kinds of organism can be extremophiles too, e.g. some plants are able to grow in really salty soil near the sea.

Adaptations in microorganisms

Microorganisms have a huge variety of adaptations so that they can live in a wide range of environments. For example, some microorganisms (e.g. bacteria) are known as **extremophiles** — they're adapted to live in seriously extreme conditions like super hot volcanic vents, in very salty lakes or at high pressure on the sea bed.

> **Tip:** A hydrothermal vent is an opening in the sea bed which sends out extremely hot water.

Example

Thermococcus litoralis is an extremophile. It's a bacterium that's found in deep sea hydrothermal vents. It's adapted to survive and reproduce at temperatures between 85 and 88 °C, which is much higher than most bacteria can tolerate.

Identifying and explaining adaptations

In the exam you might be given information about any organism and its environment and be asked to identify or explain its adaptations. Here's an example to show you how to do it...

> **Exam Tip**
> In the exam you could be asked to identify an animal's adaptations from explanations given. Therefore you need to think about what body features it needs. For example, if you're told that an adaptation helps an animal keep its balance when hopping then it will probably need to have a tail.

Example

The Weddell seal lives in cold Antarctic conditions and hunts fish in water covered by ice. Use this information and the diagram to explain how the seal is adapted to live in the Antarctic.

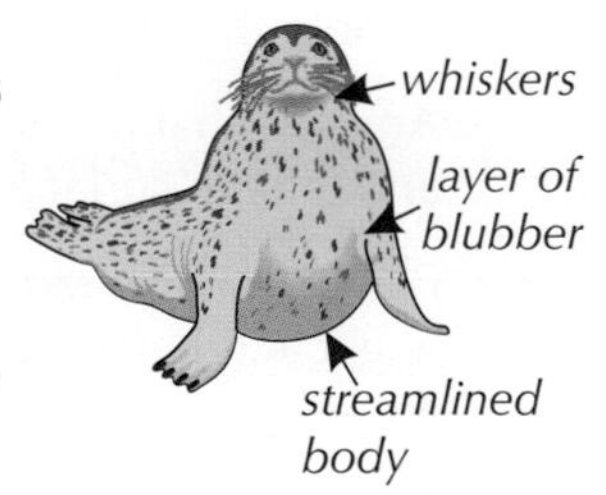

You need to look at the adaptations labelled on the diagram and think about how they allow the seal to keep warm and hunt fish, for example:

- Its whiskers help it to detect fish in the dark conditions under water.
- Its layer of blubber helps insulate it in the cold water.
- Its streamlined body helps reduce resistance from the water, so it can swim fast to catch fish.

Practice Questions — Fact Recall

Q1 What are structural adaptations?

Q2 What are functional adaptations?

Q3 Give two examples of the type of conditions in which you might find an extremophile.

4. Food Chains

It's a fact of life that organisms eat other organisms in a community. To show exactly who eats who, you can produce a food chain.

What are food chains?

Food chains show what's eaten by what in an ecosystem. Food chains always start with a **producer**. Producers make (produce) their own food using energy from the Sun.

Producers are usually green plants or algae — they make glucose (a sugar) by photosynthesis (see page 158). When a green plant produces glucose, some of it is used to make other biological molecules in the plant. These biological molecules are the plant's **biomass** — the mass of living material. Biomass can be thought of as energy stored in a plant.

Energy is transferred through living organisms in an ecosystem when organisms eat other organisms. Producers are eaten by **primary consumers**. Primary consumers are then eaten by **secondary consumers** and secondary consumers are eaten by **tertiary consumers**.

Example

Here's an example of a food chain:

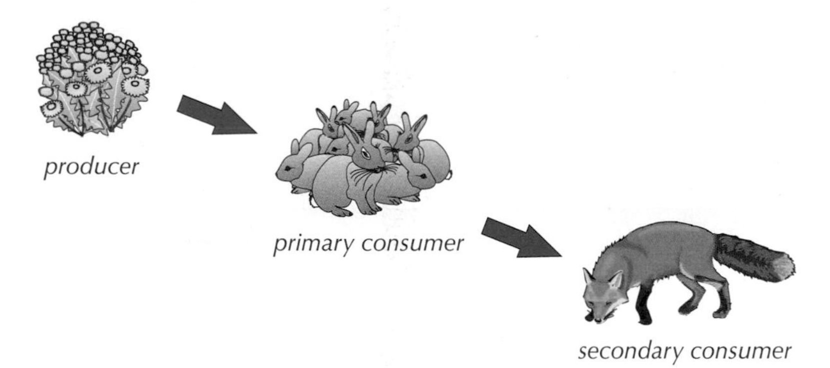

5000 dandelions feed 100 rabbits, which feed 1 fox.

Predator-prey cycles

Consumers that hunt and kill other animals are called **predators**, and their **prey** are what they eat. In a stable community containing prey and predators (as most of them do of course), the population of any species is usually limited by the amount of food available. If the population of the prey increases, then so will the population of the predators. However as the population of predators increases, the number of prey will decrease.

Learning Objectives:

- Know that food chains show the feeding relationships between different species in an ecosystem.
- Know that all food chains begin with a producer, which is a photosynthetic organism (often a green plant or algae) that makes molecules (such as glucose) and is responsible for producing all of the biomass on Earth.
- Know what primary, secondary and tertiary consumers are.
- Know that a predator is a consumer that kills and eats other animals (its prey).
- Know that the population sizes of predators and prey increase and decrease in predator-prey cycles.
- Be able to interpret graphs showing predator-prey relationships.

Specification Reference 4.7.2.1

Tip: Consumers are organisms that eat other organisms. 'Primary' means first, so primary consumers are the first consumers in a food chain. Secondary consumers are second and tertiary consumers are third.

Exam Tip
You might be asked to interpret a graph like this in the exam — so make sure you really understand what's going on and why.

Tip: Predator-prey cycles are an example of how species in a community can be interdependent. In this case, the predator relies on the prey for food.

Example

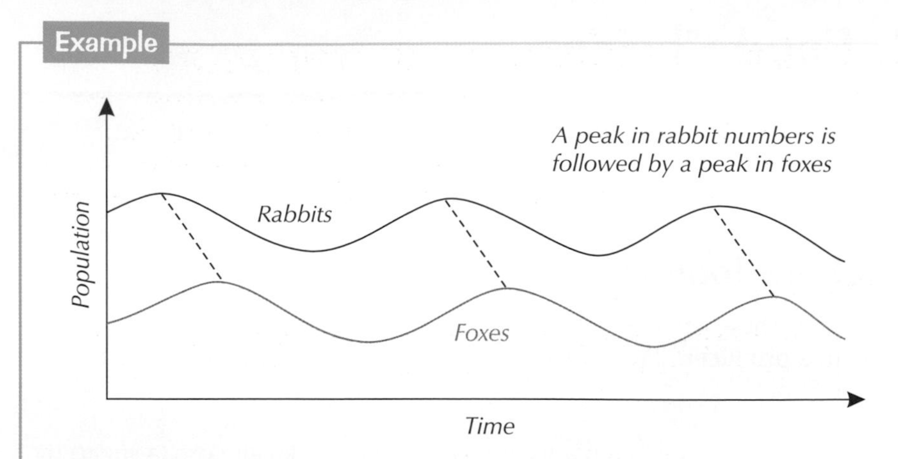

More grass means more rabbits. More rabbits means more foxes. But more foxes means fewer rabbits. Eventually fewer rabbits will mean fewer foxes again. This up and down pattern continues...

Predator-prey cycles are always out of phase with each other. This is because it takes a while for one population to respond to changes in the other population. E.g. when the number of rabbits goes up, the number of foxes doesn't increase immediately because it takes time for them to reproduce.

Practice Questions — Fact Recall

Q1 What type of organism is always found at the start of a food chain?

Q2 Name the group of organisms that eat the secondary consumers.

Q3 What are predators?

Practice Question — Application

Q1 The following food chain is observed in an African savanna.

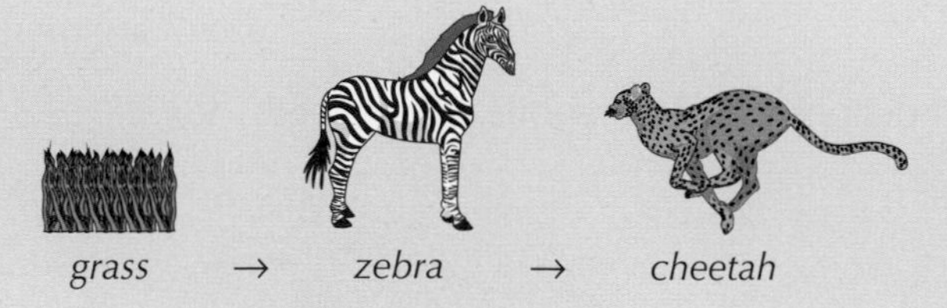

a) Name the prey organism in this food chain.

This food chain forms part of a stable community.

b) Suggest the effect that an increase in grass growth would have on the population of cheetahs. Explain your answer.

c) After a certain amount of time, the population of cheetahs would return to its previous level. Explain why this would happen.

5. Using Quadrats and Transects

Quadrats are really useful tools for investigating the population size and distribution of organisms in a particular place. You can also use them alongside transects to investigate the distribution of organisms across an area.

Learning Objectives:

- Know that ecologists use quadrats to investigate the population size and distribution of organisms.
- Understand how to use sampling to investigate how a factor affects the distribution of species (Required Practical 9).
- Understand and be able to work out the mean, mode and median of data on the abundance of organisms.
- Understand how to investigate the size of a population of a certain species in a habitat (Required Practical 9).
- Know that transects can also be used to investigate the distribution of organisms across an area.
- Be able to produce graphs of abundance data with appropriate axes scales.

Specification Reference
4.7.2.1

Distribution of organisms

As you know from page 315, a habitat is the place where an organism lives, e.g. a playing field. The distribution of an organism is where an organism is found, e.g. in a part of the playing field.

Where an organism is found is affected by environmental factors (see pages 318-319). An organism might be more common in one area than another due to differences in environmental factors between the two areas.

Examples

- In a playing field, you might find that daisies are more common in the open than under trees, because there's more light available in the open and daisies need light to survive (they use it for photosynthesis — see page 158).
- Some types of mayfly are more common in colder parts of a stream, as they can't tolerate the warmer temperatures in other parts of the stream.

There are a couple of ways to study the distribution of an organism:

1. You can measure how common an organism is in two or more sample areas (e.g. using **quadrats** — see below) and compare them.
2. You can study how the distribution changes across an area, e.g. by placing quadrats along a **transect** (see page 327).

Both of these methods give quantitative data (numbers) about the distribution.

The data you collect can be used to provide evidence for environmental change. For instance, if the distribution of organisms across an area changes over time, this could be due to changes in the environment.

Tip: Quadrats are a tool used by ecologists. Ecologists are scientists who specialise in studying the interaction between organisms and their environment.

Quadrats

A quadrat is a square frame enclosing a known area, e.g. 1 m^2 (see Figure 2).

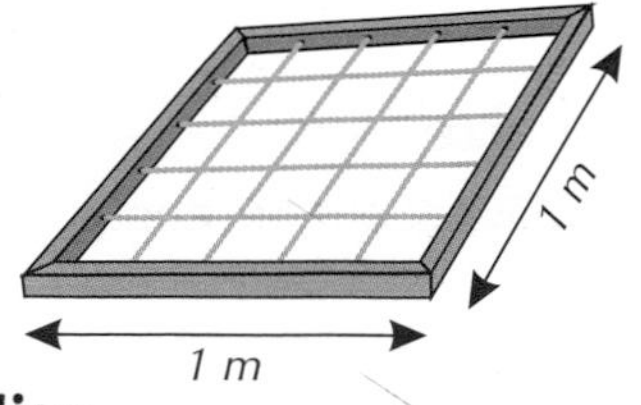

Figure 1: *A diagram of a 1 m^2 quadrat.*

Investigating distribution

REQUIRED PRACTICAL 9

You can use quadrats to study the distribution of small organisms that are slow moving or that don't move around. You can investigate the effect of a factor on the distribution of a species by comparing how common an organism is in two sample areas (e.g. shady and sunny spots in a playing field). Just follow these simple steps:

Tip: Before you start this experiment, check the area you're investigating for hazards (e.g. broken glass) and make sure you wash your hands after touching the soil.

> **Tip:** It's really important that the quadrats are placed randomly within the sample area. Taking random samples improves the validity of the study.

1. Place a 1 m^2 quadrat on the ground at a random point within the first sample area. E.g. divide the area into a grid and use a random number generator to pick coordinates (see page 373).
2. Count all the organisms within the quadrat.
3. Repeat steps 1 and 2 as many times as you can.
4. Work out the mean number of organisms per quadrat within the first sample area.

> **Tip:** If you're counting flowering plants in your quadrat, make sure you count the number of actual plants, not the number of flowers. Some plants will have more than one flower, and some might not currently have any flowers on them.

Example

Anna counted the number of daisies in 7 quadrats within her first sample area and recorded the following results:

Quadrat	1	2	3	4	5	6	7
Number of Daisies	18	22	20	23	23	25	23

Here the mean is: $\frac{\text{total number of organisms}}{\text{number of quadrats}}$

$= \frac{154}{7} =$ **22 daisies per quadrat**

Figure 2: *A student using a quadrat to gather data on the distribution of organisms.*

5. Repeat steps 1 to 4 in the second sample area.
6. Finally compare the two means. E.g. you might find 2 daisies per m^2 in the shade, and 22 daisies per m^2 (lots more) in the open field.

You might want to find and compare the mode and median of your data too.

Example

1. The mode is the number that appears most often in the data. For Anna's data above, the mode is **23 daisies per quadrat**.
2. To find the median, you first need to put your data in numerical order. The median is the middle value in your data.

   18, 20, 22, (23), 23, 23, 25

   The median of Anna's data is **23 daisies per quadrat**.

> **Tip:** There's more on means, modes and medians on page 14 of the Working Scientifically section.

Estimating population size

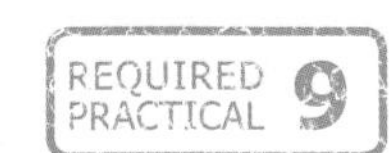

You can also use quadrats to work out the population size of an organism in one area.

Example

Students used 0.25 m^2 quadrats to randomly sample daisies on an open field. The students found a mean of 10.5 daisies per quadrat. The field had an area of 800 m^2. Estimate the population of daisies on the field.

1. Work out the mean number of organisms per m^2.

$1 \div 0.25 = 4$ ← *Because the quadrat is only $0.25\ m^2$, you first need to work out how many quadrats make up $1\ m^2$.*

$4 \times 10.5 = 42$ daisies per m^2

2. Then multiply the mean by the total area (in m^2) of the habitat.

$800 \times 42 =$ **33 600 daisies in the open field**

Tip: The population size of an organism is sometimes called its abundance.

Tip: If your quadrat has an area of $1\ m^2$, the mean number of organisms per m^2 is just the same as the mean number per quadrat.

Transects

You can use lines called transects (see Figure 3) to help find out how organisms (like plants) are distributed across an area — e.g. if an organism becomes more or less common as you move from a hedge towards the middle of a field.

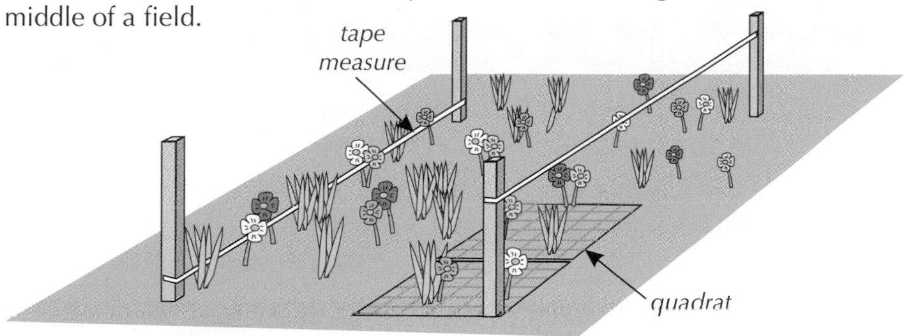

Figure 3: *A diagram showing two transects.*

Tip: Transects can be used in any type of habitat, not just fields. For example, along a beach or in a stream.

Figure 4: *Students using a quadrat along a transect.*

Here's what to do:

1. Mark out a line in the area you want to study using a tape measure.
2. Then collect data along the line. You can do this by just counting all the organisms you're interested in that touch the line. Or, you can collect data by using quadrats. These can be placed next to each other along the line or at intervals, for example, every 2 m.

Percentage cover

If it's difficult to count all the individual organisms in the quadrat (e.g. if they're grass) you can calculate the percentage cover. This means estimating the percentage area of the quadrat covered by a particular type of organism, e.g. by counting the number of little squares covered by the organisms.

Exam Tip
You need to be able to draw your own graphs to show data about the abundance of different organisms. Here's an example:

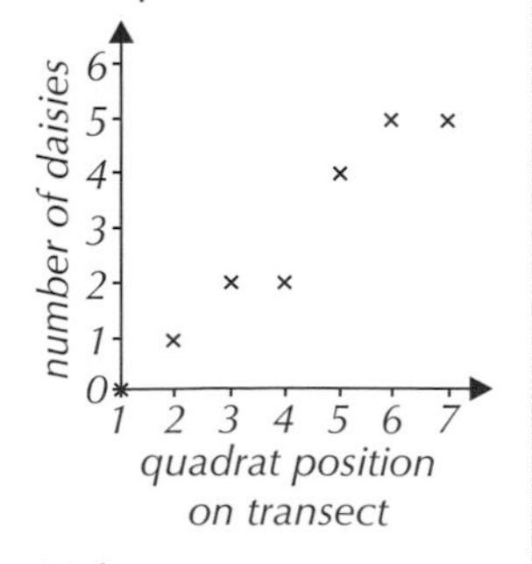

Make sure you use axes with appropriate scales. See page 18 for more on drawing graphs.

Example

MATHS SKILLS

Some students were measuring the distribution of organisms from one corner of a school playing field to another, using quadrats placed at regular intervals along a transect. Below is a picture of one of the quadrats. Calculate the percentage cover of each organism, A and B.

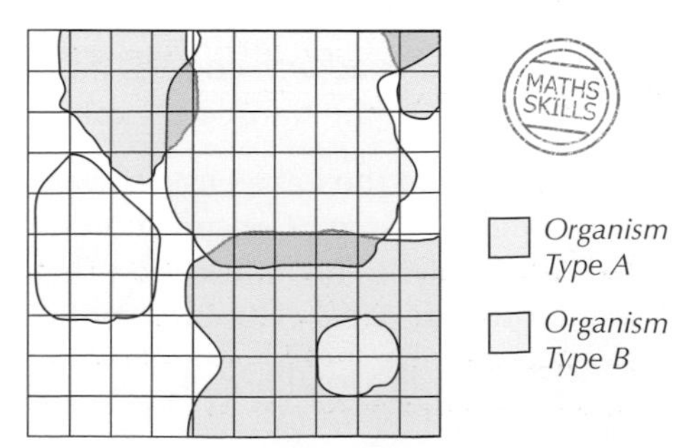

Tip: You count a square if it's more than half covered.

1. Count the number of squares covered by organism A.
   Type A = 42 squares
2. Make this into a percentage — divide the number of squares covered by the organism by the total number of squares in the quadrat (100), then multiply the result by 100.

$$\frac{42}{100} \times 100 = 0.42 \times 100 = \mathbf{42\%}$$

3. Do the same for organism B.
   Type B = 47 squares

$$\frac{47}{100} \times 100 = 0.47 \times 100 = \mathbf{47\%}$$

Practice Questions — Fact Recall

Q1 Describe how you would use random sampling with a quadrat to compare the distribution of organisms in two sample areas.

Q2 Describe one way in which a transect can be used to measure the distribution of organisms across an area.

Practice Questions — Application

Q1 Joanne read that bulrushes grow best in moist soil or in shallow water. She wanted to find out whether this is true, so she investigated the distribution of bulrushes in her garden. She used a transect (as shown in the diagram) and recorded the number of bulrushes in each 1 m^2 quadrat as shown in the table.

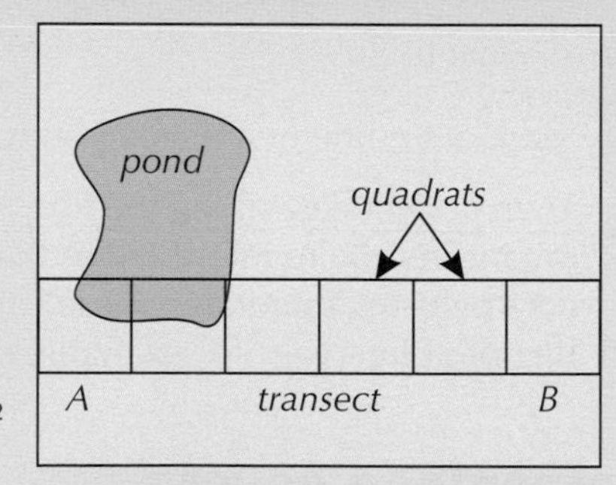

Quadrat	1	2	3	4	5	6
Number of Bulrushes	1	5	5	20	43	37

a) Calculate the mean number of bulrushes per quadrat.

b) Joanne started counting at quadrat 1 and moved along the transect in order, to quadrat 6. Assuming that the information Joanne read was correct, do you think quadrat 1 is at end A or end B of the transect? Explain your answer.

Q2 An area of the Lake District was sampled for the presence of a species of slug. Conservationists used 0.25 m^2 quadrats placed randomly in the landscape. The mean number of the slugs found in a quadrat was 2.5 and the area under investigation was 2000 m^2.

Estimate the total population of the slugs in the area.

6. Environmental Change **Higher**

The distribution of species is affected by changes in the environment.

Learning Objectives:

- H Know that environmental changes such as changes to water availability, temperature and the concentrations of atmospheric gases can affect the distribution of different species.
- H Know that environmental changes can be seasonal, geographic or related to human interaction with the ecosystem.
- H Be able to evaluate data showing the effect of environmental changes on the distribution of organisms.

Specification Reference 4.7.2.4

Environmental change and distribution

Environmental changes can cause the distribution of organisms to change. A change in distribution means a change in where an organism lives. Environmental changes that can affect organisms in this way include changes in:

Availability of water

The distribution of some animal and plant species in the tropics changes between the wet and the dry seasons — i.e. the times of year where there is more or less rainfall, and so more or less water available.

Example **Higher**

Each year in Africa, large numbers of giant wildebeest migrate, moving north and then back south as the rainfall patterns change.

Temperature

The distribution of bird species in Germany is changing because of a rise in average temperature.

Example **Higher**

The European bee-eater bird is a Mediterranean species but it's now present in parts of Germany.

Tip: H Average temperatures are usually warmer in the Mediterranean than in Germany.

The composition of atmospheric gases

The distribution of some species changes in areas where there is more air pollution.

Example **Higher**

Some species of lichen can't grow in areas where sulfur dioxide is given out by certain industrial processes.

Figure 1: *The distribution of European bee-eaters has changed due to a rise in average temperature.*

These environmental changes can be caused by seasonal factors, geographic factors or human interaction. For example, the rise in average temperature is due to global warming, which has been caused by human activity (see page 343).

Practice Questions — Fact Recall

Q1 Give three environmental changes that can affect the distribution of species.

Q2 Give three causes of environmental changes.

7. The Cycling of Materials

Materials such as water and carbon are being constantly recycled on Earth. It's really important that this happens. For example, it means that there's always enough material to make new organisms when old organisms die.

Learning Objectives:

- Know that the water on Earth is constantly recycled in a process called the water cycle, in which water follows a continuous chain of being evaporated and precipitated, and that this provides the plants and animals on land with fresh water.
- Understand that all materials in our environment are recycled to produce the building blocks required to make new organisms, and that lots of different materials are constantly cycling through both the biotic and abiotic parts of an ecosystem.
- Know that microorganisms decay dead animal and plant material, returning mineral ions to the soil for use by plants.
- Know that the carbon on Earth is constantly cycled between organisms and the atmosphere.
- Know that carbon is removed from the atmosphere by plants as CO_2, to be used in photosynthesis.
- Know that the decay of dead plant and animal material by microorganisms returns carbon to the atmosphere in the form of CO_2, which is produced when the microorganisms respire.

Specification Reference 4.7.2.2

The water cycle

The water here on planet Earth is constantly recycled. Energy from the Sun makes water **evaporate** from the land and sea, turning it into water vapour. Water also evaporates from plants — this is known as **transpiration** (see p. 92). The warm water vapour is carried upwards (as warm air rises). When it gets higher up it cools and condenses to form clouds. Water falls from the clouds as **precipitation** (usually rain, but sometimes snow or hail) onto land, where it provides fresh water for plants and animals. It then drains into the sea, before the whole process starts again.

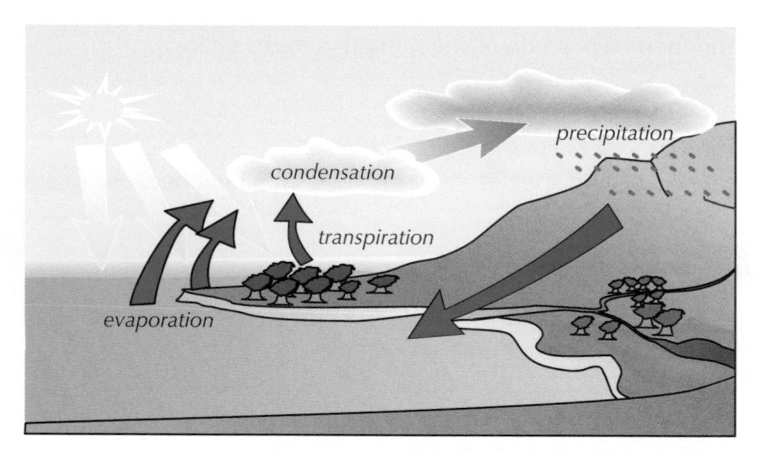

Figure 1: *Diagram showing the steps in the water cycle.*

Recycling of elements

Living things are made of materials they take from the world around them. For example, plants turn elements like carbon, oxygen, hydrogen and nitrogen from the soil and the air into the complex compounds (carbohydrates, proteins and fats) that make up living organisms. These get passed up the food chain.

These materials are returned to the environment in waste products, or when the organisms die and decay. Materials decay because they're broken down (digested) by microorganisms. This happens faster in warm, moist, aerobic (oxygen rich) conditions because microorganisms are more active in these conditions.

Decay puts the stuff that plants need to grow (e.g. mineral ions) back into the soil. In a **stable community**, the materials that are taken out of the soil and used by plants, etc. are balanced by those that are put back in. There's a constant cycle of lots of different materials happening through both the biotic and abiotic parts of an ecosystem.

The carbon cycle

Carbon is constantly being cycled — from the air, through food chains and eventually back out into the air again. The carbon cycle shows how carbon is recycled — see Figure 2.

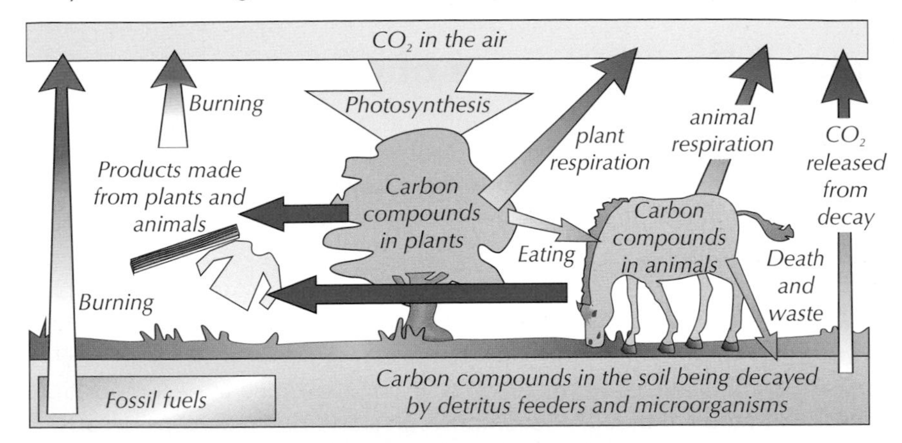

Figure 2: *The carbon cycle.*

Tip: Fossil fuels are made of decayed animal and plant matter.

Carbon is taken out of the air

The whole carbon cycle is "powered" by **photosynthesis**. CO_2 (carbon dioxide) is removed from the atmosphere by green plants and algae, and the carbon is used to make glucose, which can be turned into carbohydrates, fats and proteins that make up the bodies of the plants and algae.

Carbon moves through food chains

Some of the carbon becomes part of the fats and proteins in animals when the plants and algae are eaten. The carbon then moves through the food chain. The energy that green plants and algae get from photosynthesis is transferred up the food chain.

When plants, algae and animals die, other animals (called **detritus feeders**) and microorganisms feed on their remains. Animals also produce waste, and this too is broken down by detritus feeders and microorganisms.

Figure 3: *Woodlice are detritus feeders.*

Carbon is returned to the air

Some carbon is returned to the atmosphere as CO_2 when the plants, algae, animals (including detritus feeders) and microorganisms **respire**. Also CO_2 is released back into the air when some useful plant and animal products, e.g. wood and fossil fuels, are burnt (**combustion**).

Practice Questions — Fact Recall

Q1 How does CO_2 from the atmosphere first enter the food chain?

Q2 How is carbon returned to the atmosphere from dead leaves?

Practice Question — Application

Q1 A student said, "Some of the water in trees eventually ends up in the sea." Do you agree with this statement? Explain your answer.

8. Decay

The process of decay is essential for keeping nature's cycle of nutrients going...

Learning Objectives:

- Know how temperature, oxygen availability and water availability affect the rate of decay.
- Know that compost is made out of decomposed material and is a useful natural fertiliser that can be used in gardens and on crops.
- Know that farmers and gardeners produce compost by creating ideal conditions for the decay of waste.
- Be able to investigate how temperature affects the rate of decay of milk by measuring a change in the pH (Required Practical 10).
- Be able to convert data about decay between number and graph form.
- Be able to draw graphs with appropriate axes scales to present data on decay.
- Be able to carry out rate calculations for the decay of material.
- Know that biogas contains methane, which can be used as fuel.
- Know that biogas is produced when waste is anaerobically decayed in a biogas generator.

Specification Reference 4.7.2.3

Factors affecting the rate of decay

Microorganisms such as bacteria and fungi, as well as detritus feeders (see previous page) are the critters responsible for decomposition (decay). You'd think that organisms that feed on dead stuff wouldn't be that picky, but they're a bit like Goldilocks — everything has to be just right for them to work at their best.

Temperature

Warmer temperatures make things decompose more quickly because they increase the rate that the enzymes (see page 115) involved in decomposition work at. If it's too hot though, decomposition slows down or stops because the enzymes are destroyed and the organisms die. Really cold temperatures slow the rate of decomposition too.

Oxygen availability

Many organisms need oxygen to respire, which they need to do to survive. The microorganisms involved in anaerobic decay (such as those used to produce biogas — see page 334) don't need oxygen though.

Water availability

Decay takes place faster in moist environments because the organisms involved in decay need water to carry out biological processes.

Number of decay organisms

The more microorganisms and detritus feeders there are, the faster decomposition happens.

Compost

Compost is decomposed organic matter (e.g. food waste) that is used as a natural fertiliser for crops and garden plants. It recycles nutrients back into the soil — giving you a lovely garden and improving crop growth. Farmers and gardeners try to provide the ideal conditions for quick decay to make compost. Compost bins recreate the ideal conditions for decay — see Figure 1.

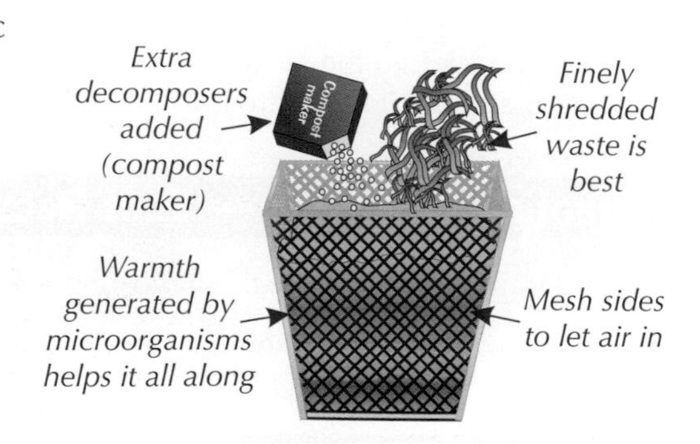

Figure 1: *A compost bin provides the right conditions for microorganisms to work.*

Investigating the rate of decay

REQUIRED PRACTICAL 10

You can investigate decay by observing the action of the enzyme lipase on a sample of milk that has been made alkaline. When lipase breaks the milk down, the pH of the milk decreases.

This practical looks at how temperature affects the rate of decay. In it, an indicator dye called phenolphthalein is used — it has a pink colour when the pH is around 10, but becomes colourless when the pH falls below 8.3. Here's what you need to do:

1. Measure out 5 cm^3 of lipase solution and add it to a test tube. Label this tube with an 'L' for lipase.
2. Measure out 5 cm^3 of milk and add it to a different test tube.
3. Add 5 drops of phenolphthalein indicator to the tube containing milk.
4. Then measure out 7 cm^3 of sodium carbonate solution and add it to the tube containing milk and phenolphthalein. This makes the solution in the tube alkaline, so it should turn pink.
5. Put both tubes into a water bath set to 30 °C and leave them to reach the temperature of the water bath. You could stick a thermometer into the milk tube to check this.
6. Once the tubes have reached 30 °C, use a calibrated dropping pipette (a dropping pipette with a scale) to put 1 cm^3 of the lipase solution into the milk tube and start a stopwatch straight away.
7. Stir the contents of the tube with a glass rod. The enzyme will start to decompose the milk.
8. As soon as the solution loses its pink colour, stop the stopwatch and record how long the colour change took in a table.

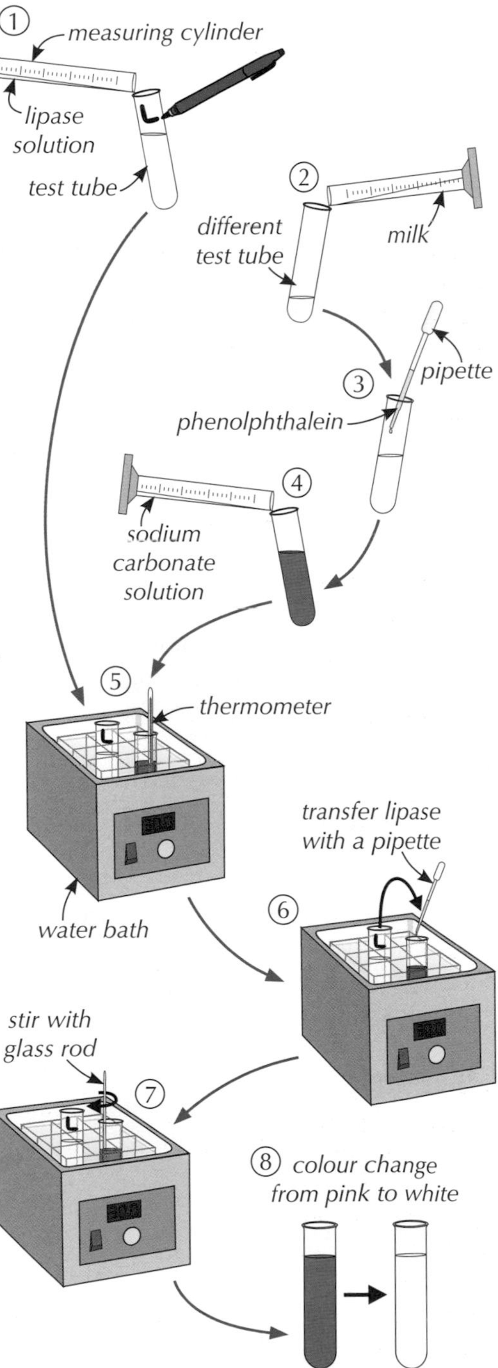

Figure 2: The process of investigating the effect of temperature on the decay of milk by lipase.

Tip: Make sure you carry out a risk assessment before you start this experiment. Avoid getting phenolphthalein, lipase or sodium carbonate solution on your skin and wear safety goggles to avoid getting them in your eyes. Make sure you keep the phenolphthalein away from sources of ignition (e.g. lit Bunsen burners).

Tip: Make sure you only ever put your thermometer in the milk and phenolphthalein tube to avoid contamination between this tube and the lipase solution.

Exam Tip
In the exam, you might need to convert data about the rate of decay between numerical form (e.g. data tables) and graphs, so make sure you can do it.

Exam Tip
You need to be able to draw your own graphs to show data about the rate of decay. Make sure you use axes with appropriate scales. Head to page 18 for more on drawing graphs.

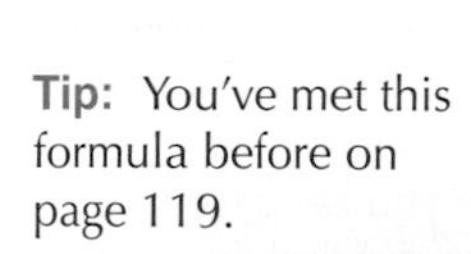

Tip: You can make a water bath capable of temperatures below room temperature by adding ice cubes to a beaker of water and measuring the temperature with a thermometer.

Tip: You've met this formula before on page 119.

9. Repeat the experiment at a range of different temperatures (e.g. 10 °C, 20 °C, 40 °C, 50 °C). Make sure you carry out the experiment three times at each temperature, then calculate the mean time taken for the colour change to occur at each temperature.
10. You can use your results to calculate the rate of decay using this formula. The units will be s^{-1} since rate is given per unit of time.

$$\text{Rate} = \frac{1000}{\text{time}}$$

Example

In one run of the experiment, the mean time taken for the colour change to occur was 40 seconds. Calculate the rate of decay.

$$\text{Rate} = \frac{1000}{\text{time}} = \frac{1000}{40} = \mathbf{25\ s^{-1}}$$

Biogas

Biogas is mainly made up of methane, which can be burned as a fuel. Lots of different microorganisms are used to produce biogas. They decay plant material and animal waste (e.g. faeces) anaerobically (without oxygen). This type of decay produces methane gas. Sludge waste from, e.g. sewage works or sugar factories, is used to make biogas on a large scale.

***Figure 3:** Biogas can also be collected from landfill sites. The waste (buried under the grey bit) releases gases as it is broken down.*

Biogas is made in a simple fermenter called a digester or generator. Biogas generators need to be kept at a constant temperature to keep the microorganisms respiring away.

Biogas can't be stored as a liquid (it needs too high a pressure), so it has to be used straight away — for heating, cooking, lighting, or to power a turbine to generate electricity.

Types of biogas generator

There are two main types of biogas generator — batch generators and continuous generators.

Batch generators make biogas in small batches. They're manually loaded up with waste, which is left to digest, and the by-products are cleared away at the end of each session. They don't have to be filled up as often as continuous generators, but they don't produce biogas at a steady rate.

***Figure 4:** A large-scale biogas production plant. The gas is stored in the white round part and the digestion takes place in the large containers at the back.*

Continuous generators make biogas all the time. Waste is continuously fed in, and biogas is produced at a steady rate. Continuous generators are more suited to large-scale biogas projects.

Figure 5 on the next page shows a simple biogas generator. Whether a generator is big or small, or for batch or continuous use, it needs to have the following:

- an inlet for waste material to be put in,
- an outlet for the digested material to be removed through,
- an outlet so that the biogas can be piped to where it is needed.

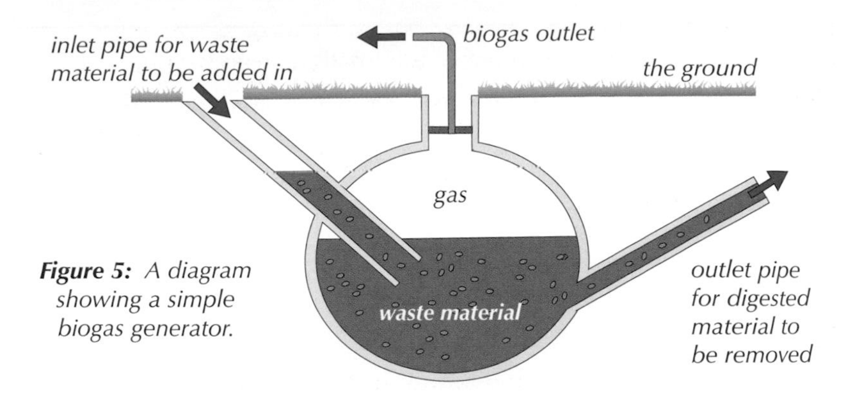

Figure 5: A diagram showing a simple biogas generator.

Figure 6: A small-scale biogas generator being used in India. Animal manure is fed into the generator to be fermented.

Practice Questions — Fact Recall

Q1 Compost is the decayed remains of animal and plant matter. Why is compost used on gardens?

Q2 What's the difference between a batch biogas generator and a continuous biogas generator?

Practice Questions — Application

Q1 Gary recycles his garden waste by making it into compost. He does this in his garden using this compost bin:

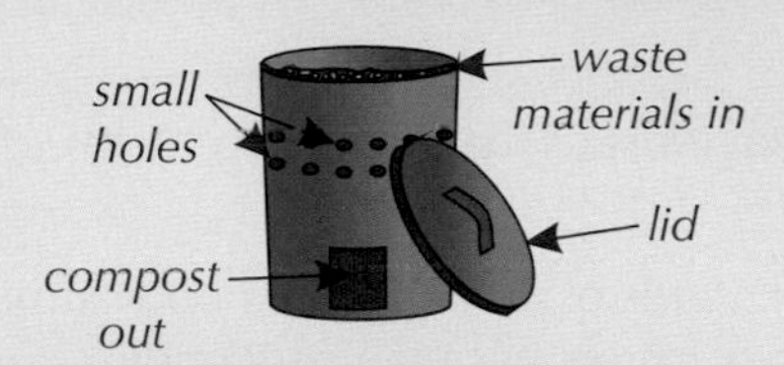

a) Suggest why the compost bin has small holes in it.

b) Gary has found that compost is made more quickly if he puts his compost bin in a sunny area of the garden. Explain why this is.

c) Gary's neighbour told him that he should always keep the lid on his compost bin to stop water vapour escaping out of the bin. Explain how the presence of water vapour would help compost to be made.

Figure 7: Compost bins can come in lots of different shapes and sizes.

Topic 7a Checklist — Make sure you know...

Competition

- [ ] That an ecosystem is the interaction of a community of living organisms (biotic) with the non-living (abiotic) parts of their environment.
- [ ] The levels of organisation that make up an ecosystem, e.g. population, community.
- [ ] That organisms rely on resources from their environment and other organisms for survival, and that organisms have to compete with each other for the limited resources.
- [ ] That plants compete for light and space, as well as water and mineral ions from the soil, and that animals compete for space, food, water and mates.

cont...

- ❑ How to identify the resources being competed for using relevent information given to you.
- ❑ How to read and interpret information about the interaction of organisms in communities from graphs, charts and tables.
- ❑ That species in a community are interdependent — they rely on each other for the resources they need, so changes in an environment can have knock-on effects for lots of different organisms.
- ❑ That a stable community is a community where all the species and environmental factors are in balance so that the population sizes remain roughly constant.

Abiotic and Biotic Factors

- ❑ These examples of abiotic (non-living) factors — moisture level, light intensity, temperature, carbon dioxide level, wind intensity and direction, oxygen level, soil pH and mineral content.
- ❑ How changes in the abiotic factors of an ecosystem can affect a community.
- ❑ These examples of biotic (living) factors — the arrival of new predators, competition from other species, the arrival of new pathogens and the availability of food.
- ❑ How changes in the biotic factors of an ecosystem can affect a community.
- ❑ How to read and interpret data about biotic and abiotic factors from graphs, charts and tables.

Adaptations

- ❑ That adaptations are features or characteristics that allow organisms to survive in the environmental conditions of their habitat.
- ❑ That adaptations can be structural (features of an organism's body structure), behavioural, or functional (the way that things work inside an organism's body).
- ❑ That extremophiles are organisms that live in extreme environments, such as where temperature, pressure or salt concentration are high (e.g. some species of bacteria live in deep sea vents).
- ❑ How to identify an animal's adaptations and understand how they help the organism to survive.

Food Chains

- ❑ That food chains show the feeding relationships between different species in an ecosystem.
- ❑ That a food chain always starts with a photosynthetic organism such as a green plant or algae, known as a producer, which makes molecules like glucose, and that producers make all of the biomass on Earth.
- ❑ That producers are eaten by primary consumers, and primary consumers are eaten by secondary consumers, which are eaten by tertiary consumers.
- ❑ That a predator is a consumer that kills and eats another animal, which is known as its prey, and that the population sizes of predators and prey go up and down in predator-prey cycles.

Using Quadrats and Transects

- ❑ That quadrats (square frames that enclose a known area) can be used by ecologists to investigate the population size and distribution of organisms in an area.

cont...

- [ ] How to investigate the effect of a factor on the distribution of a species using a quadrat and sampling techniques (Required Practical 9).
- [ ] How to calculate the mean, mode and median of data on the abundance of organisms.
- [ ] How to investigate the size of a population of a certain species in a habitat using a quadrat (Required Practical 9).
- [ ] How transects can be used to investigate the distribution of organisms across an area (e.g. how the distribution of a certain species varies across an entire field).
- [ ] How to draw graphs, using appropriate axes, to show data relating to the abundance of species.

Environmental Change

- [ ] H That environmental changes such as changes in water availability, temperature and the concentrations of atmospheric gases can affect the distribution of species.
- [ ] H That environmental changes affecting the distribution of organisms can be seasonal, geographic or related to human interaction with the ecosystem.
- [ ] H How to evaluate data showing the effect of environmental changes on organisms.

The Cycling of Materials

- [ ] That the water cycle is a continually occurring chain of evaporation and precipitation that allows the water on Earth to be constantly recycled and provides water to the plants and animals living on land.
- [ ] That there is a constant cycle of lots of different materials through both the biotic and abiotic components of an ecosystem.
- [ ] That the materials used to make the building blocks of organisms are recycled into new organisms.
- [ ] That mineral ions are returned to the soil from dead animal and plant material when microorganisms break the dead material down.
- [ ] That the carbon cycle shows how carbon is recycled between organisms and the atmosphere.
- [ ] That carbon is returned to the atmosphere as carbon dioxide when microorganisms that decay dead plant and animal material respire, and that this gas is then used by plants in photosynthesis.

Decay

- [ ] That the rate of decay is affected by temperature, oxygen availability and water availability.
- [ ] That compost is a natural fertiliser used by farmers and gardeners on crops and gardens that is made out of decomposed material, and that farmers and gardeners produce compost by creating ideal conditions for waste to decay.
- [ ] How to investigate how temperature affects the rate of decay of milk by the enzyme lipase by measuring a change in the pH (Required Practical 10).
- [ ] How to convert data about the rate of decay between numerical and graph forms, and how to draw graphs showing data related to the rate of decay with appropriate scales for the axes.
- [ ] How to carry out rate calculations for the decay of material.
- [ ] That biogas contains methane, which can be used as fuel. Biogas can be produced by decomposing waste material using microorganisms in anaerobic conditions (without oxygen) in a biogas generator.

Exam-style Questions

1 A biology class were studying the distribution of organisms on a rocky shore line. Three students each set up a transect which ran from the water's edge up the shore, as shown in the diagram. All three students counted the number of limpets in each of the 50 cm × 50 cm quadrats. The number of limpets counted by each student are shown in the table.

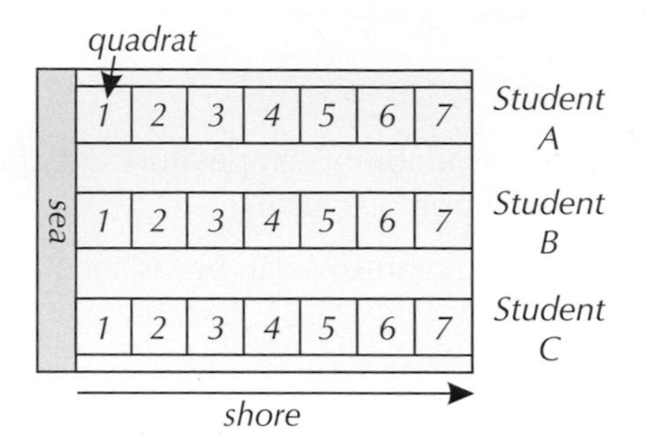

Quadrat number	1	2	3	4	5	6	7
Student A No. of Limpets	18	65	70	55	30	10	0
Student B No. of Limpets	6	72	76	41	21	18	3
Student C No. of Limpets	14	83	84	57	26	16	1

1.1 Calculate the mean number of limpets found in quadrat 4.

(2 marks)

1.2 Limpets need to stay moist in order to survive, but they are able to survive for a period of time without water by clamping down onto the surface of rocks — this helps to prevent them from drying out. Close to the water's edge, limpets have lots of competition from other species for space.

Use this information and the data above to describe the distribution of limpets along the shore line and suggest reasons for it.

(3 marks)

1.3 During the class study, another group of students investigated the distribution of a type of starfish. They placed 20 quadrats at random in each sample area of shoreline that they were interested in studying.

Why did the students place their quadrats randomly?

(1 mark)

2* The diagram below shows how a particular food chain is part of the carbon cycle.

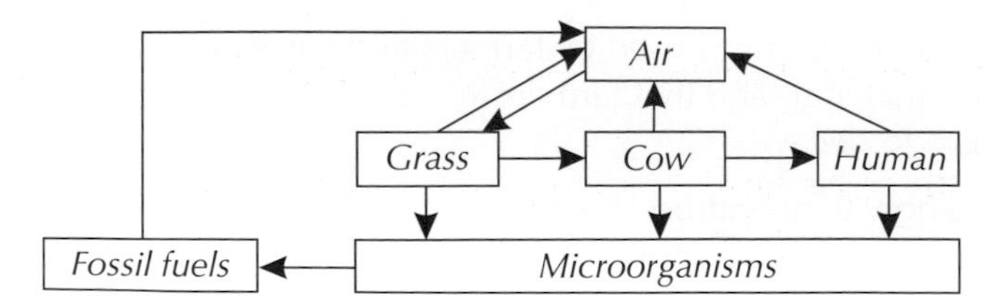

Use the information in the diagram and your own knowledge to describe the processes in which carbon is cycled between this food chain, other living organisms and the air.

(6 marks)

3 Scientists have been studying a species of wolf. The wolf lives in rocky mountains where the temperature at night can drop as low as –15 °C. Its main source of prey is rodents that live in burrows beneath the ground. The wolf has adaptations to help it survive in the rocky mountains as shown on the diagram below.

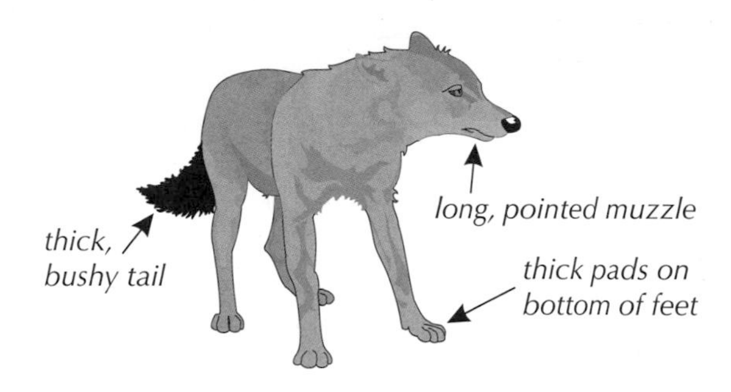

3.1 What is an adaptation?

(1 mark)

3.2 Use the information and the diagram above to suggest how the wolf is adapted to survive in the rocky mountains.

(3 marks)

Scientists have been recording the total population size of the wolf over a number of years. Their findings and some other data are shown in the table and graph below.

Year	Total population size of wolf	Total population size of one type of prey (million)	Rabies outbreak in this year
2007	401	1.96	No
2008	327	2.01	Yes
2009	330	2.09	No
2010	341	2.06	No
2011	265	2.01	Yes
2012	269	1.99	No

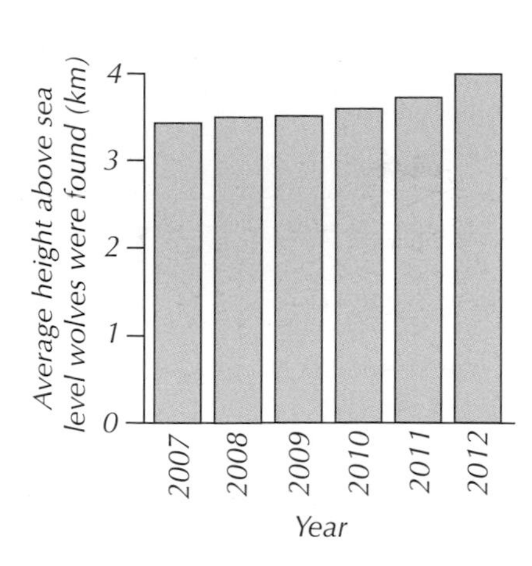

3.3 Describe the overall trend in the total population size of the wolf species studied.

(1 mark)

3.4 Suggest the biotic factor that's responsible for this trend.
Use data from the table to support your answer.

(3 marks)

3.5 Since 2007, the land higher than 3.5 km above sea level has gradually been taken over as farmland. Use data from the graph to suggest how this has affected the distribution of the wolf population.

(2 marks)

1. Biodiversity and Waste

Learning Objectives:
- Know that biodiversity is the variety of different species of organisms on Earth, or within an ecosystem.
- Know that high biodiversity is important because it results in stable ecosystems, where one species is less dependent on another single species for the resources it needs.
- Know that maintaining biodiversity is vital for the survival of the human species.
- Know that lots of things that humans do reduce biodiversity and it's only relatively recently that we've started to put measures in place to stop this from continuing.
- Understand that the human population is growing very quickly and the standard of living is getting higher, which means that more resources are being used and more waste is being produced.
- Know that, unless dealt with properly, the increasing amounts of waste and chemicals being produced can lead to increasing amounts of pollution.
- Know how pollution can occur in water, land and air, and how this can reduce biodiversity.

Specification References
4.7.3.1, 4.7.3.2

How humans impact the environment makes it onto the news a lot nowadays — everything we do has an effect, and it's not usually good...

What is biodiversity?

Biodiversity is the variety of different species of organisms on Earth, or within an ecosystem.

High biodiversity is important. It makes sure that ecosystems (see page 315) are stable. Different species are interdependent — they depend on each other for things like shelter and food. Different species can also help to maintain the right physical environment for each other (e.g. the acidity of the soil). A high biodiversity means that one species is less likely to have to rely on a single different species for the resources and physical environment that it needs. This is because there's a higher chance that more than one species in the ecosystem can help provide the same resources and conditions.

For the human species to survive, it's important that a good level of biodiversity is maintained. Lots of human actions, including waste production (see next page) and deforestation (see p. 346), as well as global warming (see page 343) are reducing biodiversity. However, it's only recently that we've started taking measures to stop this from continuing.

The increasing human population

There are currently over seven billion people in the world — the population is rising very quickly, and it's not slowing down (see Figure 1).

This is mostly due to modern medicine and farming methods, which have reduced the number of people dying from disease and hunger. This is great for all of us humans, but it means we're having a bigger effect on the environment we live in.

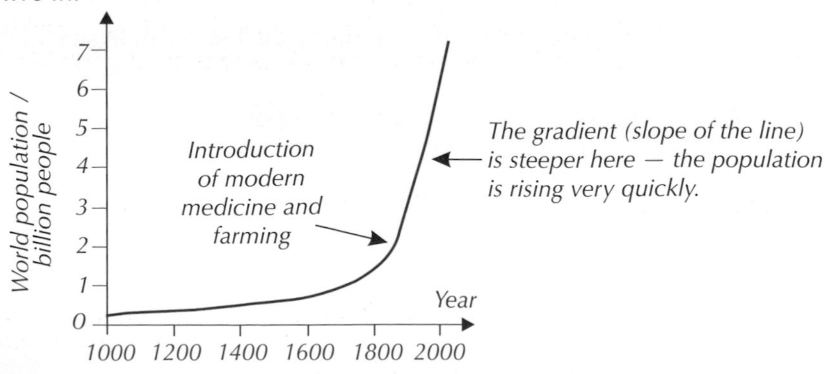

Figure 1: *Graph showing the increasing population of the world.*

Demands on the environment

When the Earth's population was much smaller, the effects of human activity were usually small and local. Nowadays though, our actions can have a far more widespread effect. Our increasing population puts pressure on the environment, as we take the resources we need to survive.

People around the world are also demanding a higher **standard of living** (and so demand luxuries to make life more comfortable — cars, computers, etc.). So we use more raw materials (e.g. oil to make plastics), but we also use more energy for the manufacturing processes. This all means we're taking more and more resources from the environment more and more quickly.

Unfortunately, many raw materials are being used up quicker than they're being replaced. So if we carry on like we are, one day we're going to run out.

Waste

As we make more and more things we produce more and more waste, including waste chemicals. And unless this waste is properly handled, more harmful pollution will be caused. Pollution affects water, land and air and kills plants and animals, reducing biodiversity.

Tip: Fertilisers are chemicals added to soil to provide nutrients for crop growth. Pesticides are chemicals that kill pests (like insects) that eat and damage the crops. Herbicides kill weeds that compete with the crops for nutrients, water, etc.

Water

Sewage and toxic chemicals from industry can pollute lakes, rivers and oceans, affecting the plants and animals that rely on them for survival (including humans). And the chemicals used on land to help grow crops (e.g. fertilisers, pesticides and herbicides) can be washed into water.

Example

If too much fertiliser is added to a field and it rains, then the excess fertiliser can get washed into nearby rivers and lakes. The nutrients in the fertiliser allow lots of algae to grow, causing an algal bloom (see Figure 2). The bloom blocks out the light, so plants in the water can't photosynthesise and die. Microorganisms feeding on the dead plants use up the oxygen in the water, which can eventually lead to the death of fish and other animals.

Figure 2: *Algal bloom in a pond. The algae is the thick yellow-green material that you can see covering the surface of the water.*

Land

We use toxic chemicals for farming (e.g. pesticides and herbicides). We also bury nuclear waste underground, and we dump a lot of household waste in landfill sites, where it doesn't break down so easily (and some doesn't break down at all).

Air

Smoke and acidic gases released into the atmosphere can pollute the air.

Tip: Finding land to dump and bury waste (landfill sites) will become more of a problem as the population grows and we produce more waste.

Examples

- Burning some fossil fuels releases sulfur dioxide. When sulfur dioxide mixes with clouds it forms dilute sulfuric acid, which falls as acid rain. Acid rain can damage buildings and kill plants.

Tip: Fossil fuels are fuels formed from the remains of organisms that died a very, very long time ago.
The three main types of fossil fuel are oil, gas and coal.

- Carbon dioxide (CO_2) is also released when fossil fuels are burnt. The increasing level of CO_2 is contributing to global warming and climate change (see page 343).

Practice Questions — Fact Recall

Q1 Explain why it's important to have high biodiversity.

Q2 True or false? The population of the world is decreasing, but people's standard of living is increasing.

Q3 How can the chemicals used to grow crops cause water pollution?

Q4 How do humans cause land pollution?

Practice Question — Application

Q1 The graph below shows the amount of waste recycled and the amount sent to landfill in one city for three separate years.

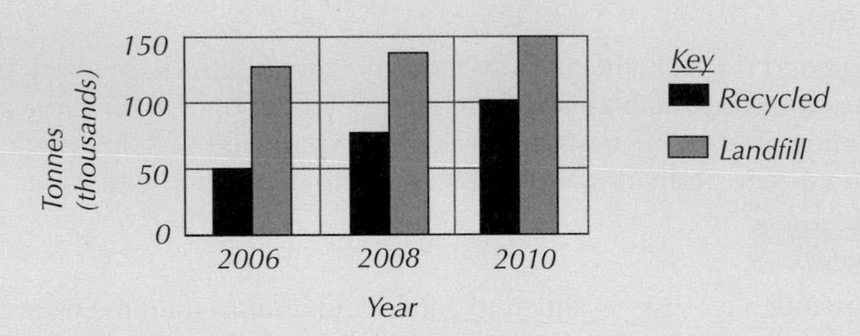

a) Describe the trends shown by the graph.

b) What percentage of the total waste shown in the graph for 2010 was recycled?

c) Suggest why the amount of space taken up by landfill sites around the world is increasing.

d) Suggest one reason why it is important to recycle products such as metal cans.

Figure 3: Landfill sites take up room, pollute the land around them and are ugly (they cause visual pollution).

Tip: Metal is a raw material.

2. Global Warming

You've probably heard the term 'global warming' thrown about. Now it's time to get down to the details of what it really means and what causes it.

Global warming

The temperature of the Earth is a balance between the energy it gets from the Sun and the energy it radiates back out into space. Gases in the atmosphere naturally act like an insulating layer. They absorb most of the energy that would normally be radiated out into space, and re-radiate it in all directions (including back towards the Earth). This increases the temperature of the planet. It's called the greenhouse effect — see Figure 1.

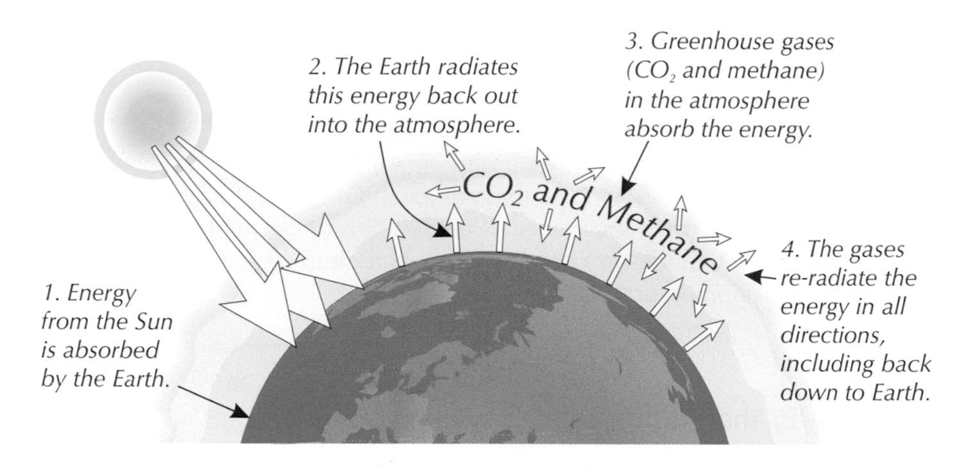

Figure 1: *Diagram illustrating the greenhouse effect.*

If this didn't happen, then at night there'd be nothing to keep any energy in, and we'd quickly get very cold indeed. But recently we've started to worry that this effect is getting a bit out of hand...

There are several different gases in the atmosphere which help keep the energy in. They're called "greenhouse gases", and the main ones whose levels we worry about are carbon dioxide (CO_2) and methane — because the levels of these two gases are rising quite sharply.

The Earth is gradually heating up because of the increasing levels of greenhouse gases — this is global warming. Global warming is a type of **climate change** and causes other types of climate change, e.g. changing rainfall patterns.

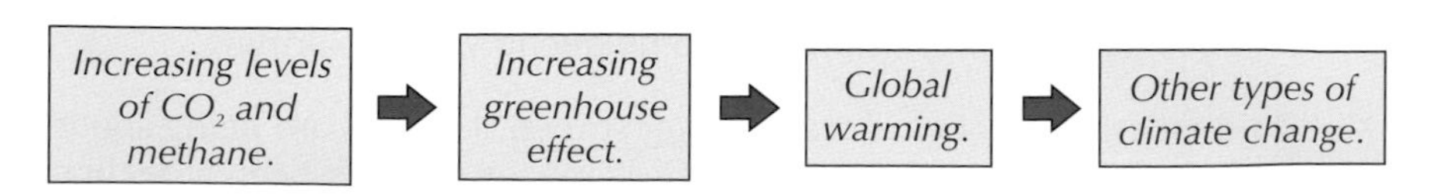

Figure 2: *A flow chart summarising the causes and effects of an increasing greenhouse effect.*

Learning Objectives:

- Know that the increasing levels of carbon dioxide and methane in the atmosphere are leading to global warming.
- Be able to recall some of the biological consequences that global warming can lead to.

Specification Reference
4.7.3.5

Tip: Scientists now agree that human activities are causing global warming — our activities are causing more greenhouse gases to be released. E.g. deforestation is causing more CO_2 to be released (see page 346). Burning fossil fuels is also causing more CO_2 to be released.

The consequences of global warming

Global warming could be a big problem for us. An increase of even a few degrees could have serious consequences for our planet, for example:

Rising sea level leading to habitat loss

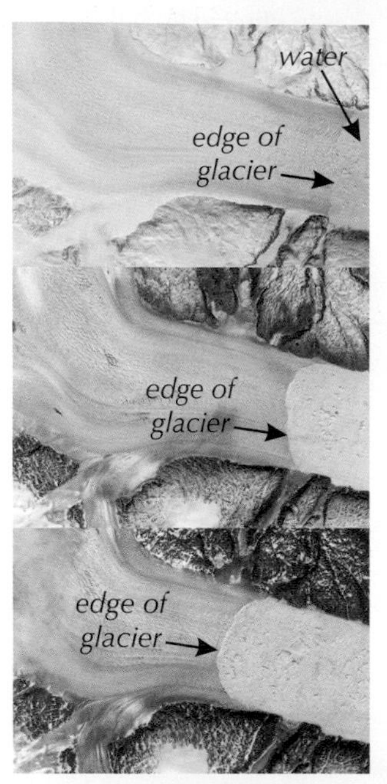

Figure 3: *Satellite images of the Helheim glacier in 2001 (top), 2003 (middle) and 2005 (bottom).*

As the sea gets warmer, it expands, causing sea level to rise. Sea level has risen a little bit over the last 100 years. If it keeps rising it'll be bad news for people and animals living in low-lying places like the Netherlands, East Anglia and the Maldives. It will lead to flooding, resulting in the loss of habitats (where organisms live).

Higher temperatures also make ice melt. Water that's currently 'trapped' on land (as ice in glaciers and at the poles) runs into the sea, causing sea level to rise even more.

Examples

- The coastline of Louisiana (a southern state of the USA) has seen a sea level rise of at least 8 inches in the last 50 years.
- The Helheim glacier (a thick mass of ice in Greenland) has retreated (melted) by 4 km in four years. We know this because of satellite images like the ones in Figure 3. The glacier had stayed in roughly the same position up until 2001.

Tip: A decrease in rainfall could lead to drought.

Changes in the distribution of organisms

The distribution of organisms is just where they're usually found within a certain area. E.g. in the UK, red squirrels are found in the north of England and in Scotland. Wild plants and animals are found in places where they can survive — where it's the right temperature and they have access to the right resources like water, food and shelter. As the average global temperature goes up and other changes to the weather occur (e.g. more or less rainfall), the distribution of many wild animal and plant species may change.

Some species may become more widely distributed, e.g. species that need warmer temperatures may spread further as the conditions they thrive in exist over a wider area. Other species may become less widely distributed, e.g. species that need cooler temperatures may have smaller ranges as the conditions they thrive in exist over a smaller area.

Examples

- The Comma is a species of butterfly that was until recently only found in England and Wales. Its distribution is extending further north though, into Scotland, and this is thought to be due to global warming.
- An increasing number of warm-water striped dolphins are being spotted in the waters around England. At the same time there's been a decline in spottings of the cold-water white-beaked dolphins. This could be because the sea around England is getting warmer.

Figure 4: *A Comma butterfly.*

Changes to migration patterns

Lots of birds and other animals migrate — they move to different areas at different times of the year. For example, swallows come to the UK in summer but they spend their winters in hotter countries. Global warming could lead to changes in migration patterns. This includes changes to where organisms migrate (e.g. some birds may migrate further north, as more northern areas are getting warmer), when they migrate, and possibly if they migrate at all.

Tip: Global warming is also affecting when animals and plants do other things, e.g. they might breed or flower at different times of the year.

Example

Studies have shown that some birds living in Africa over winter have started to migrate back to their European summer breeding grounds earlier than ever before. This could be directly due to temperature change, or because the temperature is causing the birds' food sources to appear at earlier times.

Less biodiversity

Biodiversity (see page 340) could be reduced — if some species are unable to adapt to a change in the climate, then they won't survive and the species will become extinct. For example, if it's too hot for them where they usually live but they can't survive in other habitats because there isn't the right food source, then they might die out.

Practice Questions — Fact Recall

Q1 True or False? Decreasing levels of greenhouse gases are contributing to an increase in the average global temperature.

Q2 Name the two main greenhouse gases we are worried about.

Practice Questions — Application

Q1 Since the 1950s, a species of bird has started to become more common in a small island country. The population of the species on the island started increasing as the average global temperature rose due to global warming. Scientists have suggested that it is the effect of global warming that has changed the distribution of the species.

a) Explain how global warming can lead to changes in the distribution of organisms.

b) Besides causing distribution changes, suggest one other way in which global warming might affect birds found in the country.

3. Deforestation and Land Use

The expanding human population means that we need more development. More development means that we need to use more land. Sometimes, that can lead to problems. Another page of bad news, I'm afraid...

Learning Objectives:

- Know that humans use land for things like building, quarrying, farming and waste disposal, which means that there's less land for other animals and plants.
- Know that humans cut down large areas of tropical forest to clear the land for livestock, rice crops and crops to make biofuels from.
- Know that humans sometimes destroy peat bogs and other areas of peat to use the peat in compost for gardeners.
- Know that the destruction of areas of peat results in more CO_2 being released into the atmosphere (when the peat is burnt or broken down by microorganisms).
- Know why the destruction of areas of peat reduces biodiversity.

Specification References
4.7.3.3, 4.7.3.4

Using land

Humans reduce the amount of land available to other animals and plants. The four main human activities that do this are:

- Building
- Quarrying
- Farming
- Dumping waste

Sometimes, the way we use land has a bad effect on the environment — for example, if it requires deforestation or the destruction of habitats like peat bogs and other areas of peat.

Deforestation

Deforestation is the cutting down of forests. This causes big problems when it's done on a large-scale, such as cutting down rainforests in tropical areas. It's done for various reasons, including:

- To clear land for farming (e.g. cattle or rice crops) to provide more food.
- To grow crops from which biofuels (such as those based on ethanol) can be produced.

Problems caused by deforestation

Deforestation leads to a variety of issues:

Less biodiversity

The more species, the greater the biodiversity. Habitats like tropical rainforests can contain a huge number of different species, so when they are destroyed there is a danger of many species becoming extinct — reducing biodiversity.

Tip: Head back to page 340 for more on biodiversity.

More carbon dioxide released into the atmosphere

As you saw on page 343, an increasing level of carbon dioxide in the atmosphere is contributing to global warming. Deforestation increases the amount of carbon dioxide released into the atmosphere, as carbon dioxide is released when trees are burnt to clear land — see Figure 1 on the next page. (Carbon in wood doesn't contribute to atmospheric pollution until it's released by burning.) Microorganisms feeding on bits of dead wood also release carbon dioxide as a waste product of respiration.

Tip: Remember, carbon dioxide (CO_2) is a greenhouse gas.

Less carbon dioxide taken in

Cutting down loads of trees means that the amount of carbon dioxide removed from the atmosphere during photosynthesis is reduced — see Figure 1. Trees 'lock up' some of the carbon that they absorb during photosynthesis in their wood, which can remove it from the atmosphere for hundreds of years. Removing trees means that less is locked up.

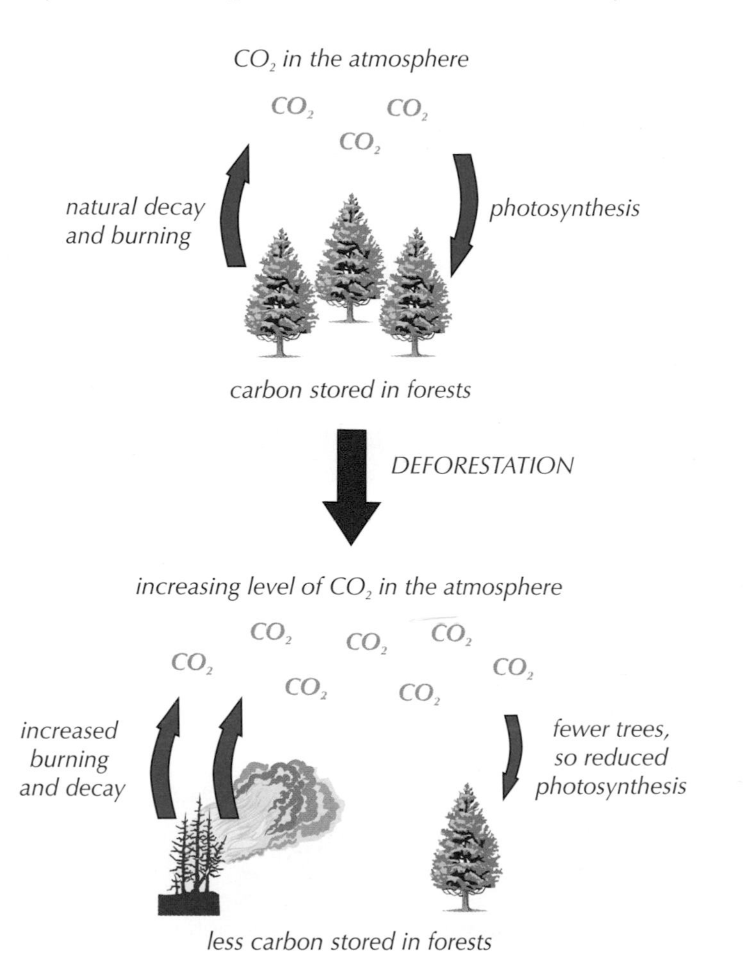

Figure 1: *The amount of carbon dioxide in the atmosphere should stay fairly constant (top cycle). But deforestation leads to an increase in the atmospheric carbon dioxide level (bottom cycle).*

Tip: The destruction of peat bogs also leads to the release of more CO_2 (see next page).

The destruction of peat bogs

Bogs are areas of land that are acidic and waterlogged. Plants that live in bogs don't fully decay when they die, because there's not enough oxygen. The partly-rotted plants gradually build up to form **peat** (a brown, soil-like material — see Figure 2). So the carbon in the plants is stored in the peat instead of being released into the atmosphere.

Figure 2: *Some peat.*

However, peat bogs are often drained so that the area can be used as farmland, or the peat is cut up and dried to use as fuel. It's also sold to gardeners as compost. Peat is being used faster than it forms.

Figure 3: *A peat bog that's had some peat harvested from it. The brown trench is where peat blocks have been dug up and taken away.*

When peat is drained, it comes into more contact with air and some microorganisms start to decompose it. These microorganisms use oxygen and substances found in the peat as reactants for respiration. They release carbon dioxide as a product of respiration, contributing to global warming (see page 343). Carbon dioxide is also released when peat is burned as a fuel.

Destroying the bogs also destroys (or reduces the area of) the habitats of some of the animals, plants and microorganisms that live there, so reduces biodiversity.

Practice Questions — Fact Recall

Q1 Give three things that humans use land for that reduces the amount of land available to other animals and plants.

Q2 What is deforestation?

Practice Questions — Application

Q1 The graph shows how much forest was cleared each year in one region of a rainforest and the use of the area after clearance.

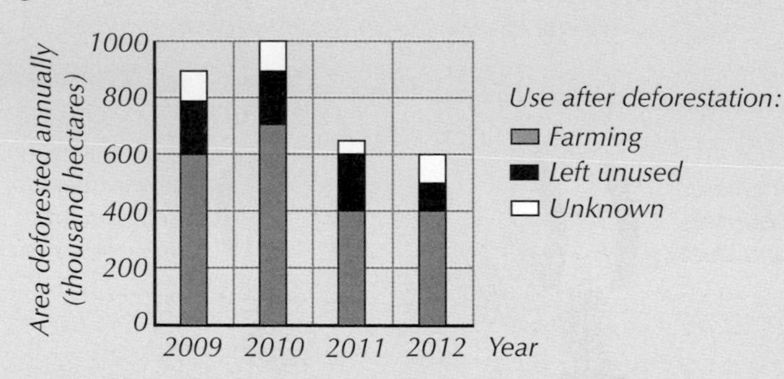

a) How much forest was cleared in this region in 2009?

b) i) How much land was used for farming after clearance in 2012?

ii) What kind of farming might be taking place on the cleared land?

Q2 Dave needs some compost for his vegetable plot. A peat-free variety is being advertised as an environmentally-friendly option. Explain why the manufacturers are making this claim.

Tip: The bar chart in Q1 is a composite bar chart — it's just like a normal one except the bars are placed on top of each other instead of side by side. Here's the information for 2011 shown in composite form (left) and normal form (right):

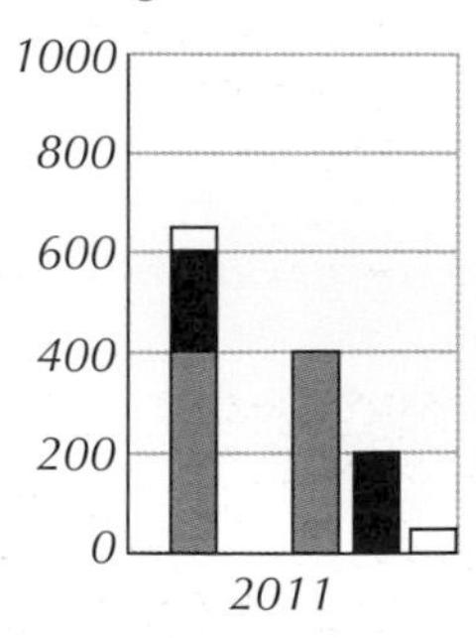

4. Maintaining Ecosystems and Biodiversity

You learned about biodiversity back on p. 340 and ecosystems on p. 315. It's important that we protect and maintain them. These pages tell you about some of the methods being used, as well as some of the issues involved.

Protection programmes

It's important that biodiversity is maintained at a high enough level to make sure that ecosystems are stable.

In some areas, programmes have been set up by concerned citizens and scientists to minimise damage by human activities to ecosystems and biodiversity.

You need to know some examples of these programmes:

Example

- Breeding programmes have been set up to help prevent endangered species from becoming extinct. These are where animals are bred in captivity to make sure the species survives if it dies out in the wild. Individuals can sometimes be released into the wild to boost or re-establish a population.
- Programmes to protect and regenerate rare habitats like mangroves (areas of mangrove trees, which grow in the mud in coastal regions that flood twice a day with the tide and have roots partly above ground), heathland (an open landscape made up mainly of short shrubs like heather) and coral reefs. Protecting these habitats helps to protect the species that live there — preserving the ecosystem and biodiversity in the area.
- There are programmes to reintroduce hedgerows and field margins around fields on farms where only a single type of crop is grown. Field margins are areas of land around the edges of fields where wild flowers and grasses are left to grow. Hedgerows and field margins provide a habitat for a wider variety of organisms than could survive in a single crop habitat.
- Some governments have introduced regulations and programmes to reduce the level of deforestation taking place and the amount of carbon dioxide being released into the atmosphere by businesses. This could reduce the increase of global warming (see page 343).
- People are encouraged to recycle to reduce the amount of waste that gets dumped in landfill sites. This could reduce the amount of land taken over for landfill, leaving ecosystems in place.

Learning Objectives:

- Know that programmes have been set up to reduce the effects of human activities on ecosystems and biodiversity.
- Know the following examples of programmes designed to maintain ecosystems and biodiversity:
  - Breeding programmes,
  - Programmes to protect and regenerate rare habitats,
  - Programmes to reintroduce field margins and hedgerows to single-crop fields,
  - Programmes to reduce deforestation and carbon dioxide emissions,
  - Recycling programmes.
- Know examples of the positive and negative effects of human interaction with ecosystems, and be able to evaluate their effect on biodiversity.

Specification Reference 4.7.3.6

Figure 1: *There are breeding programmes in place to protect various endangered animals, such as the Mauritius Olive White-eye shown in this photo.*

Tip: You need to be able to describe some of the positive and negative ways in which humans interact with ecosystems and understand how they impact biodiversity. You can read about some of the negatives on pages 341-348. Page 349 covers some of the positives.

Conflicting pressures of maintaining biodiversity

Maintaining biodiversity isn't as simple as you would hope. There are lots of conflicting pressures that have to be taken into account:

Examples

- Protecting biodiversity costs money. For example, governments sometimes pay farmers a subsidy to reintroduce hedgerows and field margins to their land. It can also cost money to keep a watch on whether the programmes and regulations designed to maintain biodiversity are being followed. There can be conflict between protecting biodiversity and saving money — money may be prioritised for other things.

- Protecting biodiversity may come at a cost to local people's livelihood. For example, reducing the amount of deforestation is great for biodiversity, but the people who were previously employed in the tree-felling industry could be left unemployed. This could affect the local economy if people move away with their family to find work.

- There can be conflict between protecting biodiversity and protecting our food security. Sometimes certain organisms are seen as pests by farmers (e.g. locusts and foxes) and are killed to protect crops and livestock so that more food can be produced. As a result, however, the food chain and biodiversity can be affected.

- Development is important, but it can affect the environment. Many people want to protect biodiversity in the face of development, but sometimes land is in such high demand that previously untouched land with high biodiversity has to be used for development, e.g. for housing developments on the edge of towns, or for new agricultural land in developing countries.

Figure 2: *This photo shows a large area of rainforest being cleared for use as agricultural land.*

Practice Questions — Fact Recall

Q1 What is the purpose of a breeding programme?

Q2 Outline and explain what a farmer who only grows one crop might do to protect biodiversity.

Q3 Give one reason why people are encouraged to recycle as much waste as possible.

Topic 7b Checklist — Make sure you know...

Biodiversity and Waste

- [ ] The meaning of 'biodiversity' — the variety of species of organisms on Earth, or within an ecosystem.
- [ ] The importance of biodiversity in maintaining stable ecosystems, which often involve lots of interdependent species — e.g. species relying on each other for things like food, shelter and keeping the physical environment right.
- [ ] That high biodiversity means that one species is less dependent on another single species, because there is more likely to be multiple species in the ecosystem that can help provide the resources and physical environment that it needs.
- [ ] That for the survival of the human species, it's important that biodiversity is maintained at a good level.
- [ ] That many human activities lead to a reduction in biodiversity, and that actions to combat this have only been put into place relatively recently.
- [ ] That the rapidly growing human population and the increase in the standard of living is putting a strain on resources and producing increasing amounts of waste.
- [ ] How the increasing amounts of waste cause pollution of water, land and air, and that this can kill plants and animals, which can result in a reduction in biodiversity.

Global Warming

- [ ] That increasing levels of carbon dioxide and methane, which trap energy inside the Earth's atmosphere, are contributing to global warming, causing the average global temperature to rise.
- [ ] Examples of how global warming can have biological consequences for the planet.

Deforestation and Land Use

- [ ] That humans use lots of land for activities like building, quarrying, farming and dumping waste, and that this reduces the amount of land available to other species.
- [ ] That deforestation of large areas of tropical forest is sometimes carried out by humans to make the land available for things like farming cattle and growing crops for food, or for growing crops for biofuel production.
- [ ] That humans often destroy peat bogs and other areas of peat, e.g. to use the peat in garden compost.
- [ ] That when areas of peat are destroyed, there is an increase in the amount of carbon dioxide released into the atmosphere due to the burning and decomposition of peat, and biodiversity is also reduced.

Maintaining Ecosystems and Biodiversity

- [ ] That there are programmes to help reduce the effects of human activities on ecosystems, such as breeding programmes for endangered species, protection and regeneration of rare habitats, programmes to introduce margins and hedgerows around fields, schemes to reduce deforestation and carbon dioxide emissions, and recycling programmes.
- [ ] Examples of both positive and negative ways that humans interact with ecosystems, and how these can affect the level of biodiversity in an ecosystem.

Exam-style Questions

1 **Figure 1** shows the edge of a mangrove forest where it meets the water. Mangrove forests are ecosystems found in tropical and subtropical coastal areas. The trees in mangrove forests form branching root systems that are partly exposed above the ground. Mangrove forests frequently flood with sea water as the tide comes in. They're important habitats for a huge variety of organisms, including fish, crabs, birds and even Bengal tigers.

Enormous areas of mangrove forest have been destroyed to make way for large-scale shrimp farming. This has badly affected biodiversity in the areas involved.

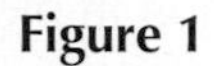

Figure 1

1.1 Explain what is meant by biodiversity.

(1 mark)

1.2 Some projects now aim to protect the biodiversity of mangrove forests by protecting the remaining areas of mangrove forest from being used for shrimp farming.

Suggest **two** pressures that could affect the success of these projects.

(2 marks)

1.3 The reason that shrimp farming has increased is not due to the increasing requirement for food to feed the growing human population.

Suggest a reason for why the size of the shrimp farming industry has increased.

(2 marks)

Global warming is a major issue facing modern scientists and society as a whole. It is caused by increasing levels of certain gases.

1.4 Explain why clearing areas of mangrove forest could contribute to global warming.

(4 marks)

1.5 Explain how global warming could reduce biodiversity.

(2 marks)

1. Trophic Levels

You might recognise some of this stuff about food chains from page 323. Unfortunately, now you need to learn a bit more about it. Enter the exciting world of trophic levels...

Learning Objectives:
- Know that each trophic level in a food chain is assigned a number based on its position in the food chain, beginning with plants and algae at level 1.
- Know the differences between the different trophic levels of an ecosystem.
- Know that trophic level 1 contains the producers.
- Know that trophic level 2 contains the primary consumers, which are herbivores.
- Know that trophic level 3 contains the secondary consumers, which are carnivores.
- Know that trophic level 4 contains carnivores called tertiary consumers.
- Know that apex predators have no predators, so are at the top of the food chain.
- Understand how decomposers break down dead plant and animal material into small soluble food molecules.

Specification Reference 4.7.4.1

What are trophic levels?

Trophic levels are the different stages of a food chain. They consist of one or more organisms that perform a specific role in the food chain. Trophic levels are named after their location in the food chain using numbers. The first level is called trophic level 1. Each level after that is numbered in order based on how far along the food chain the organisms in the trophic level are. For example, the organisms second in line in the food chain belong to trophic level 2, organisms third in line are in trophic level 3, and so on. You need to know the differences between the different trophic levels:

Trophic level 1

Trophic level 1 contains producers. Producers are the organisms at the starting point of a food chain, e.g. plants and algae. They're called producers because they make their own food by photosynthesis using energy from the Sun.

Trophic level 2

Trophic level 2 contains primary consumers. Herbivores that eat the plants and algae are primary consumers. Herbivores eat only plants and algae.

Trophic level 3

Trophic level 3 contains secondary consumers. Carnivores that eat the primary consumers are secondary consumers. Carnivores are meat eaters.

Trophic level 4

Trophic level 4 contains tertiary consumers. Carnivores that eat other carnivores (the secondary consumers) are tertiary consumers. Carnivores that have no predators are at the top of the food chain, so they're always in the highest trophic level. They're known as apex predators.

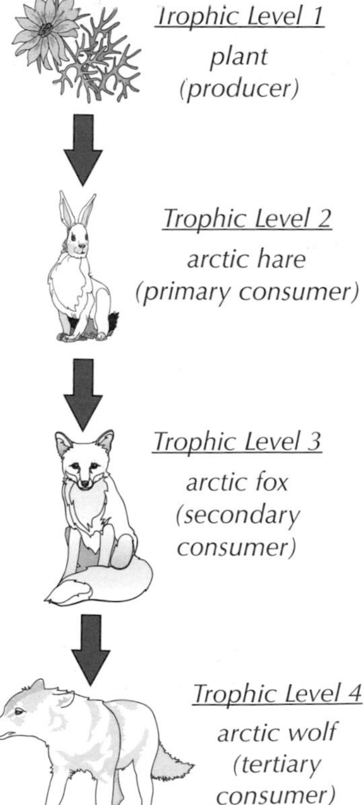

Figure 1: *A food chain demonstrating the different trophic levels.*

Tip: There can be more than four trophic levels in a food chain, but there are only usually 4 or 5 because so much energy is lost from the food chain at each trophic level (see page 358).

Decomposers

Decomposers such as bacteria and fungi play an important role in ecosystems. They decompose any dead plant or animal material left in an environment. They can do this by secreting (releasing) enzymes that break the dead stuff down into small soluble food molecules. These then diffuse into the microorganisms. This process also releases nutrients into the environment, which the producers need in order to grow.

Tip: There's more on decomposition on page 332.

Practice Questions — Fact Recall

Q1 What are trophic levels?

Q2 What are apex predators?

Q3 What role do decomposers play in an ecosystem?

Practice Question — Application

Q1 The organisms below are all part of the same food chain.

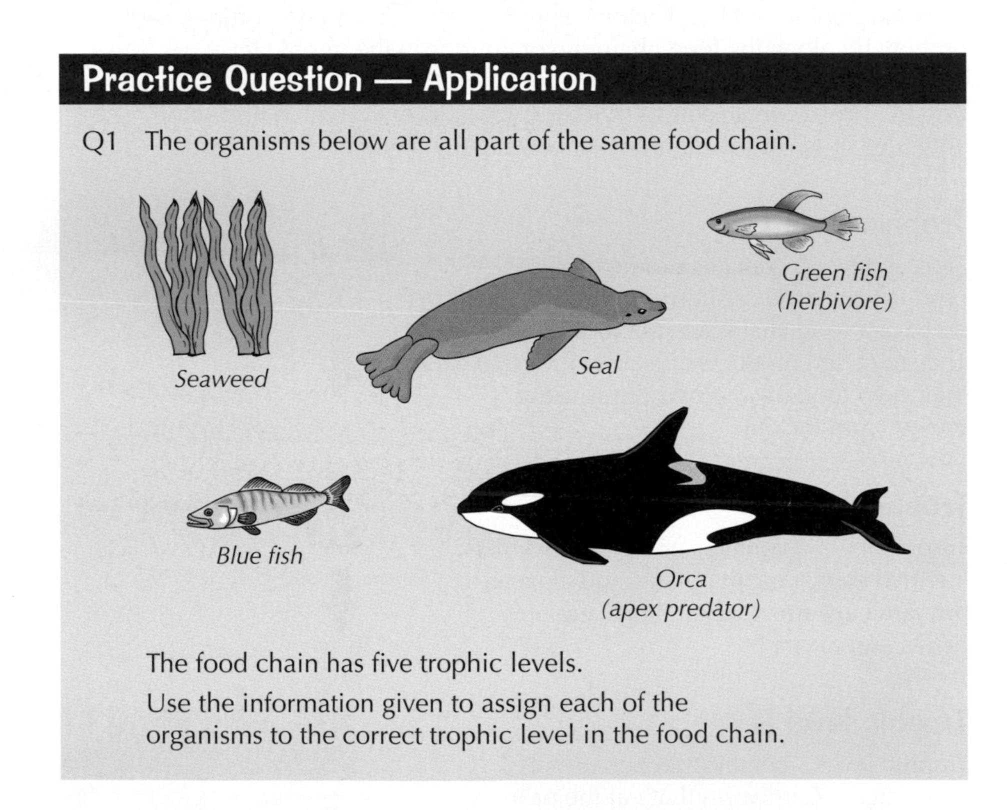

The food chain has five trophic levels.

Use the information given to assign each of the organisms to the correct trophic level in the food chain.

2. Pyramids of Biomass

They're perhaps slightly less exciting than the Egyptian pyramids, but pyramids of biomass are an important topic. They give information about the amount of biomass at each trophic level in a food chain.

Learning Objectives:
- Know how the relative amounts of biomass present in each trophic level of a food chain can be represented in a pyramid of biomass.
- Understand how the number of organisms in each trophic level is affected by the loss of biomass at each trophic level.
- Know that the bottom level of a pyramid of biomass always represents trophic level 1.
- Be able to draw pyramids of biomass using given data.
- Be able to describe what pyramids of biomass show.

Specification References
4.7.4.2, 4.7.4.3

What are pyramids of biomass?

Pyramids of biomass show the relative mass of each trophic level. There's less energy and less biomass every time you move up a stage (trophic level) in a food chain. So there are usually fewer organisms every time you move up a level too, as there's less energy available to support them.

Example

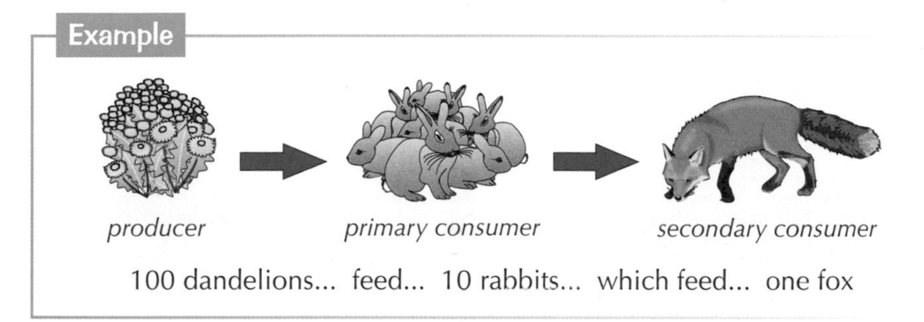

100 dandelions... feed... 10 rabbits... which feed... one fox

This isn't always true though — for example, if 500 fleas are feeding on the fox, the number of organisms has increased as you move up to that stage in the food chain. So a better way to look at the food chain is often to think about biomass instead of number of organisms. You can use information about biomass to construct a pyramid of biomass to represent the food chain.

Tip: Biomass just means the mass of living material.

Each bar on a pyramid of biomass shows the relative mass of living material at a trophic level — basically how much all the organisms at each level would "weigh" if you put them all together. So the one fox above would have a big biomass and the hundreds of fleas would have a very small biomass. Biomass pyramids are practically always pyramid-shaped.

Tip: See page 358 for the reasons why biomass is lost at each trophic level.

Example

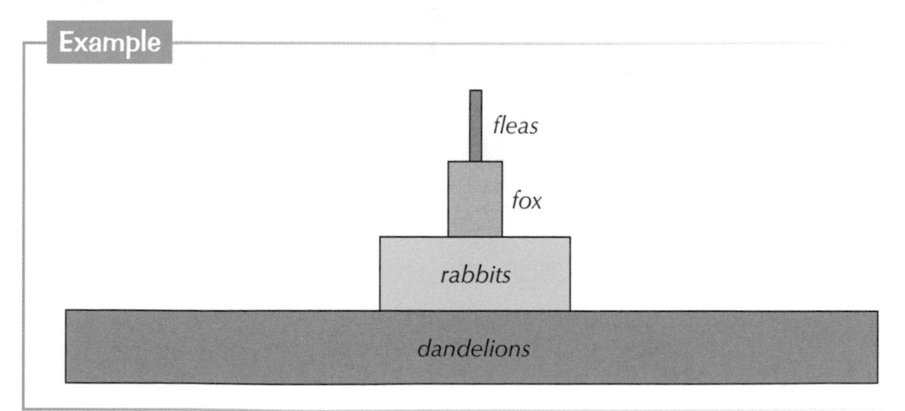

The big bar along the bottom of the pyramid shows trophic level 1. It always represents the producer (e.g. plants or algae). The next bar will be the primary consumer (the animal that eats the producer), then the secondary consumer (the animal that eats the primary consumer) and so on up the food chain.

Constructing a pyramid of biomass

You need to be able to construct pyramids of biomass. Luckily it's pretty simple — they'll give you all the information you need to do it in the exam.

Example

A great tit weighing 20 g feeds on caterpillars weighing 190 g in total, which in turn feed on a rose bush weighing 2000 g. Draw and label a pyramid of biomass to represent this food chain.

Here's what you need to remember when drawing a pyramid like this:

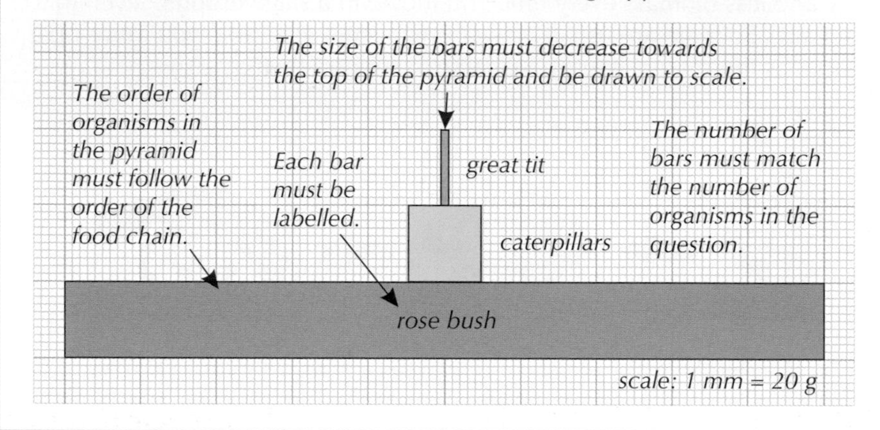

Exam Tip
If you're asked to construct a pyramid of biomass from scratch, you may need to decide on a suitable scale for the bars. Whatever scale you pick, the largest bar needs to fit in the space you're provided with. In this example, each small square on the graph paper measures 1 mm^2. The largest bar is 86 mm across, so that makes each mm square across worth 0.5 kg. All of your bars have to use the same scale.

Exam Tip
The space you have to write your answer in the exam might not be graph paper like it is here. If this is the case, you need to be really careful and use a ruler to measure the bars so that you can draw them accurately to scale.

Interpreting a pyramid of biomass

It's easy to look at pyramids of biomass and explain what they show about the food chain — just remember, the biomass at each stage should be drawn to scale.

Example

Look at the pyramid of biomass below:

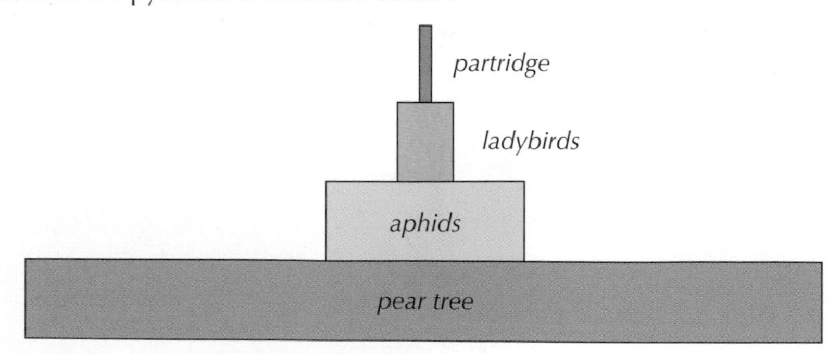

Even if you know nothing about the natural world, you're probably aware that a tree is quite a bit bigger than an aphid. So what's going on here is that lots (probably thousands) of aphids are feeding on a great big tree. Quite a lot of ladybirds are then eating the aphids, and a few partridges are eating the ladybirds.

Biomass and energy are still decreasing as you go up the levels — it's just that one tree can have a very big biomass, and can fix a lot of the Sun's energy using all those leaves.

Tip: Remember, pyramids of biomass represent food chains, so if the number of organisms in one trophic level changes (e.g. if their population size decreases), this will affect their biomass and the biomass of organisms elsewhere in the food chain.

Tip: Plants fix the Sun's energy during photosynthesis (the process by which plants make their own food).

You could also be expected to do a bit of maths, such as working out the ratio of the biomass at different levels.

Example

You might be given a pyramid of biomass like this one and be asked to work out the ratio of the biomass of aphids to the biomass of the pear tree.

So you need to count how many squares wide the aphid and pear tree bars are and then work out the ratio. Like this:

aphid bar = 5 squares

pear tree bar = 50 squares

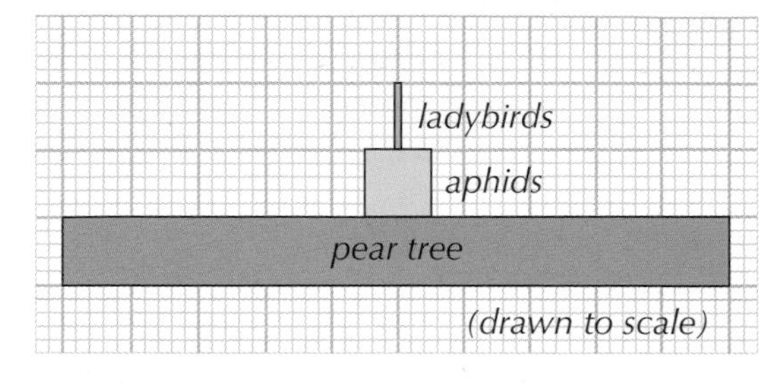

So the ratio of the aphids' biomass to the pear tree's biomass is 5:50, which is the same as 1:10.

Tip: There's more about working out ratios on page 377.

Practice Questions — Fact Recall

Q1 What is biomass?

Q2 Why is a pyramid of biomass nearly always pyramid shaped?

Q3 Which trophic level is always shown by the bottom bar on a pyramid of biomass?

Practice Questions — Application

Q1 Look at this pyramid of biomass.

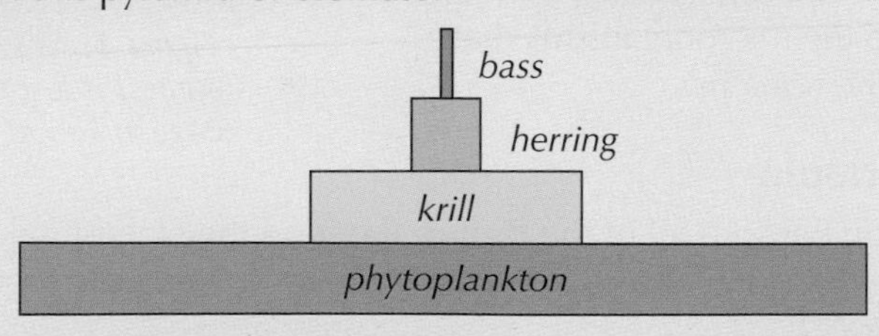

a) Which organism is the producer?

b) Which organism has the smallest biomass?

c) Why is the bar representing the herring smaller than the bar representing the krill?

3. Biomass Transfer

Biomass is transferred along a food chain, but lots is lost along the way...

Learning Objectives:
- Know that plants and algae are the producers in a food chain, and that they transfer approximately 1% of the energy from the light that hits them for use in photosynthesis.
- Know that only about 10% of the biomass in a trophic level is passed on to the next trophic level in a food chain.
- Know that biomass is lost at each trophic level because some of the ingested material isn't absorbed.
- Know that lots of glucose (obtained from biomass) is used by organisms for respiration.
- Know that biomass is also lost because some of the material that is absorbed is converted into other substances that are lost as waste.
- Be able to calculate the efficiency of biomass transfer from one trophic level to the next.

Specification Reference 4.7.4.3

How is biomass transferred?

Energy from the Sun is the source of energy for nearly all life on Earth. Producers, such as green plants and algae, use energy transferred by light from the Sun to make food (glucose) during photosynthesis. Of the energy that hits these producers, only about 1% is transferred for photosynthesis. Some of the glucose is used by the plants and algae to make biological molecules. These biological molecules make up the plant's biomass — the mass of living material. Biomass stores energy.

Biomass is transferred through a food chain in an ecosystem when organisms eat other organisms. However, not much biomass gets transferred from one trophic level to the next. In fact, only about 10% of the biomass is passed on to the next level.

How is biomass lost?

Biomass is lost at each stage of the food chain. It's lost for a number of different reasons:

Uneaten material

Organisms don't always eat every single part of the organism they're consuming. For example, some material that makes up plants and animals is inedible (e.g. bone). This means that not all the biomass can be passed to the next stage of the food chain. Also some organisms die before they're eaten, so their remains are left to decay and their energy doesn't get passed along the food chain (instead the energy gets passed to the microorganisms that break down the remains).

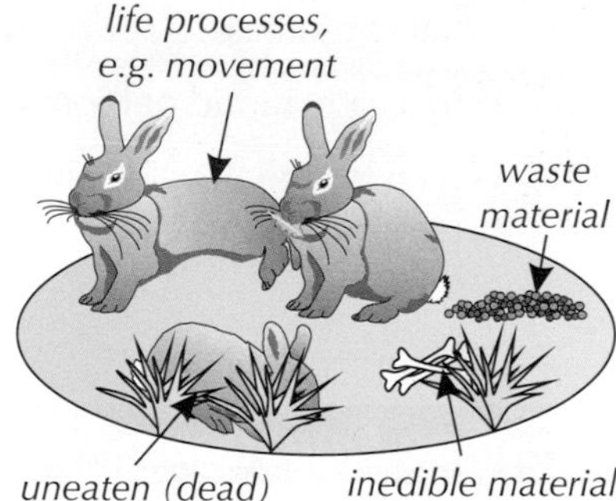

Figure 1: *Ways that biomass is lost at each stage in a food chain.*

Waste products

Organisms don't absorb all of the stuff in the food they ingest (take in). The stuff that they don't absorb is egested (released) as faeces (poo).

Some ingested biomass is converted into other substances that are lost as waste. For example, organisms use a lot of glucose (obtained from the biomass) in respiration to provide energy for movement and keeping warm, etc. rather than to make more biomass. This is especially true for mammals and birds, whose bodies must be kept at a constant temperature which is normally higher than their surroundings. This process produces lots of waste carbon dioxide and water as by-products. Urea is another waste substance, which is released in urine with water when the proteins in the biomass are broken down.

Tip: There's more on microorganisms and decay on page 332.

Tip: Think of all the food (energy) you've eaten in your life. If someone ate you they wouldn't get all that energy — most of it will have been used up just keeping you alive or has passed out of your body as waste products.

Biomass calculations

You can work out how much biomass has been lost at each level by taking away the biomass that is available at that level from the biomass that was available at the previous level.

You can also calculate the **efficiency of biomass transfer** between trophic levels using this formula:

$$\text{efficiency} = \frac{\text{biomass transferred to the next level}}{\text{biomass available at the previous level}} \times 100$$

Exam Tip
Always double check that you're using the formula the right way round, otherwise you could be chucking away relatively easy marks.

Example

MATHS SKILLS

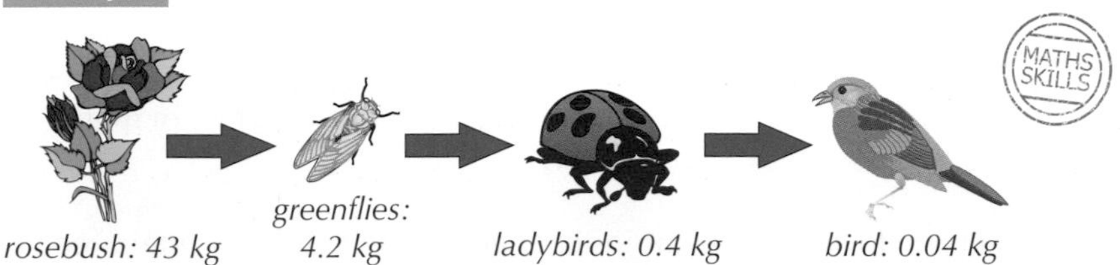

The numbers show the amount of biomass available to the next level. So 43 kg is the amount of biomass available to the greenflies, and 4.2 kg is the amount available to the ladybirds.

Biomass lost at the 1st trophic level

= 43 kg – 4.2 kg

= **38.8 kg**

Efficiency of biomass transfer at the 1st trophic level

$= \frac{4.2 \text{ kg}}{43 \text{ kg}} \times 100$

= **9.8%**

Exam Tip
In the exam, the question might talk about a food chain that you've never seen before. You don't need to worry about what organisms are involved — the only information you need is the amount of biomass at each trophic level.

Practice Questions — Fact Recall

Q1 True or false? Only about 10% of the biomass at one trophic level is passed on to the next.

Q2 Some biomass is lost between trophic levels when biomass is converted into substances that are lost as waste, such as urea. Give two other ways that biomass is lost between trophic levels.

Practice Question — Application

Q1 The diagram shows the biomass at each level of a food chain.

Trophic level 1 *57 kg* → *Trophic level 2* *5.8 kg* → *Trophic level 3* *0.56 kg* → *Trophic level 4* *0.055 kg*

Calculate the efficiency of biomass transfer between the 3rd and 4th trophic levels.

4. Food Security and Farming

In the UK, it's easy to take the food available to us for granted. But on a global scale, the supply of plentiful and nutritious food to feed the world's population isn't as secure as we would like. We've also got to think about how our food is produced and whether our food supply is sustainable.

Learning Objectives:

- Know what is meant by 'food security'.
- Be able to give examples of biological factors that affect food security.
- Know that we must find sustainable methods to feed the entire world population.
- Know that fish stocks must be maintained at a sustainable level that allows breeding to occur, otherwise some species will completely disappear from the oceans.
- Understand how fishing quotas and regulating the size of fishing nets could conserve fish stocks.
- Know that food production from livestock can be made more efficient by reducing the transfer of energy from the livestock to the environment, and how this can be done.
- Know that the growth of animals can be increased by feeding them high-protein foods.

Specification References 4.7.5.1, 4.7.5.2, 4.7.5.3

Factors affecting food security

Food security is having enough food to feed the population. There's a wide range of things that can threaten food security:

Examples

- The world population keeps increasing, with the birth rate of many developing countries rising quickly.
- As diets in developed countries change, the demand for certain foods to be imported from developing countries can increase. This means that already scarce food resources can become more scarce.
- Farming can be affected by new pests and pathogens (e.g. bacteria and viruses) or changes in the environmental conditions (e.g. a lack of rain). This can result in the loss of crops and livestock, and can lead to widespread famine.
- The high input costs of farming (e.g. the price of seeds, machinery and livestock) can make it too expensive for people in some countries to start or maintain food production, meaning that there sometimes aren't enough people producing food in these areas to feed the people.
- In some parts of the world, there are conflicts that affect the availability of food and water.

Sustainable methods of food production are needed so that enough food can be made to feed everyone now and in the future. Sustainable production means making enough food without using resources faster than they renew.

Overfishing

Overfishing (catching too many fish) is reducing fish stocks in the oceans. This means there's less fish for us to eat, the ocean's food chains are affected and some species of fish may disappear altogether in some areas — for example, cod are at risk of disappearing from the north west Atlantic.

To tackle this problem, we need to maintain fish stocks at a level where the fish continue to breed. This is sustainable food production. Fish stocks can be maintained (conserved) in several ways.

Tip: Fish stocks are the numbers of fish of different species we think are left in the seas and oceans.

Fishing quotas

There are limits on the number and size of fish that can be caught in certain areas. This prevents certain species from being overfished.

Net size

There are different limits of the mesh size of the fish net, depending on what's being fished. This is to reduce the number of 'unwanted' and discarded fish — the ones that are accidently caught, e.g. shrimp caught along with cod. Using a bigger mesh size will let the 'unwanted' species escape. It also means that younger fish will slip through the net, allowing them to reach breeding age.

Efficiency of food production

Food production can be made more efficient. Limiting the movement of livestock and keeping them in a temperature-controlled environment reduces the transfer of energy from livestock to the environment. This makes farming more efficient as the animals use less energy moving around and controlling their own body temperature. This means that more energy is available for growth, so more food can be produced from the same input of resources.

Figure 1: *Chickens being farmed in barns. The temperature can be controlled, so less energy is wasted by the chickens.*

Examples

- Livestock like calves and chickens can be factory farmed. This involves raising them in small pens.
- Fish can also be factory farmed in underwater cages where their movement is restricted.
- Some animals are also fed high-protein food to further increase their growth.

The positives and negatives of factory farming

Factory farming helps us improve the efficiency of food production. This is useful — it means cheaper food for us and generally better standards of living for farmers. It also helps to feed an increasing human population. But it also has disadvantages:

Tip: Factory farmed chickens are also called battery chickens.

- Some people think that forcing animals to live in unnatural and uncomfortable conditions is cruel. There's a growing demand for farmed products produced with animal welfare taken into account.

Example

Food labelled as 'free range' lets consumers know that the animals used to produce it were free to roam around outdoors. Organic meat production also has high animal welfare standards and the animals aren't given growth hormones. Organic and free range farming is less efficient than intensive farming, so organic and free range foods cost more than if they were produced using intensive farming methods. But some people are willing to pay this extra cost.

- The crowded conditions on factory farms create a favourable environment for the spread of diseases.

Examples

- Avian flu is a serious concern on poultry farms. It's a highly contagious disease between birds, so is easily spread between poultry on factory farms. It can be fatal for some types of domesticated bird such as chickens, and infected flocks are often destroyed, meaning that farmers face losing lots of livestock in the event of an outbreak on their farm.
- Foot-and-mouth disease is another highly infectious disease that can easily spread in the conditions on factory farms. It can affect sheep, pigs and cattle, and animals that become infected can't be eaten by humans, so farmers with infected livestock lose lots of money.

- It requires lots of energy to maintain the temperature-controlled environment. This often means using power from fossil fuels — which we wouldn't be using if the animals were grazing in their natural environment.

Choosing what methods to use for farming can involve conflict and has to involve compromise if we are going to keep feeding the increasing population.

Practice Questions — Fact Recall

Q1 What is meant by 'food security'?

Q2 Describe two methods of conserving fish stocks.

Q3 a) Describe the conditions used to factory farm cows.

b) Explain why these conditions make food production more efficient.

Practice Questions — Application

Q1 Although food security is generally good in developed countries, food security in developing countries is often under threat. Describe two factors that can threaten food security in developing countries.

Q2 In a shop, a box of 6 eggs from chickens that were free to roam outside (free range eggs) costs £1.80 and a box of 6 factory farmed ones costs £1.

a) Why do the free range eggs cost more to buy?

b) Suggest why someone might still choose to buy the free range eggs instead of the factory farmed ones.

5. Biotechnology

Biotechnology is an ever-expanding area of science that uses the power of biology to produce useful things. These pages cover a few examples of its uses.

Mycoprotein

As the population of the world increases, it'll be important to find new food sources to add to those that currently exist. Using modern biotechnology techniques, large amounts of microorganisms can be cultured (grown) industrially under controlled conditions in large vats for use as a food source.

A fairly modern food source that is becoming increasingly popular is mycoprotein. Mycoprotein means protein from fungi. It's used to make high-protein meat substitutes for vegetarian meals, e.g. Quorn™. A fungus called ***Fusarium*** is the main source of mycoprotein. It is grown in large vats on glucose syrup, which acts as food for the fungus.

The fungus respires aerobically, so oxygen is supplied, together with nitrogen (as ammonia) and other minerals. The mixture is also kept at the right temperature and pH. Once ready, the fungal biomass is harvested, purified and dried to make the mycoprotein. It's then processed further by adding flavourings and other ingredients.

Producing human insulin

Genetic engineering is transferring a useful gene from one organism to another (see page 294). Bacteria can be genetically engineered to make human insulin:

1. A plasmid (a loop of DNA) is removed from a bacterium.
2. The insulin gene is cut out of a human chromosome using a restriction enzyme. Restriction enzymes recognise specific sequences of DNA and cut the DNA at these points. The cut leaves one of the DNA strands with unpaired bases — this is called a 'sticky end'.
3. The plasmid is cut open using the same restriction enzyme — leaving the same sticky ends.
4. The plasmid and the human insulin gene are mixed together.
5. Ligase (an enzyme) is added. This joins the sticky ends together to produce recombinant DNA (two different bits of DNA stuck together).
6. The recombinant DNA is inserted into a bacterium.
7. The modified bacterium is grown in a vat under controlled conditions. You end up with millions of bacteria that produce insulin. The insulin can be harvested and purified to treat people with diabetes.

Figure 3 on the next page summarises these steps.

Learning Objectives:

- Know that we need to find new food sources to meet the demands of a growing population.
- Know that we can now culture large amounts of microorganisms in industrial vats under controlled conditions to use as a source of food.
- Know that mycoprotein, made from the fungus *Fusarium*, is an example of a food source made from microorganisms, and know how it is produced.
- Know that human insulin can be produced by genetically modified bacteria in order to treat diabetes.
- Know that genetically modified (GM) crops could be used to produce more food or food with better nutritional content, e.g. golden rice.

Specification Reference 4.7.5.4

Figure 1: *A selection of food products made from mycoprotein.*

Figure 2: *Genetically engineered bacteria being grown in an industrial vat.*

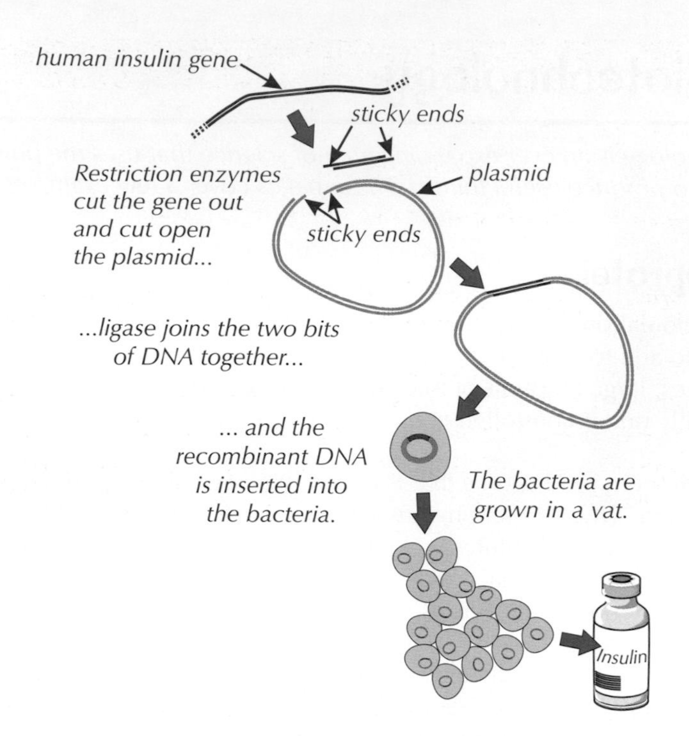

Figure 3: *The steps involved in producing human insulin using genetically engineered bacteria.*

Genetically modified (GM) crops

Many people in the world today don't have enough food to eat (or the diet they have isn't varied). This mostly happens in developing countries — like those in Africa and parts of Asia. Crops can be genetically modified to give them certain characteristics. Modifying certain crops could potentially help to solve the issues of food shortage and malnutrition, but not everyone agrees that GM crops are a good idea.

Benefits of genetically modifying crops

On the plus side, GM crops can increase the amount of food that a crop provides (its yield). For example, crops can be produced that are resistant to pests or are able to grow better in drought conditions.

People living in developing countries often lack nutrients in their diet. GM crops could be engineered to contain the nutrient that's missing. For example, 'golden rice' has been engineered to produce a chemical that's converted in the body into vitamin A.

Concerns about genetically modifying crops

Many people argue that people go hungry because they can't afford to buy food, not because there isn't any food about. So they argue that you need to tackle poverty first. Also, there are fears that countries may become dependent on companies who sell GM seeds. Additionally, it is sometimes poor soil that is the main reason why crops fail in a certain area, and even GM crops won't survive.

Practice Questions — Fact Recall

Q1 What is mycoprotein?

Q2 Describe how mycoprotein is produced.

Q3 How can human insulin be produced by bacteria?

Practice Question — Application

Q1 Scientists noticed that many children in a developing country were suffering from a disease caused by the deficiency of a certain vitamin. The children's diet involved large quantities of corn. Suggest how genetic modification could be used to help prevent the disease.

Topic 7c Checklist — Make sure you know...

Trophic Levels

- [ ] That the different stages of a food chain are called trophic levels and that each trophic level is assigned a number related to its position in the food chain, e.g. the first trophic level is called trophic level 1 and the third trophic level is called trophic level 3.
- [ ] How the trophic levels in an ecosystem differ from each other.
- [ ] That trophic level 1 always contains the producers (e.g. plants and algae) of a food chain.
- [ ] That trophic level 2 contains the primary consumers — the herbivores that eat the producers.
- [ ] That trophic level 3 contains the secondary consumers — the carnivores that eat the primary consumers.
- [ ] That trophic level 4 contains the tertiary consumers — the carnivores that eat the secondary consumers.
- [ ] That the predators at the top of the food chain (so have no predators that eat them) are called apex predators.
- [ ] That dead plant and animal material is broken down by decomposers, which secrete enzymes to break the material down into smaller soluble molecules that can then diffuse into the decomposers.

Pyramids of Biomass

- [ ] That pyramids of biomass show the relative amounts of biomass in each trophic level of a food chain.
- [ ] That the number of organisms at each trophic level of a food chain is affected by the loss of biomass at each trophic level.
- [ ] That in a pyramid of biomass, the bottom level always represents trophic level 1.
- [ ] How to construct a pyramid of biomass accurately from relevant data.
- [ ] How to interpret pyramids of biomass and describe what they show.

cont...

Biomass Transfer

- [ ] That plants and algae are the producers in a food chain, and that they transfer approximately 1% of the energy from the light that hits them for use in photosynthesis to make glucose.
- [ ] That when organisms in one trophic level eat the organisms from the previous trophic level, only about 10% of the biomass is passed on from the previous trophic level.
- [ ] That biomass is lost between trophic levels because some of the biomass ingested by an organism isn't absorbed so is egested as faeces.
- [ ] That lots of the glucose obtained from biomass is used by organisms for respiration.
- [ ] Some ingested biomass is converted into other substances that are released as waste — e.g. CO_2 and water are waste products of respiration, and urea is a waste product of the breakdown of proteins, which is released with water in urine.
- [ ] How to work out the efficiency of biomass transfer between two trophic levels in a food chain.

Food Security and Farming

- [ ] That food security is having enough food to feed the population.
- [ ] That food security can be affected by rising birth rates, changes in the diets of people in developed countries (meaning that developing countries export more of their already scarce food resources), issues with farming caused by new pests, pathogens, changing environmental conditions and high input costs, and conflicts.
- [ ] That fish stocks must be maintained at a sustainable level that allows breeding to occur, to avoid the risk of certain species of fish disappearing from oceans, e.g. cod from the north west Atlantic.
- [ ] That introducing fishing quotas (limits on the number of fish that can be caught) and regulating the width of the mesh used in fishing nets (to avoid accidentally catching unwanted species and young fish) could help to conserve fish stocks.
- [ ] That modern farming methods can be used to increase the efficiency of food production by reducing the amount of energy that is transferred from livestock to the environment (by limiting the animals' movement and controlling the temperature of their environment).
- [ ] That high-protein foods can be fed to animals to increase their growth.

Biotechnology

- [ ] That we can use modern biotechnology to culture large amounts of microorganisms for use as food.
- [ ] That mycoprotein, made from the fungus *Fusarium*, is an example of a food that can be made from microorganisms, and that the fungus is grown in aerobic conditions in large industrial vats on glucose syrup, which it uses as a food source.
- [ ] That bacteria can be genetically modified to make human insulin that is then purified for medical use to treat diabetes.
- [ ] That genetically modified (GM) crops could be used to address food shortages (by increasing crop yield) and to improve the nutritional content of crops.
- [ ] That golden rice is an example of a GM crop that has been successfully produced to improve the nutritional value of rice.

Exam-style Questions

1 **Figure 1** shows a pyramid of biomass.
It represents a food chain involving algae, fish and birds.

Figure 1

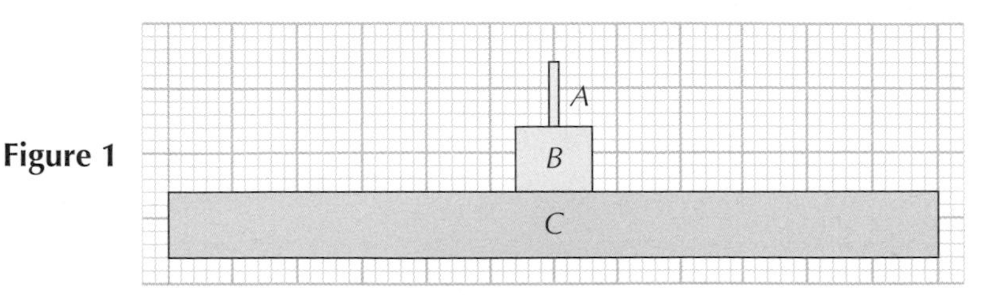

1.1 Explain why the Sun is needed at the start of all food chains.

(4 marks)

1.2 Which bar on the pyramid (A-C) represents the algae? Explain your answer.

(1 mark)

1.3 Calculate the ratio of bar C to bar B.

(2 marks)

1.4 Neither of the ratios in the pyramid of biomass (bar C to bar B, or bar B to bar A) are 1:1 because biomass is lost as you go up a food chain.
Explain how biomass is lost from a food chain.

(3 marks)

2 **Figure 2** shows chickens in a factory farm.

Figure 2

2.1 Explain how factory farming can improve the efficiency of producing chickens for food.

(3 marks)

2.2 Fish can also be factory farmed. Describe how this is carried out.

(1 mark)

2.3 Factory farming can be controversial.
Suggest **two** disadvantages of factory farming methods.

(2 marks)

1. Measuring Substances

As part of GCSE Biology, you'll have to do at least ten practicals, called Required Practical Activities. You'll need to know how to use various pieces of apparatus and carry out different scientific techniques. And not only do you need to carry out the practicals, you could also be asked about them in the exams. Luckily, all the Required Practical Activities are covered in this book, and the next few pages cover some of the techniques that you'll need to know about.

Tip: The Required Practical Activities in this book are marked with a big stamp like this...

The practicals that you do in class might be slightly different to the ones in this book (as it's up to your teacher exactly what method you use), but they'll cover the same principles and techniques.

Measuring the mass of a solid

To weigh a solid, start by putting the container you are weighing your substance into on a balance. Set the balance to exactly zero and then weigh out the correct amount of your substance. Easy peasy.

Measuring temperature

You can use a thermometer to measure the temperature of a solution. Make sure that the bulb of the thermometer is completely submerged in the solution and that you wait for the temperature to stabilise before you take your initial reading. Read off the scale on the thermometer at eye level to make sure it's correct.

Tip: When you're reading off a scale, write down the value of the graduation that the amount is closest to.

Measuring the volume of a liquid

There's more than one way to measure the volume of a liquid. Whichever method you use, always read the volume from the bottom of the meniscus (the curved upper surface of the liquid) when it's at eye level.

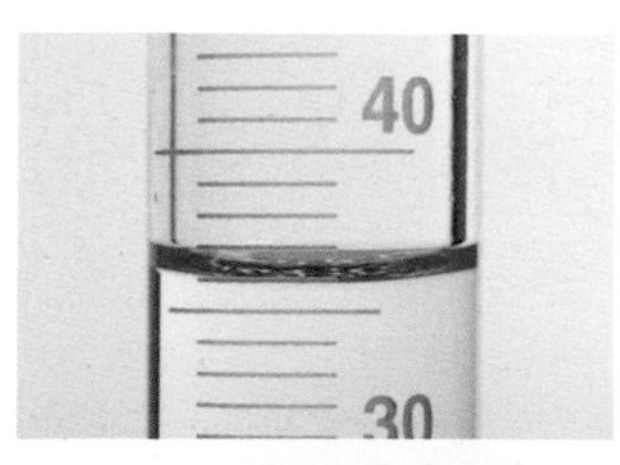

Figure 1: *The meniscus of a fluid in a measuring cylinder, viewed at eye level.*

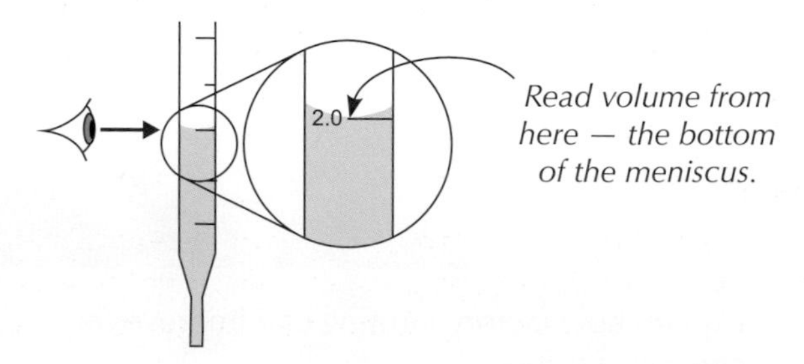

Figure 2: *The technique for correctly measuring the volume of a liquid.*

Measuring cylinders

Measuring cylinders come in all different sizes. Make sure you choose one that's the right size for the measurement you want to make. It's no good using a huge 1 dm^3 cylinder to measure out 2 cm^3 of a liquid — the graduations will be too big, and you'll end up with massive errors. It'd be much better to use one that measures up to 10 cm^3.

Tip: You could use a dropping pipette if you just wanted to transfer a couple of drops of liquid and didn't need it to be particularly accurate, for example, if you were adding an indicator to a mixture (see next page).

Pipettes

Pipettes are used to suck up and transfer volumes of liquid between containers. Dropping pipettes are used to transfer drops of liquid. Graduated pipettes are used to transfer accurate volumes. A pipette filler (see Figure 3) is attached to the end of a graduated pipette, to control the amount of liquid being drawn up.

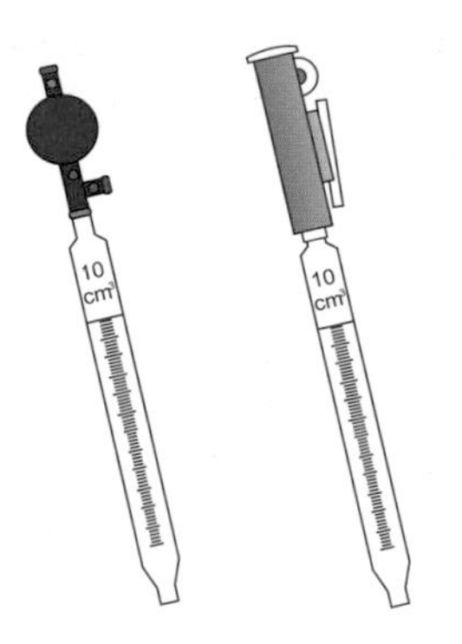

Figure 3: Graduated pipettes with two different types of pipette filler attached.

Measuring the volume of a gas

There are a few different ways of measuring the volume of a gas. However you do it, it's key that you make sure that the equipment is set up so that none of the gas can escape, otherwise your results won't be accurate.

Gas syringes

To accurately measure the volume of gas, you should use a gas syringe (see Figure 4).

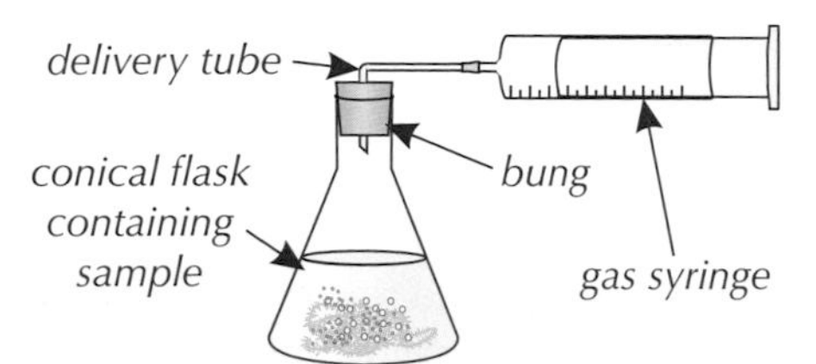

Figure 4: *A gas syringe attached to a conical flask.*

Displacement of water

As an alternative to using a gas syringe, you can use an upturned measuring cylinder filled with water. The gas will displace the water so you can read the volume off the scale.

Other methods

Other methods to measure the amount of gas include counting the bubbles produced or measuring the length of a gas bubble drawn along a tube (see page 167). These methods are less accurate, but will give you relative amounts of gas to compare results.

Measuring the pH of a substance

The method you should use to measure pH depends on your experiment:

Indicator dyes

Indicators are dyes that change colour depending on whether they're in an acid or an alkali. You use them by adding a couple of drops of the indicator to the solution you're interested in. Universal indicator is a mixture of indicators that changes colour gradually as pH changes — see Figure 5. It's useful for estimating the pH of a solution based on its colour.

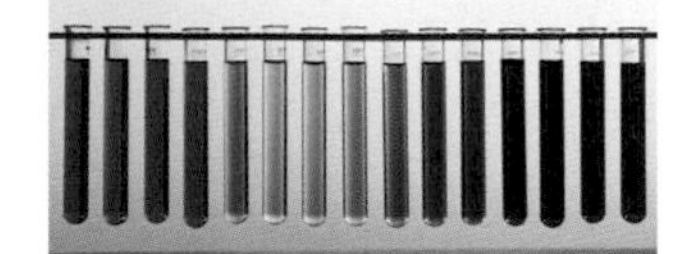

Figure 5: *A set of test tubes containing liquids at a range of pH levels with universal indicator added.*

Indicator paper

Indicator paper is useful if you don't want to colour the entire solution that you're testing. It changes colour depending on the pH of the solution it touches. You can also hold a piece of damp indicator paper in a gas sample to test its pH.

Tip: Blue litmus paper turns red in acidic conditions and red litmus paper turns blue in alkaline conditions.

pH meters

pH meters have a digital display that gives an accurate value for the pH of a solution.

Measuring time

You should use a stopwatch to time experiments. These measure to the nearest 0.1 s so are pretty sensitive. Make sure you start and stop the stopwatch at exactly the right time.

2. Heating Substances and Using Potometers

Lots of scientific experiments involve heating one or more of the substances involved. The method of heating used depends on the substance being heated and what temperature it needs to be heated to. Potometers are handy pieces of apparatus that are used to measure how much water a plant takes up.

Bunsen burners

Tip: Some things take a long time to cool down, and you can't necessarily tell by looking whether they're hot or cold. So, after heating equipment, you should always handle it with tongs so you don't get burnt.

Bunsen burners are good for heating things quickly. You can easily adjust how strongly they're heating. But you need to be careful not to use them if you're heating flammable substances as the flame means the substance would be at risk of catching fire.

To use a Bunsen burner, you should first connect it to a gas tap, and check that the hole is closed. Place it on a heat-proof mat. Next, light a splint and hold it over the Bunsen burner. Now, turn on the gas. The Bunsen burner should light with a yellow flame. The more open the hole is, the more strongly the Bunsen burner will heat your substance. Open the hole to the amount you want. As you open the hole more, the flame should turn more blue.

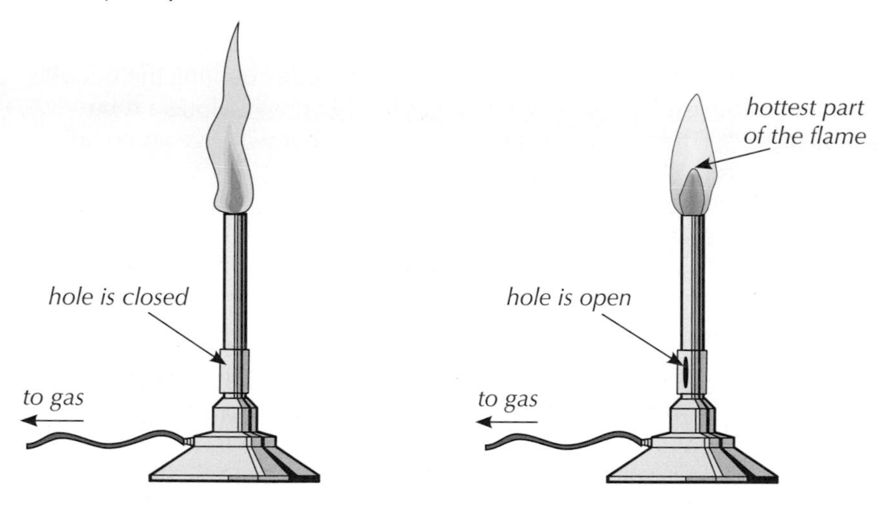

Figure 1: *A Bunsen burner with the hole closed (left) and the hole open (right).*

If your Bunsen burner is alight but not heating anything, make sure you close the hole so that the flame becomes yellow and clearly visible. Use the blue flame to heat things. If you're heating a vessel in the flame, hold it at the top (e.g. with tongs) and point the opening away from yourself (and others). If you're heating something over the flame (e.g. a beaker of water), you should put a tripod and gauze over the Bunsen burner before you light it, and place the vessel on this.

Figure 2: *An electric heater.*

Electric heaters

Electric heaters are often made up of a metal plate that can be heated to a specified temperature. The vessel containing the substance you want to heat is placed on top of the hot plate. The vessel is only heated from below, so you'll usually have to stir the substance inside to make sure it's heated evenly.

Water baths

A water bath is a container filled with water that can be heated to a specific temperature. A simple water bath can be made by heating a beaker of water over a Bunsen burner and monitoring the temperature with a thermometer. However, it is difficult to keep the temperature of the water constant.

An electric water bath will monitor and adjust the temperature for you. To use one, start by setting the temperature on the water bath, and allow the water to heat up. To make sure it has reached the right temperature, use a thermometer. Place the vessel containing your substance in the water bath using a pair of tongs. The level of the water outside the vessel should be just above the level of the substance inside the vessel. The substance will then be warmed to the same temperature as the water.

As the substance in the vessel is surrounded by water, the heating is very even. Water boils at 100 °C though, so you can't use a water bath to heat something to a higher temperature than this — the water won't get hot enough.

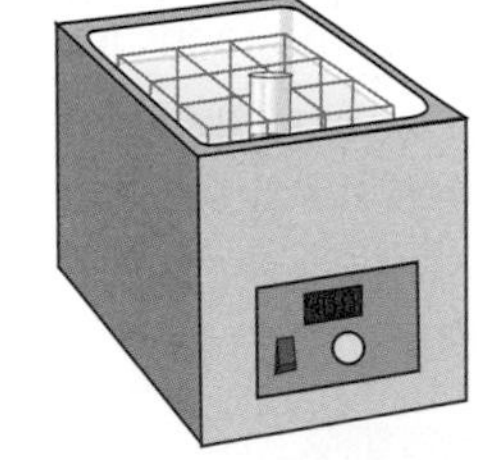

Figure 3: *A typical water bath.*

Measuring water uptake with a potometer

A potometer is a special piece of apparatus used to measure the water uptake by a plant. Here's how to set one up:

1. Cut a shoot underwater to prevent air from entering the xylem. Cut it at a slant to increase the surface area available for water uptake.
2. Assemble the potometer in water and insert the shoot under water, so no air can enter.
3. Remove the apparatus from the water but keep the end of the capillary tube submerged in a beaker of water.
4. Check that the apparatus is watertight and airtight.
5. Dry the leaves, allow time for the shoot to acclimatise and then shut the tap.
6. Remove the end of the capillary tube from the beaker of water until one air bubble has formed, then put the end of the tube back into the water.
7. A potometer can be used to estimate the transpiration rate of a plant. There's more about this on page 93.

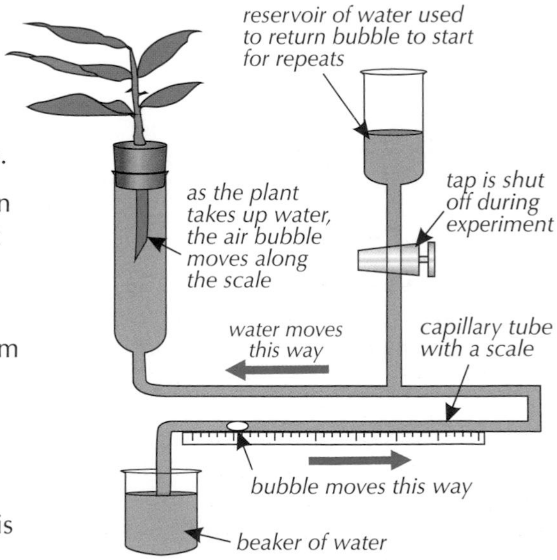

reservoir of water used to return bubble to start for repeats
as the plant takes up water, the air bubble moves along the scale
tap is shut off during experiment
water moves this way
capillary tube with a scale
bubble moves this way
beaker of water

Figure 4: *A potometer set up and ready to use.*

Tip: If there are air bubbles in the apparatus or the plant's xylem, it will affect your results. That's why the whole process of setting up a potometer is done underwater.

Tip: There are different types of potometer available. They won't all look like the one shown in Figure 4.

Figure 5: *Students using a potometer with a slightly different design to the one shown in Figure 4.*

3. Cell Size and Scale Bars

Knowing the real size of a cell can be handy when you're making a drawing of it. You can use that measurement to draw a scale bar for your diagram.

Measuring the size of a single cell

When viewing cells under a microscope, you might need to work out their size. To work out the size of a single cell:

- Place a clear, plastic ruler on top of your microscope slide. Clip the ruler and slide onto the stage.
- Select the objective lens that gives an overall magnification of × 100.
- Adjust the focus to get a clear image of the cells.
- Move the ruler so that the cells are lined up along 1 mm. Then count the number of cells along this 1 mm sample.
- 1 mm = 1000 μm. So to calculate the length of a single cell in μm, divide 1000 μm by the number of cells in the sample.

Tip: You can read all about using a microscope on page 26.

Example

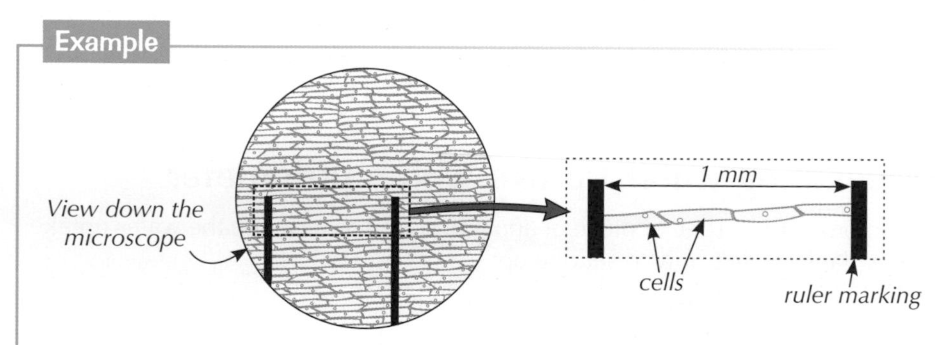

Figure 1: *Using a ruler to measure cells.*

In Figure 1, there are 4 cells in 1 mm. So the length of a single cell is:
1000 ÷ 4 = **250 μm**

Finding the length of a scale bar

If you draw a diagram of a cell you've observed under a microscope, you might want to include a scale bar. Once you know the size of one cell, you can use it to calculate how long your scale bar should be.

To draw a 500 μm scale bar, just use this formula:

$$\text{scale bar length (μm)} = \frac{\text{drawn length of cell (μm)} \times 500}{\text{actual length of cell (μm)}}$$

Tip: A scale bar is a line added to a diagram to show its relative real size. So if a 100 μm scale bar is 1 mm long, it means every 1 mm in the diagram has a real length of 100 μm.

Tip: For a scale bar with a different scale, just replace the 500 in the formula with the number of micrometers you want the scale bar to represent.

Example

If a drawing of one of the cells above is 23 mm long, the length of the scale bar should be:

$$\frac{23\,000 \text{ μm} \times 500}{250 \text{ μm}} = 46\,000 \text{ μm} = \mathbf{46 \text{ mm}}$$

drawing of cell
scale bar
500 μm

4. Sampling

When you're investigating the population size of a certain organism, or its distribution in an area, using sampling can save you an awful lot of time and effort. Read on for more...

What is sampling?

When you're investigating a population, it's generally not possible to count every single organism in the population. This means that you need to take samples of the population you're interested in. The sample data will be used to draw conclusions about the whole population, so it's important that it accurately represents the whole population. To make sure a sample represents a population, it should be random.

Tip: If a sample doesn't represent the population as a whole, it's said to be biased.

If you're interested in the distribution of an organism in an area, or its population size, you can take population samples in the area you're interested in using quadrats or transects (see pages 325-327).

If you only take samples from one part of the area, your results will be biased — they may not give an accurate representation of the whole area. To make sure that your sampling isn't biased, you need to use a method of choosing sampling sites in which every site has an equal chance of being chosen. For example:

Tip: Head to pages 326-327 to see how sampling can be used to estimate the total population of a species in an area.

Example

MATHS SKILLS

If you're looking at plant species in a field...

1. Divide the field into a grid.
2. Label the grid along the bottom and up the side with numbers or letters.
3. Use a random number generator (on a computer or calculator) to select coordinates, e.g. (2,6).
4. Take your samples at these coordinates.

<u>*Non-random sampling*</u>

Only looks at a small part of the field.

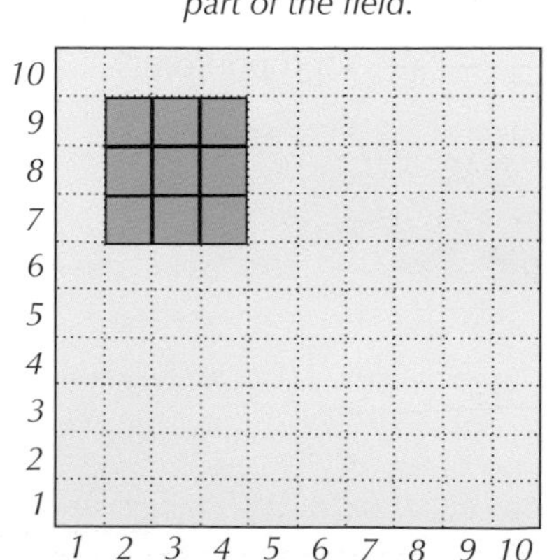

<u>*Random sampling*</u>

Randomly selects squares from all over the field.

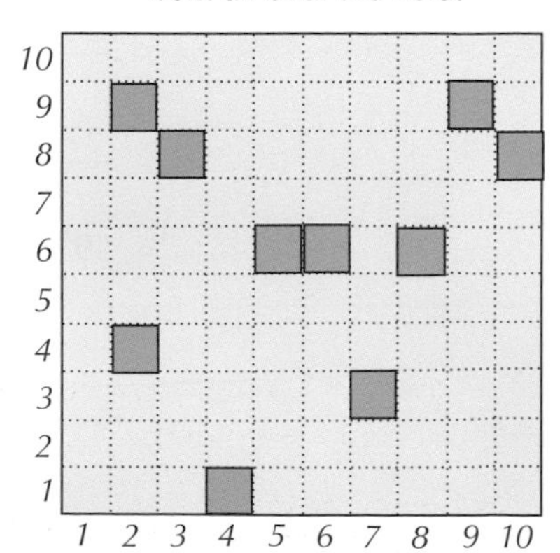

Figure 1: *Many scientific calculators have a handy random number generator built in.*

Collecting health data using sampling

As mentioned on the previous page, it's not practical (or even possible) to study an entire human population. You need to use random sampling to choose members of the population you're interested in.

Example

A health professional is investigating how many people diagnosed with Type 2 diabetes in a particular country also have heart disease.

1. All the people who have been diagnosed with Type 2 diabetes in the country of interest are identified by hospital records. In total, there are 270 196 people.
2. These people are assigned a number between 1 and 270 196.
3. The sample size is decided — e.g. 4250 people.
4. Then a random number generator is used to choose the sample group (e.g. it selects the individuals #72 063, #11 822, #193 123, etc.)
5. The proportion of people in the sample that have heart disease can be used to estimate the total number of people with Type 2 diabetes that also have heart disease:
   - Find the number of people in the sample that have heart disease too — here it's 935 people.
   - Work out what proportion of the sample that number is — you can calculate it as a percentage.

$$\text{proportion (\%)} = \frac{\text{number of people sampled with heart disease}}{\text{total number of people sampled}}$$

$$= \frac{935}{4250} \times 100$$

$$= \mathbf{22\%}$$

   - Then use your proportion to estimate the total number of people in the population with both Type 2 diabetes and heart disease.

$$= \frac{\text{total population size}}{100} \times \text{proportion (\%)}$$

$$= \frac{270\,196}{100} \times 22\%$$

$$= \mathbf{59\,443\ people}$$

Tip: See page 378 for more on percentages.

Tip: Some calculators have a percentage button, which you can use instead of doing this second calculation.

5. Safety and Ethics

Science can be quite dangerous at times, so it's really important that you keep yourself (and others) safe in the lab. Some experiments can also involve ethical issues that you must deal with respectfully and responsibly.

Working safely

Make sure you're working safely in the lab. Before you start any experiment, make sure you know about any safety precautions to do with your method or the chemicals you're using. You need to follow any instructions that your teacher gives you carefully. The chemicals you're using may be hazardous — for example, they might be flammable (catch fire easily), or they might irritate or burn your skin if it comes into contact with them.

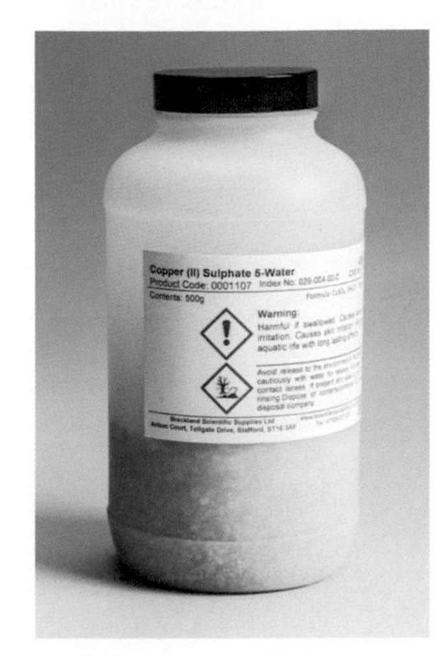

Figure 1: *You'll often see hazard labels on bottles of chemicals. These tell you in what way a chemical is dangerous.*

Make sure that you're wearing sensible clothing when you're in the lab (e.g. open shoes won't protect your feet from spillages). When you're doing an experiment, you should wear a lab coat to protect your skin and clothing. Depending on the experiment, you may need to also wear safety goggles and gloves.

Figure 2: *Diagram showing lab safety gear.*

You also need to be aware of general safety in the lab, e.g. keep anything flammable away from lit Bunsen burners, don't directly touch any hot equipment, handle glassware carefully so it doesn't break, etc.

Working ethically

You need to think about ethical issues in your experiments.

Any organisms involved in your investigations need to be treated safely and ethically. Animals need to be treated humanely — they should be handled carefully and any wild animals captured for studying (e.g. during an investigation of the distribution of an organism) should be returned to their original habitat. Any animals kept in the lab should also be cared for in a humane way, e.g. they should not be kept in overcrowded conditions.

If you are carrying out an experiment involving other students (e.g. investigating the effect of caffeine on reaction time), they should not be forced to participate against their will or feel pressured to take part.

Tip: See page 191 for more on investigating the effect of caffeine on reaction time.

Maths Skills

Exam Tip
Around 10% of the marks in the exams will test mathematical skills. That's a lot of marks so it's definitely worth making sure you're up to speed.

Maths skills for GCSE Biology

Maths crops up quite a lot in GCSE Biology so it's really important that you've mastered all the maths skills you'll need before sitting your exams. Maths skills are covered throughout this book but here's an extra little section, just on maths, to help you out.

1. Calculations

Calculations are the cornerstone of maths in science. So being able to carry them out carefully is pretty important.

Standard form

You need to be able to work with numbers that are written in **standard form**. Standard form is used for writing very big or very small numbers with lots of zeros in a more convenient way. Standard form must always look like this:

This number must always be between 1 and 10. → $A \times 10^n$ ← *This number is the number of places the decimal point moves.*

Examples

- 1 000 000 can be written as 1×10^6.
- 0.017 can be written as 1.7×10^{-2}.

Tip: When you're writing a measurement in standard form, make sure you keep the same number of significant figures. E.g. 0.00400 cm^3 = 4.00×10^{-3} cm^3. This'll make sure that you don't lose any accuracy.

You can write numbers in standard form by moving the decimal point. Which direction the decimal point moves, and how many places it moves, is described by *'n'*. If the decimal point has moved to the left, *'n'* is positive. If the decimal point has moved to the right, *'n'* is negative.

Example

Here's how to write out 0.000056 in standard form.

1. Move the decimal point to give the smallest number you can between 1 and 10 — this is *'A'*.

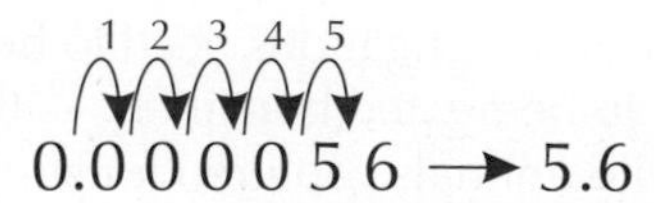

2. Count the number of places the decimal point has moved. In this example, the decimal point has moved five places to the right.
3. Write that number in the place of *'n'*. Remember, if the decimal point has moved to the left, *'n'* is positive. If the decimal point has moved to the right (like in this example), *'n'* is negative.
4. So 0.000056 is the same as 5.6×10^{-5}.

Tip: Double check you've got it right by doing the multiplication — you should end up with the number you started with. So for this example, you'd check $5.6 \times 10^{-5} = 0.000056$. It's easy to do this on your calculator — see the next page for how to type it in.

The key things to remember with numbers in standard form are...

- When 'n' is positive, the number is big. The bigger 'n' is, the bigger the number is.
- When 'n' is negative, the number is small. The smaller 'n' is (the more negative), the smaller the number is.
- When 'n' is the same for two or more numbers, you need to look at the start of each number to work out which is bigger. For example, 4.5 is bigger than 3.0, so 4.5×10^5 is bigger than 3.0×10^5.

There's a special button on your calculator for using standard form in a calculation — it's the 'Exp' button. So if, for example, you wanted to type in 2×10^7, you'd only need to type in: '2' 'Exp' '7'. Some calculators may have a different button that does the same job, for example it could say 'EE' or '$\times 10^x$' instead of 'Exp' — see Figure 1.

Figure 1: *The 'Exp' or '$\times 10^x$' button is used to input standard form on calculators.*

Using ratios

Ratios can be used to compare quantities.

Example

An organism with a surface area to volume ratio of 2 : 1 would theoretically have a surface area twice as large as its volume.

Ratios are usually written like this:

A colon separates one quantity from the other.

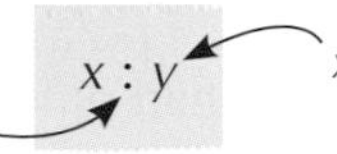

x and y stand for the quantities that you want to compare.

Ratios are usually most useful in their simplest (smallest) form.
To simplify a ratio, divide each side by the same number. It's in its simplest form when there's nothing left you can divide by to give a whole number ratio.

Tip: When you divide one side of a ratio by an amount, you have to divide the other side by the same number for the ratio to keep the same proportions.

Example

MATHS SKILLS

To simplify the ratio 28 : 36, divide both sides by 4.

You get **7 : 9**. ← *You can't divide 7 and 9 by the same number to get two whole numbers, so this must be the ratio's simplest form.*

To compare two ratios, it's best to get the number on the right-hand side of each ratio to be 1. Then you can easily see which ratio is the largest.
To get a ratio of X : Y in the form X : 1, divide both sides by Y.

Example

Organism A has a surface area to volume ratio of 6 : 1.
Organism B has a surface area to volume ratio of 30 : 9.

To compare the two ratios, you need to write the ratio 30 : 9 in the form of X : 1. To do this, just divide both sides by 9:

30 ÷ 9 = 3.3... = 3.3   30 : 9 → 3.3 : 1   *9 ÷ 9 = 1*

So the ratio of Organism B is equal to **3.3 : 1**. 3.3 is smaller than 6, so Organism A has the larger surface area to volume ratio.

Tip: If you're not sure what number to divide by to simplify a ratio, start by trying to divide both sides by a small number, e.g. 2 or 3, then check to see if you can simplify your answer further. E.g. you could simplify 28 : 36 by dividing each side by 2 to get 14 : 18. But you could simplify it further by dividing by 2 again to get 7 : 9. You can't simplify the ratio any further, so it's in its simplest form.

Calculating percentages

Percentages are another way of comparing quantities. They come in handy when you want to compare amounts from different-sized samples.

To give the amount X as a percentage of total amount Y, you need to divide X by Y, then multiply by 100.

Exam Tip
Rather than having to calculate a percentage, you might be asked to give your answer as a fraction, i.e. X/Y. To simplify a fraction, divide the top and bottom of the fraction by the same number. E.g. 4/40 = (4 ÷ 4)/(40 ÷ 4) = 1/10. To get the fraction as simple as possible, you might have to do this more than once.

Example

The amount of biomass available in the first level of a food chain is 40 kg. 4 kg of this biomass is transferred to the next level of the food chain. What percentage of the original biomass is transferred?

1. You want to give 4 kg as a percentage of 40 kg, so divide 4 by 40:

$$4 \div 40 = 0.1$$

2. Multiply this amount by 100:

$$0.1 \times 100 = \mathbf{10\%}$$

Calculating percentage change

When investigating the change in a variable, you may want to compare results that didn't have the same initial value. For example, you may want to compare the change in mass of potato cylinders left in different concentrations of sugar solution that had different initial masses (see p. 55). One way to do this is to calculate the percentage change.

To calculate it you use this equation:

Tip: A positive value for percentage change indicates an increase and a negative value indicates a decrease.

$$\text{percentage (\%) change} = \frac{\text{final value} - \text{original value}}{\text{original value}} \times 100$$

Example

A student is investigating the effect of the concentration of sugar solution on potato cells. She records the mass of potato cylinders before and after placing them in sugar solutions of different concentrations. The table below shows some of her results.

MATHS SKILLS

Potato cylinder	Concentration (mol/dm^3)	Mass at start (g)	Mass at end (g)
1	0.0	7.5	8.7
2	1.0	8.0	6.8

Which potato cylinder had the largest percentage change?

1. Stick each set of results into the equation:

$$\% \text{ change} = \frac{\text{final value} - \text{original value}}{\text{original value}} \times 100$$

potato cylinder 1: $\frac{8.7 - 7.5}{7.5} \times 100 = 16\%$

potato cylinder 2: $\frac{6.8 - 8.0}{8.0} \times 100 = -15\%$

2. Compare the results.

16% is greater than 15%, so potato cylinder 1 (in the 0.0 mol/dm^3 sugar solution) had the largest percentage change.

Tip: The mass at the start is the original value and the mass at the end is the final value.

Tip: A decrease in mass gives a negative value for the percentage change.

Estimating values

Estimating can help you to check your answer is roughly correct. To estimate the answer to a calculation, just round everything off to nice, convenient numbers so that you get a simple calculation that you can do in your head.

Most of the time, you should round the numbers to just one significant figure.

Example

A student calculated the number of stomata per mm^2 on ten different leaves.

The results are shown in the table below.

Leaf	1	2	3	4	5	6	7	8	9	10
Number of stomata/mm^2	245	287	322	185	316	299	178	250	332	420

The student calculates that the mean number of stomata per mm^2 is 283. You could estimate the mean of the data like this:

1. Round the numbers to one significant figure.

200 300 300 200 300 300 200 300 300 400

2. Do the calculation with the rounded numbers.

(200 + 300 + 300 + 200 + 300 + 300 + 200 + 300 + 300 + 400) ÷ 10

= 2800 ÷ 10 = **280**

From the estimated calculation, 283 is a sensible answer. If your answer was massively different, you would know that your calculation had gone wrong somewhere.

Exam Tip
By working out an estimate in the exam, you can check that you haven't made any silly mistakes while calculating your answer (like missing out a decimal place along the way) and it's a lot quicker than typing the whole thing into your calculator again.

2. Algebra

Every so often, biology can involve a bit of rearranging equations and substituting values into equations. It can be easy to make simple mistakes though, so here are a few things to remember...

Algebra symbols

Here's a reminder of some of the symbols that you will come across:

Symbol	Meaning
=	equal to
<	less than
<<	much less than
>	greater than
>>	much greater than
$\propto$	proportional to
~	roughly equal to

Tip: An example of using 'proportional to' can be found on p. 169. There are two types of proportion — direct and inverse. When two values are directly proportional to each other, if one increases, the other increases at the same rate. When two values are inversely proportional to each other, if one increases, the other decreases at the same rate.

Using equations

Equations can show relationships between variables. To rearrange an equation, make sure that whatever you do to one side of the equation you also do to the other side.

Tip: The word formula is sometimes used instead of the word equation.

Example

You can find the magnification of something using the equation: magnification = image size ÷ real size

You can rearrange this equation to find the image size by multiplying each side by the real size: image size = magnification × real size.

Tip: There's more about this equation for magnification on p. 27.

To use an equation, you need to know the values of all but one of the variables. Substitute the values you do know into the equation, and do the calculation to work out the final variable. Always make sure the values you put into an equation have the right units.

Tip: Converting all the values into the correct units <u>before</u> putting them into the equation stops you making silly mistakes.

Example

Calculating a BMI is one way to find whether a person is a healthy weight. It's calculated by this equation: $\text{BMI} = \frac{\text{weight (kg)}}{\text{height}^2\text{ (m)}}$

A man weighs 84 kg and is 2.0 m tall. Calculate his BMI.

In this scenario, the weight is 84 kg and the height is 2.0 m. The units for both of these values are the right ones for the equation, so you can just go ahead and substitute them in:

$$\text{BMI} = \frac{\text{weight (kg)}}{\text{height}^2\text{ (m)}} = \frac{84}{2.0^2} = \frac{84}{4} = \mathbf{21}$$

MATHS SKILLS

To make sure your units are correct, it can help to write down the units on each line of your calculation.

3. Graphs

Results are often presented using graphs. They make it easier to see relationships between variables and can be used to calculate other quantities.

Finding the intercept of a graph

The y-intercept of a graph is the point at which the line of best fit crosses the y-axis. Meanwhile, the x-intercept is the point at which the line of best fit crosses the x-axis.

Example

A graph can be plotted to show the percentage change in mass of several pieces of potato against the different concentrations of sucrose solution that they were left in overnight.

Tip: See pages 55-56 for more on how to carry out this experiment and analyse the results.

MATHS SKILLS

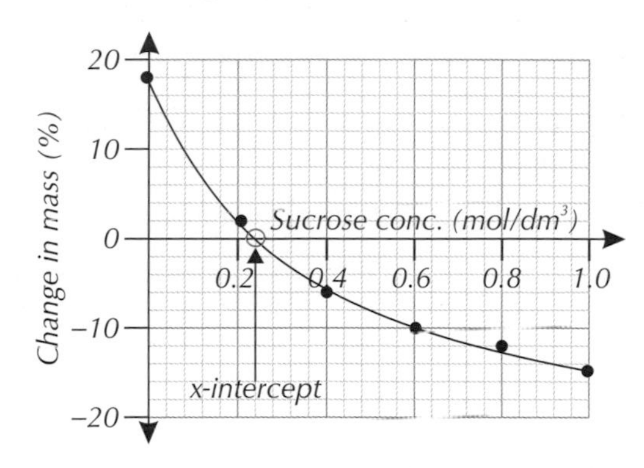

The approximate concentration of the solution in the potato cells, which is equal to the x-intercept, is **0.24 mol/dm³**.

Linear graphs

A linear graph is a straight line graph. This means that one of the variables plotted on the axes increases or decreases in relation to the other, which is shown by a straight line. An example of a linear graph is shown in Figure 1.

Tip: If one variable increases in proportion with the other (so if x is doubled, y is doubled, etc.), the graph shows a directly proportional relationship.

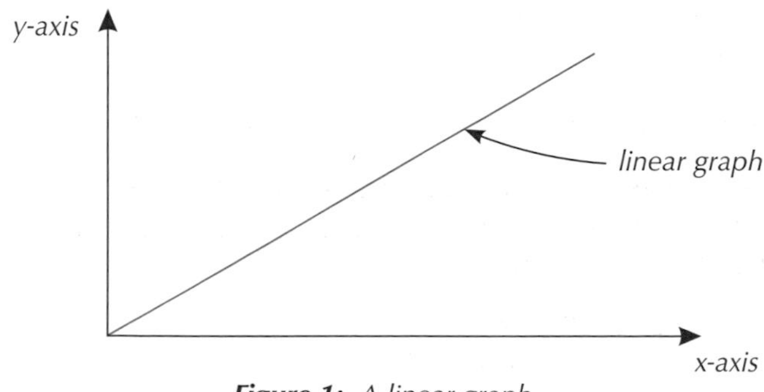

Figure 1: *A linear graph.*

Tip: The gradient of a graph tells you how quickly the dependent variable changes if you change the independent variable.

Tip: When using this equation to find a rate, x should always be the time.

Tip: The units for the gradient are the units for y divided by the units for x.

Finding the rate

Rate is a measure of how much something is changing over time. Calculating a rate can be useful when analysing your data, e.g. you might want to the find the rate of a reaction. You can find the rate from a graph that shows a variable changing over time by finding the **gradient** (how steep it is).

For a linear graph, you can calculate the rate by finding the gradient of the line, using the equation:

$$\text{Gradient} = \frac{\text{Change in } y}{\text{Change in } x}$$

Change in y is the change in value on the y-axis and **change in x** is the change in value on the x-axis.

Example

To find the rate at which oxygen is produced in the graph below:

1. Pick two points on the line that are easy to read and a good distance apart.
2. Draw a vertical line down from one point and a horizontal line across from the other to make a triangle.

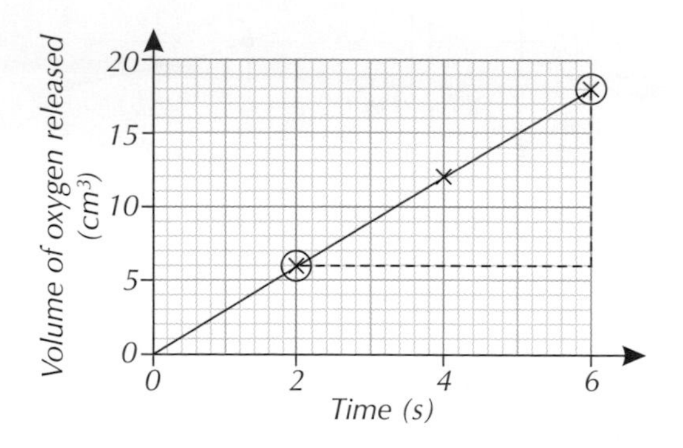

3. Use the scales on the axes to work out the length of each line. The vertical side of the triangle is the change in y and the horizontal side of the triangle is the change in x.

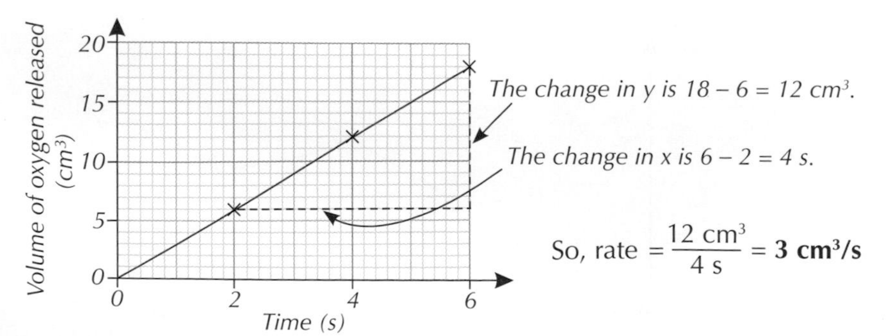

So, rate $= \frac{12 \text{ cm}^3}{4 \text{ s}} = \mathbf{3 \text{ cm}^3/\text{s}}$

$y = mx + c$

The y-intercept and the gradient of a linear graph can be used to work out the equation of the graph line. The equation of a straight line is given by:

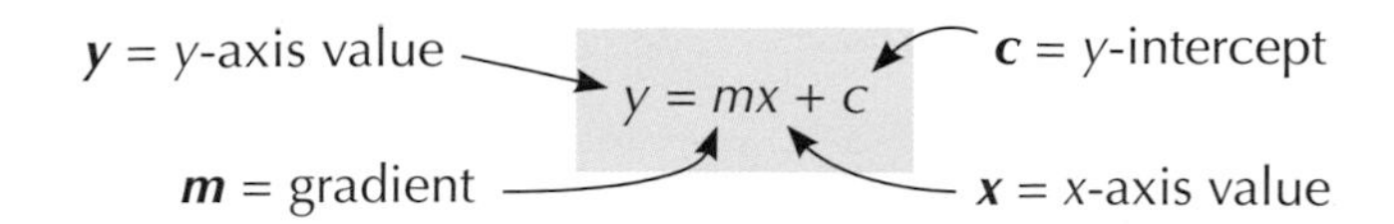

The equation of the line shows the relationship between the two variables. To work out the equation, you just need to plug in the values for m and c.

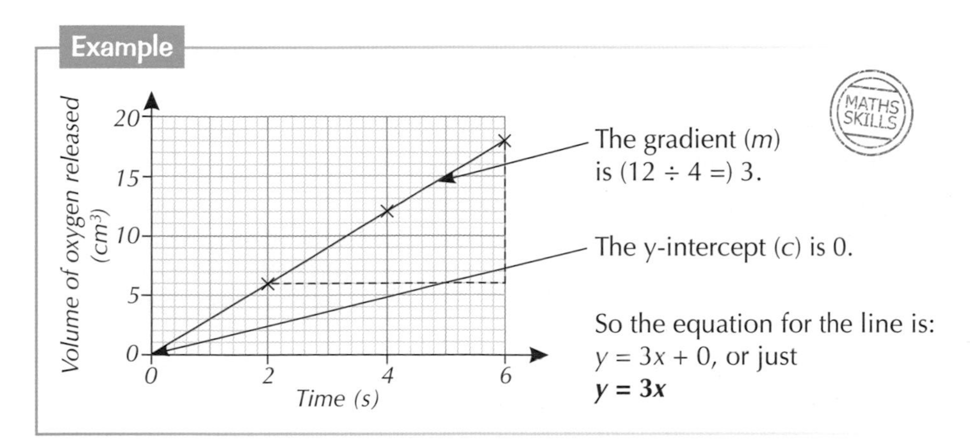

Tip: mx is just the same as writing '$m \times x$'.

Tip: If the straight line passes through the origin of the graph (the point at which the x-axis and y-axis meet), then the y-intercept is just zero.

Tip: You can check your equation is correct using the values of x and y from a point on the graph. Just pop the value for x into the equation and it should give you the value of y.

1. The Exams

Unfortunately, to get your GCSE you'll need to sit some exams. And that's what this page is about — what to expect in your exams.

Exam Tip
Make sure you have a good read through these pages. It might not seem all that important now but you don't want to get any surprises just before an exam.

Assessment for GCSE Biology

To get your GCSE in Biology you'll have to do some exams that test your knowledge of Biology, your understanding of the Required Practical Activity experiments and how comfortable you are with Working Scientifically. You'll also be tested on your maths skills.

All the content that you need to know is in this book. All the Required Practical Activities are also covered in detail and clearly labelled, examples that use maths skills are marked up, and there are even dedicated sections on Working Scientifically (pages 2-22), Maths Skills (pages 376-383) and Practical Skills (pages 368-375).

Grading

When you sit your exams, you'll be given a grade between 1 and 9 based on your results. 9 is the highest grade, and 1 is the lowest. Which grades you can get will depend on which set of exams you sit — Foundation tier, or Higher tier.

If you take the Foundation tier exams, you can get a grade between 1 and 5, with 5 being the maximum grade you can get. If you sit the Higher tier exams, you can get a grade between 3 and 9.

The exams

You'll sit two separate exams at the end of Year 11 — remember that you could be asked questions on the Required Practical Activities and Working Scientifically in either of them. You're allowed to use a calculator in both of your GCSE Biology exams, so make sure you've got one.

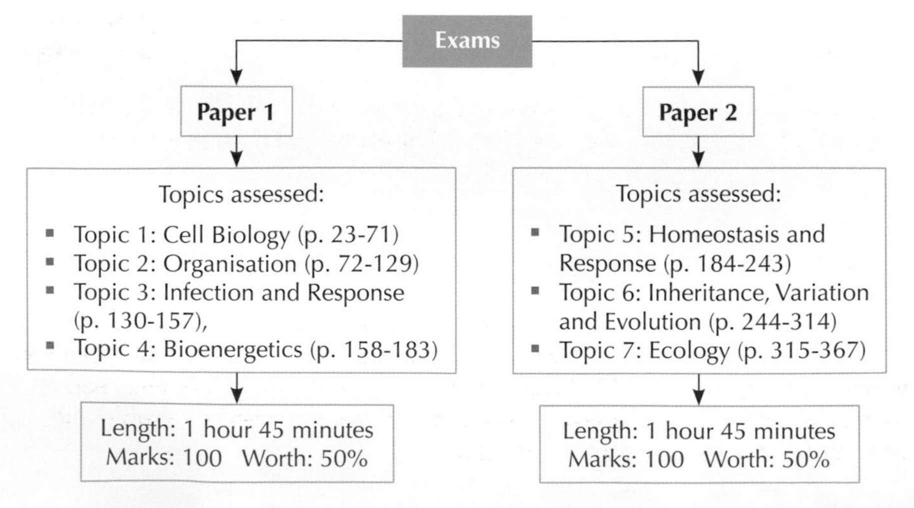

Exam Tip
Use these topic lists as a guide, but be aware that you may be expected to use basic biology knowledge from different parts of the course in either exam.

2. Exam Technique

Knowing the science is vitally important when it comes to passing your exams. But having good exam technique will also help. So here are some handy hints on how to squeeze every mark you possibly can out of those examiners.

Time management

Good time management is one of the most important exam skills to have — you need to think about how much time to spend on each question. Check out the length of your exams (you'll find them on the previous page and on the front of your exam papers). These timings give you about 1 minute per mark. Try to stick to this to give yourself the best chance to get as many marks as possible.

Don't spend ages struggling with a question if you're finding it hard to answer — move on. You can come back to it later when you've bagged loads of other marks elsewhere. Also, you might find that some questions need a lot of work for only a few marks, while others are much quicker — so if you're short of time, answer the quick and easy questions first.

Exam Tip
You shouldn't really be spending more time on a 1 mark question than on a 4 mark question. Use the marks available as a rough guide for how long each question should take to answer.

Example

The questions below are both worth the same number of marks but require different amounts of work.

1.1 What name is given to the tissue that covers the leaf?

(1 mark)

2.1 By reading the information above and looking at the diagram, suggest **one** way in which the structure of a motor neurone cell helps it to carry out its function.

(1 mark)

Question 1.1 only asks you to write down the name of a tissue — if you can remember it this shouldn't take you too long.

Question 2.1 asks you to make a suggestion about something you don't already know the answer to. Since you need to interpret some information and a diagram, and then apply your existing scientific knowledge to work out your answer, it may take you a bit longer than just writing down a name.

So, if you're running out of time it makes sense to do questions like 1.1 first and come back to questions like 2.1 if you've got time at the end.

Exam Tip
Don't forget to go back and do any questions that you left the first time round — you don't want to miss out on marks because you forgot to do the question.

Reading the question

Make sure you always read the whole question carefully. It can be easy to look at a question and read what you're expecting to see, rather than what it's actually asking you. Read it through before you start answering, and read it once again when you've finished, to make sure your answer is sensible and matches up to what the question is asking.

Exam Tip
The amount of space given for your answer should also give you an idea about how much you need to write.

Remember to pay attention to the marks available too. They can often give you a sense of how much work is needed to answer the question. If it's just a one or two mark question, it'll often only need a single word or phrase as an answer, or a very simple calculation. Questions worth more marks are likely to be longer questions, which need to be clearly structured and will involve writing a short paragraph or a more complicated calculation.

Making educated guesses

Make sure you answer all the questions that you can — don't leave any blank if you can avoid it. If a question asks you to tick a box, circle something or draw lines between boxes, you should never, ever leave it blank, even if you're short on time. It only takes a second or two to answer these questions, and even if you're not absolutely sure what the answer is you can have a good guess.

Example

Look at the question below.

1.1 Which of the following are hormones involved in the menstrual cycle? Tick **two** boxes.

Insulin	☐	Oestrogen	☐
Testosterone	☐	Luteinising hormone	☐

(2 marks)

Say you knew that oestrogen is involved in the menstrual cycle, and that insulin helps to regulate blood sugar level, but weren't sure about the other two hormones.

You can tick oestrogen — you know it's involved in the menstrual cycle. You can leave the insulin box blank because you know it's involved in regulating blood sugar level. That leaves you with testosterone and luteinising hormone. If you're not absolutely sure which is involved in the menstrual cycle and which isn't, just have a guess. You won't lose any marks if you get it wrong and there's a 50% chance that you'll get it right.

Exam Tip
If you're asked, for example, to tick two boxes, make sure you only tick two. If you tick more than two, you won't get the marks even if some of your answers are correct.

3. Question Types

If all questions were the same, exams would be mightily boring. So really, it's quite handy that there are lots of different question types. Here are just a few...

Command words

Command words are just the bits of a question that tell you what to do. You'll find answering exam questions much easier if you understand exactly what they mean, so here's a brief summary of the most common ones:

Command word:	What to do:
Give / Name / Write down	Give a brief one or two word answer, or a short sentence.
Complete	Write your answer in the space given. This could be a gap in an equation or table, or you might have to finish a diagram.
Describe	Write about what something's like, e.g. describe the trend in a set of results.
Explain	Make something clear, or give the reasons why something happens. The points in your answer need to be linked together, so you should include words like because, so, therefore, due to, etc.
Calculate	Use the numbers in the question to work out an answer.
Suggest	Use your scientific knowledge to work out what the answer might be.
Compare	Give the similarities and differences between two things.
Determine	Use data or information you're given in the question to work something out, e.g. you might be asked to determine the rate of a reaction from a graph.
Estimate	Use rounded numbers to work out an approximate answer.

Exam Tip
When you're reading an exam question, you might find it helpful to underline the command words. It can help you work out what type of answer to give.

Exam Tip
It's easy to get describe and explain mixed up, but they're quite different. For example, if you're asked to describe some data, just state the overall pattern or trend. If you're asked to explain data, you'll need to give reasons for the trend.

Exam Tip
Some questions might ask you to both describe and explain something — so make sure you do both things.

Some questions will also ask you to answer 'using the information provided' (e.g. a graph, table or passage of text) — if so, you must refer to the information you've been given or you won't get the marks.

Required Practical Activities

The Required Practical Activities are ten specific experiments that you need to cover during your lessons. At least 15% of the total marks in your exams will be for questions that test your understanding of these experiments and the techniques involved in them. There are a lot of different types of question you could be asked on these experiments. Here are some basic areas you might be asked about:

- Carrying out the experiment — e.g. planning or describing a method, describing how to take measurements or use apparatus.

Exam Tip
The Required Practical Activity questions are likely to have some overlap with Working Scientifically, so make sure you've brushed up on pages 2-22.

- Risk assessment — e.g. describing or explaining dangers which can arise from the experiment, or safety precautions which should be taken.
- Understanding variables — e.g. identifying control, dependent and independent variables.
- Data handling — e.g. plotting graphs or doing calculations using some sample results provided.
- Analysing results — e.g. making conclusions based on sample results.
- Evaluating the experiment — e.g. making judgements on the quality of results, identifying where mistakes have been made in the method, suggesting improvements to the experiment.

Required Practical Activity questions won't be pointed out to you in the exam, so you'll need to make sure you know the practicals inside out, and can recognise them easily. For an example of a question testing your understanding of a practical, see page 129.

Levels of response questions

Some questions are designed to assess your ability to present and explain scientific ideas in a logical and coherent way, as well as your scientific knowledge. These questions often link together different topics and are worth more marks than most other question types. You'll be told which questions these are on the front of your exam paper.

This type of question is marked using a 'levels of response' mark scheme. Your answer is given a level depending on the number of marks available and its overall quality and scientific content. Here's an idea of how the levels may work out for a 6 mark question:

Exam Tip
You're likely to get level of response questions with command words like 'Explain', 'Evaluate' or 'Describe'.

Example

- An answer that has no relevant information and makes no attempt to answer the question receives no marks.
- A Level 1 answer usually makes one or two correct statements, but does not fully answer the question. For instance, when asked to explain how the features of an organism make it adapted to its environment, it might state one or two correct facts about the organism's features, but not attempt to link them to the environment. These answers receive 1 or 2 marks.
- A Level 2 answer usually makes a number of correct statements, with explanation, but falls short of fully answering the question. It may miss a step, omit an important fact, or not be organised as logically as it should be. These answers receive 3 or 4 marks.
- A Level 3 answer will answer the question fully, in a logical fashion. It will make a number of points that are explained and related back to the question. Any conclusions it makes will be supported by evidence in the answer. These answers receive 5 or 6 marks.

Exam Tip
It might be useful to write a quick plan of your answer in the spare space of your paper. This can help you get your thoughts in order, so you can write a logical, coherent answer. But remember to cross your plan out after you've written your answer so it doesn't get marked.

Exam Tip
Make sure your writing is legible — you don't want to lose marks just because the examiner couldn't read your handwriting.

Make sure you answer the question fully, and cover all points indicated in the question. You also need to organise your answer clearly — the points you make need to be in a logical order. Use specialist scientific vocabulary whenever you can. For example, if you're describing cell division, you'd need to use scientific terms like mitosis and chromosomes. Obviously you need to use these terms correctly — it's no good knowing the words if you don't know what they actually mean.

There are some exam-style questions that use this type of mark scheme in this book (marked up with an asterisk, *). You can use them to practise writing logical and coherent answers. Use the worked answers given at the back of this book to mark what you've written. The answers will tell you the relevant points you could've included, but it'll be down to you to put everything together into a full, well-structured answer.

Exam Tip
Make sure your writing style is appropriate for an exam. You need to write in full sentences and use fairly formal language.

Comparing and evaluating information

In the exam, you may be given some information to read and then be asked to make some comparisons.

Example

Drug X and drug Y are both weight-loss drugs. You may be given some information on how drug X and drug Y work, how effective they are and what side effects they have. You could then be asked to compare the drugs directly, or give the advantages of one compared to the other.

To compare the two types of drug, you need to do more than just pick out relevant information from the question and repeat it in your answer — you need to make clear comparisons between the two.

Example

If the information tells you that people taking drug X lose 1 kilogram per week on average and that people taking drug Y lose 1.5 kilograms per week on average, you need to say in your answer that "people taking drug Y lose **more** weight per week on average than people taking drug X". It's even better to say that "people taking drug Y lose **0.5 kilograms more** weight per week on average than people taking drug X".

Evaluating information

If you are asked to evaluate two things, you need to weigh up their advantages and disadvantages.

The question may ask you to write a conclusion too, e.g. make an overall judgement about which drug is best. If so, you must include a conclusion in your answer and you must back it up with evidence from the question.

Calculations

In the GCSE Biology exams, around 10% of the total marks will come from questions that test your maths skills. Questions that involve a calculation can seem a bit scary. But they're really not that bad. Here are some tips to help you out...

Figure 1: *A calculator. Under the pressure of an exam it's easy to make mistakes in calculations, even if they're really simple ones. So don't be afraid to put every calculation into the calculator.*

- Show your working — this is the most important thing to remember. It only takes a few seconds more to write down what's in your head and it might stop you from making silly errors and losing out on easy marks. You won't get a mark for a wrong answer, but you could get marks for the method you used to work out the answer.
- Check your answer — a good way to do this is to work backwards through your calculation. You should also think about whether your answer seems sensible — if it's a much bigger or smaller number than you were expecting, you might have gone wrong somewhere.

Example

A potato chip weighs 10.2 g at the start of the experiment. At the end of the experiment it weighs 11.6 g. To calculate the change in mass, you'd subtract 10.2 g from 11.6 g — which works out as 1.4 g. You'd expect the answer to be around 1 g, so this seems sensible. To check it though, just add 1.4 to 10.2 — you should end up with 11.6.

- Sometimes you'll be asked to pick some numbers out of a table or read values off a graph to use in your calculation. If so, always read the question carefully so you know exactly what figures you need to use. Make sure you read the headings in the table carefully too (or the axes on the graph) to make sure you understand what's being shown.

Exam Tip
These aren't the only calculations you could be asked to do in an exam — they're just examples of the sort of thing that's likely to come up.

In the exam, you should be prepared to do things like estimate a value, calculate a percentage or a mean, put numbers into equations that you've been given, and also be able to rearrange these equations.

Answers

Topic 1 — Cell Biology

Topic 1a — Cell Structure and Cell Division

1. Cell Structure

Page 24 — Fact Recall Questions

Q1 the genetic material / the nucleus

Q2 In the cytoplasm.

Q3 aerobic respiration

Q4 A cell wall, a permanent vacuole and chloroplasts.

Q5 A small ring of DNA.

2. Microscopy

Page 29 — Application Questions

Q1 magnification = image size ÷ real size
= 18 ÷ 0.002 = **× 9000**

Q2 image size = magnification × real size
= 400 × 0.08 = **32 mm**

Q3 real size = image size ÷ magnification
= 10.5 ÷ 1500 = 0.007 mm
0.007 mm × 1000 = **7 µm**

The question asks for the answer to be in µm, so to convert mm to µm you need to multiply by 1000.

3. Cell Differentiation and Specialisation

Page 32 — Fact Recall Questions

Q1 The process by which a cell changes to become specialised for its job.

Q2 In most animal cells, the ability to differentiate is lost at an early stage, whereas lots of plant cells don't ever lose this ability.

Q3 a) To carry electrical signals from one part of the body to another.

b) It is long in order to cover more distance and it has branched connections at each end to connect to other nerve cells (to form a network throughout the body).

Q4 They contain lots of mitochondria.

Page 32 — Application Questions

Q1 E.g. you'd expect to find a lot of ribosomes because they make proteins.

The function of a gastric chief cell is to secrete proteins — and ribosomes are where proteins are made in the cell. So it makes sense that you'd find a lot of ribosomes in a gastric chief cell.

Q2 E.g. the folds in the cell membrane give the cells a large surface area for absorbing food molecules efficiently. Lots of mitochondria provide the energy from respiration needed to absorb food molecules.

This question is pretty tricky — you need to think about the unusual shape of the cell and look at the things it contains. Cells that need to absorb things tend to have a large surface area (e.g. like red blood cells and palisade leaf cells), so look for anything about the cell that increases its surface area. You also know that the cell needs energy — so look for mitochondria, which are the site of respiration (a process which transfers energy).

4. Chromosomes

Page 34 — Fact Recall Questions

Q1 in the nucleus

Q2 DNA

Q3 a) short sections of DNA/a chromosome (that control our characteristics)

b) controlling the development of different characteristics

Page 34 — Application Questions

Q1 a) X

b) Z

c) Y

Q2 Because genes are too small.

5. Mitosis

Page 37 — Fact Recall Questions

Q1 Any two from: e.g. growth / development / replacing damaged cells.

Q2 once

Page 37 — Application Questions

Q1 a) genetically identical

Mitosis produces cells that are genetically identical to the parent cells.

b) two

If a cell has two sets of chromosomes it means that it will have two copies of each chromosome.

Q2 (6 ÷ 198) × 1500 = 45.45...
45.45... ÷ 60 = **0.76 hours**

6 out of 198 cells are in the last stage of mitosis. This suggests that the proportion of time the cells spend in this stage must be 6/198th of the cell cycle. Once you've worked out the time taken in minutes, don't forget to convert your answer into hours as the question asks.

6. Binary Fission

Page 39 — Application Questions

Q1 a) 6 hours × 60 = 360 minutes
360 ÷ 24 = 15 divisions
2^{15} = 32 768 = **33 000 cells (2 s.f.)**

b) 33 000 = **3.3×10^4 cells**

Q2 a) $128 = 2 \times 2 \times 2 \times 2 \times 2 \times 2 \times 2 = 2^7$
so a population of 128 cells is produced after 7 divisions.
7 divisions would take 7 × 30 minutes = 210 minutes.
210 minutes ÷ 60 = **3.5 hours**

b) $64 = 2 \times 2 \times 2 \times 2 \times 2 \times 2 = 2^6$
The bacterial cell has divided 6 times in 270 minutes, so the mean division time is 270 ÷ 6 = **45 minutes.**

7. Culturing Microorganisms

Page 43 — Application Questions

Q1 Using the scale, 4 mm = 1 cm.

Antibiotic	Diameter (cm)	Radius (cm)	Area (cm²)
A	13 ÷ 4 = 3.25	1.625	$\pi \times (1.625)^2$ = 8.29... = 8
B	4 ÷ 4 = 1	0.5	$\pi \times (0.5)^2$ = 0.78... = 1

Don't forget to halve the diameter to get the radius before calculating the area.

Q2 There would be no inhibition zone around the paper disc holding the antibiotic.

Q3 Out of the antibiotics tested, antibiotic A is the more effective antibiotic against the strain of bacteria used.

8. Stem Cells

Page 46 — Fact Recall Questions

Q1 An undifferentiated cell that has the potential to differentiate into different types of cell.

Q2 any type of cell

Q3 E.g. in the bone marrow.

Q4 They would be genetically identical to the patient, so they won't be rejected by the patient's body.

Q5 meristem

Page 46 — Application Questions

Q1 a) Stem cells could be made to differentiate into neurones, which could replace the damaged/dead neurones.

b) E.g. producing nerve cells to replace damaged tissue in people with paralysis. / Producing insulin-producing cells for people with diabetes.

Q2 Because the embryos leftover from fertility clinics will be destroyed anyway.

Q3 E.g. because the embryo still has the potential to develop into a human life before this point.

Pages 49-50 — Cell Structure and Cell Division Exam-style Questions

1.1 the nucleus ***(1 mark)***

1.2 two ***(1 mark)***

1.3 mitosis ***(1 mark)***

2.1 D ***(1 mark)***

2.2 A eukaryotic cell, e.g. because it has a nucleus/ mitochondria/chloroplasts / it doesn't have plasmids/ circular strand of DNA in the cytoplasm ***(1 mark).***

2.3 A cell that performs a specific function ***(1 mark).***

2.4 stem cells ***(1 mark)***

2.5 E.g. it has lots of chloroplasts / its chloroplasts are mainly at the top of the cell (nearer the light) / it's thin so lots of them can be packed in at the top of the leaf ***(1 mark).***

3.1 4 × 60 = 240 minutes
240 ÷ 48 = 5 divisions
$2^5 = 2 \times 2 \times 2 \times 2 \times 2$ = **32 cells** ***(2 marks for correct answer, otherwise 1 mark for correct working.)***

3.2 genetically identical ***(1 mark)***

3.3 binary fission ***(1 mark)***

3.4 E.g. warm environment ***(1 mark)***, lots of nutrients ***(1 mark)***

3.5 Width of image = 14 mm
2.8 μm ÷ 1000 = 0.0028 mm
14 ÷ 0.0028 = **× 5000** ***(2 marks for correct answer, otherwise 1 mark for correct working.)***

4.1 Any two from: e.g. heat/sterilise the agar jelly ***(1 mark)*** / sterilise the inoculating loop before using it, e.g. by passing it through a flame ***(1 mark)*** / sterilise the Petri dish before using it ***(1 mark).***

4.2 Antibiotic 5 ***(1 mark)*** because this disc has the biggest inhibition zone around it ***(1 mark).*** This means that more bacteria were killed/unable to grow around antibiotic 5 than around any of the other antibiotics / the bacteria were least resistant to antibiotic 5 ***(1 mark).***

It stands to reason that the best antibiotic for getting rid of the infection is the one that kills the most bacteria.

4.3 E.g. the size of the paper discs ***(1 mark)***, the concentration of the antibiotics ***(1 mark).***

4.4 E.g. it shows that the lack of growth around the antibiotic discs is due to the effect of the antibiotics and not something else ***(1 mark).***

Topic 1b — Transport in Cells

1. Diffusion

Page 53 — Fact Recall Questions

Q1 Diffusion is the spreading out of particles from an area of higher concentration to an area of lower concentration.

Q2 E.g. oxygen, glucose, amino acids, water

Q3 The rate of diffusion is increased because the particles have more energy, so they move faster.

Page 53 — Application Questions

Q1 B

Q2 Inside respiring cells. The carbon dioxide molecules must be diffusing from an area of higher concentration (inside respiring cells) to an area of lower concentration (the bloodstream).

Q3 a) The smoke particles diffuse from where there is a higher concentration (near the stage), to where there is a lower concentration (at the opposite end of the hall).

b) There are already some smoke particles in the air the second time the smoke machine is set off, whereas there weren't the first time. This means there is a smaller difference in the concentration of the smoke particles at each end of the hall, so the rate of diffusion is slower the second time.

2. Osmosis

Page 57 — Fact Recall Questions

Q1 The movement of water molecules across a partially permeable membrane from a region of higher water concentration to a region of lower water concentration.

Q2 If a cell is short of water, the surrounding solution will usually have a higher concentration of water molecules/be more dilute than the solution inside the cell. This means water molecules will move by osmosis from the surrounding solution into the cell.

Page 57 — Application Question

Q1 a) i) They got longer/increased in length (by an average of 3.1 mm in the 0.00 M solution and an average of 2.1 mm in the 0.25 M solution).

ii) The solutions contained a higher concentration of water molecules/were more dilute than the fluid inside the potato cells. So water molecules moved into the potato cells by osmosis and the potato cylinders increased in size.

Cells that take on water by osmosis will also increase in mass. So you could vary this experiment slightly by measuring the change in mass of the potato cylinders.

b) Accept any answer between – 3 mm and – 4 mm.

c) i) the length of the potato cylinders

The dependent variable is the variable you measure — in this case potato cylinder length.

ii) the concentration of the sugar solution

The independent variable is the variable you change.

d) Any two from: e.g. the volume of each solution / the temperature each solution was kept at / the time the cylinders were left for / the type of sugar used.

3. Active Transport

Page 59 — Fact Recall Questions

Q1 The movement of particles against a concentration gradient (i.e. from an area of lower concentration to an area of higher concentration) using energy transferred during respiration.

Q2 false

Cells can absorb ions from very dilute solutions using active transport.

Q3 They are essential for healthy growth.

Q4 The concentration of mineral ions inside the root hair cells is greater than in the soil / they are taking up mineral ions against the concentration gradient.

Q5 E.g. glucose/sugar

4. Exchange Surfaces

Page 62 — Fact Recall Questions

Q1 a) diffusion, active transport and osmosis

b) diffusion and active transport

Q2 Any three from: e.g. they may be thin. / They may have a large surface area. / They may have lots of blood vessels. / They may be ventilated.

Page 63 — Application Questions

Q1 a) Any two from: e.g. it may be thin, so oxygen and carbon dioxide have a short distance to diffuse/a short diffusion pathway. / It may have a large surface area so lots of oxygen/carbon dioxide can diffuse at once. / It may have lots of blood vessels to get gases into and out of the blood quickly.

b) Because elephants have a much smaller surface area to volume ratio than earthworms. This means that not enough oxygen can diffuse from their outside surface to supply their entire volume, so they need a specialised gas exchange surface for the oxygen to diffuse across efficiently.

Q2 a) A — $6 \times (1 \times 1) = \mathbf{6\ \mu m^2}$

B — $2 \times (0.5 \times 0.5) + 4 \times (2 \times 0.5) = 0.5 + 4$
$= \mathbf{4.5\ \mu m^2}$

C — $2 \times (0.2 \times 0.4) + 2 \times (1 \times 0.4) + 2 \times (1 \times 0.2)$
$= 0.16 + 0.8 + 0.4 = \mathbf{1.36\ \mu m^2}$

b) A — volume: $1 \times 1 \times 1 = \mathbf{1\ \mu m^3}$

B — volume: $2 \times 0.5 \times 0.5 = \mathbf{0.5\ \mu m^3}$

C — volume: $1 \times 0.4 \times 0.2 = \mathbf{0.08\ \mu m^3}$

c) A — **6 : 1**

B — 4.5 : 0.5
$4.5 \div 0.5 = 9$, $0.5 \div 0.5 = 1$
so ratio is **9 : 1**

C — 1.36 : 0.08
$1.36 \div 0.08 = 17$, $0.08 \div 0.08 = 1$
so ratio is **17 : 1**

d) C, because it has the largest surface area to volume ratio.

5. Exchanging Substances

Page 65 — Fact Recall Questions

Q1 E.g. they have a large surface area. / They have a moist lining for dissolving gases. / They have thin walls. / They have a good blood supply.

Q2 They increase the surface area of the small intestine so that nutrients can be absorbed more quickly.

Page 67 — Fact Recall Questions

Q1 By diffusion through the stomata.

Q2 E.g. oxygen

Q3 It has a flattened shape and there are air spaces inside the leaf.

Q4 They increase the surface area of the gills available for diffusion of gases (and so increase the rate of diffusion).

Q5 It maintains a large concentration gradient of oxygen between the water and the blood because it flows in the opposite direction to water. The concentration of oxygen in the water is always higher than in the blood, so as much oxygen as possible diffuses from the water into the blood.

Pages 70-71 — Transport in Cells Exam-style Questions

1.1 surface area:
$0.6 \times 0.6 = 0.36$
$3.0 \times 0.6 = 1.8$
$(2 \times 0.36) + (4 \times 1.8) = 7.92\ \mu m^2$
volume:
$0.6 \times 0.6 \times 3.0 = 1.08\ \mu m^3$
ratio:
$7.92 \div 1.08 = 7.3333... = 7$
so the ratio is **7 : 1** ***(3 marks for correct answer, otherwise 1 mark for correct surface area and 1 mark for correct volume.)***

1.2 Because it is a single-celled organism ***(1 mark)***, so it has a large surface area to volume ratio ***(1 mark)***. This means it can exchange enough substances across its cell membrane/outer surface to meet its needs ***(1 mark)***.

1.3 small holes in the surface of a leaf ***(1 mark)***

1.4 It is a lower concentration than inside the leaf ***(1 mark)***.

1.5 E.g. the flattened shape of the leaf increases its surface area for diffusion into and out of the leaf ***(1 mark)***. / There are air spaces inside the leaf to increase the surface area available for diffusion into and out of the cells ***(1 mark)***.

1.6 gills ***(1 mark)***

1.7 It would decrease the rate of diffusion of oxygen ***(1 mark)*** because it would reduce the difference in concentration of oxygen between the water and the blood in the gills / reduce the concentration gradient ***(1 mark)***.

2.1 Glucose may be absorbed by active transport when the concentration of glucose molecules in the small intestine is lower than the concentration of glucose molecules in the blood ***(1 mark)***. Respiration is needed to provide energy for this process ***(1 mark)***.

2.2 The damage to the villi would decrease the surface area of the small intestine ***(1 mark)***, meaning less glucose and other nutrients would be absorbed ***(1 mark)***.

2.3 Osmosis ***(1 mark)***. The water moves from an area of higher concentration (in the large intestine) to an area of lower concentration (in the bloodstream) ***(1 mark)***.

3.1 Water moves by osmosis from a region of higher water concentration to region of a lower water concentration ***(1 mark)***. The potato has a lower water concentration than the water in the dish, so water moves from the dish into the potato cells ***(1 mark)***. The potato has a higher water concentration than the sugar/well, so water moves from the potato into the well and dissolves the sugar/ creates a sugar solution ***(1 mark)***.

3.2 There's no sugar in the well in the potato in experiment B, so no water is drawn out of the potato into the well at the top ***(1 mark)***. This means the difference in the concentration of the solution in the potato cells and in the dish is smaller ***(1 mark)***, so the movement of water molecules by osmosis is reduced ***(1 mark)***.

Topic 2 — Organisation

Topic 2a — Tissues, Organs and Organ Systems

1. Cell Organisation

Pages 74-75 — Fact Recall Questions

Q1 cells

Q2 A group of similar cells that work together to carry out a particular function.

Q3 A group of different tissues that work together to perform a certain function.

Q4 organ system

Q5 A is the salivary glands, which produce digestive juices. B is the liver, which produces bile. C is the stomach, where food is digested. D is the pancreas, which produces digestive juices. E is the small intestine, where food is digested and soluble food molecules are absorbed. F is the large intestine, which absorbs water from undigested food, leaving faeces.

Page 75 — Application Questions

Q1 A tissue, because it consists of a group of similar cells that work together to perform a particular function.

Q2 a) It contracts, to move the fertilised egg cell along the fallopian tube to the uterus.

b) An organ, because it consists of a group of different tissues that work together to perform a certain function.

c) egg cell, muscular tissue, uterus, reproductive system

2. The Lungs

Page 78 — Fact Recall Question

Q1 a) trachea

b) bronchi

c) i) alveolus

ii) Because this is where gas exchange occurs. Oxygen diffuses out of part C/the alveoli into the blood being carried by the capillaries. At the same time, carbon dioxide diffuses out of the blood in the capillaries into part C/the alveoli.

Page 78 — Application Question

Q1 breaths per minute = 366 ÷ 9
= 40.666...
= **41 breaths per minute**

If your answer doesn't come out as a whole number, you'll need to round it up because you can't take part of a breath.

3. Circulatory System — The Heart

Page 82 — Fact Recall Questions

Q1 To transport food and oxygen to every cell in the body and to carry waste products such as carbon dioxide and urea to where they can be removed from the body.

Q2 One circuit pumps deoxygenated blood from the heart to the lungs, and then oxygenated blood from the lungs back to the heart. The other circuit pumps oxygenated blood from the heart to the rest of the body, and then deoxygenated blood from the rest of the body back to the heart.

Q3 right atrium, right ventricle, left atrium, left ventricle

Q4 a) vena cava and pulmonary vein
b) pulmonary artery and aorta

Q5 Blood flows through the pulmonary vein and the vena cava into the atria. The atria contract, pushing blood into the ventricles. The ventricles then contract, forcing the blood into the pulmonary artery and aorta, and out of the heart.

Page 82 — Application Questions

Q1 Because the coronary arteries supply the heart with its own supply of oxygenated blood. When this supply is blocked, the heart might not get all the oxygen it needs to function.

Q2 a) It is controlled by a group of cells in the right atrium wall that act as a pacemaker. These cells produce a small electric impulse which spreads to the surrounding muscle cells, causing them to contract.
b) A pacemaker is a little device that's implanted under the skin and has a wire going to the heart. It produces an electric current to keep the heart beating regularly.

Q3 Similarity:
E.g. both circulatory systems involve the blood going to an organ to pick up oxygen. / Both systems involve a heart pumping blood around the body.
Difference:
E.g. the circulatory system of a fish is just one circuit, whereas that of a human is two separate circuits. / The circulatory system of a fish picks up oxygen at the gills, whereas that of a human picks up oxygen at the lungs.

4. Circulatory System — The Blood Vessels

Page 85 — Fact Recall Questions

Q1 An artery has thick walls compared to its lumen. The walls have thick layers of muscle and elastic fibres.

Q2 a capillary

Q3 Capillaries carry blood very close to the body cells. The substances needed by the body cells diffuse out of the blood, through the walls of capillaries, and into the body cells.

Q4 Any two from: e.g. a vein has thinner walls than an artery. / A vein has a bigger lumen than an artery. / A vein has valves whereas an artery doesn't.

Page 85 — Application Questions

Q1 A vein, because veins carry blood into the heart.

Q2 A. Arteries carry blood at a higher pressure than veins and their walls are more muscular as a result. Blood vessel A carries blood at a higher pressure than blood vessel B, and so is more likely to be an artery and have more muscle in its walls.

Q3 rate of blood flow = 1075 ÷ 5
= 215 ml/min
Yes, the patient may have a blockage in this artery because the rate of blood flow is much less than the average for this artery.

5. Circulatory System — The Blood

Page 87 — Fact Recall Questions

Q1 a) haemoglobin
b) It carries oxygen.

Q2 To defend against microorganisms that cause disease.

Q3 False
White blood cells do have a nucleus. Red blood cells don't (so they have more room for haemoglobin).

Q4 Small fragments/pieces of cells.

Page 88 — Application Questions

Q1 Platelets, because platelets help the blood to clot at the site of a wound.

Q2 Haemoglobin is needed to carry oxygen around the body in red blood cells, so without enough haemoglobin, organs won't get enough oxygen.

Q3 Fay, because the number of white blood cells in her blood is below the normal range / lower than the number of white blood cells in Imogen's blood. White blood cells help to defend the body against microorganisms, so with fewer white blood cells in her blood, Fay is more likely to get an infection.

6. Plant Cell Organisation

Page 90 — Fact Recall Questions

Q1 false

Q2 palisade mesophyll tissue

Q3 They allow gases to diffuse in and out of cells.

Q4 In the growing tips of shoots and roots.

Q5 So that light can pass through to the palisade layer, which contains lots of chloroplasts for photosynthesis.

7. Transpiration and Translocation

Page 92 — Fact Recall Questions

Q1 a) food substances / dissolved sugars
b) the leaves

Q2 water and mineral ions

Q3 Water is transported up a plant in the transpiration stream. Water escapes from the leaves by evaporation and diffusion. This creates a slight shortage of water in the leaves, which causes more water to be drawn up the xylem and into the leaves. This in turn causes more water to be drawn up the xylem from the roots.

Remember, the transpiration stream only goes on in the xylem — it doesn't involve the phloem.

8. Transpiration and Stomata

Page 95 — Fact Recall Questions

Q1 The transpiration rate increases.

Q2 The uptake of water by a plant.

Q3 a) guard cells
b) When the plant is short of water, the guard cells lose water and become flaccid, making the stomata close.

Page 96 — Application Questions

Q1 a) They have lost water by evaporation and diffusion (through their stomata).
b) The plant next to the fan. Diffusion is quicker when air flow is greater. The movement of the fan will increase air flow.
c) E.g. instead of using a fan he could leave one plant in a warmer/colder place than the other plant, e.g. in an airing cupboard/refrigerator.

Q2 a) As temperature increases, the distance moved by the bubble in 20 minutes/the transpiration rate also increases. This is because at higher temperatures, water evaporates from the leaves more quickly. This means that water is drawn into the xylem from the glass tube more quickly, so the bubble moves further along the tube in a given time.
b) rate of transpiration = 26 ÷ 20
= **1.3 mm/min**
c) E.g. they could have tested the transpiration rate at more temperatures/a wider range of temperatures.

One feature of a valid experiment is that it answers the original question, in this case: how does temperature affect transpiration rate? Testing the transpiration rate at more temperatures gives them a better idea of this.

d) You would expect the bubble to have moved less than 26 mm in 20 minutes. Coating the underside of some of the leaves with nail varnish would have blocked some of the plant's stomata and reduced the amount of water that was able to escape from the leaves. Therefore the rate of transpiration would have been slower, so the bubble would have moved a shorter distance in 20 minutes.

Pages 99-100 — Tissues, Organs and Organ Systems Exam-style Questions

1.1 C ***(1 mark)***

1.2 C ***(1 mark)***

2.1 A — vena cava ***(1 mark)***, B — right ventricle ***(1 mark)***, C — left atrium ***(1 mark)***.

2.2 X, because veins have thinner walls than arteries ***(1 mark)***.

3 Red blood cells ***(1 mark)***. They have a biconcave shape to give them a large surface area for absorbing oxygen ***(1 mark)***. They contain a red pigment called haemoglobin, which carries the oxygen ***(1 mark)***. They don't have a nucleus, allowing for more room for haemoglobin, meaning that they can carry more oxygen ***(1 mark)***.

4.1 epidermal tissue ***(1 mark)***

4.2 Palisade mesophyll tissue ***(1 mark)***. Having lots of chloroplasts suggests that the cells are adapted for photosynthesis ***(1 mark)*** and it is in the palisade mesophyll tissue that most photosynthesis takes place ***(1 mark)***.

4.3 A decrease in the concentration of carbon dioxide in the air spaces outside the leaf cells would reduce the rate of diffusion of carbon dioxide into the cells ***(1 mark)***. This is because the difference between the concentration of carbon dioxide inside the cells and outside the cells would be smaller ***(1 mark)***.

5.1 A plant loses water through its stomata ***(1 mark)***. Evaporation/water loss is quickest in hot, dry conditions ***(1 mark)***. So closing the stomata when it is hot and dry and only opening them when it is cooler, will reduce water loss from the plant ***(1 mark)***.

5.2 Trapping water vapour close to the surface of the leaf will reduce the difference in the concentration of water molecules between the inside and outside of the leaf ***(1 mark)***. This will slow down evaporation/water loss from the plant ***(1 mark)***.

Plants in dry conditions want to conserve as much water as possible because there's so little available in the environment.

Topic 2b — Health and Disease

1. Introduction to Health and Disease

Page 102 — Fact Recall Questions

Q1 Health is the state of physical and mental wellbeing.

Q2 Communicable diseases can be spread between people or between people and animals, whereas non-communicable diseases cannot.

Q3 Whether or not you have a good, balanced diet, the amount of stress that you're under and your life situation.

Page 102 — Application Question

Q1 People with high-stress jobs had on average 3 more colds per year than people with low-stress jobs. This could be due to the fact that being under stress can have an effect on physical health, which may increase the likelihood of catching a cold.

2. Cardiovascular Disease

Page 106 — Fact Recall Questions

Q1 Disease of the heart or blood vessels.

Q2 The arteries that supply blood to the heart muscle get blocked by fatty deposits, causing them to become narrow and restrict blood flow.

Q3 a) A tube inserted inside arteries to widen them and keep them open.

b) They can be inserted into the coronary arteries to make sure blood can pass through to the heart muscles. This keeps the person's heart beating and keeps the person alive.

Page 106 — Application Question

Q1 a) E.g. they are less likely to be rejected by the body's immune system than donor hearts.

b) Any two from: e.g. surgery to fit an artificial heart can lead to bleeding and infection. / Artificial hearts don't work as well as healthy natural ones — parts of the heart could wear out or the electrical motor could fail. / Blood doesn't flow through artificial hearts as smoothly as through natural hearts, which can cause blood clots and lead to strokes. / A patient receiving an artificial heart has to take drugs to thin their blood and make sure clots don't occur, which can cause problems with bleeding if they're hurt in an accident. / Having an artificial heart in the body may be uncomfortable for the patient.

c) E.g. artificial heart two may be more convenient for the patient, as they will be able to move around more freely than if they had to carry a battery pack around with them. Also, artificial heart two doesn't need to have wires coming out of the patient's body, unlike artificial heart one. This may reduce the risk of infection compared to artificial heart one, as there are no openings in the skin where microorganisms could enter the body. However, artificial heart two has only been implanted into 14 patients, whereas artificial heart one has been implanted into 1100 patients. So there's a greater chance that there may be problems with the use of artificial heart two, which have not yet been identified due to the small number of patients it has been used for. Overall, artificial heart two may be more convenient for the patient and have a lower risk of infection than artificial heart one, but there's a greater risk of unexpected problems occurring.

Remember, if you're asked to evaluate the use of something in the exam, you should give arguments for and against its use and use any relevant facts you're given to support the points you make. It's a good idea to sum up your evaluation at the end too.

3. Risk Factors for Non-Communicable Diseases

Page 108 — Fact Recall Questions

Q1 Something linked to an increase in the likelihood that a person will develop a certain disease during their lifetime.

Q2 The presence of substances in the environment and aspects of a person's lifestyle.

Q3 E.g. nationally, people from deprived areas are more likely to smoke, have a poor diet and not exercise. This means the incidence of cardiovascular disease, obesity and Type 2 diabetes is higher in those areas.

Q4 Any two from: e.g. cardiovascular disease / lung disease / lung cancer

Q5 E.g. obesity

Q6 Something that causes cancer.

4. Cancer

Page 110 — Fact Recall Questions

Q1 No

Benign tumours aren't cancerous but malignant ones are.

Q2 Malignant tumours grow and spread to neighbouring healthy tissues. Cells can break off and spread to other parts of the body by travelling in the bloodstream. The malignant cells then invade healthy tissues elsewhere in the body and form secondary tumours. Malignant tumours are dangerous and can be fatal — they are cancers.

Page 110 — Application Question

Q1 a) Robyn's BMI = 77 ÷ (1.78 × 1.78) = **24.3**

Lauren's BMI = 75 ÷ (1.55 × 1.55) = **31.2**

b) Lauren, because she is obese, and obesity is a risk factor for cancer.

Pages 113-114 — Health and Disease Exam-style Questions

1.1 D ***(1 mark)***

1.2 malignant tumour ***(1 mark)***

1.3 Because there was more than one similar tumour, it suggests that cells from the original tumour had spread to another part of the body in the bloodstream ***(1 mark)*** to form a secondary tumour ***(1 mark)***. Only malignant/cancerous tumours can do this ***(1 mark)***.

1.4 C ***(1 mark)***

1.5 Statins reduce the amount of 'bad' cholesterol present in the bloodstream ***(1 mark)***. This slows down the rate of fatty deposits forming inside arteries ***(1 mark)***.

2.1 The damage may cause the valve tissue to stiffen, so it won't open properly ***(1 mark)***, or it might cause the valve to become leaky, allowing blood to flow in both directions rather than just forward ***(1 mark)***. This means that blood doesn't circulate as effectively as normal ***(1 mark)***.

2.2 Mechanical heart valves are man-made ***(1 mark)***, whereas biological heart valves are taken from humans or other mammals ***(1 mark)***.

2.3 The patient could be fitted with an artificial heart ***(1 mark)***.

3.1 (3 ÷ 154) × 100 = **1.9%** ***(1 mark)***

3.2 Stents can be inserted into narrowed arteries/arteries with restricted blood flow ***(1 mark)*** to help keep them open/allow blood to flow more freely ***(1 mark)***. People who have had a heart attack are likely to have narrowed coronary arteries, so putting stents in these arteries can prevent another heart attack ***(1 mark)***.

3.3 E.g. yes, because the results show that patients are less likely to die after having a drug-eluting stent inserted compared to a bare-metal stent ***(1 mark)***. Also renarrowing of the artery is much less likely in patients treated with a drug-eluting stent (9%) compared to those treated with a bare-metal stent (21%) ***(1 mark)***. The trial was fairly large (307 patients) which means the results are more likely to be representative of the whole population ***(1 mark)***.

Topic 2c — Enzymes and Digestion

1. Enzymes

Page 117 — Fact Recall Questions

Q1 Because they increase the speed of a biological reaction without being changed or used up in the reaction.

Q2 active site

Q3 false

Different enzymes work best at different pHs.

Page 117 — Application Question

Q1 It should slow down the rate of reaction. This is because heating hexokinase up to a high temperature/50 °C will probably cause the bonds in hexokinase to break and the active site to lose its shape. This would mean that glucose will no longer be able to fit into hexokinase and the reaction won't be catalysed.

For questions like this you just need to apply your own knowledge — e.g. that enzymes lose their shape at high temperatures and that enzymes need their unique shape to work — to the specific enzyme named in the question.

2. Investigating Enzymatic Reactions

Page 119 — Application Question

Q1 a) By adding a buffer solution to the boiling tube containing the enzyme and starch mixture.

b) The iodine solution would remain browny-orange rather than changing to blue-black.

c) $\text{Rate} = \frac{1000}{\text{time}} = \frac{1000}{120}$

$= \mathbf{8.3\ s^{-1}}$ (2 s.f.)

3. Enzymes and Digestion

Page 122 — Fact Recall Questions

Q1 false

Digestive enzymes catalyse the breakdown of big molecules into smaller molecules, e.g. protease enzymes catalyse the breakdown of proteins into amino acids.

Q2 amylase

Q3 In the stomach, the pancreas and the small intestine.

Q4 lipases

Q5 Bile neutralises the stomach acid and so creates the ideal alkaline conditions for enzymes in the small intestine to work in.

Page 123 — Application Question

Q1

	Amylase	Proteases	Lipases	Bile
Made where?	B, F, G	D, F, G	F, G	C
Work(s) where?	A, G	D, G	G	G

Make sure you learn where digestive enzymes and bile are produced and where they work — you can pick up easy marks with this information in the exam. However, it's easy to get things mixed up, so watch out. E.g. bile is made in the liver but stored in the gall bladder — make sure you don't get those two organs mixed up.

4. Food Tests

Page 126 — Fact Recall Questions

Q1 Benedict's solution

Q2 The solution would change from browny-orange to black or blue-black.

Q3 blue

Page 126 — Application Question

Q1 A pestle and mortar should be used to grind up the piece of food into little bits. The ground up food should then be transferred into a beaker and mixed with some distilled water. A glass rod should be used to give it a good stir to dissolve some of the food in the water. 5 cm^3 of the solution should then be transferred into a test tube and a pipette used to add 3 drops of Sudan III stain solution to the test tube. The test tube should then be gently shaken. If the food sample contains lipids, the mixture will separate out into two layers. The top layer will be bright red. If no lipids are present, no separate red layer will form at the top of the liquid.

Pages 128-129 — Enzymes and Digestion Exam-style Questions

1.1 C *(1 mark)*

1.2 To act as a biological catalyst. / To speed up the rate of a reaction (in a living organism) without being changed or used up itself *(1 mark)*.

1.3 How to grade your answer:

Level 0: There is no relevant information. *[No marks]*

Level 1: There is a brief description of the function of at least one type of digestive enzyme, with mention of where it is made or where it functions. *[1 to 2 marks]*

Level 2: There is a description of the function of at least two types of digestive enzyme, with mention of where they are made and where they work. *[3 to 4 marks]*

Level 3: There is a detailed description of the function of several types of digestive enzyme, including where each of them is made and where each of them works in the digestive system. *[5 to 6 marks]*

Here are some points your answer may include:

Carbohydrases are digestive enzymes that convert carbohydrates into simple sugars.

Amylase is a carbohydrase that catalyses the conversion of starch into sugars. Amylase is made in the salivary glands, the pancreas and the small intestine. It works in the mouth and the small intestine.

Protease enzymes are digestive enzymes that catalyse the conversion of proteins into amino acids. Proteases are made in the stomach, the pancreas and the small intestine. They work in the stomach and the small intestine.

Lipase enzymes are digestive enzymes that catalyse the conversion of lipids into glycerol and fatty acids. Lipases are made in the pancreas and the small intestine. They work in the small intestine.

2.1 Because every enzyme has an active site with a unique shape that usually only fits the substrate involved in a single reaction *(1 mark)*. If the substrate doesn't match the enzyme's active site, then the reaction won't be catalysed *(1 mark)*.

2.2 A, because higher temperatures can increase the rate of an enzyme-controlled reaction *(1 mark)*.

2.3 Because the temperature was too high *(1 mark)*, so the enzyme became denatured *(1 mark)* and so could no longer catalyse the reaction *(1 mark)*.

3.1 A *(1 mark)*

3.2 That glucose is not present in the sample *(1 mark)*.

3.3 She should have ground up the food in a pestle and mortar *(1 mark)*. The ground up food should then have been transferred to a beaker and distilled water added *(1 mark)*. The mixture should then have been stirred to dissolve some of the food *(1 mark)*. Finally, the solution should have been filtered through a funnel lined with filter paper *(1 mark)*.

3.4 Iodine solution *(1 mark)*. If the sample contained starch, the solution would have changed from a browny-orange *(1 mark)* to black or blue-black *(1 mark)*.

4 The photograph shows that in the test tubes with hydrochloric acid only and pepsin only (test tubes 1 and 2), the meat hasn't been fully digested *(1 mark)*. However, in the test tube with both pepsin and hydrochloric acid (test tube 3) the meat sample has been completely digested *(1 mark)*. This suggests that pepsin requires acidic conditions/low pH to function at its best *(1 mark)*.

Topic 3 — Infection and Response

1. Communicable Disease

Page 131 — Application Questions

Q1 They produce toxins that damage your cells and tissues.

Q2 The infected student might cough or sneeze, releasing droplets containing the virus into the air. Someone else in the classroom could breathe these in and become infected.

2. Viral Diseases

Page 132 — Fact Recall Questions

Q1 In the air through droplets from an infected person's sneeze or cough.

Q2 E.g. a red skin rash and a fever.

Q3 By sexual contact or by exchanging bodily fluids.

Q4 antiretroviral drugs

Q5 E.g. tomatoes

Q6 In plants with tobacco mosaic virus, parts of the leaves become discoloured. This means that the plant can't carry out photosynthesis as well, so growth is affected.

Photosynthesis produces glucose, which, amongst other things, is used by a plant for growth.

3. Fungal and Protist Diseases

Page 133 — Fact Recall Questions

Q1 Purple or black spots will develop on the leaves, and the leaves may turn yellow and drop off.

Q2 Because this can prevent people from being bitten by mosquitoes when they're asleep, meaning that the protist that causes malaria isn't passed on.

4. Bacterial Diseases and Preventing Disease

Page 135 — Fact Recall Questions

Q1 Any three from: e.g. fever / stomach cramps / vomiting / diarrhoea

Q2 E.g. eating chicken that caught the disease when it was alive / eating food that has been contaminated by being prepared in unhygienic conditions.

Q3 It is treated with antibiotics.

Q4 The vectors spread the disease to humans, and so by destroying the vectors, the disease cannot be passed on.

Page 135 — Application Questions

Q1 It will prevent them from passing the disease on to anyone else.

Q2 The nurse may come into contact with pathogens for many diseases. By washing his hands, this will prevent pathogens being passed on to other people that he comes into contact with.

5. Fighting Disease

Page 137 — Fact Recall Questions

Q1 It acts as a barrier to pathogens. It secretes antimicrobial substances which kill pathogens.

Q2 They consume pathogens using phagocytosis. They produce antitoxins. They produce antibodies.

Page 137 — Application Questions

Q1 E.g. hairs and mucus in your nose trap particles that could contain the SARS-CoV. / The trachea and bronchi secrete mucus to trap the SARS-CoV. / The trachea and bronchi are lined with cilia. These waft mucus containing the SARS-CoV up to the back of the throat where it can be swallowed.

Q2 a) John should now be immune to measles. If he is infected with the measles virus again, his white blood cells should rapidly produce the antibodies against it — meaning he won't get ill.

b) Antibodies are specific to a particular type of antigen. John has never had German measles before, so he won't have made antibodies against the antigens on the German measles virus. This means it will take his white blood cells time to produce antibodies against the virus, during which he could get ill.

6. Fighting Disease — Vaccination

Page 139 — Application Question

Q1 a) E.g. the vaccination will involve injecting a small amount of the dead or inactive whooping cough pathogen. The pathogen will carry antigens which will trigger the white blood cells to produce antibodies to attack the pathogen. If the vaccinated person is infected with the whooping cough pathogen at a later date, their white blood cells should be able to rapidly mass-produce antibodies to kill off the pathogen.

b) If a large percentage of the population are vaccinated it means that the people who aren't vaccinated (i.e. very young babies) are less likely to catch the disease because there are fewer people to pass it on.

7. Fighting Disease — Drugs

Page 142 — Fact Recall Questions

Q1 false

Q2 Antibiotics are drugs that kill or prevent the growth of bacteria. E.g. penicillin / methicillin.

Q3 Because different antibiotics kill different types of bacteria.

Q4 Viruses reproduce using your own body cells, so it's difficult to develop drugs that destroy the virus without killing the body's cells.

Q5 To slow down the rate of development of resistant strains.

Q6 aspirin

Q7 Alexander Fleming. He noticed that one of his Petri dishes of bacteria also had mould on it and the area around the mould was free of the bacteria. He found that the mould on the Petri dish was producing a substance that killed the bacteria — this substance was penicillin.

Page 142 — Application Question

Q1 a) Because antibiotics don't kill viruses.

b) E.g. because it won't tackle the underlying cause of Chloe's flu (it will only help to relieve her symptoms).

8. Developing Drugs

Page 145 — Fact Recall Questions

Q1 Human cells, human tissues and animals.

Q2 Any two from: e.g. to find out whether the drug works. / To find out about the drug's toxicity. / To find out the optimum/best dose of the drug.

Q3 a) healthy volunteers

b) It is very low.

Q4 A placebo is a substance that's like the drug being tested but doesn't do anything.

Q5 A double-blind trial is a clinical trial where neither the doctor nor the patients know who has been given the drug and who has been given the placebo, until the results of the trial have been gathered.

It's helpful to remember that 'double' means 'two' — so in a double-blind trial, there are two groups of people (doctors and patients) who don't know who receives the drug or who receives the placebo.

Page 145 — Application Questions

Q1 a) E.g. a capsule without paracetamol

b) E.g. an inhaler without steroids

c) E.g. an injection without cortisone

Q2 a) i) No, because it was a double-blind trial.

ii) E.g. a pill without any weight-loss drug.

b) E.g. Group 2 was included in the trial to make sure that the new drug, Drug X, worked as well as/ better than other, similar weight-loss drugs already available on the market (like Drug Y). / To see how Drug X compared to Drug Y.

c) That, on average, people taking Drug X in this trial lost 4 lbs more than those taking the placebo, but 3 lbs less than those taking Drug Y.

9. Monoclonal Antibodies

Page 146 — Fact Recall Questions

Q1 A mouse B-lymphocyte and a tumour cell.

Q2 The hybridoma cells are cloned to get lots of identical cells that all produce the same antibodies (monoclonal antibodies). These are collected and purified.

10. Monoclonal Antibody Uses

Page 150 — Fact Recall Questions

Q1 a hormone / HCG

Q2 cancer

Q3 They can cause side effects, e.g. fever, vomiting and low blood pressure.

Page 150 — Application Questions

Q1 Monoclonal antibodies could be made that will target a specific chemical in a banned drug. Blood samples from the athletes could be tested using the monoclonal antibodies — if the chemical is present, the monoclonal antibodies will bind to it and can be detected using the fluorescent dye.

Q2 The blue beads are always carried in the flow of urine up the test stick. If the test has worked, the antibodies attached to the blue beads will bind to the antibodies in the control window, turning it blue.

11. Plant Diseases and Defences

Page 152 — Fact Recall Questions

Q1 They suffer from chlorosis and have yellow leaves.

Q2 To identify the particular pathogen that is causing a plant disease.

Q3 Any two from: e.g. some plants have thorns or hairs. / Some plants have leaves that droop or curl when something touches them. / Some plants mimic other organisms or objects to trick other organisms.

Pages 156-157 — Infection and Response Exam-style Questions

1.1 E.g. repeating episodes of fever ***(1 mark)***.

1.2 protist ***(1 mark)***

1.3 The vaccination involves being injected with a small amount of dead or inactive typhoid pathogens ***(1 mark)***. These carry antigens ***(1 mark)*** which cause your white blood cells to produce antibodies to attack them ***(1 mark)***. If live pathogens of the same type appear after that, the white blood cells can rapidly mass-produce antibodies to kill off the pathogen ***(1 mark)***.

2.1 nitrate ion deficiency ***(1 mark)***

2.2 Any three from: e.g. spots on the leaves / patches of decay / abnormal growths / malformed stems or leaves / discolouration ***(1 mark for each correct answer)***.

2.3 fungus ***(1 mark)***

2.4 It may have been carried in water ***(1 mark)*** or by the wind ***(1 mark)***.

2.5 The plant should be treated with a fungicide ***(1 mark)*** and stripped of its affected leaves ***(1 mark)***. The leaves should then be destroyed ***(1 mark)***.

3.1 It produces hydrochloric acid which kills pathogens ***(1 mark)***.

3.2 The trachea and bronchi secrete mucus to trap pathogens ***(1 mark)***. They are also lined with cilia ***(1 mark)***. These waft the mucus up to the back of the throat where it can be swallowed ***(1 mark)***.

3.3 They can produce toxins that damage your cells and tissues ***(1 mark)***.

3.4 E.g. *Salmonella* food poisoning / gonorrhoea ***(1 mark)***

3.5 White blood cells ***(1 mark)*** consume pathogens by phagocytosis ***(1 mark)*** and also produce antitoxins ***(1 mark)*** and antibodies ***(1 mark)***.

4.1 A microorganism that causes disease ***(1 mark)***.

4.2 In 1980, there were just over 4 million reported measles cases ***(1 mark)***. Between 1980 and 2014 the number of reported measles cases dropped to around 0.3 million (after rising to a peak of 4.5 million in around 1981) ***(1 mark)***. The estimated vaccine coverage was around 15% of the population in 1980 ***(1 mark)***. Between 1980 and 2014 it increased to around 85% ***(1 mark)***.

The question asks you to use data from the graph to support your answer — so you must include some figures to get the marks. The graph has three different axes, which can make things a bit tricky. Take your time and work out what each one shows before answering the question.

4.3 E.g. the data suggests that as the estimated percentage vaccination coverage increased, the number of reported measles cases decreased — this supports the case for vaccinating people against measles ***(1 mark)***. However it doesn't prove that the increase in vaccination coverage definitely caused the decrease in measles cases, since other factors may have been at work ***(1 mark)***. It also doesn't tell us anything about the side effects of the vaccine ***(1 mark)***.

5.1 Drugs were originally extracted from plants and microorganisms ***(1 mark)***. They are now synthesised by chemists in labs in the pharmaceutical industry ***(1 mark)***.

5.2 Because live animals have an intact small intestine and circulatory system ***(1 mark)***. Testing on just cells or tissues wouldn't show whether the drug has the desired effect on the absorption of nutrients from the small intestine to the bloodstream because these body systems would not be intact ***(1 mark)***.

5.3 How to grade your answer:

Level 0: There is no relevant information. *[No marks]*

Level 1: There is a brief explanation of what happens during a clinical trial, but no information about placebos or double-blind testing. *[1 to 2 marks]*

Level 2: There is some explanation of what happens during a clinical trial, including some information about placebos or double-blind testing. *[3 to 4 marks]*

Level 3: There is a clear and detailed explanation of what happens during a clinical trial, including clear and detailed information about placebos and double-blind trials. *[5 to 6 marks]*

Here are some points your answer may include:

The drug is first tested on healthy volunteers. This is to make sure that it doesn't have any harmful side effects when the body is working normally.

A very low dose of the drug is given at first, and this is gradually increased.

If the results of the tests on healthy volunteers are good, the drug is tested on people suffering from the illness.

The optimum dose is found — this is the dose of drug that is the most effective and has few side effects.

To test how well the drug works, a placebo may be used.

Patients are randomly put into two groups — one is given the new drug, the other is given a placebo. This is so the doctor can see the actual difference the drug makes — it allows for the placebo effect.

Clinical trials are often double-blind — neither the doctor or the patients in the study know who is getting the drug or the placebo.

This is so the doctors monitoring the patients and analysing the results aren't subconsciously influenced by their knowledge.

Topic 4 — Bioenergetics

1. The Basics of Photosynthesis

Page 159 — Fact Recall Questions

Q1 glucose

Q2 carbon dioxide

Q3 oxygen

Q4 a) chloroplasts

b) It absorbs light.

Q5 carbon dioxide + water → glucose + oxygen

Q6 glucose

Page 159 — Application Questions

Q1 Plant C. It received the most hours of sunlight, so it will have photosynthesised for longer. As photosynthesis produces glucose, it will have produced the most glucose.

Q2 3, because this type of plant cell contains the most chloroplasts. Chloroplasts contain chlorophyll, which is needed for plants to photosynthesise.

2. How Plants Use Glucose

Page 161 — Fact Recall Questions

Q1 respiration

Q2 To strengthen the cell walls.

Q3 nitrate ions

Q4 As fats and oils.

Q5 Starch is insoluble, so it doesn't draw in water and cause the cells to swell up.

3. The Rate of Photosynthesis

Page 165 — Fact Recall Questions

Q1 a) A factor which stops photosynthesis from happening any faster.

b) E.g. light intensity, carbon dioxide level, temperature.

Q2 true

Pages 165-166 — Application Questions

Q1

Environmental conditions	Most likely limiting factor
Outside on a cold winter's day.	temperature
In an unlit garden at 1:30 am, in the UK, in summer.	light
On a windowsill on a warm, bright day.	carbon dioxide concentration

Q2 a) Because before point X, increasing the light intensity increases the rate of photosynthesis.

b) Because carbon dioxide concentration is limiting the rate of photosynthesis in Flask A. Flask A has a lower carbon dioxide concentration than Flask B but all the other variables that could affect the rate of photosynthesis are the same for both flasks. Therefore the reason why the rate of photosynthesis levels off at a lower level in Flask A, is most likely to be because of the lower carbon dioxide concentration in this flask.

4. Investigating Photosynthesis Rate

Page 168 — Application Question

Q1 $25 \div 10 = 2.5$ cm

2.5 cm $\div 3$ min $= 0.833... =$ **0.8 cm/min (1 s.f.)**

5. The Inverse Square Law

Page 169 — Application Question

Q1 a) light intensity $= 1 \div d^2$,

$1 \div 15^2 =$ **0.004 a.u.**

b) light intensity $= 1 \div d^2$,

$1 \div 30^2 =$ **0.001 a.u.**

6. Artificially Controlling Plant Growth

Page 172 — Fact Recall Questions

Q1 E.g. so that they can create the ideal conditions for photosynthesis. This means that their plants photosynthesise faster, so the rate of plant growth is also increased.

Q2 e.g. carbon dioxide concentration

Page 172 — Application Questions

Q1 a) i) 30 ÷ 8 = **3.75 cm per week**.

ii) 17.5 ÷ 8 = **2.19 cm per week**.

Graphs like this with two y-axes can be tricky — just take your time and make sure you're reading the value off the correct axis.

b) Carbon dioxide concentration. Light, temperature and carbon dioxide concentration can all affect the rate of photosynthesis, which affects growth rate. The graph shows that there was very little difference in the temperature of each greenhouse throughout the experiment, and both greenhouses were exposed to the same amount of light. Therefore it's most likely to be the extra carbon dioxide produced by the paraffin heater in Greenhouse A which caused the higher average growth rate of plants in this greenhouse.

c) E.g. the cost of running the heaters.

7. Aerobic Respiration

Page 174 — Fact Recall Questions

Q1 Respiration using oxygen. / The process of transferring energy from glucose using oxygen.

Q2 It transfers energy to the environment.

Q3 CO_2 and H_2O

To answer this question, you first have to identify what the products of aerobic respiration are — water and carbon dioxide. Then you need to write down the correct symbols.

Q4 E.g. mammals use energy to build larger molecules from smaller ones, to contract muscles and to keep warm.

Page 174 — Application Question

Q1 a) oxygen

b) water

8. Anaerobic Respiration

Page 176 — Fact Recall Questions

Q1 The body uses anaerobic respiration during vigorous exercise when it can't get enough oxygen to the muscles for aerobic respiration.

We're not respiring by anaerobic respiration all the time — just when increased muscle activity means we can't get oxygen to our muscles fast enough for them to respire aerobically.

Q2 An oxygen debt is the amount of extra oxygen the body needs after exercise to react with the build up of lactic acid and remove it from the cells.

Q3 In the liver, lactic acid is converted to glucose.

Q4 glucose → ethanol + carbon dioxide

Q5 a) fermentation

b) E.g. making bread and making alcoholic drinks.

9. Exercise

Page 178 — Application Question

Q1 E.g. her breathing rate and breath volume will increase.

10. Metabolism

Page 179 — Fact Recall Questions

Q1 One molecule of glycerol and three fatty acids.

Q2 amino acids

Q3 urea

Q4 Metabolism is the sum of all of the reactions that happen in a cell or the body.

Pages 182-183 — Bioenergetics
Exam-style Questions

1.1 in the chloroplasts ***(1 mark)***

1.2 carbon dioxide + water → glucose + oxygen ***(1 mark)***

1.3 B. It took the shortest time for all of the discs to be floating ***(1 mark)*** suggesting that photosynthesis was happening the fastest in this condition ***(1 mark)***.

1.4 Increasing the light intensity will increase the rate of photosynthesis up to a point ***(1 mark)***. However, past this point increasing the light intensity will have no further effect on the rate of photosynthesis ***(1 mark)*** as other limiting factors may come into play, such as carbon dioxide concentration or temperature ***(1 mark)***.

1.5 It should take less than 18 minutes because the rate of photosynthesis should be faster ***(1 mark)***, as there is more carbon dioxide available in the solution for photosynthesis ***(1 mark)***.

1.6 How to grade your answer:

Level 0: There is no relevant information. ***[No marks]***

Level 1: There is a brief description of one or two ways that glucose can be used by plants. ***[1 to 2 marks]***

Level 2: There is some description of three or more ways that glucose can be used by plants. ***[3 to 4 marks]***

Level 3: There is a clear and detailed description of four or more ways that glucose can be used by plants. ***[5 to 6 marks]***

Here are some points your answer may include:

Some of the glucose is used for respiration.

Some glucose is converted into cellulose, which is used to make strong cell walls.

Some glucose is combined with nitrate ions from the soil to make amino acids, which can be joined together to make proteins.

Some glucose is converted into starch for storage.

Some glucose is converted into lipids (fats and oils) for storage.

2.1 (151 + 163 + 154) / 3 = **156 beats per minute**
(2 marks for correct answer, otherwise 1 mark for correct working.)

Make sure that you always include units in your answer — in this case, the units are beats per minute.

2.2 His heart rate increased to increase blood flow to the muscles ***(1 mark)*** to provide more oxygen for respiration ***(1 mark)***.

2.3 E.g. making larger molecules from smaller ones ***(1 mark)***. Keeping the body warm/at a steady temperature ***(1 mark)***.

3.1 glucose + oxygen → carbon dioxide + water
(1 mark each for glucose and oxygen on the left-hand side of the equation, 1 mark each for carbon dioxide and water on the right.)

3.2 Anaerobic respiration forms less ATP than aerobic respiration, which shows that it transfers less energy ***(1 mark)***. This is because glucose is not completely oxidised in anaerobic respiration ***(1 mark)***.

Topic 5 — Homeostasis and Response

Topic 5a — The Nervous System

1. Homeostasis

Page 185 — Fact Recall Questions

Q1 Homeostasis is the regulation of the conditions inside your body (and cells) to maintain a stable internal environment, in response to changes in both internal and external conditions.

Q2 Any two from: e.g. temperature, blood glucose content, water content of the body.

Q3 A change in the environment.

Q4 the coordination centre

Page 185 — Application Question

Q1 Receptors detect that the water content of the body is too low. They send this information to the coordination centre, which processes the information and organises a response from the effectors. The effectors respond to increase the amount of water in the body / bring the water content back to its optimum level.

2. The Nervous System

Page 187 — Fact Recall Questions

Q1 glands

Q2 E.g. motor neurones, relay neurones

Page 187 — Application Question

Q1 a) i) The sound of the cat moving.
ii) Receptors in the dog's ears (that are sensitive to sound).
iii) sensory neurone
b) i) motor neurone
ii) They will contract.

3. Synapses and Reflexes

Page 190 — Fact Recall Questions

Q1 Chemicals diffuse across the gap between the two neurones, which sets off an electrical impulse in the next neurone.

Q2 A reflex is a fast, automatic response to a stimulus.

Q3 No

Reflexes are automatic — you don't have to think about them, so they don't pass through conscious parts of the brain.

Q4 a relay neurone

Q5 The muscle will contract.

Page 190 — Application Question

Q1 a) pain
b) A muscle in the leg. It contracts, moving the foot away from the source of the pain (the pin).
c) Stimulus → Receptor → Sensory neurone → Relay neurone → Motor neurone → Effector → Response

4. Investigating Reaction Time

Page 192 — Application Questions

Q1 It would increase reaction time. / It would make the reaction time slower.

Q2 a) E.g. the person being tested should sit with their arm resting on the edge of a table. The ruler should be held vertically between their thumb and forefinger, with the zero end of the ruler level with their thumb and finger. The ruler should then be dropped (without any warning). The person being tested should try to catch the ruler as soon as they see it fall. Their reaction time should be measured by the number on the ruler where it's caught, at the top of the thumb. This test should then be repeated and the mean distance (i.e. reaction time) calculated.
b) E.g. amount of caffeine consumed / amount of sleep / age.

5. The Brain

Page 194 — Fact Recall Questions

Q1 cerebral cortex

Q2 at the back of the brain

Q3 E.g. by studying patients with brain damage, by electrically stimulating the brain, by using MRI scans.

Page 194 — Application Question

Q1 the cerebellum

For questions like this, you need to look at the information to see what the symptoms are. You can then link the symptoms to which skill is being affected, and then link the skill to the part of the brain responsible for it.

6. The Eye

Page 197 — Fact Recall Questions

Q1 sclera

Q2 cornea

Q3 a) It contains light receptor cells that are sensitive to light intensity and colour.

b) It carries impulses from the receptors on the retina to the brain.

Q4 a) The circular muscles in the iris contract and the radial muscles relax, so the pupil becomes smaller / the iris becomes larger.

b) Bright light can damage the retina. The iris reflex to bright light quickly reduces the amount of light that can enter the eye, so the retina is protected.

Q5 A

Page 197 — Application Questions

Q1 A — iris
B — cornea
C — retina

Q2 a) X — ciliary muscle
Y — suspensory ligaments

b) The ciliary muscles (X) would contract and because of this the suspensory ligaments (Y) would slacken.

c) The changes would cause the lens to become more curved/fat, which would increase the amount by which it refracts light, so the light is refracted by the right amount to focus on the retina.

7. Correcting Vision Defects

Page 199 — Fact Recall Questions

Q1 Behind the retina.

Q2 A concave lens / a lens which curves inwards.

Q3 a) hyperopia

b) myopia

Q4 A thin lens that sits on the surface of the eye and is shaped to compensate for the fault in focussing.

Q5 E.g. laser eye surgery / replacement lens surgery

Page 199 — Application Question

Q1 The glasses have a concave lens which refracts the light so that the rays focus on the retina.

8. Controlling Body Temperature

Page 202 — Fact Recall Questions

Q1 Receptors in the thermoregulatory centre that are sensitive to the temperature of the blood flowing through the brain and receptors in the skin that are sensitive to skin temperature.

Q2 a) They dilate.

b) More blood flows close to the surface of the skin, which makes it easier for energy to be transferred from the skin to the environment.

Q3 Temperature receptors detect that the core body temperature is too low. The receptors send impulses to the thermoregulatory centre, which processes the information and sends impulses to the effectors. The effectors produce a response that reduces the amount of heat lost from the body (so the body warms up).

Page 202 — Application Question

Q1 a) E.g. hairs on her skin will stand up. / The amount of sweat she produces will decrease. / The blood vessels supplying her skin capillaries will constrict/get narrower. / She will shiver/her muscles will contract automatically.

b) E.g. the hairs on her skin standing up trap an insulating layer of air next to her skin, reducing the amount of energy transferred to the environment. / The lack of sweat decreases the amount of cooling that occurs from sweat evaporating from her skin. / The constriction of blood vessels supplying the skin capillaries reduces the amount of blood that flows close to the surface of the skin, meaning that less energy is transferred from the skin to the environment. / Her shivering/the automatic contraction of her muscles helps to raise body temperature as it increases the rate of respiration, which transfers some energy to warm the body.

Pages 205-206 — The Nervous System Exam-style Questions

1.1 The muscle in the upper arm ***(1 mark)***.

1.2 sensory neurone ***(1 mark)***

1.3 It would get faster ***(1 mark)***. The presence of the drug increases the amount of chemical released at the synapses, so it would take less time for an impulse to be triggered in the next neurone ***(1 mark)***.

1.4 E.g. reflexes are fast ***(1 mark)***, so we quickly respond to danger, decreasing our chances of injury ***(1 mark)***. Reflexes are automatic ***(1 mark)***, so we don't have to waste time thinking about our response, which reduces our chance of injury ***(1 mark)***.

2.1 retina ***(1 mark)***

2.2 cerebellum ***(1 mark)***

In the diagram, the muscle is an effector, so you need to think about which part of the brain is involved in muscle coordination.

2.3 before exercise: 4 + 7 = 11 people ***(1 mark)***
after exercise: 4 + 9 + 6 = 19 people ***(1 mark)***
19 – 11 = 8 people
So 8 more people had a reaction time of less than 0.3 s after exercise than before exercise ***(1 mark)***.

2.4 E.g. the reaction times of the group of participants after exercise were generally faster than their reaction times before exercise ***(1 mark)***.

3.1 The thermoregulatory centre ***(1 mark)*** in the brain contains receptors that are sensitive to the temperature of the blood flowing through the brain ***(1 mark)***. The thermoregulatory centre also receives inputs about skin temperature from receptors in the skin (via nervous impulses) ***(1 mark)***.

3.2 So that it remains at the optimum temperature for enzyme action / cell function ***(1 mark)***.

3.3 E.g. more sweat will be produced by his sweat glands ***(1 mark)***. Sweat transfers energy from the skin to the environment as it evaporates, helping to reduce body temperature ***(1 mark)***. The blood vessels supplying his skin capillaries will dilate ***(1 mark)*** so more blood will flow close to the surface of the skin, making it easier for energy to be transferred from the blood to the environment ***(1 mark)***.

Topic 5b — The Endocrine System

1. Hormones

Page 209 — Fact Recall Questions

Q1 By the blood (plasma).

Q2 false

They only affect particular organs, called target organs.

Q3 (endocrine) glands

Q4 E.g. oestrogen

Q5 nerves

2. Controlling Blood Glucose

Page 214 — Fact Recall Questions

Q1 the pancreas

Q2 Insulin makes body cells take up more glucose from the blood.

Q3 the pancreas

Q4 It causes glycogen to be converted into glucose which then enters the blood.

Q5 A condition where the pancreas produces little or no insulin, which means blood glucose can rise to a dangerous level.

Q6 E.g. eating a carbohydrate-controlled diet and exercising regularly.

Page 214 — Application Question

Q1 a) Glucagon, because the subject's blood glucose level rises following the injection. Glucagon causes glycogen to be converted back into glucose, which enters the blood, causing the blood glucose level to rise.

b) i) insulin

ii) The blood glucose level would fall following the injection, as insulin causes body cells to take up more glucose from the blood, causing the blood glucose level to fall.

3. Controlling Water Content

Page 217 — Fact Recall Questions

Q1 selective reabsorption

Q2 Any two from: e.g. glucose / water / ions

Q3 E.g. urea / water / (excess) ions

Q4 Any two from: e.g. in the urine / in the sweat / in the air that we breathe out.

Q5 false

The body loses more water as sweat when you are hot, so more water needs to be taken into the body to balance this loss.

Q6 osmosis

Pages 217-218 — Application Questions

Q1

Person	Largest volume of urine	Smallest volume of urine	Most concentrated urine	Most dilute urine
Katherine		✓	✓	
Cate				
Julia	✓			✓

Q2 a) i) (volume of water reabsorbed ÷ volume of water filtered) × 100
= (178.2 ÷ 180) × 100 = **99%**

ii) It would increase. The person would lose more water in sweat, so the kidney would reabsorb more water in order to balance this loss.

iii) A small amount of concentrated urine would be produced.

b) 800 millimoles. All of the glucose filtered by the kidney should be reabsorbed in a person with healthy kidneys.

c) 720 – 500 = **220 millimoles per day**

If a greater amount of a substance is filtered by the kidneys than is reabsorbed, then the remainder will end up in the urine.

4. Kidney Failure

Page 221 — Fact Recall Questions

Q1 false

People with kidney failure need to have dialysis regularly to keep the concentrations of dissolved substances in the blood at normal levels.

Q2 E.g. proteins

Q3 false

There are waiting lists for organs such as kidneys. People needing a kidney transplant may have to wait a long time for a suitable donor organ to become available. They'll need to have dialysis while they are waiting.

Pages 221-222 — Application Questions

Q1 a) So that useful sodium ions won't be lost from the blood during dialysis, as they won't diffuse across the barrier into the dialysis fluid.

b) i) Urea is a waste product which is carried to the kidneys in the blood.

ii) As there is some urea in the plasma there needs to be less in the dialysis fluid for urea to be removed from the blood. If there is no urea in the dialysis fluid at the start of the dialysis session, then there will be a concentration gradient between the blood and the dialysis fluid, allowing urea to leave the blood via diffusion.

c) E.g. there would be a similar amount. Glucose isn't a waste product — it's a substance that the body needs. If there are similar amounts in both the blood and the dialysis fluid then glucose won't be lost from the blood during dialysis.

Q2 a) Any three from: e.g. there are long waiting lists for donor kidneys. / There is a risk of rejection even if a kidney with a matching tissue-type is found. / The patient has to take drugs which suppress the immune system, which makes them vulnerable to other illnesses and infections. / A kidney transplant is a major operation, so it can be risky.

b) Any two from: e.g. dialysis isn't pleasant and each session takes 3-4 hours. / She will no longer be at risk of infections that can be picked up during dialysis. / She won't be at risk of blood clots from dialysis. / She won't have to be as careful about what she eats. / She won't have to limit her fluid intake.

Page 224 — The Endocrine System Exam-style Questions

1.1 B ***(1 mark)***

1.2 insulin ***(1 mark)***

1.3 People with Type 1 diabetes produce little or no insulin ***(1 mark)***. Injecting insulin stops their blood glucose level from rising to a dangerous level ***(1 mark)*** as insulin causes the body's cells to take up glucose from the blood ***(1 mark)***.

1.4 A pancreas transplant could be an effective treatment for Type 1 diabetes because the transplanted pancreas could take over the role of the patient's faulty pancreas and produce the right amount of insulin ***(1 mark)***.
A pancreas transplant would be ineffective in treating Type 2 diabetes because the patient's pancreas already produces enough insulin ***(1 mark)***, but their body's cells don't respond properly to the hormone ***(1 mark)***.

1.5 E.g. the patient should eat a carbohydrate-controlled diet / limit their intake of foods rich in simple carbohydrates ***(1 mark)***, as simple carbohydrates cause the blood glucose level to rise rapidly ***(1 mark)***.
The patient should try to do regular exercise ***(1 mark)*** as this will remove glucose from their blood ***(1 mark)***.

2.1 A ***(1 mark)***

2.2 Excess amino acids are broken down/deaminated by the liver ***(1 mark)***. This forms ammonia ***(1 mark)***, which is converted into urea ***(1 mark)***.

2.3 100 – 99.5 = 0.5%
(575 ÷ 100) × 0.5 = **2.88 g per day**
(2 marks for correct answer, otherwise 1 mark for correct working.)

Topic 5c — Animal and Plant Hormones

1. Puberty and the Menstrual Cycle

Page 228 — Fact Recall Questions

Q1 oestrogen

Q2 e.g. stimulates sperm production.

Q3 It causes the egg to mature.

Q4 It stimulates the release of an egg at around the middle (day 14) of the menstrual cycle. / It stimulates ovulation.

Q5 oestrogen and progesterone

Q6 The pituitary gland and the ovaries.

Page 228 — Application Question

Q1 a) LH (luteinising hormone). LH is the hormone responsible for stimulating the release of an egg. The concentration of the hormone on the graph increases just before the middle of the cycle (day 14) — the time at which an egg is normally released.

b) the pituitary gland

c) LH is needed to stimulate the release of an egg. This woman's LH level peaks at a much lower level than the other woman's, suggesting that she may not be releasing an egg during her menstrual cycle. This could be the reason why she's struggling to have children.

2. Contraceptives

Page 230 — Fact Recall Questions

Q1 It inhibits FSH production, and so prevents egg maturation and therefore release.

Q2 progesterone

Q3 E.g. condom / diaphragm.

3. Increasing Fertility

Page 232 — Fact Recall Questions

Q1 To stimulate the maturation of multiple eggs.

Q2 The eggs are fertilised in the lab using the male's sperm.

Q3 Any two from: e.g. IVF treatment can lead to multiple births / the success rate is low / it is emotionally stressful / it is physically stressful.

Q4 E.g. it results in unused embryos, which are potential human lives, being destroyed.

Page 233 — Application Question

Q1 a) (7768 ÷ 17 765) × 100 = **43.73% (4 s.f.)**

b) (41.74/100) × 17 892 = **7468**

c) IVF often involves the transfer of more than one embryo into the uterus, which results in an increased chance of multiple pregnancies, and therefore more multiple pregnancies in those undergoing IVF.

4. Thyroxine and Adrenaline

Page 235 — Fact Recall Questions

Q1 E.g. thyroxine helps to regulate basal metabolic rate and stimulates protein synthesis for growth and development.

Q2 the adrenal glands

Page 235 — Application Question

Q1 a) adrenaline

b) It increases the supply of oxygen and glucose to cells in his brain and muscles, so that his body is ready for 'fight or flight'.

5. Plant Hormones

Page 238 — Fact Recall Questions

Q1 The growth response of a plant to gravity.

Q2 More auxin accumulates on the lower side of the root. As the extra auxin inhibits cell elongation in roots, the cells on the top side of the root elongate faster, and the root bends downwards towards gravity.

6. Uses of Plant Hormones

Page 241 — Fact Recall Questions

Q1 auxins

Q2 gibberellins

Q3 Ethene stimulates fruit to ripen.

Page 241 — Application Question

Q1 a) For this type of plant, rooting powder B is the best type of rooting powder to use for the first three weeks of growth. This is because it's more effective at increasing root length each week for the first three weeks after planting, than rooting powder A.

It's important to remember that the results only apply to this particular type of plant. You also can't say what will happen after the three week period Dan recorded his results for.

b) It was a control. / To show that the results were likely to be due to the presence of the rooting powder and nothing else.

Page 243 — Animal and Plant Hormones Exam-style Questions

1.1 C — because the plant is growing towards it ***(1 mark)***.

1.2 Sample X because it has been taken from the side that has grown more / the side in the shade ***(1 mark)***. Auxin makes the cells in plant shoots elongate ***(1 mark)***, so the side with more elongation must contain the most auxin ***(1 mark)***.

1.3 Rooting powders contain auxins ***(1 mark)***. These promote root development, helping the new plant to grow ***(1 mark)***.

2.1 C ***(1 mark)***

2.2 How to grade your answer:

Level 0: There is no relevant information. ***[No marks]***

Level 1: There is some description of how the level of thyroxine is controlled by negative feedback, but it is incomplete or points are not clearly linked. ***[1 to 2 marks]***

Level 2: There is a detailed description of how the level of thyroxine is controlled by negative feedback, with points that are clearly linked together. ***[3 to 4 marks]***

Here are some points your answer may include:

The amount of thyroxine in the blood is kept at the right level by a negative feedback system.

When the level of thyroxine in the blood rises higher than normal, the secretion of thyroid stimulating hormone (TSH) from the pituitary gland is inhibited.

This means the amount of thyroxine released from the thyroid gland is reduced.

So the level in the blood falls back towards normal.

When the level of thyroxine in the blood is lower than normal, the secretion of TSH from the pituitary gland is stimulated again.

This means the amount of thyroxine released from the thyroid gland is increased.

So the level in the blood rises back towards normal.

2.3 1 day = 24 hours
$1860 \div 24 = 77.5$
$77.5 \times 6 =$ **465 kcal**
(2 marks for correct answer, otherwise 1 mark for correct working.)

Topic 6 — Inheritance, Variation and Evolution

Topic 6a — DNA and Reproduction

1. DNA

Page 245 — Fact Recall Questions

Q1 In the nucleus.

Q2 (long molecules of coiled up) DNA

Q3 Because they are made up of two strands coiled together in a double helix shape.

Q4 By telling the cell what order to put amino acids in.

Q5 The entire set of genetic material in an organism.

2. The Structure of DNA

Page 247 — Fact Recall Questions

Q1 a) One sugar molecule, one phosphate molecule and one base.

b) The sugar and phosphate molecules.

Q2 three

Q3 Each sequence of three bases codes for an amino acid. The amino acids are joined together to make various proteins, depending on the order of the gene's bases.

Q4 They switch genes on and off, and therefore control whether or not a gene is expressed/used to make a protein.

Page 247 — Application Questions

Q1 serine, threonine, asparagine, glutamic acid

Q2 A G A T G C T T A C T A

A always pairs with T, and C always pairs with G.

3. Protein Synthesis

Page 248 — Fact Recall Questions

Q1 On ribosomes in the cell cytoplasm.

Q2 They bring amino acids to the ribosomes in the correct order.

Q3 It folds into a unique shape which allows the protein to perform its task.

4. Mutations

Page 250 — Fact Recall Questions

Q1 True

Q2 False

Page 250 — Application Question

Q1 The mutation will change the sequence of DNA bases in the gene. This can change the sequence of amino acids, and therefore the protein that is coded for. If an altered protein is coded for, the shape of the active site may change, meaning that the substrate may no longer bind to it.

5. Reproduction

Page 252 — Fact Recall Questions

Q1 meiosis

Q2 egg and sperm cells

Q3 one

Page 252 — Application Question

Q1 a) Asexual reproduction. There is only a single parent cell and no fusion of gametes. The offspring are genetically identical/clones.

b) Sexual reproduction. There are two parents, gametes have fused together and offspring show genetic variation.

Remember, sexual reproduction doesn't always involve sexual intercourse — it's the fusion of gametes that's important. Both animals and plants can reproduce sexually.

c) Sexual reproduction. There are two parents and the offspring has characteristics of both parents, suggesting that it has a mixture of genes from both parents.

d) Asexual reproduction. There is only one parent and no fusion of gametes. The offspring are genetically identical to/clones of the mother snake.

6. Meiosis

Page 254 — Fact Recall Questions

Q1 one

Q2 In the reproductive organs.

Q3 two

Q4 four

Q5 False

Q6 By repeatedly dividing by mitosis.

Page 254 — Application Questions

Q1 a) 39

b) 38

c) 32

Q2 Before the first division, the cell duplicates its DNA. So when the first division happens, cells will have one full set of chromosomes again.

Duplicating the DNA creates X-shaped chromosomes, where each 'arm' of the chromosome has the same DNA.

7. More on Reproduction

Page 256 — Fact Recall Questions

Q1 It increases the chance of a species surviving a change in the environment because it's likely that variation will have led to some of the offspring being able to survive in the new environment. These offspring are more likely to breed successfully and pass the genes for the beneficial characteristics on.

Q2 Because there only needs to be one parent, asexual reproduction uses less energy than sexual reproduction as organisms don't need to find a mate. / Because there only needs to be one parent, asexual reproduction is faster than sexual reproduction because organisms don't need to find a mate.

Q3 When it is in the human host.

Q4 It produces runners that new identical strawberry plants grow from.

Pages 259-260 — DNA and Reproduction Exam-style Questions

1.1 Sexual, because it involves the joining of male and female gametes ***(1 mark)***.

1.2 E.g. it produces variation in the offspring ***(1 mark)***. This increases the chance of the species surviving a change in the environment because it's likely that variation will have led to some of the offspring being able to survive in the new environment ***(1 mark)***. The individuals that have a better chance of survival are more likely to breed successfully ***(1 mark)*** and pass the genes for the characteristics on ***(1 mark)***.

1.3 They have received genetic material/chromosomes from both the male pig and the female pig ***(1 mark)*** and it is this genetic material/the chromosomes that control the offspring's characteristics ***(1 mark)***.

1.4 So that when two gametes join at fertilisation, the full number of chromosomes is reached (two copies of each) ***(1 mark)***.

1.5 mitosis ***(1 mark)***

2.1 The human genome is the entire set of genetic material in a human ***(1 mark)***.

The question asks specifically about the 'human genome', so make sure your answer is specific to humans.

2.2 Any two from, e.g. it allows scientists to identify genes in the genome that are linked to different types of disease. / Knowing which genes are linked to inherited diseases will help us to understand them better and could help us to develop effective treatments for them. / Scientists can look at genomes to trace the migration of certain populations of people around the world ***(2 marks)***.

3.1 How to grade your answer:

Level 0: There is no relevant information. ***[No marks]***

Level 1: One or both terms are described clearly but there is no detail on how they are linked together. ***[1 to 2 marks]***

Level 2: Both terms are described fully and the link between them is made clear. ***[3 to 4 marks]***

Here are some points your answer may include:

Genes are small sections of DNA that code for a particular sequence of amino acids which are put together to make a specific protein.

Chromosomes are very long structures made up of DNA.

Genes are found on chromosomes.

3.2 nucleus ***(1 mark)***

3.3 a sugar molecule ***(1 mark)***

3.4 four ***(1 mark)***

Remember, amino acids are coded for by sequences of three bases.

3.5 T A G T C G G A T C A A ***(2 marks for all bases correct, 1 mark for 8 or more bases correct.)***

3.6 The mutation might change which amino acids are coded for ***(1 mark)***. Because amino acids are joined in a specific order to produce a particular protein, a change in the amino acid sequence might produce a different protein ***(1 mark)***.

Topic 6b — Genetics

1. X and Y Chromosomes

Page 264 — Fact Recall Questions

Q1 one

Q2 false

Half carry X chromosomes, and half carry Y chromosomes.

Page 264 — Application Questions

Q1

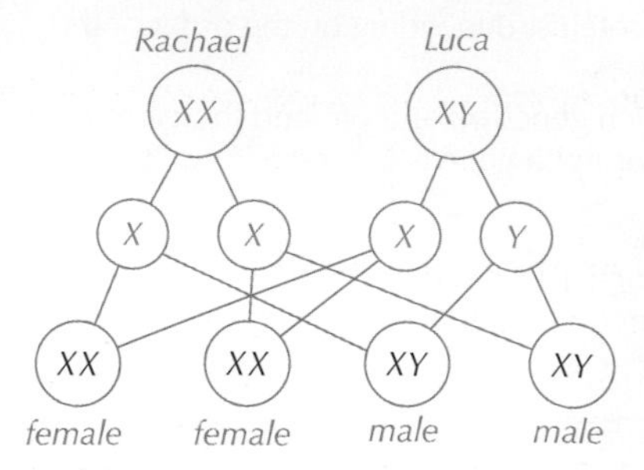

Q2 0.5 or 50% or 1:1 or 1 in 2 or ½.

Q3

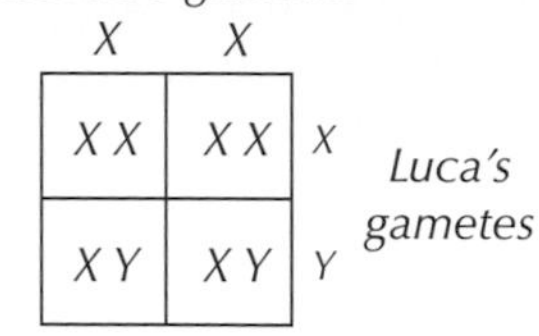

2. Alleles and Genetic Diagrams

Page 270 — Fact Recall Questions

Q1 E.g. mouse fur colour / red-green colour blindness in humans

Q2 a) Different versions of the same gene.

b) When an organism has two alleles for a particular gene that are the same, e.g. TT.

c) The allele for the characteristic that's shown even if two different alleles for the same gene are present.

d) The characteristics you have.

Page 270 — Application Questions

Q1 a) heterozygous

b) ss

c) SS

Q2 a) a rough coat

b) i) *gametes' alleles*

	R	*r*
R	*RR*	*Rr*
r	*Rr*	*rr*

(rows: *gametes' alleles*)

ii) 0.25 or 25% or 1:3 or 1 in 4 or ¼.

3. Inherited Disorders

Page 274 — Fact Recall Questions

Q1 cell membrane

Q2 recessive

You need to have two recessive alleles for cystic fibrosis to have the disease.

Q3 E.g. to see if the embryo carries an inherited disorder.

Page 274 — Application Questions

Q1 a) Ff

b) ff

Q2 a) Because there are no carriers of the disease, just sufferers.

b) a purple circle

c) i) dd

ii) Dd

You know from the diagram that Kate has polydactyly, so she must have at least one copy of the dominant allele 'D'. You also know that she doesn't have two copies of the dominant allele, as her father Clark doesn't have polydactyly, so she couldn't have inherited a second 'D' allele from him. Therefore her genotype must be Dd.

d) i) *Kate's alleles*

	D	*d*
d	*Dd*	*dd*
d	*Dd*	*dd*

Aden's alleles

ii) 0.5 or 50% or 1:1 or 1 in 2 or ½

4. The Work of Mendel

Page 277 — Fact Recall Questions

Q1 false

Mendel found that the hereditary units are passed on unchanged from parents to offspring.

Q2 No one knew about genes, DNA or chromosomes when Mendel was still alive, so the significance of his work wasn't realised until after he had died.

These days people recognise Mendel as one of the founding fathers of genetics. He was just a bit ahead of his time. Poor Mendel.

Q3 1 — B. Mendel's research on pea plants.

2 — C. The observation of chromosomes during cell division.

3 — A. The structure of DNA being determined.

Page 278 — Application Question

Q1

Seed colour of pea plant	Type of hereditary unit		
	Just green	Just yellow	Both green and yellow
Green	✓	✗	✗
Yellow	✗	✓	✓

The yellow hereditary unit is dominant, so if the pea plant has both yellow and green hereditary units its seeds will be yellow. If the plant just has yellow hereditary units its seeds will also be yellow. For its seeds to be green, the plant has to have two of the green hereditary units.

Pages 280-281 — Genetics Exam-style Questions

1 C ***(1 mark)***

2.1 They have extra fingers or toes ***(1 mark)***.

2.2 DD or Dd ***(1 mark)***

2.3 A ***(1 mark)***

2.4

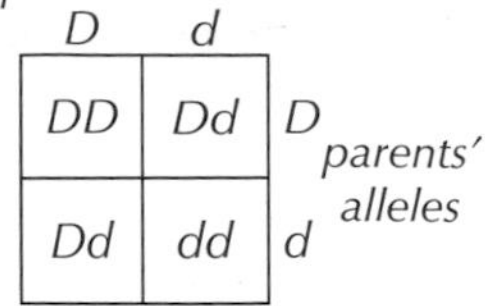

DD and Dd offspring have polydactyly.

3:1 of polydactyly offspring:unaffected offspring

(1 mark for correctly identifying the parents' genotypes, 1 mark for correctly identifying possible genotypes of the offspring, 1 mark for correctly identifying possible phenotypes of the offspring, 1 mark for correct ratio.)

3.1

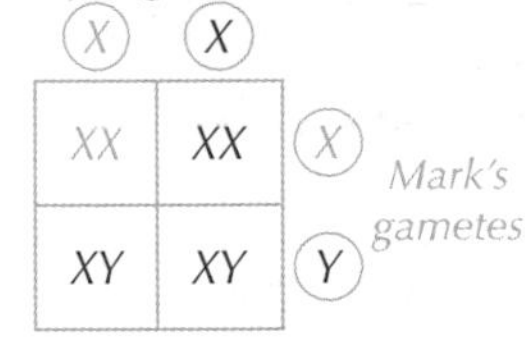

(1 mark for correctly identifying Kaye and Mark's missing gametes, 1 mark for correctly identifying the possible gamete combinations of the offspring.)

3.2 XY ***(1 mark)***

Males have the XY chromosome combination, females have the XX chromosome combination.

3.3 E.g.

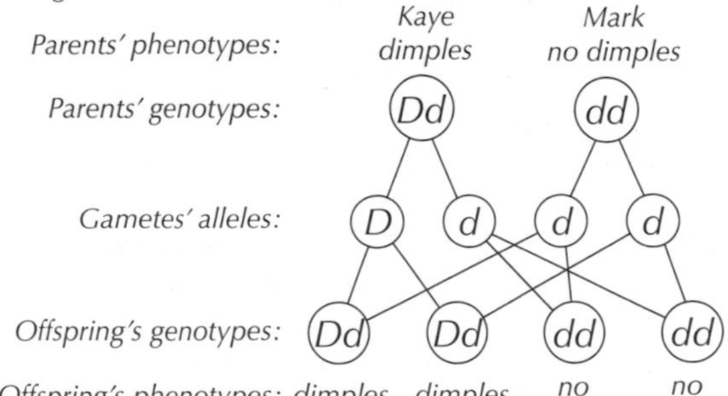

The baby has a 0.5 / 50% / 1:1 / 1 in 2 / ½ chance of having dimples. ***(1 mark for correctly identifying Kaye's genotype, 1 mark for correctly identifying Mark's genotype, 1 mark for correctly identifying possible genotypes of the offspring, 1 mark for correctly identifying possible phenotypes of the offspring, 1 mark for identifying the correct probability of the baby having dimples.)***

You could have drawn a Punnett square to answer this question instead.

4.1 Cystic fibrosis is an inherited disorder ***(1 mark)***, which affects the cell membranes ***(1 mark)***.

4.2 Homozygous, because he has a recessive disorder ***(1 mark)***.

4.3 Ff ***(1 mark)***. For a child to have cystic fibrosis, both parents must carry the cystic fibrosis allele ***(1 mark)***, as it is a recessive disorder ***(1 mark)***.

Tony is a carrier so his alleles must be Ff. As their child has cystic fibrosis (ff) it must have inherited a cystic fibrosis allele from both Tony and Bex, as it's a recessive disorder. Bex doesn't have cystic fibrosis, so she must be a carrier (Ff).

4.4 Any two from: e.g. there may come a point where everyone wants to screen their embryos so they can pick the most 'desirable' one ***(1 mark)***. / It implies that people with genetic problems are 'undesirable', which could increase prejudice ***(1 mark)***. / Some people think destroying the rejected embryos is unethical as they could have developed into humans ***(1 mark)***. / There's a risk of miscarriage with embryo screening ***(1 mark)***. / If an inherited disorder is diagnosed, it could lead to a termination/abortion ***(1 mark)***. / Screening embryos is expensive ***(1 mark)***.

4.5 E.g.

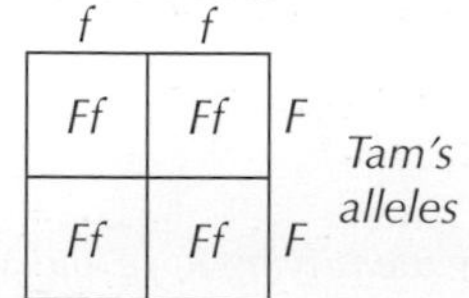

No, because to have cystic fibrosis you need the genotype ff. None of the offspring are ff. Ivana and Tam's offspring will only have the genotype Ff, so they could only be carriers, not have the disorder.

(1 mark for correctly identifying the Ivana's genotype, 1 mark for correctly identifying Tam's genotype, 1 mark for correctly identifying possible genotypes of the offspring, 1 mark for correctly explaining the answer by identifying possible phenotypes of the offspring.)

Topic 6c — Evolution and Classification

1. Variation

Page 285 — Fact Recall Questions

Q1 False

Q2 a) A change to the sequence of bases in DNA.

b) Most mutations have no effect on the protein the gene codes for, so most mutations have no effect on the organism's phenotype.

c) Mutations can result in a new phenotype being seen in a species.

Page 285 — Application Questions

Q1 a) genes only

b) both

c) genes only

Q2 a) both

b) environment only

Q3 E.g. how often you train / how hard you train / how well you are coached / how good the facilities you train at are / how good your diet is.

You're not expected to know the answer to this question, just to make sensible suggestions.

Q4 The results from Study 1 suggest that genes must influence IQ because identical twins have the same genes and non-identical twins have different genes. The results from Study 2 suggest that environment also influences IQ. This is because identical twins have exactly the same genes, which means that differences in their IQs must be down to differences in their environment/the way they were brought up.

2. Evolution and Extinction

Page 288 — Fact Recall Questions

Q1 over 3 billion years ago

Q2 They evolved from simple organisms.

Q3 a) (Charles) Darwin

b) Observations made on a huge round-the-world trip, along with experiments, discussions and new knowledge of fossils and geology.

c) The process by which species evolve.

Q4 The species doesn't exist any more.

Page 289 — Application Questions

Q1 E.g. the original rats showed variation — some were resistant to warfarin, others weren't / a genetic variant appeared (due to a mutation) which made some rats resistant to warfarin. The warfarin-resistant rats were better suited to the environment (because they weren't killed by the warfarin), so they were more likely to survive and breed successfully. This meant that the genetic variant for warfarin-resistance was more likely to be passed on to the next generation.

You could get asked to explain the selection of pretty much any characteristic in the exam — make sure you can apply the key points of Darwin's theory to any context.

Q2 a) By around 3.5 – 2.1 = **1.4 cm** (accept any answer between 1.3 and 1.5 cm).

b) E.g. the reindeer population in 1810 would have shown variation — some would have had shorter fur and some longer fur (due to a mix of genetic variants in the population). The reindeer with shorter fur would have been better suited to the new, warmer environment they found themselves in (as they would have been less likely to overheat), so they would have been more likely to survive and breed successfully. This meant that the genetic variant for short fur was more likely to be passed on to the next generation. This genetic variant became more common in the population, eventually reducing the average fur length.

3. Ideas About Evolution

Page 291 — Application Questions

Q1 It shows that the acquired characteristic of clipped flight feathers is not passed on from the parent birds to their offspring.

Q2 a) E.g. Lamarck may have argued that if an anteater used its tongue a lot to reach into ant nests, then its tongue would get longer. This acquired characteristic would then be passed on to the next generation and the anteater's offspring would have been born with long tongues.

b) E.g. tongue lengths in anteaters used to vary — some were long and some were short (due to a mix of genetic variants in the population) / a mutation occasionally produced a genetic variant that caused some anteaters to be born with a long tongue. Long-tongued anteaters were better adapted to their environment/could get more food with their long tongues, so were more likely to survive and reproduce. So the genetic variant for a long tongue was more likely to be passed on to the next generation and eventually all anteaters were born with long tongues.

4. Selective Breeding

Page 293 — Fact Recall Questions

Q1 E.g. high meat production / high milk production / good disease resistance in crops / a good temperament in dogs / big or unusual flowers in plants.

Q2 From your existing stock, select the organisms which have the characteristics you're after. Breed them with each other. Select the best of the offspring, and breed them together. Continue this process over several generations, and the desirable trait gets stronger and stronger. Eventually, all the offspring will have the characteristic.

Q3 E.g. it reduces the gene pool, which known as inbreeding. Inbreeding can cause health problems, such as genetic defects and susceptibility to diseases.

5. Genetic Engineering

Page 296 — Fact Recall Questions

Q1 'Cutting out' and transferring a gene responsible for a desirable characteristic from one organism's genome into another organism, so that it also has the desired characteristic.

Q2 It means that the organism will develop with the characteristic coded for by the gene.

Q3 E.g. producing human insulin.

Q4 a) genetically modified

b) E.g. insects, herbicides

Page 296 — Application Questions

Q1 Enzymes would be used to cut out the *Bt* crystal protein gene from the *B. thuringiensis* DNA. The gene would be inserted into a vector (e.g. a virus or bacterial plasmid). The vector would then be introduced to the target plant, and the *Bt* crystal protein gene would be inserted into its cells.

Q2 E.g. if insects ate the crop, they'd be poisoned by the protein. If less of the crop was eaten by insects, this could improve crop yield.

Q3 E.g. it could affect the number of weeds/flowers/insects that live in and around the crop. / People might develop allergies to the crop. / The gene for the *Bt* protein might get into the natural environment affecting, e.g., weeds.

6. Cloning

Page 300 — Fact Recall Questions

Q1 a) Any two from: e.g. they can be produced quickly. / They can be produced cheaply. / Lots of ideal offspring/offspring with known characteristics can be produced.

b) tissue culture

Q2 Embryo transplant. Adult cell cloning.

Page 300 — Application Questions

Q1 By giving it an electric shock.

Q2 No. The egg cell from the domestic cat will have had its nucleus/genetic material removed before the black-footed cat nucleus was inserted. So the embryo will only contain genetic material from the black-footed cat.

Q3 a) E.g. there are more domestic cats around than black-footed cats, so using domestic cats to carry the embryos may mean more black-footed kittens can be produced.

b) E.g. using skin cells from adults that have died as well as those that are still alive may increase the number of possible 'parents', so more black-footed cats can be produced.

The idea behind using adult cell cloning to help save endangered species is that it's a way of producing lots of offspring relatively rapidly — you need to apply this idea to the question to make sensible suggestions here.

Q4 E.g. the cloned animals may also not be as healthy as normal ones. Cloning could also give the animals a reduced gene pool, meaning that if a new disease appears, they could all be wiped out.

7. Fossils

Page 303 — Fact Recall Questions

Q1 The remains of an organism from many years ago that are found in a rock.

Q2 That organisms lived ages ago and how today's species have evolved over time.

Q3 Early organisms were soft-bodied and soft tissue tends to decay away completely, without forming fossils. Plus fossils that did form millions of years ago may have been destroyed by geological activity, e.g. the movement of tectonic plates. So the fossil record is incomplete.

Page 303 — Application Question

Q1 A E.g. an organism (or organisms) burrowed into soft material. The material later hardened around the burrows to form casts in the rock.

B E.g. the hard shell of the snail decayed away slowly. It was gradually replaced by minerals as it decayed, forming a rock-like substance shaped like the shell.

C E.g. a leaf became buried in a soft material. The material later hardened around the leaf, which decayed, leaving a cast of itself in the rock.

D E.g. the mammoth died and it's body became trapped in frozen ground. It was too cold for decay microbes to work, so the mammoth's body was preserved.

8. Speciation

Page 305 — Fact Recall Questions

Q1 The development of a new species.

Q2 Because they have a wide range of alleles.

Q3 When individuals from different populations have changed so much that they can no longer interbreed to produce fertile offspring.

Q4 1858

Q5 On the Origin of Species

Q6 His work on warning colours in animals.

9. Antibiotic-Resistant Bacteria

Page 307 — Fact Recall Questions

Q1 Because they reproduce rapidly.

Q2 E.g. MRSA

Q3 The rate of development is slow, and it's a very costly process.

Page 307 — Application Questions

Q1 E.g. because it's important for doctors to avoid over-prescribing antibiotics in order to slow down the development of antibiotic resistance. James' infection is only mild.

Q2 E.g. people with a *Streptococcus pneumoniae* infection may have been treated with penicillin. A mutation may have given rise to a genetic variant that caused some of these bacteria to be resistant to the penicillin, meaning that only the non-resistant bacteria would have been killed. The individual resistant bacteria would have survived and reproduced, increasing the population of penicillin-resistant *Streptococcus pneumoniae* bacteria.

10. Classification

Page 310 — Application Question

Q1 a) i) the lion

ii) the snow leopard

b) the jaguar

c) yes

The lion and the snow leopard are two of the most distantly related organisms on this tree, but they still share a common ancestor:

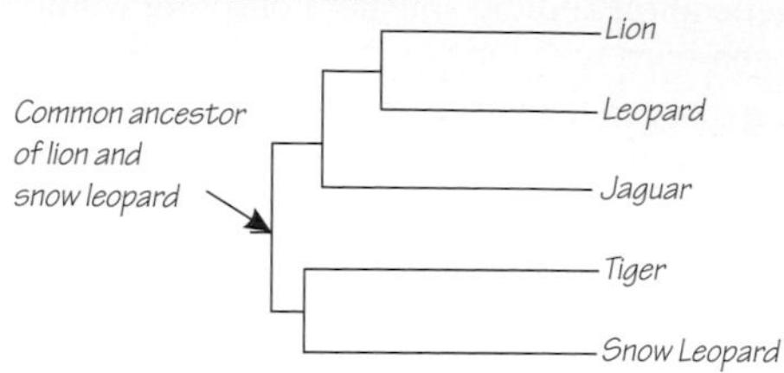

Pages 313-314 — Evolution and Classification Exam-style Questions

1.1 C ***(1 mark)***

1.2 turtles ***(1 mark)***

1.3 The original milk snake population showed variation — some had similar colouring to the coral snake, others didn't. / A mutation produced a genetic variant which caused some milk snakes to have similar colouring to the coral snake ***(1 mark)***. Milk snakes that looked like the coral snake were less likely to be eaten ***(1 mark)*** and so more likely to reproduce and pass on the genetic variant that causes coral snake colouring to the next generation ***(1 mark)***.

1.4 It went against common religious beliefs at the time ***(1 mark)***. Darwin couldn't explain how useful characteristics appeared or were passed on (as he didn't know about genes) ***(1 mark)***. There wasn't enough evidence to convince many scientists ***(1 mark)***.

2.1 Populations of the original squirrel species became separated/isolated by the formation of the canyon ***(1 mark)***. Each population showed variation because they had a wide range of alleles ***(1 mark)***. In each population, individuals with characteristics that made them better adapted to their environment were more likely to survive and breed successfully ***(1 mark)***. So the alleles that controlled beneficial characteristics were more likely to be passed onto the next generation ***(1 mark)***. Eventually individuals from the different populations changed so much that they were unable to interbreed to produce fertile offspring ***(1 mark)***.

Remember, the main steps leading to speciation are: isolation, genetic variation in the isolated populations, natural selection causing different characteristics and therefore different alleles to become more common in each population, populations change and are unable to interbreed.

2.2 Individuals from the bird species were able to fly across the canyon ***(1 mark)***, so there were no isolated populations ***(1 mark)***.

Something that's a physical barrier for one species won't necessarily be a barrier for another. It depends on things like the size of the organisms involved, how they move around, how many of them there are, etc.

2.3 The hard shells won't have decayed easily so they will have lasted a long time when buried ***(1 mark)***. As the shells did decay, they will have been gradually replaced by minerals, forming a rock-like substance shaped like the original shells ***(1 mark)***.

2.4 Soft tissue tends to decay away completely without forming fossils ***(1 mark)***.

2.5 E.g. by geological activity / they may have been crushed by the movements of tectonic plates ***(1 mark)***.

Topic 7 — Ecology

Topic 7a — Organisms and Their Environment

1. Competition

Page 317 — Fact Recall Questions

Q1 The populations of different species living in a habitat.

Q2 The interaction of a community of living organisms with the abiotic parts of their environment.

Q3 Space (territory), food, water and mates.

Q4 A community in which all the species and environmental factors are in balance so that the population sizes are roughly constant.

2. Abiotic and Biotic Factors

Page 320 — Fact Recall Questions

Q1 Any three from: e.g. moisture level / light intensity / temperature / carbon dioxide level / wind intensity / wind direction / oxygen level / soil pH / soil mineral content.

Q2 Any three from: e.g. new predators arriving / competition from other species / new pathogens / availability of food.

Page 320 — Application Question

Q1 a) i) It decreased.

ii) an abiotic factor

Remember, it's the oxygen concentration that changed, and oxygen isn't alive — making it an abiotic factor.

b) The population size of the herons is likely to have decreased because there were fewer fish to eat.

Here the environmental change was caused by a biotic factor — the availability of food.

3. Adaptations

Page 322 — Fact Recall Questions

Q1 Features of an organism's body structure that allow it to live in the conditions of its environment.

Q2 Things that go on inside an organism's body that can be related to processes like reproduction and metabolism that allow it to live in the conditions of its environment.

Q3 Any two from: e.g. very hot conditions / very salty conditions / high pressure conditions.

4. Food Chains

Page 324 — Fact Recall Questions

Q1 a producer

Q2 tertiary consumers

Q3 Consumers that hunt and kill other animals.

Page 324 — Application Question

Q1 a) zebra

b) The cheetah population will increase, because an increase in grass means more food for the zebra population, so the zebra population will increase. This means there is more food for the cheetahs, so their population will grow.

c) As the cheetah population increases, the zebra population will decrease because there are more cheetahs to eat them. This will lead to the cheetah population decreasing as the amount of food available decreases.

5. Using Quadrats and Transects

Page 328 — Fact Recall Questions

Q1 E.g. you would place the quadrat on the ground at a random position in the first sample area and count the number of the organisms within the quadrat. You would then repeat this many times. Next you would repeat this whole process in the second sample area. Finally you would work out the mean number of organisms per quadrat or the population size in each sample area and compare the results.

Q2 A line can be marked out across the area you want to study and all of the organisms that touch the line can be counted. / A line can be marked out across the area you want to study and data can be collected using quadrats placed along the line.

Page 328 — Application Questions

Q1 a) $1 + 5 + 5 + 20 + 43 + 37 = 111$

$111 \div 6 =$ **18.5**

Remember the mean is the average you get by adding together all the values in the data and dividing it by the number of values that you have.

b) End B, as the amount of bulrushes is lower here and you would expect there to be fewer bulrushes further away from the pond as they prefer moist soil or shallow water.

Q2 First calculate how many of the quadrats used by the conservationists would fit in 1 m^2: $1 \div 0.25 = 4$
Then work out how many slugs there would be in 1 m^2: $4 \times 2.5 = 10$
Finally, work out how many slugs there would be in 2000 m^2: $10 \times 2000 =$ **20 000 slugs in the area**

6. Environmental Change

Page 329 — Fact Recall Questions

Q1 E.g. changes in the availability of water, temperature and the composition of atmospheric gases/pollution.

Q2 Seasonal factors, geographic factors and human interaction.

7. The Cycling of Materials

Page 331 — Fact Recall Questions

Q1 CO_2 is removed from the atmosphere by green plants and algae during photosynthesis, and the carbon is used to make glucose. The glucose can be turned into carbohydrates, fats and proteins that make up the bodies of the plants and algae.

Q2 Dead leaves are fed on by detritus feeders and microorganisms. These organisms release CO_2 when they respire.

Page 331 — Application Question

Q1 Yes. Water evaporates from plants in transpiration and cools and condenses to form clouds as it gets higher up. Water falls from the clouds as precipitation onto land. It then drains into the sea.

8. Decay

Page 335 — Fact Recall Questions

Q1 It acts as a natural fertiliser that returns nutrients back into the soil.

Q2 Batch generators make biogas in small batches, whereas continuous generators make biogas all the time.

When you're asked to talk about differences or make comparisons, make sure you use words like 'whereas', 'however', 'on the other hand'...

Page 335 — Application Questions

Q1 a) To let oxygen in — many microorganisms which break down waste material need oxygen for respiration.

b) Sunny areas of the garden will have higher temperatures. Microorganisms work best in warm conditions, so compost is made faster.

c) The presence of water vapour would increase the moisture inside the compost bin and, as microorganisms need water to carry out biological processes, it would help compost to be made more quickly.

To answer these questions you just need to remember the conditions that microorganisms need to work best and apply that knowledge to the question.

Pages 338-339 — Organisms and Their Environment Exam-style Questions

1.1 $55 + 41 + 57 = 153$
$153 \div 3 =$ **51**
(2 marks for correct answer, otherwise 1 mark for correct working.)

1.2 The number of limpets increases as you move away from the water's edge, and then begins to decrease after the position of quadrat 3 ***(1 mark)***. The low number of limpets in quadrats closest to the water's edge could be due to competition for space from other organisms ***(1 mark)***. The decrease in the number of limpets after quadrat 3 could be due to there being less water available further from the water's edge, which increases the limpets' chance of drying out ***(1 mark)***.

1.3 To increase the validity of their results ***(1 mark)***.

2 How to grade your answer:

Level 0: There is no relevant information. ***[No marks]***

Level 1: There is a brief description of one or two steps in the carbon cycle, but no names of any processes are given. ***[1 to 2 marks]***

Level 2: There is a description of three or four steps in the carbon cycle, but not all of the processes are named. ***[3 to 4 marks]***

Level 3: There is a detailed description of five or six steps in the carbon cycle with named processes. ***[5 to 6 marks]***

Here are some points your answer may include:

The grass absorbs carbon dioxide from the air in photosynthesis.

The grass uses this carbon to make glucose/carbohydrates/fats/proteins.

This carbon is passed onto the cows when the cows eat the grass, and onto humans when the humans eat the cows, so it moves through the food chain.

Dead organisms/waste materials are broken down/decayed by detritus feeders/microorganisms.

All the organisms/the grass/cows/humans/microorganisms respire and release carbon dioxide into the atmosphere.

Combustion/burning of fossil fuels/wood also releases carbon dioxide into the air.

3.1 A characteristic/feature that helps an organism to survive in the conditions in its natural environment ***(1 mark)***.

3.2 It has a thick bushy tail to help it keep warm at night when temperatures drop very low ***(1 mark)***. It has a long, pointed muzzle so it can catch prey through the narrow entrance of a burrow ***(1 mark)***. It has thick pads on the bottom of its feet to protect them as it's moving over rocky ground ***(1 mark)***.

You might not have come across any of these adaptations before, but just use the information you're given about the rocky mountains and think sensibly about how the features would help the wolf survive there.

3.3 It is decreasing ***(1 mark)***.

3.4 The outbreak of rabies is responsible for the trend ***(1 mark)***. In the years when there were rabies outbreaks the population size fell ***(1 mark)***. For example, the numbers fell from 401 to 327 in 2008 / 341 to 265 in 2011 when there was also a rabies outbreak ***(1 mark)***.

There's a lot of data to look at in the table so think carefully about what you're looking for. The numbers of prey don't really change much throughout the study so that's unlikely to have caused the change in the number of wolves, so it's sensible to look at the outbreaks of rabies.

3.5 It has made the wolf population move to higher ground ***(1 mark)*** from an average height of 3.4 km above sea level in 2007 to an average height of 4 km above sea level in 2012 ***(1 mark)***.

Topic 7b — Human Impacts on the Environment

1. Biodiversity and Waste

Page 342 — Fact Recall Questions

Q1 It makes sure that ecosystems are stable.

Q2 false

Both the population of the world and the general standard of living are increasing.

Q3 They can be washed from the land into water.

Q4 By using toxic chemicals for farming (e.g. pesticides and herbicides), by burying nuclear waste and by creating landfill sites for other waste.

Page 342 — Application Question

Q1 a) Both the amount of waste recycled and the amount sent to landfill has increased each year. But the amount recycled each year has increased by more than the amount sent to landfill (an extra 50 000 tonnes of waste was recycled in 2010 compared to 2006, but only 20 000 tonnes more was sent to landfill).

b) (100 000 ÷ 250 000) × 100 = **40%**

To find the total amount of waste in 2010, you need to add together the amount recycled (100 000) and the amount sent to landfill (150 000), which equals 250 000. You can then work out the percentage of this figure that was recycled.

c) The population of the world is increasing and so we are producing more waste, some of which ends up in landfill sites. Also, as the standard of living around the world increases we create even more waste. So the amount of space taken up by landfill sites, where waste is dumped, increases.

d) E.g. as the human population and standard of living increase, we use more raw materials (e.g. metal). It's important to recycle products, so that we don't run out of raw materials. / As the human population and standard of living increase, we produce more waste. It's important to recycle products as it reduces the amount of waste sent to landfill and therefore the amount of space taken up by landfill sites.

2. Global Warming

Page 345 — Fact Recall Questions

Q1 False

It's <u>increasing</u> levels of these gases that are contributing to an increase in the average global temperature.

Q2 carbon dioxide and methane

Page 345 — Application Question

Q1 a) Organisms are found in places where they can survive. If the average global temperature increases the distribution of organisms may change. E.g. animals that require cooler temperatures may only exist in a smaller area and animals that require warmer temperatures may become more widely distributed.

b) E.g. global warming could cause a change in birds' migration patterns, so birds that come to the country for only part of the year could be around for a shorter/longer period of time or it might cause birds to arrive and leave at slightly different times of the year. / If some bird species fail to adapt to the changes in their environment caused by global warming, then they might die out. / Birds that live, feed or breed near the coast might have problems if the sea level rises because of global warming.

3. Deforestation and Land Use

Page 348 — Fact Recall Questions

Q1 Any three from: building / quarrying / farming / dumping waste

Q2 The cutting down of forests.

Page 348 — Application Questions

Q1 a) 900 thousand hectares

To find the total amount of forest cleared, you need to read off the value for the full height of the bar. Don't forget to include the units in your answer either.

b) i) 400 thousand hectares

ii) E.g. cattle farming, growing rice crops or growing crops to create biofuels.

Q2 Peat bogs store carbon instead of releasing it into the atmosphere as carbon dioxide. If peat bogs are destroyed (e.g. to use the peat in compost), this carbon dioxide will be released, contributing to global warming. So peat-free compost helps to prevent the destruction of peat bogs, and so could be considered an environmentally-friendly option.

4. Maintaining Ecosystems and Biodiversity

Page 350 — Fact Recall Questions

Q1 They help protect biodiversity by preventing endangered species from becoming extinct.

Q2 They might reintroduce hedgerows and field margins. These protect biodiversity by providing a habitat for a wider variety of organisms around the field than would otherwise survive in the single crop habitat.

Q3 E.g. to reduce the amount of waste that gets dumped in landfill sites, reducing the amount of land taken over for landfill and so leaving ecoystems in place.

Page 352 — Human Impacts on the Environment Exam-style Questions

1.1 The variety of different species of organisms on Earth, or within an ecosystem ***(1 mark)***.

1.2 E.g. it would cost money to keep a watch on whether people are following the programme. The local economy could be harmed if people are forced to move away with their family to find work because the jobs that would have been available from shrimp production are not available ***(2 marks)***.

1.3 E.g. because people around the world are demanding a higher standard of living ***(1 mark)***, and shrimp could be seen as a luxury food ***(1 mark)***.

1.4 The removal of the trees will mean that less carbon dioxide will be removed from the atmosphere by the trees in photosynthesis ***(1 mark)***. More carbon dioxide will be released into the atmosphere if the trees are burnt to clear the land ***(1 mark)*** and by the respiration of microorganisms feeding on bits of dead wood ***(1 mark)***. This will result in an increased level of carbon dioxide in the atmosphere, contributing to global warming ***(1 mark)***.

1.5 Global warming is a type of climate change, and if some species are unable to adapt to the change in the climate ***(1 mark)***, then they won't survive and the species will become extinct ***(1 mark)***.

Topic 7c — Biomass, Food and Biotechnology

1. Trophic Levels

Page 354 — Fact Recall Questions

Q1 The different stages of a food chain, which consist of one or more organisms that perform a specific role in the food chain.

Q2 Carnivores that are at the top of the food chain because they have no predators.

Q3 They decompose/break down dead plant or animal material left in an environment.

Page 354 — Application Question

Q1 Trophic level 1 – Seaweed
Trophic level 2 – Green fish
Trophic level 3 – Blue fish
Trophic level 4 – Seal
Trophic level 5 – Orca

2. Pyramids of Biomass

Page 357 — Fact Recall Questions

Q1 Biomass is the mass of living material.

Q2 Each bar of a pyramid of biomass represents the biomass of one trophic level of a food chain. Biomass nearly always decreases as you move up a food chain, so the bars of the pyramid will get smaller nearer the top.

Q3 trophic level 1

Page 357 — Application Questions

Q1 a) phytoplankton
b) bass
c) Because biomass has been lost between the krill trophic level and the herring trophic level in the food chain.

3. Biomass Transfer

Page 359 — Fact Recall Questions

Q1 True

Q2 E.g. organisms don't always eat every single part of the organism they're consuming. Also, organisms don't absorb all of the stuff in the food they ingest. What they don't absorb is egested as faeces.

Page 359 — Application Question

Q1 $\frac{0.055\text{ kg}}{0.56\text{ kg}} \times 100$
$= \mathbf{9.8\%}$

4. Food Security and Farming

Page 362 — Fact Recall Questions

Q1 Food security is having enough food to feed the population.

Q2 E.g. by using fishing quotas. These limit the number and size of fish that can be caught in certain areas, preventing certain species from being overfished. By controlling mesh size of the fish nets. Using a bigger mesh size will reduce the number of 'unwanted' and discarded fish, by allowing unwanted species to escape and younger fish to slip through the net, allowing them to reach breeding age.

Q3 a) They can be raised in small pens in a temperature-controlled environment.
b) The conditions reduce the transfer of energy from the cows to the environment.

Page 362 — Application Questions

Q1 Any two from: The world population keeps increasing, with the birth rate of many developing countries rising quickly. / As diets in developed countries change, the demand for certain foods to be imported from developing countries can increase. This means that already scarce food resources can become more scarce. / Farming can be affected by new pests and pathogens (e.g. bacteria and viruses) or changes in the environmental conditions (e.g. a lack of rain). This can result in the loss of crops and livestock, and can lead to widespread famine. / The high input costs of farming (e.g. the price of seeds, machinery and livestock) can make it too expensive for people in some countries to start or maintain food production, meaning that there sometimes aren't enough people producing food in these areas to feed the people. / In some parts of the world, there are conflicts that affect the availability of food and water.

Q2 a) Because the chickens are free to roam about, they will lose more energy than the factory farmed chickens. This makes free range farming a less efficient method of farming, so the eggs cost more to produce and more to buy.

b) E.g. some people think it's cruel to force animals to live in intensive farming conditions. Free range animals live in more natural environments and have room to move around. / The animals need to be kept warm to reduce the energy they lose as heat. This often means using power from fossil fuels — which we wouldn't be using if the animals were grazing in their natural environment. This increases air pollution, so people may choose free range eggs to avoid this form of pollution.

5. Biotechnology

Page 365 — Fact Recall Questions

Q1 Protein from fungi used to make high-protein meat substitutes for vegetarian meals.

Q2 A fungus called *Fusarium* is grown in large vats on glucose syrup, which acts as food for the fungus. The fungus respires aerobically, so oxygen is supplied, together with nitrogen and other minerals. The mixture is also kept at the right temperature and pH. Once ready, the fungal biomass is harvested, purified and dried to make the mycoprotein.

Q3 A bacterium can be genetically modified to produce human insulin by transferring the gene that codes for human insulin to the bacterium and growing it in a vat under controlled conditions to create millions of bacteria that produce insulin. The insulin can be harvested and purified.

Page 365 — Application Question

Q1 The corn crops could be genetically engineered to contain the vitamin that the children are deficient in.

Page 367 — Biomass, Food and Biotechnology Exam-style Questions

1.1 It provides the energy ***(1 mark)*** needed for producers/ green plants and algae to photosynthesise ***(1 mark)*** and produce biomass ***(1 mark)***, which is passed on throughout the food chain by organisms eating each other ***(1 mark)***.

1.2 C, because algae are at the start of the food chain/are producers ***(1 mark)***.

1.3 Width of bar B = 6 squares.
Width of bar C = 60 squares.
So ratio of bar C to bar B = 60:6 = **10:1**
(2 marks for correct answer, otherwise 1 mark for correctly finding the widths of bar B and bar C).

1.4 Biomass is lost when organisms don't eat every single part of the organism they're consuming because some parts are inedible ***(1 mark)***. It is also lost when some of the material ingested by an organism is not absorbed and is egested as faeces ***(1 mark)***. Some ingested biomass is converted into substances that are lost as waste ***(1 mark)***.

2.1 Factory farming limits the movement of livestock and keeps them in a temperature-controlled environment ***(1 mark)***. This makes farming more efficient as the animals use less energy moving around and controlling their own body temperature ***(1 mark)***. This means that more energy is available for growth, so more food can be produced from the same input of resources ***(1 mark)***.

2.2 They are farmed in underwater cages where their movement is restricted ***(1 mark)***.

2.3 Any two from: e.g. some people think that forcing animals to live in unnatural and uncomfortable conditions is cruel. / The crowded conditions on factory farms create a favourable environment for the spread of diseases, like avian flu and foot-and-mouth disease. / The animals need to be kept warm to reduce the energy they lose as heat. This often means using power from fossil fuels — which we wouldn't be using if the animals were grazing in their natural environment. ***(2 marks)***

Glossary

A

Abdomen (humans)
The lower part of the body, ending at the hips.

Abiotic factor
A non-living factor of the environment.

Accommodation
The ability of focusing on near or distant objects by changing the shape of the lens in the eye.

Accurate result
A result that is very close to the true answer.

Active transport
The movement of particles against a concentration gradient (i.e. from an area of lower concentration to an area of higher concentration) using energy transferred during respiration.

Adaptation
A feature or characteristic that helps an organism to survive in the conditions of its natural environment.

Adrenaline
A hormone secreted by the adrenal glands that is released in response to stressful situations in order to prepare the body for 'fight or flight'.

Adult cell cloning
A method of cloning animals, which involves taking the nucleus from an adult body cell and inserting it into an unfertilised egg cell that has had its nucleus removed.

Aerobic respiration
The reactions involved in breaking down glucose using oxygen, to transfer energy. Carbon dioxide and water are produced.

Allele
An alternative version of a gene.

Alveolus
A tiny air sac in the lungs, where gas exchange occurs.

Amino acid
A small molecule that is a building block of proteins.

Amylase
A digestive enzyme that catalyses the breakdown of starch into sugars, in the mouth and small intestine.

Anaerobic respiration
The incomplete breakdown of glucose, which produces lactic acid in humans, and CO_2 and ethanol in plants and yeast. It takes place in the absence of oxygen.

Anomalous result
A result that doesn't seem to fit with the rest of the data.

Anti-diuretic hormone (ADH)
A hormone that stimulates the kidney tubules to become more permeable, so that the kidneys reabsorb more water.

Antibiotic
A drug used to kill or prevent the growth of bacteria.

Antibiotic resistance
When bacteria aren't killed by an antibiotic.

Antibody
A protein produced by white blood cells in response to the presence of an antigen (e.g. on the surface of a pathogen).

Antigen
A molecule on the surface of a cell. A foreign antigen triggers white blood cells to produce antibodies.

Antitoxin
A protein produced by white blood cells which counteracts the toxins produced by invading bacteria.

Aorta
A blood vessel (artery) which transports blood from the heart to the rest of the body (excluding the lungs).

Archaea (three-domain system)
A domain in the three-domain system of classification that consists of primitive bacteria, often found in extreme places.

Artery
A blood vessel that carries blood away from the heart.

Artificial blood product
A product used as a substitute for normal blood, e.g. to replace lost blood volume if someone has lost a lot of blood.

Artificial heart
A mechanical device that's put into a person to pump blood if their own heart fails.

Asexual reproduction
Where organisms reproduce by mitosis to produce genetically identical offspring.

Atrium
A chamber of the heart into which blood enters from either the pulmonary vein or the vena cava.

Auxin
A plant hormone that controls the growth of a plant in response to different stimuli.

B

Bacteria (three-domain system)
A domain in the three-domain system of classification that consists of true bacteria, e.g. E.coli and Staphylococcus.

Bacterium
A single-celled microorganism without a 'true' nucleus. Some bacteria are able to cause disease.

Behavioural adaptation
The way an organism behaves that helps it survive in its environment.

Bias
Prejudice towards or against something.

Bile
A fluid that is made in the liver, stored in the gall bladder and released into the small intestine. It aids digestion by creating alkaline conditions in the small intestine and by emulsifying fats.

Binary fission
A type of simple cell division carried out by prokaryotic cells whereby the cell makes copies of its genetic material, before splitting into two daughter cells.

Binomial system
The system used in classification for naming organisms using a two-part Latin name.

Biodiversity
The variety of different species of organisms on Earth, or within an ecosystem.

Biogas generator
A fermenter used to produce biogas.

Biological heart valve
A valve taken from a human or other mammal to replace a faulty heart valve.

Biomass
The mass of living material.

Biotic factor
A living factor of the environment.

Blood
A tissue which transports substances around the body in the circulatory system.

Blood cholesterol level
The level of cholesterol (a fatty substance) in the blood.

Capillary
A type of blood vessel involved in the exchange of materials at tissues.

Carbohydrase
A type of digestive enzyme that catalyses the breakdown of starch into sugars in the mouth and small intestine.

Carbon cycle
The continuous cycle of carbon from the air, through food chains and back into the air.

Cardiovascular disease
Disease of the heart or blood vessels.

Carrier
A person who carries the allele for an inherited disorder, but who doesn't have any symptoms of the disorder.

Catalyst
A substance that increases the speed of a reaction, without being changed or used up in the reaction.

Categoric data
Data that comes in distinct categories (e.g. sex — male and female, etc.).

Cell cycle
The process that all body cells from multicellular organisms use to grow and divide.

Cell membrane
A membrane surrounding a cell, which holds it all together and controls what goes in and out.

Cell wall
A structure surrounding some cell types, which gives strength and support.

Cellulose
A molecule which strengthens cell walls in plants and algae.

Central Nervous System (CNS)
The brain and spinal cord. It's where reflexes and actions are coordinated.

Cerebellum
An area at the back of the brain responsible for muscle coordination.

Cerebral cortex
The outer layer of the brain responsible for things like consciousness, intelligence, memory and language.

Chamber (of the heart)
An area of the heart (atrium or ventricle) through which blood is pumped.

Chlorophyll
A green substance found in chloroplasts which absorbs light for photosynthesis.

Chloroplast
A structure found in plant cells and algae, which contains chlorophyll. Chloroplasts are the site of photosynthesis.

Chromosome
A long molecule of DNA found in the nucleus, which carries genes.

Circulatory system
A system which uses blood to transport materials around the body.

Climate change
A change in things like temperature or weather patterns in a part of the world or across the whole world. E.g. global warming is a type of climate change.

Clinical trial
A set of drug tests on human volunteers.

Clone
An organism that is genetically identical to another organism.

Cloning
Making a genetically identical copy of another organism.

Communicable disease
A disease that can spread between individuals.

Community
The populations of different species living in a habitat.

Continuous data
Numerical data that can have any value within a range (e.g. length, volume or temperature).

Contraceptive
A method of preventing pregnancy that can be hormonal or non-hormonal.

Control experiment
An experiment that's kept under the same conditions as the rest of the investigation, but doesn't have anything done to it.

Control group
A group that matches the one being studied, but the independent variable isn't altered. It's kept under the same conditions as the group in the experiment.

Control variable
A variable in an experiment that is kept the same.

Coordination centre
An organ (e.g. the brain, spinal cord or pancreas) that processes information from receptors and organises a response from the effectors.

Coronary artery
A blood vessel which supplies blood to the heart muscle.

Coronary heart disease
A disease in which the coronary arteries are narrowed by the build up of fatty deposits.

Correlation
A relationship between two variables.

Culture (of microorganisms)
A population of one type of microorganism that's been grown under controlled conditions.

Cutting (plants)
A small piece of a plant (usually with a new bud on) that can be taken and grown into a new plant.

Cystic fibrosis
An inherited disorder of the cell membranes caused by a recessive allele.

Cytoplasm
A gel-like substance in a cell where most of the chemical reactions take place.

Decay
The breakdown of dead organisms.

Deficiency disease
A disease caused by a lack of a certain nutrient.

Deforestation
The cutting down of forests (large areas of trees).

Dependent variable
The variable in an experiment that is measured.

Dialysis fluid
A fluid that has the same concentration of glucose and ions as blood plasma. It is used in kidney dialysis machines.

Diaphragm (thorax)
The muscle separating the abdomen from the thorax.

Differentiation
The process by which a cell becomes specialised for its job.

Diffusion
The spreading out of particles from an area of higher concentration to an area of lower concentration.

Discrete data
Numerical data that can only take certain values with no in-between value (e.g. number of people).

Distribution
Where organisms are found in a particular area.

DNA (deoxyribonucleic acid)
The molecule in cells that stores genetic information.

Dominant allele
The allele for the characteristic that's shown by an organism if two different alleles are present for that characteristic.

Double-blind trial
A clinical trial where neither the doctors nor the patients know who has received the drug and who has received the placebo until all the results have been gathered.

E

Ecosystem
The interaction of a community of living organisms with the abiotic parts of their environment.

Effector
Either a muscle or gland which responds to nervous impulses.

Efficacy
Whether something, e.g. a drug, works or not.

Electron microscope
A microscope that uses electrons to form an image.

Embryo transplant (cloning)
A method of cloning animals, where an embryo is created and then split, before any cells become specialised, to produce clones. The clones are then implanted into the uteruses of host mothers.

Embryonic screening
Genetic analysis of either a cell taken from an embryo before it's implanted into the uterus during IVF, or a sample of cells taken from the placenta, in order to check that the embryo doesn't carry any inherited disorders.

Endothermic reaction
A reaction where energy is transferred from the environment.

Enzyme
A protein that acts as a biological catalyst.

Epidemic
A big outbreak of disease.

Epidermal tissue
A type of plant tissue which covers the whole plant.

Ethene
A gas produced by plants that controls cell division and stimulates ripening.

Eukaryota (three-domain system)
A domain in the three-domain system of classification that consists of a broad range of organisms including fungi, plants, animals and protists.

Eukaryotic cell
A complex cell, such as a plant or animal cell.

Evaporation
The process by which a liquid turns into a gas.

Evolution
The changing of the inherited characteristics of a population over time.

Exchange surface
A specialised surface in an organism used for the exchange of materials, e.g. gases and dissolved substances.

Excretion
The removal of waste products from the body.

Exothermic reaction
A reaction that transfers energy to the environment.

Extinction
When no living individuals of a species remain.

Extremophile
An organism that's adapted to live in seriously extreme conditions.

F

Factory farming
When animals are grown in a temperature-controlled environment where their movement is limited, so they waste little energy keeping warm or moving.

Fair test
A controlled experiment where the only thing that changes is the independent variable.

Family tree (genetics)
A diagram which shows how a characteristic (or disorder) is inherited in a group of related people.

Fermentation
The process of anaerobic respiration in yeast cells.

Fertilisation
The fusion of male and female gametes during sexual reproduction.

Fertility
The ability to conceive a child.

Fishing quota
A limit on the number and size of fish that can be caught in a certain area.

Follicle stimulating hormone (FSH)
A hormone produced by the pituitary gland involved in the menstrual cycle. It causes eggs to mature in the ovaries and stimulates the ovaries to produce oestrogen.

Food security
Having enough food to feed the population.

Fossil
The remains of an organism from many years ago, which is found in rock.

Fossil record
The history of life on Earth preserved as fossils.

Functional adaptation
Something that goes on inside an organism's body that helps the organism to survive in its environment.

Fungus
A microorganism that can cause disease, and that produces spores that can be spread to other organisms.

G

Gamete
A sex cell, e.g. an egg cell or a sperm cell in animals.

Gene
A short section of DNA, found on a chromosome, which contains the instructions needed to make a protein (and so controls the development of a characteristic).

Genetic engineering
The process of cutting out a useful gene from one organism's genome and inserting it into another organism's cell(s).

Genetic variant
A different form of a gene (produced by a mutation).

Genetically modified (GM) crop
A crop which has had its genes modified through genetic engineering.

Genome
All of the genetic material in an organism.

Genotype
What alleles you have, e.g. Tt.

Geotropism
See gravitropism.

Gibberellin
A type of plant hormone that stimulates seed germination, stem growth and flowering.

Gland
The place where hormones are produced and secreted from.

Global warming
The rise in the average global temperature.

Glucagon
A hormone produced and secreted by the pancreas when blood glucose level is too low. It causes glycogen to be converted back into glucose, increasing the blood glucose level.

Glycogen
A molecule that acts as a store of glucose in liver and muscle cells.

Gonorrhoea
A sexually transmitted bacterial disease.

Gravitropism
The growth of a plant in response to gravity. Also known as geotropism.

Greenhouse effect
When gases (called greenhouse gases) in the atmosphere absorb energy radiated by the Earth and re-radiate it back down towards the Earth, helping to keep it warm.

Guard cell
A cell found on either side of a stoma, which controls the stoma's size.

Habitat
The place where an organism lives.

Haemoglobin
A red pigment found in red blood cells which carries oxygen.

Hazard
Something that has the potential to cause harm (e.g. fire, electricity, etc.).

Heterozygous
Where an organism has two alleles for a particular gene that are different.

HIV
A virus that attacks the immune system cells so that eventually it can't cope with other infections or cancers. It causes AIDS.

Homeostasis
The regulation of conditions inside your body (and cells) to maintain a stable internal environment, in response to changes in both internal and external conditions.

Homozygous
Where an organism has two alleles for a particular gene that are the same.

Hormone
A chemical messenger which travels in the blood to activate target cells.

Hybridoma
A cell made by fusing a mouse lymphocyte with a tumour cell, which is used to produce lots of monoclonal antibodies.

Hyperopia
Long-sightedness.

Hypothesis
A possible explanation for a scientific observation.

Immunity
The ability of the white blood cells to respond quickly to a pathogen.

***In vitro* fertilisation (IVF)**
The artificial fertilisation of eggs in the lab.

Inbreeding
When closely related animals or plants are bred together.

Independent variable
The variable in an experiment that is changed.

Inherited disorder
A disorder caused by a faulty allele, which can be passed on to an individual's offspring.

Insulin
A hormone produced and secreted by the pancreas when blood glucose level is too high. It causes the body's cells to take up more glucose from the blood, reducing the blood glucose level.

Interdependence
Where, in a community, each species depends on other species for things such as food, shelter, pollination and seed dispersal.

Kidney
The organ responsible for producing urine. The kidneys play an important role in the regulation of water and ion content in the body, and in the removal of waste products such as urea.

Kidney dialysis
A way of artificially filtering the blood to remove waste products and keep the concentration of dissolved ions in the blood at normal levels. It is used to treat patients with kidney failure.

Kidney transplant
Where a damaged kidney is replaced with a healthy donor kidney in patients with kidney failure.

Lactic acid
The product of anaerobic respiration that builds up in muscle cells.

Large intestine
An organ in the mammalian digestive system where water from undigested food is absorbed, producing faeces.

Limiting factor
A factor which prevents a reaction from going any faster.

Linnaean system
The system of classification devised by Carl Linneaus that groups living things according to the characteristics and structures that make them up.

Lipase
A type of digestive enzyme tha catalyses the breakdown of li fatty acids and glycerol, in th intestine.

Liver
An organ found in the mammalian digestive system with several important roles in the body, e.g. the production of bile and the conversion of glucose into glycogen.

Lung
A gas exchange organ in mammals.

Luteinising hormone (LH)
A hormone produced by the pituitary gland, which stimulates egg release around the middle of the menstrual cycle.

M

Malaria
A disease caused by a protist and spread by mosquitoes, which causes repeating episodes of fever.

Mean (average)
A measure of average found by adding up all the data and dividing by the number of values there are.

Measles
A viral disease that causes a red skin rash and fever.

Mechanical heart valve
A man-made heart valve used to replace a faulty heart valve.

Median (average)
The middle value in a set of data when they're in order of size.

Medulla
An area at the top of the spinal cord and base of the brain, that controls unconscious activities e.g. breathing and heartbeat.

Meiosis
A type of cell division where a cell divides twice to produce four genetically different gametes. It occurs in the reproductive organs.

Menstrual cycle
The monthly sequence of events in which the female body releases an egg and prepares the uterus (womb) in case it receives a fertilised egg.

Meristem tissue
Tissue found at the growing tips of plant shoots and roots that is able to differentiate.

Metabolism
All the chemical reactions that happen in a cell or the body.

Methane
A greenhouse gas.

Mitochondria
Structures in a cell which are the site of most of the reactions for aerobic respiration.

Mitosis
A type of cell division where a cell reproduces itself by splitting to form two identical offspring.

Mode (average)
The most common value in a set of data.

Monoclonal antibodies
Antibodies produced from lots of clones of a single white blood cell, which will only target one specific protein antigen.

Motor neurone
A nerve cell that carries electrical impulses from the CNS to effectors.

MRI scanner
A tube-like machine that can be used to produce a very detailed picture of the brain's structures.

MRSA (methicillin-resistant *Staphylococcus aureus*)
A strain of bacteria that is resistant to the powerful antibiotic methicillin.

Multicellular organism
An organism made up of more than one cell.

Muscle fatigue
Where muscles become tired and can't contract efficiently.

Mutation
A random change in an organism's DNA.

Mycoprotein
Protein produced by fungi, in particular *Fusarium*.

Myopia
Short-sightedness.

N

Natural selection
The process by which species evolve.

Negative correlation
When one variable decreases as another variable increases.

Negative feedback
A mechanism that restores a level back to optimum in a system.

Nervous system
The organ system in animals that allows them to respond to changes in their environment.

Neurone
A nerve cell. Neurones transmit information around the body, including to and from the CNS.

Non-communicable disease
A disease that cannot spread between individuals.

Nucleotide
A repeating unit in DNA and RNA that consists of a sugar molecule, a phosphate molecule and a base.

Nucleus (of a cell)
A structure found in animal and plant cells which contains genetic material that controls the activities of the cell.

Obesity
A condition defined as being 20% or more over the maximum recommended body mass.

Oestrogen
A hormone produced by the ovaries which is involved in the menstrual cycle. It's found in some oral contraceptives.

Optimum dose (in drug testing)
The dose of a drug that is most effective and has few side effects.

Optimum level (in the body)
A level of something (e.g. water, ions or glucose) that enables the body to work at its best.

Oral contraceptive
A hormone-containing pill taken by mouth in order to reduce fertility and therefore decrease the chance of pregnancy.

Organ
A group of different tissues that work together to perform a certain function.

Organ system
A group of organs working together to perform a particular function.

Osmosis
The movement of water molecules across a partially permeable membrane from a region of higher water concentration to a region of lower water concentration.

Ovary
An organ in the female body which stores and releases eggs. It is also a gland and secretes the hormone oestrogen.

Oxygen debt
The amount of extra oxygen your body needs after exercise to react with the build up of lactic acid and remove it from cells.

Oxyhaemoglobin
A molecule formed when haemoglobin combines with oxygen.

P

Painkiller
A drug that relieves pain.

Palisade mesophyll tissue
Leaf tissue where most photosynthesis happens.

Pancreas
An organ (and gland) in the mammalian digestive system which produces digestive juices. It also produces insulin and glucagon to control blood sugar level.

Partially permeable membrane
A membrane with tiny holes in it, which lets some molecules through it but not others.

Pathogen
A microorganism that causes disease, e.g. a bacterium, virus, protist or fungus.

Peat bog
An area of land that is acidic and waterlogged, so plants don't fully decompose when they die, producing peat.

Peer review
The process where scientists check each other's results and explanations to make sure they're scientific.

Penicillin
A type of antibiotic.

Permanent vacuole (plant cells)
A structure in plant cells that contains cell sap.

Phagocytosis
The process by which white blood cells engulf foreign cells and digest them.

Phenotype
The characteristics you have, e.g. blue eyes.

Phloem
A type of plant tissue which transports dissolved sugars around the plant.

Photosynthesis
The process by which plants use energy to convert carbon dioxide and water into glucose and oxygen.

Phototropism
The growth of a plant in response to light.

Pituitary gland
A gland located in the brain that is responsible for secreting various hormones.

Placebo (in drug testing)
A substance that's like the drug being tested but doesn't do anything.

Plasma
The liquid component of blood, which transports the contents of the blood around the body.

Platelet
A small fragment of a cell found in the blood, which helps blood to clot at a wound.

Polydactyly
An inherited disorder caused by a dominant allele that results in a person being born with extra fingers or toes.

Positive correlation
When one variable increases as another variable increases.

Potometer
Apparatus used to measure water uptake by a plant.

Precise result
A result that is close to the mean.

Preclinical trial
A set of drug tests on human cells and tissues, and live animals that occurs before clinical trials.

Predator
An animal that hunts and kills other animals.

Prediction
A statement based on a hypothesis that can be tested.

Prey
An animal that is hunted and killed by another animal.

Primary consumer
An organism in a food chain that feeds on a producer.

Producer
An organism at the start of a food chain that makes its own food using energy from the Sun.

Progesterone
A hormone produced by the ovaries, which is involved in the menstrual cycle. It's found in some oral contraceptives.

Prokaryotic cell
A small, simple cell, e.g. a bacterium.

Protease
A type of digestive enzyme that catalyses the breakdown of proteins into amino acids, in the stomach and small intestine.

Protein
A large biological molecule made up of long chains of amino acids.

Protist
A pathogen that is often transferred to other organisms by a vector, which doesn't get the disease itself.

Pulmonary artery
A blood vessel (artery) which transports blood out of the heart to the lungs.

Pulmonary vein
A blood vessel (vein) which transports blood into the heart from the lungs.

Punnett square
A type of genetic diagram.

Pyramid of biomass
A diagram to represent the biomass at each stage of a food chain.

Q

Quadrat
A square frame enclosing a known area which can be used to study the distribution of organisms.

R

Random error
A difference in the results of an experiment caused by things like human error in measuring.

Range
The difference between the smallest and largest values in a set of data.

Receptor
A group of cells which are sensitive to a stimulus. E.g. light receptor cells in the eye are sensitive to light.

Recessive allele
An allele whose characteristic only appears in an organism if there are two copies present.

Reflex
A fast, automatic response to a stimulus.

Reflex arc
The passage of information in a reflex.

Relay neurone
A nerve cell that carries electrical impulses from sensory neurones to motor neurones.

Repeatable result
A result that will come out the same if the experiment is repeated by the same person using the same method and equipment.

Reproducible result
A result that will come out the same if someone different does the experiment, or a slightly different method or piece of equipment is used.

Resolution (microscopes)
The ability to distinguish between two parts. A higher resolution gives a sharper image.

Ribcage
A set of bones in the thorax, which protect the lungs.

Ribosome
A structure in a cell, where proteins are made.

Risk
The chance that a hazard will cause harm.

Risk factor
Something that is linked to an increased likelihood that a person will develop a certain disease.

Root hair cell
A cell on the surface of a plant root, which absorbs water and mineral ions.

Rooting powder
A powder containing auxins, which can be applied to plant cuttings to assist root development.

Rose black spot
A fungus that causes purple or black spots to develop on the leaves of rose plants. The leaves can then turn yellow and drop off.

S

Salivary gland
An organ (and gland) in the mammalian digestive system which produces digestive juices.

Salmonella
A type of bacteria that causes food poisoning.

Sampling
Taking a number of organisms from a population to study.

Secondary consumer
An organism in a food chain that eats a primary consumer.

Selective breeding (artificial selection)
When humans artificially select the plants or animals that are going to breed so that the genes for particular characteristics remain in the population.

Selective weedkiller
A weedkiller that contains auxins. It kills weeds (unwanted plants), without affecting the growth of crops.

Sense organ
An organ which contains receptors that detect stimuli, e.g. the eye.

Sensory neurone
A nerve cell that carries electrical impulses from the receptors in the sense organs to the CNS.

Sex chromosome (humans)
One of the 23rd pair of chromosomes — together they determine whether an individual is male or female.

Sexual reproduction
Where two gametes combine at fertilisation to produce a genetically different new individual.

SI unit
A unit of measurement recognised as standard by scientists worldwide.

Single gene cross
Where you cross two parents to look at the inheritance of just one characteristic controlled by a single gene.

Small intestine
An organ in the mammalian digestive system where food is digested and soluble food molecules are absorbed.

Specialised cell
A cell which performs a specific function.

Speciation
The development of a new species.

Species
A group of similar organisms that can reproduce to give fertile offspring.

Spongy mesophyll tissue
Tissue in a leaf that contains air spaces.

Stable community
A community where all the species and environmental factors are in balance so that the population sizes are roughly constant.

Standard of living
A measure of things like what access people have to food, education, health care and luxury goods like TVs, cars and computers.

Starch
An insoluble carbohydrate used as a store of glucose in plants.

Statins
A group of medicinal drugs that are used to decrease the risk of heart and circulatory disease.

Stem cell
An undifferentiated cell which has the ability to become one of many different types of cell or more stem cells.

Stent
A wire mesh tube that's inserted inside an artery to help keep it open.

Sterilisation (avoiding contamination)
The process of destroying microorganisms (such as bacteria) on an object.

Stimulus
A change in the environment.

Stoma
A tiny hole in the surface of a leaf.

Stomach
An organ in the mammalian digestive system where food is digested.

Structural adaptation
A feature of an organism's body structure that helps it survive in its environment.

Sulfur dioxide
A gas released by burning fossil fuels, which can cause acid rain if it mixes with rain clouds in the atmosphere.

Sustainable food production
Producing food in a way that means we have enough to eat now and in the future without using resources faster than they renew.

Synapse
The connection between two neurones.

Systematic error
An error that is consistently made every time throughout an experiment.

Target cell
A particular cell in a particular place, which is affected by a hormone.

Tertiary consumer
An organism in a food chain that eats a secondary consumer.

Testosterone
The main male reproductive hormone, produced by the testes.

Theory
A hypothesis which has been accepted by the scientific community because there is good evidence to back it up.

Therapeutic cloning
A type of cloning where the embryo is made to have the same genetic information as the patient.

Thermoregulatory centre
An area of the brain which controls and monitors body temperature.

Thorax (humans)
The upper part of the body, excluding the arms and head. Contains the lungs.

Three-domain system
The system of classification based on evidence gathered from new chemical analysis techniques, like RNA sequence analysis.

Thyroxine
A hormone secreted by the thyroid gland that regulates metabolism and stimulates protein synthesis for growth and development.

Tissue
A group of similar cells that work together to carry out a particular function. It can include more than one type of cell.

Tissue culture (plants)
A method of cloning plants in which a few plant cells are put on a growth medium containing hormones and allowed to grow into new plants.

Tobacco mosaic virus
A virus that affects many species of plants, e.g. tomatoes, and causes a mosaic pattern on the leaves.

Toxicity
How harmful something is, e.g. a drug.

Toxin
A poison. Toxins are often produced by bacteria.

Transect
A line which can be used to study the distribution of organisms across an area.

Translocation
The movement of dissolved sugars around a plant.

Transpiration stream
The movement of water from a plant's roots, through the xylem and out of the leaves.

Trial run
A quick version of an experiment that can be used to work out the range of variables and the interval between the variables that will be used in the proper experiment.

Trophic level
A stage in a food chain.

Tumour
A growth of abnormal cells.

Type 1 diabetes
A condition where the pancreas produces little or no insulin, which means blood glucose can rise to a dangerous level.

Type 2 diabetes
A condition in which the body is unable to control blood sugar level.

U

Uncertainty
The amount of error your results might have.

Urea
A waste product produced from the breakdown of amino acids in the liver.

Uterus
The main female reproductive organ and where the embryo develops during pregnancy. (Another word for the womb.)

Vaccination
The injection of dead or inactive microorganisms to provide immunity against a particular pathogen.

Valid result
A result that is repeatable, reproducible and answers the original question.

Valve (in the circulatory system)
A structure within the heart or a vein which prevents blood from flowing in the wrong direction.

Variable
A factor in an investigation that can change or be changed (e.g. temperature or concentration).

Variation
The differences that exist between individuals.

Vasoconstriction
When blood vessels supplying the skin capillaries constrict (get narrower).

Vasodilation
When blood vessels supplying the skin capillaries dilate (get wider).

Vector (in genetic engineering)
Something used to transfer DNA into a cell, e.g. a virus or a bacterial plasmid.

Vein
A blood vessel that carries blood to the heart.

Vena cava
A blood vessel (vein) which transports blood into the heart from the rest of the body (excluding the lungs).

Ventricle
A chamber of the heart which pumps blood out of the heart through either the pulmonary artery or the aorta.

Virus
A disease-causing agent about 1/100th of the size of a bacterial cell. Can only replicate within host body cells.

Water bath
Apparatus that uses water heated to a specific temperature to heat samples contained in vessels to that same temperature.

Water cycle
The continuous cycle of water by evaporation and precipitation.

White blood cell
A cell which forms part of the blood and part of the immune system, helping to defend the body against disease.

Xylem
A type of plant tissue which transports water and mineral ions around the plant.

Zero error
A type of systematic error caused by using a piece of equipment that isn't zeroed properly.

Acknowledgements

Data Acknowledgements

Data used to construct the table on the use of stents on page 114 reprinted from the Journal of the American College of Cardiology, Vol 49/19. H Vernon Anderson et al. Drug-Eluting Stents for Acute Myocardial Infarction, pgs 1931-1933. © 2007, with permission from Elsevier.

Data used to construct the measles graph on page 157 from http://www.who.int/immunization/monitoring_surveillance/burden/vpd/surveillance_type/active/big_measles_global_coverage.jpg accessed March 2016.

Data used to produce the IVF table and question on page 233 from the Human Fertilisation and Embryology Authority (HFEA).

Data on pregnancy rates used in question on page 233 from the Human Fertilisation and Embryology Authority (HFEA).

Data used to construct the big cat evolutionary tree on page 310 reprinted from Molecular Phylogenetics and Evolution, Vol 56. Brian W. Davis, Gang Li, William J. Murphy. Supermatrix and species tree methods resolve phylogenetic relationships within the big cats, Panthera (Carnivora: Felidae). Pages 64-76. © 2010, with permission from Elsevier.

Photograph Acknowledgements

Cover Photo **Scott Sinklier/AGStockUSA**/Science Photo Library, p 3 Science Photo Library, p 5 iStock.com/**LourdesPhotography**, p 6 **Tony Craddock**/Science Photo Library, p 8 iStock.com/**muslian** (Fig. 1), p 8 **Tony McConnell**/Science Photo Library (Fig. 2), p 9 **Tek Image**/Science Photo Library, p 12 iStock.com/**tunart**, p 21 **Adam Hart-Davis**/Science Photo Library, p 23 **Alfred Pasieka**/Science Photo Library, p 24 **Dr. Martha Powell, Visuals Unlimited**/Science Photo Library, p 25 **Biophoto Associates**/Science Photo Library (Fig. 1 top), p 25 **National Cancer Institute**/Science Photo Library (Fig. 1 bottom), p 27 **Kevin & Betty Collins, Visuals Unlimited**/Science Photo Library, p 31 **Steve Gschmeissner**/Science Photo Library (Fig. 3), p 31 **Steve Gschmeissner**/Science Photo Library (Fig. 4), p 31 **Dr Keith Wheeler**/Science Photo Library (Fig. 6), p 33 **Sovereign, ISM**/Science Photo Library (Fig. 2), p 33 **Power And Syred**/Science Photo Library (Fig. 4), p 34 **Manfred Kage**/Science Photo Library, p 36 **Steve Gschmeissner**/Science Photo Library (all images on page), p 37 **Herve Conge, ISM**/Science Photo Library, p 38 **Dr. Tony Brain**/Science Photo Library, p 40 **CNRI**/Science Photo Library, p 41 **Michael Gabridge/Visuals Unlimited, Inc.**/Science Photo Library, p 43 **John Durham**/Science Photo Library, p 45 **Pascal Goetgheluck**/Science Photo Library, p 46 iStock.com/**Epitavi**, p 51 **Andrew Lambert Photography**/Science Photo Library, p 61 iStock.com/**EcoPic** (Fig. 1 top), p 61 iStock.com/**GlobalP** (Fig. 1 bottom), p 63 **E. R. Degginger**/Science Photo Library (left), p 63 **Tom & Pat Leeson**/Science Photo Library (right), p 64 **Biophoto Associates**/Science Photo Library, p 65 **Eye Of Science**/Science Photo Library, p 67 iStock.com/**petrescudaniel**, p 72 **Steve Gschmeissner**/Science Photo Library, p 73 **Dr Keith Wheeler**/Science Photo Library, p 76 **Zephyr**/Science Photo Library, p 82 iStock.com/**Suljo**, p 83 **Steve Gschmeissner**/Science Photo Library, p 84 **CNRI**/Science Photo Library, p 86 **Steve Gschmeissner**/Science Photo Library, p 87 **Power And Syred**/Science Photo Library (Fig. 3), p 87 **Biophoto Associates**/Science Photo Library (Fig. 4), p 87 **Susumu Nishinaga**/Science Photo Library (Fig. 5), p 87 **Antonia Reeve**/Science Photo Library (Fig. 6), p 89 **Eye Of Science**/Science Photo Library, p 91 **Dr. Richard Kessel & Dr. Gene Shih/Visuals Unlimited, Inc.**/Science Photo Library (Fig. 2), p 91 **Biophoto Associates**/Science Photo Library (Fig. 4), p 94 **Dr Jeremy Burgess**/Science Photo Library (Fig. 2 top), p 94 **Dr Jeremy Burgess**/Science Photo Library (Fig. 2 bottom), p 95 **Ted Kinsman**/Science Photo Library, p 100 **Power And Syred**/Science Photo Library (Fig. 3), p 100 **Eye Of Science**/Science Photo Library (Fig. 4), p 103 **Dr P. Marazzi**/Science Photo Library, p 104 **Hank Morgan**/Science Photo Library, p 105 **CNRI**/Science Photo Library, p 109 **James Stevenson**/Science Photo Library, p 110 **CNRI**/Science Photo Library, p 113 iStock.com/**Willsie**, p 116 **Clive Freeman, The Royal Institution**/Science Photo Library, p 124 **Martyn F. Chillmaid**/Science Photo Library, p 125 **Andrew Lambert Photography**/Science Photo Library (Fig. 4), p 125 **Andrew Lambert Photography**/Science Photo Library (Fig. 6), p 129 **Martyn F. Chillmaid**/Science Photo Library, p 130 **Juergen Berger**/Science Photo Library (Fig. 2 top), p 130 **AMI Images**/Science Photo Library (Fig. 2 bottom), p 131 **Dr Keith Wheeler**/Science Photo Library (Fig. 3), p 131 **Steve Gschmeissner**/Science Photo Library (Fig. 4 top), p 131 Science Photo Library (Fig. 4 bottom), p 132 **Norm Thomas**/Science Photo Library, p 133 **Ian Gowland**/Science Photo Library, p 135 **CDC**/Science Photo Library, p 136 **Juergen Berger**/Science Photo Library (Fig. 1), p 136 **Biology Media**/Science Photo Library (Fig. 2), p 139 **Saturn Stills**/Science Photo Library, p 141 **Scott Camazine**/Science Photo Library, p 143 **James King-Holmes**/Science Photo Library, p 144 **St. Bartholomew's Hospital**/Science Photo Library, p 147 **Mark Thomas**/Science Photo Library, p 149 **Dr K. Sikora**/Science Photo Library, p 151 **UK Crown Copyright Courtesy Of Fera**/Science Photo Library, p 152 **Geoff Kidd**/Science Photo Library (Fig. 2 top), p 152 **Vaughan Fleming**/Science Photo Library (Fig. 2 bottom), p 158 **Biophoto Associates**/Science Photo Library, p 160 **Biophoto Associates**/Science Photo Library (Fig. 1), p 160 **Angel Fitor**/Science Photo Library (example box left), p 160 **Victor De Schwanberg**/Science Photo Library (example box right), p 161 **The Picture Store**/Science Photo Library, p 163 Science Photo Library, p 167 **E. R. Degginger**/Science Photo Library, p 170 **Angel Fitor**/Science Photo Library (Fig. 1), p 170 iStock.com/**36clicks** (Fig. 3), p 174 **Samuel Ashfield**/Science Photo Library, p 176 **Sputnik**/Science Photo Library, p 178 **BSIP, Laurent/B. Hop Ame**/Science Photo Library (Fig. 2), p 178 iStock.com/**milla1974** (Fig. 3), p 186 **Sovereign, ISM**/Science Photo Library, p 188 **Thomas**

Deerinck, **NCMIR**/Science Photo Library, p 189 **PH. Gerbier**/Science Photo Library, p 192 iStock.com/**Spanishalex**, p 193 **Du Cane Medical Imaging Ltd**/Science Photo Library, p 198 **Patrick Dumas/Look At Sciences**/Science Photo Library, p 199 iStock.com/**andreea-cristina** (Fig. 4), p 199 **Pascal Goetgheluck**/Science Photo Library (Fig. 5), p 201 iStock.com/**toos**, p 202 **Robert Markus**/Science Photo Library, p 207 **Scott Camazine**/Science Photo Library, p 211 **Coneyl Jay**/Science Photo Library, p 219 **Life In View**/Science Photo Library, p 220 **Cordelia Molloy**/Science Photo Library, p 221 **Dr. Barry Slaven/Visuals Unlimited, Inc.**/Science Photo Library, p 226 **Professors P. M. Motta & J. Van Blerkom**/Science Photo Library, p 229 **Cordelia Molloy**/Science Photo Library, p 230 **Gustoimages**/Science Photo Library, p 231 **Zephyr**/Science Photo Library, p 232 **Zephyr**/Science Photo Library, p 236 **Martin Shields**/Science Photo Library, p 237 **Martin Shields**/Science Photo Library, p 239 **Geoff Kidd**/Science Photo Library, p 245 **Lawrence Berkeley National Laboratory**/Science Photo Library, p 251 **Eye Of Science**/Science Photo Library, p 252 **Wim Van Egmond**/Science Photo Library, p 253 **Adrian T Sumner**/Science Photo Library, p 255 iStock.com/**eurobanks**, p 256 **Alan and Linda Detrick**/Science Photo Library (Fig. 2), p 256 iStock.com/**Richard Griffin** (Fig. 3), p 262 Science Photo Library, p 263 iStock.com/**kutipie**, p 265 iStock.com/**MGJdeWit** (Fig. 1 left), p 265 iStock.com/**ivosar** (Fig. 1 right), p 266 **Wally Eberhart, Visuals Unlimited**/Science Photo Library, p 271 **Sovereign**, **ISM**/Science Photo Library, p 273 **Pascal Goetgheluck**/Science Photo Library, p 275 Science Photo Library, p 276 **Bob Gibbons**/Science Photo Library, p 282 **Kate Jacobs**/Science Photo Library (Fig. 1), p 282 **Coneyl Jay**/Science Photo Library (example box), p 288 **Science Source**/Science Photo Library, p 290 Science Photo Library (Fig. 1), p 290 **Sheila Terry**/Science Photo Library (Fig. 2), p 296 **Alfred Pasieka**/Science Photo Library (Fig. 2 top), p 296 **Bill Barksdale/AGStockUSA**/Science Photo Library (Fig. 2 bottom), p 297 **Sinclair Stammers**/Science Photo Library, p 298 **Gerard Peaucellier, ISM**/Science Photo Library, p 299 **James King-Holmes**/Science Photo Library, p 300 **Tony Camacho**/Science Photo Library, p 301 **Sinclair Stammers**/Science Photo Library (Fig. 1), p 301 **Natural History Museum, London**/Science Photo Library (Fig. 2), p 302 **Dirk Wiersma**/Science Photo Library (Fig. 3), p 302 **John Reader**/Science Photo Library (Fig. 4), p 302 **Alfred Pasieka**/Science Photo Library (Fig. 5), p 302 **Silkeborg Museum, Denmark/Munoz-Yague**/Science Photo Library (Fig. 6), p 303 **Josie Iselin, Visuals Unlimited**/Science Photo Library (top left), p 303 **Herve Conge, ISM**/Science Photo Library (top right), p 303 **Herve Conge, ISM**/Science Photo Library (bottom left), p 303 **Philippe Plailly**/Science Photo Library (bottom right), p 313 **Ken M. Highfill**/Science Photo Library, p 314 iStock.com/**kojihirano**, p 317 iStock.com/**bogdanhoria** (Fig. 1 top), p 317 iStock.com/**Aleksander** (Fig. 1 bottom), p 319 **John Devries**/Science Photo Library, p 321 iStock.com/**JohnPitcher**, p 326 **Martyn F. Chillmaid**/Science Photo Library, p 327 **Martyn F. Chillmaid**/Science Photo Library, p 329 iStock.com/**Gelpi**, p 331 **Dr Jeremy Burgess**/Science Photo Library, p 334 **Pascal Goetgheluck**/Science Photo Library (Fig. 3), p 334 **James King-Holmes**/Science Photo Library (Fig. 4), p 335 **Prof. David Hall**/Science Photo Library (Fig. 6), p 335 **Gustoimages**/Science Photo Library (Fig. 7), p 341 **Michael Marten**/Science Photo Library, p 342 **Robert Brook**/Science Photo Library, p 344 **NASA**/Science Photo Library (Fig. 3), p 344 **Dr. John Brackenbury**/Science Photo Library (Fig. 4), p 347 **Mark A. Schneider**/Science Photo Library, p 348 **Ailsa M Allaby**/Science Photo Library, p 349 **Peter Skinner**/Science Photo Library, p 350 **Matteis/Look At Sciences**/Science Photo Library, p 352 iStock.com/**bobhackettphotos**, p 361 **Alex Bartel**/Science Photo Library, p 363 **Cordelia Molloy**/Science Photo Library, p 364 **Sputnik**/Science Photo Library, p 367 iStock.com/**Baris Karadeniz**, p 368 **GIPhotoStock**/Science Photo Library, p 369 **Andrew Lambert Photography**/Science Photo Library, p 370 **Martyn F. Chillmaid**/Science Photo Library, p 371 Science Photo Library, p 372 iStock.com/**Thawatchai Tumwapee**, p 373 **Christopher McGarry**, p 375 **Martyn F. Chillmaid**/Science Photo Library, p 377 **David Maliphant**, p 390 **Photostock-Israel**/Science Photo Library.

Every effort has been made to locate copyright holders and obtain permission to reproduce sources. For those sources where it has been difficult to trace the originator of the work, we would be grateful for information. If any copyright holder would like us to make an amendment to the acknowledgements, please notify us and we will gladly update the book at the next reprint. Thank you.

Index

E

F

G

H

I

K